UNIVERSITY
D0272066

Multivariate Statistical Inference

Probability and Mathematical Statistics

A Series of Monographs and Textbooks

Editors ***Z. W. Birnbaum*** ***E. Lukacs***

University of Washington *Bowling Green State University*
Seattle, Washington *Bowling Green, Ohio*

Thomas Ferguson. Mathematical Statistics: A Decision Theoretic Approach. 1967

Howard Tucker. A Graduate Course in Probability. 1967

K. R. Parthasarathy. Probability Measures on Metric Spaces. 1967

P. Révész. The Laws of Large Numbers. 1968

H. P. McKean, Jr. Stochastic Integrals. 1969

B. V. Gnedenko, Yu. K. Belyayev, and A. D. Solovyev. Mathematical Methods of Reliability Theory. 1969

Demetrios A. Kappos. Probability Algebras and Stochastic Spaces. 1969

Ivan N. Pesin. Classical and Modern Integration Theories. 1970

S. Vajda. Probabilistic Programming. 1972

Sheldon M. Ross. Introduction to Probability Models. 1972

Robert B. Ash. Real Analysis and Probability. 1972

V. V. Fedorov. Theory of Optimal Experiments. 1972

K. V. Mardia. Statistics of Directional Data. 1972

H. Dym and H. P. McKean. Fourier Series and Integrals. 1972

Tatsuo Kawata. Fourier Analysis in Probability Theory. 1972

Fritz Oberhettinger. Fourier Transforms of Distributions and Their Inverses: A Collection of Tables. 1973

Paul Erdös and Joel Spencer. Probabilistic Methods in Combinatorics. 1973

K. Sarkadi and I. Vincze. Mathematical Methods of Statistical Quality Control. 1973

Michael R. Anderberg. Cluster Analysis for Applications. 1973

W. Hengartner and R. Theodorescu. Concentration Functions. 1973

Kai Lai Chung. A Course in Probability Theory, Second Edition. 1974

L. H. Koopmans. The Spectral Analysis of Time Series. 1974

L. E. Maistrov. Probability Theory: A Historical Sketch. 1974

William F. Stout. Almost Sure Convergence. 1974

E. J. McShane. Stochastic Calculus and Stochastic Models. 1974

Robert B. Ash and Melvin F. Gardner. Topics in Stochastic Processes. 1975

Avner Friedman, Stochastic Differential Equations and Applications, Volume 1, 1975; Volume 2. 1976

Roger Cuppens. Decomposition of Multivariate Probabilities. 1975

Eugene Lukacs. Stochastic Convergence, Second Edition. 1975

H. Dym and H. P. McKean. Gaussian Processes, Function Theory, and the Inverse Spectral Problem. 1976

N. C. Giri. Multivariate Statistical Inference. 1977

Multivariate Statistical Inference

NARAYAN C. GIRI

DEPARTMENT OF MATHEMATICS
UNIVERSITY OF MONTREAL
MONTREAL, QUEBEC, CANADA

ACADEMIC PRESS New York San Francisco London 1977

A Subsidiary of Harcourt Brace Jovanovich, Publishers

TO MY MOTHER

and

THE MEMORY OF MY FATHER

COPYRIGHT © 1977, BY ACADEMIC PRESS, INC.
ALL RIGHTS RESERVED.
NO PART OF THIS PUBLICATION MAY BE REPRODUCED OR TRANSMITTED IN ANY FORM OR BY ANY MEANS, ELECTRONIC OR MECHANICAL, INCLUDING PHOTOCOPY, RECORDING, OR ANY INFORMATION STORAGE AND RETRIEVAL SYSTEM, WITHOUT PERMISSION IN WRITING FROM THE PUBLISHER.

ACADEMIC PRESS, INC.
111 Fifth Avenue, New York, New York 10003

United Kingdom Edition published by
ACADEMIC PRESS, INC. (LONDON) LTD.
24/28 Oval Road, London NW1

Library of Congress Cataloging in Publication Data

Giri, Narayan C Date
Multivariate statistical inference.

(Probability and mathematical statistics series ;)
1. Multivariate analysis. I. Title.
QA278.G56 519.5′35 76-27441
ISBN 0–12–285650–3

222,134

AMS (MOS) 1970 Subject Classifications: 62H10, 62H15, 62H20, 62H25, and 62H30

PRINTED IN THE UNITED STATES OF AMERICA

B/1550

Contents

Preface

This book is an up-to-date presentation of both the theoretical and applied aspects of multivariate analysis, in particular multivariate normal distributions using the invariance approach. It is written for readers with knowledge of mathematics and statistics at the undergraduate level. Various aspects are explained with live data from applied areas. In conformity with the general nature of introductory textbooks in multivariate analysis, we have tried to include many examples and motivations relevant to specific topics. The material presented here is developed from the lecture notes of a year-long graduate course on multivariate analysis by the author, presented several times at the University of Montreal and in part at the Indian Statistical Institute (Calcutta), Forschungsinstitut der Mathematik (Zurich), Cornell University, the University of Arizona, Sir George Williams University (Montreal), Indian Institute of Technology (Kanpur), Indian Institute of Technology (Kharagpur), Gauhati University (India), and the Institute of Agricultural Research Statistics (India). Each chapter contains numerous related problems and complete references. The exercises at the end of each chapter also contain analogous results for the complex multivariate normal population.

Invariance is the mathematical term for symmetry with respect to certain transformations. The notion of invariance in statistical inference is of old origin. The unpublished work on Hunt and Stein toward the end of World War II has given very strong support to the applicability and meaningfulness of this notion in the framework of the general class of statistical tests. It is now established as a very powerful tool for proving the optimality of many statistical test procedures. It is a generally accepted principle that if a problem with a unique solution is invariant under a certain transformation, then the solution should be invariant under that transformation. Another compelling reason for discussing multivariate analysis through invariance is that most of the commonly used test procedures are based on the likelihood ratio test principle. Under a mild restriction on the parametric space and the probability density function under consideration, the likelihood ratio test is almost invariant with respect to the group of transformations leaving the testing problem invariant.

The selection and presentation of material to cover the wide field of multivariate analysis have not been easy. In this I have been guided by my own experience teaching graduate and undergraduate courses in statistics at various levels and by my own experience conducting and guiding research in this field for the past fifteen years. We have presented the essential tools of multivariate analysis and have discussed their theoretical basis, enabling the readers to equip themselves for further research and consultation work in this field.

Chapter I contains some special results regarding characteristic roots and vectors, and partitioned submatrices of real and complex matrices. It also contains some special theorems on real and complex matrices useful in multivariate analysis.

Chapter II deals with the theory of groups and related results that are useful for the development of invariant statistical test procedures. It also includes the Jacobians of some specific transformations that are useful for deriving multivariate sampling distributions.

Chapter III is devoted to basic notions of multivariate distributions and the principle of invariance in statistical testing of hypotheses. The interrelationships between invariance and sufficiency, invariance and unbiasedness, invariance and optimum tests, and invariance and most stringent tests are examined.

Chapter IV deals with the study of the real multivariate normal distribution through the probability density function and through a simple characterization. The second approach simplifies the multivariate theory and enables suitable generalizations from the univariate theory without further analysis. It also contains some characterizations of the real multivariate normal distribution, concentration ellipsoid and axes, regression, and multiple and partial correlation. The analogous results for the complex multivariate normal distribution are also included.

The maximum likelihood estimators of the parameters of the multivariate normal distribution and their optimum properties form the subject matter of Chapter V.

Chapter VI contains a systematic derivation of basic multivariate sampling distributions for the real case. Complex analogues of these results are included as problems.

Tests and confidence regions of mean vectors of multivariate normal populations with known and unknown covariance matrices and their optimum properties are dealt with in Chapter VII.

Chapter VIII is devoted to a systematic derivation of tests concerning covariance matrices and mean vectors of multivariate normal populations and to the study of their optimum properties.

Chapter IX contains a modern treatment of discriminant analysis. A brief history of discriminant analysis is also included.

Chapter X contains different covariance models and their analysis for the multivariate normal distribution. Principal components, factor models, canonical correlations, and time series are included here.

We feel that it will be appropriate to spread the material of the book over two three-hour, one-semester basic courses on multivariate analysis.

Acknowledgments

If the reader finds the book useful, the credit is entirely due to my teachers and colleagues, especially Professor C. M. Stein of Stanford University and Professor J. Kiefer of Cornell University under whose influence I have learned to appreciate the multivariate analysis of the present century.

Preparation and revision of the manuscript would not have been an easy task without the help of Dr. B. K. Sinha and Dr. A. K. Bhattacharji, who helped me by reading the entire manuscript with great care and diligence and offering valuable suggestions at various stages. The suggestions of the reviewers were also very helpful in improving the presentation. I would like to express my gratitude and thanks to them.

This book was written with the direct and indirect help of many people. I owe a great debt to my parents and my brothers and sisters for their help and encouragement. Had it not been for them, I probably would not have been able to complete even my secondary school.

My wife Nilima, daughter Nabanita, and son Nandan have been very helpful and patient during the preparation of the book. I gratefully acknowledge their assistance.

I wish to express my appreciation to the National Research Council of Canada and to the Ministry of Education, Government of Quebec for financial assistance for the preparation of the manuscript. I would also like to express my gratitude to the secretaries of the Department of Mathematics, University of Montreal, for an excellent job in typing the manuscript. Finally, I would like to express my sincere thanks to Professor R. Clearoux and Professor M. Ahmad for computational help.

CHAPTER I

Vector and Matrix Algebra

1.0 INTRODUCTION

The study of multivariate analysis requires knowledge of vector and matrix algebra, some basic results of which are considered in this chapter. Some of these results are stated herein without proof; proofs can be obtained from Giri (1974), Mac Lane and Birkoff (1967), Markus and Mine (1967), Perlis (1952), or any textbook on matrix algebra.

1.1 VECTORS

A vector is an ordered p-tuple $x_1, \ldots, x_p$ and is written as

$$\mathbf{x} = \begin{pmatrix} x_1 \\ \vdots \\ x_p \end{pmatrix}.$$

Actually it is called a p-dimensional column vector. For brevity we shall simply call it a p-vector or a vector. The transpose of $\mathbf{x}$ is given by $\mathbf{x}' = (x_1, \ldots, x_p)$. If all components of a vector are zero, it is called the null vector

$\mathbf{0}$. Geometrically a p-vector represents a point $A = (x_1, \ldots, x_p)$ or the directed line segment $\overline{OA}$ with the point A in the p-dimensional Euclidean space E^p. The set of all p-vectors is denoted by V^p. Obviously $V^p = E^p$ if all components of the vectors are real numbers. For any two vectors $\mathbf{x} = (x_1, \ldots, x_p)'$ and $\mathbf{y} = (y_1, \ldots, y_p)'$ we define the vector sum $\mathbf{x} + \mathbf{y} = (x_1 + y_1, \ldots, x_p + y_p)'$ and scalar multiplication by a constant a by

$$a\mathbf{x} = (ax_1, \ldots, ax_p)'.$$

Obviously vector addition is an associative and commutative operation, i.e., $\mathbf{x} + \mathbf{y} = \mathbf{y} + \mathbf{x}$, $(\mathbf{x} + \mathbf{y}) + \mathbf{z} = \mathbf{x} + (\mathbf{y} + \mathbf{z})$ where $\mathbf{z} = (z_1, \ldots, z_p)'$, and scalar multiplication is a distributive operation, i.e., for constants a, b, $(a + b)\mathbf{x} = a\mathbf{x} + b\mathbf{x}$. For $\mathbf{x}, \mathbf{y} \in V^p$, $\mathbf{x} + \mathbf{y}$ and $a\mathbf{x}$ also belong to V^p. Furthermore, for scalar constants a, b, $a(\mathbf{x} + \mathbf{y}) = a\mathbf{x} + a\mathbf{y}$ and $a(b\mathbf{x}) = b(a\mathbf{x}) = ab\mathbf{x}$.

The quantity $\mathbf{x}'\mathbf{y} = \mathbf{y}'\mathbf{x} = \sum_1^p x_i y_i$ is called the *scalar product* of two vectors $\mathbf{x}$, $\mathbf{y}$ in V^p. The scalar product of a vector $\mathbf{x} = (x_1, \ldots, x_p)'$ with itself is denoted by $\|\mathbf{x}\|^2 = \mathbf{x}'\mathbf{x}$, where $\|\mathbf{x}\|$ is called the *norm* of $\mathbf{x}$. Some geometrical significances of the norm are

(i) $\|\mathbf{x}\|^2$ is the square of the distance of the point $\mathbf{x}$ from the origin in E^p,

(ii) the square of the distance between two points $(x_1, \ldots, x_p)$, $(y_1, \ldots, y_p)$ is given by $\|\mathbf{x} - \mathbf{y}\|^2$,

(iii) the angle θ between two vectors $\mathbf{x}$, $\mathbf{y}$ is given by $\cos\theta = (\mathbf{x}/\|\mathbf{x}\|)'(\mathbf{y}/\|\mathbf{y}\|)$.

DEFINITION 1.1.1 *Orthogonal vectors* Two vectors $\mathbf{x}$, $\mathbf{y}$ in V^p are said to be orthogonal if and only if $\mathbf{x}'\mathbf{y} = \mathbf{y}'\mathbf{x} = 0$. A set of vectors in V^p is orthogonal if the vectors are pairwise orthogonal.

Geometrically two vectors $\mathbf{x}$, $\mathbf{y}$ are orthogonal if and only if the angle between them is 90°. An orthogonal vector $\mathbf{x}$ is called an orthonormal vector if $\|\mathbf{x}\|^2 = 1$.

DEFINITION 1.1.2 *Projection of a vector* The projection of a vector $\mathbf{x}$ on $\mathbf{y}$ $(\neq \mathbf{0})$, both belonging to V^p, is given by $\|\mathbf{y}\|^{-2}(\mathbf{x}' \cdot \mathbf{y})\mathbf{y}$.

If $\overline{OA} = \mathbf{x}$, $\overline{OB} = \mathbf{y}$, and P is the foot of the perpendicular from the point A on OB, then $\overline{OP} = \|\mathbf{y}\|^{-2}(\mathbf{x}' \cdot \mathbf{y})\mathbf{y}$ where O is the origin of E^p. For two orthogonal vectors $\mathbf{x}$, $\mathbf{y}$ the projection of $\mathbf{x}$ on $\mathbf{y}$ is zero.

DEFINITION 1.1.3 A set of vectors $\boldsymbol{\alpha}_1, \ldots, \boldsymbol{\alpha}_k$ in V^p is said to be linearly independent if none of the vectors can be expressed as a linear combination of the others.

Thus if $\boldsymbol{\alpha}_1, \ldots, \boldsymbol{\alpha}_k$ are linearly independent, then there does not exist a set of scalar constants $c_1, \ldots, c_k$ not all zero such that $c_1\boldsymbol{\alpha}_1 + \cdots + c_k\boldsymbol{\alpha}_k = \mathbf{0}$. It may be verified that a set of orthogonal vectors in V^p is linearly independent.

DEFINITION 1.1.4 *Vector space spanned by a set of vectors* Let $\boldsymbol{\alpha}_1, \ldots, \boldsymbol{\alpha}_k$ be a set of k vectors in V^p. Then the vector space V spanned by $\boldsymbol{\alpha}_1, \ldots, \boldsymbol{\alpha}_k$ is the set of all vectors which can be expressed as linear combinations of $\boldsymbol{\alpha}_1, \ldots, \boldsymbol{\alpha}_k$ and the null vector $\mathbf{0}$.

Thus if $\boldsymbol{\alpha}, \boldsymbol{\beta} \in V$, then for scalar constants a, b, $a\boldsymbol{\alpha} + b\boldsymbol{\beta}$ and $a\boldsymbol{\alpha}$ also belong to V. Furthermore, since $\boldsymbol{\alpha}_1, \ldots, \boldsymbol{\alpha}_p$ belong to V^p, any linear combination of $\boldsymbol{\alpha}_1, \ldots, \boldsymbol{\alpha}_k$ also belongs to V^p and hence $V \subset V^p$. So V is a *linear subspace* of V^p.

DEFINITION 1.1.5 *Basis of a vector space* A basis of a vector space V is a set of linearly independent vectors which span V.

In V^p the unit vectors $\boldsymbol{\varepsilon}_1 = (1, 0, \ldots, 0)'$, $\boldsymbol{\varepsilon}_2 = (0, 1, 0, \ldots, 0)', \ldots, \boldsymbol{\varepsilon}_p = (0, \ldots, 0, 1)'$ form a basis of V^p.

If $\boldsymbol{\alpha}_1, \ldots, \boldsymbol{\alpha}_k$ span V, then a subset of $\boldsymbol{\alpha}_1, \ldots, \boldsymbol{\alpha}_k$ forms a basis of V.

THEOREM 1.1.1 Every vector space V has a basis and two bases of V have the same number of elements.

THEOREM 1.1.2 Let the vector space V be spanned by the vectors $\boldsymbol{\alpha}_1, \ldots, \boldsymbol{\alpha}_k$. Any element $\boldsymbol{\alpha} \in V$ can be uniquely expressed as $\boldsymbol{\alpha} = \sum_1^k c_i\boldsymbol{\alpha}_i$ for scalar constants $c_1, \ldots, c_k$, not all zero, if and only if $\boldsymbol{\alpha}_1, \ldots, \boldsymbol{\alpha}_k$ is a basis of V.

DEFINITION 1.1.6 *Coordinates of a vector* If $\boldsymbol{\alpha}_1, \ldots, \boldsymbol{\alpha}_k$ is a basis of a vector space V and if $\boldsymbol{\alpha} \in V$ is uniquely expressed as $\boldsymbol{\alpha} = \sum_1^k c_i\boldsymbol{\alpha}_i$ for scalar constants $c_1, \ldots, c_k$, then the coefficient c_i of the vector $\boldsymbol{\alpha}_i$ is called the ith coordinate of $\boldsymbol{\alpha}$ with respect to the basis $\boldsymbol{\alpha}_1, \ldots, \boldsymbol{\alpha}_k$.

DEFINITION 1.1.7 *Rank of a vector space* The number of vectors in a basis of a vector space V is called the rank or the dimension of V.

1.2 MATRICES

DEFINITION 1.2.1 *Matrix* A real matrix A is an ordered rectangular array of elements a_{ij} (reals)

$$A = \begin{pmatrix} a_{11} & \cdots & a_{1q} \\ \vdots & & \vdots \\ a_{p1} & \cdots & a_{pq} \end{pmatrix} \tag{1.1}$$

and is written as $A_{p \times q} = (a_{ij})$.

A matrix with p rows and q columns is called a matrix of dimension $p \times q$ (p by q), the number of rows always being listed first. If $p = q$, we call it a square matrix of dimension p. A p-dimensional column vector is a matrix of dimension $p \times 1$. Two matrices of the same dimension $A_{p \times q}$, $B_{p \times q}$ are said to be equal (written as $A = B$) if $a_{ij} = b_{ij}$ for $i = 1, \ldots, p$, $j = 1, \ldots, q$. If all $a_{ij} = 0$, then A is called a null matrix and is denoted by boldface zero, $\mathbf{0}$. The transpose of a $p \times q$ matrix A is a $q \times p$ matrix A':

$$A' = \begin{pmatrix} a_{11} & \cdots & a_{p1} \\ \vdots & & \vdots \\ a_{1q} & \cdots & a_{pq} \end{pmatrix} \tag{1.2}$$

and is obtained by interchanging the rows and columns of A. Obviously $(A')' = A$. A square matrix A is said to be *symmetric* if $A = A'$ and is *skew symmetric* if $A = -A'$. The diagonal elements of a skew symmetric matrix are zero. In what follows we shall use the notation "A of dimension $p \times q$" instead of $A_{p \times q}$.

For any two matrices $A = (a_{ij})$ and $B = (b_{ij})$ of the same dimension $p \times q$ we define the matrix sum $A + B$ as a matrix $(a_{ij} + b_{ij})$ of dimension $p \times q$. The matrix $A - B$ is to be understood in the same sense as $A + B$ where the plus $(+)$ is replaced by the minus $(-)$ sign. Clearly $(A + B)' = A' + B'$, $A + B = B + A$, and for any three matrices A, B, C, $(A + B) + C = A + (B + C)$. Thus the operation *matrix sum* is commutative and associative.

For any matrix $A = (a_{ij})$ and a scalar constant c, the scalar product cA is defined by $cA = Ac = (ca_{ij})$. Obviously $(cA)' = cA'$, so scalar product is a distributive operation.

The *matrix product* of two matrices $A_{p \times q} = (a_{ij})$ and $B_{q \times r} = (b_{ij})$ is a matrix $C_{p \times r} = AB = (c_{ij})$ where

$$c_{ij} = \sum_{k=1}^{q} a_{ik} b_{kj}, \qquad i = 1, \ldots, p, \quad j = 1, \ldots, r. \tag{1.3}$$

The product AB is defined if the number of columns of A is equal to the number of rows of B and in general $AB \neq BA$. Furthermore $(AB)' = B'A'$. The matrix product is distributive and associative provided the products are defined, i.e., for any three matrices A, B, C,

(i) $A(B + C) = AB + AC$ (distributive),
(ii) $(AB)C = A(BC)$ (associative).

DEFINITION 1.2.2 *Diagonal matrix* A square matrix A is said to be a diagonal matrix if all its off-diagonal elements are zero.

DEFINITION 1.2.3 *Identity matrix* A diagonal matrix whose diagonal elements are unity is called an identity matrix and is denoted by I.

For any square matrix A, $AI = IA = A$.

DEFINITION 1.2.4 *Triangular matrix* A square matrix $A = (a_{ij})$ with $a_{ij} = 0, j < i$, is called an upper triangular matrix. If $a_{ij} = 0$ for $j > i$, then A is called a lower triangular matrix.

DEFINITION 1.2.5 *Orthogonal matrix* A square matrix A is said to be orthogonal if $AA' = A'A = I$.

Associated with any square matrix $A = (a_{ij})$ of dimension $p \times p$ is a unique scalar quantity $|A|$, or det A, called the determinant of A which is defined by

$$|A| = \sum_{\pi} \delta(\pi) a_{1\pi(1)} a_{2\pi(2)} \cdots a_{p\pi(p)}, \tag{1.4}$$

where π runs over all $p!$ permutations of column subscripts $(1, 2, \ldots, p)$, and $\delta(\pi) = 1$ if the number of inversions in $\pi(1), \ldots, \pi(p)$ from the standard order $1, \ldots, p$ is even and $\delta(\pi) = -1$ if the number of such inversions is odd. The number of inversions in a particular permutation is the total number of times in which an element is followed by numbers which would ordinarily precede it in the standard order $1, 2, \ldots, p$. From Chapter III on we shall consistently use the symbol det A for the determinant and reserve $|\;|$ for the absolute value symbol.

DEFINITION 1.2.6 *Minor and cofactor* For any square matrix $A = (a_{ij})$ of dimension $p \times p$, the minor of the element a_{ij} is the determinant of the matrix formed by deleting the ith row and the jth column of A. The quantity $(-1)^{i+j} \times$ the minor of a_{ij} is called the cofactor of a_{ij} and is symbolically denoted by A_{ij}.

The determinant of a submatrix (of A) of dimension $i \times i$ whose diagonal elements are also the diagonal elements of A is called a principal minor of order i. The set of leading principal minors is a set of p principal minors of orders $1, 2, \ldots, p$, respectively, such that the matrix of principal minor of order i is a submatrix of the matrix of the principal minor of order $i + 1$, $i = 1, \ldots, p$.

It is easy to verify that for any square matrix $A = (a_{ij})$ of dimension $p \times p$

$$|A| = \sum_{j=1}^{p} a_{ij} A_{ij} = \sum_{i=1}^{p} a_{ij} A_{ij}, \tag{1.5}$$

and for $j \neq j'$, $i \neq i'$,

$$\sum_{i=1}^{p} a_{ij} A_{i'j'} = \sum_{j=1}^{p} a_{ij} A_{i'j'} = 0. \tag{1.6}$$

Furthermore, if A is symmetric, then $A_{ij} = A_{ji}$ for all i, j. For a triangular or a diagonal matrix A of dimension $p \times p$ with diagonal elements a_{ii},

$|A| = \prod_{i=1}^{p} a_{ii}$. If any two columns or rows of A are interchanged, then $|A|$ changes its sign, and $|A| = 0$ if two columns or rows of A are equal or proportional.

DEFINITION 1.2.7 *Nonsingular matrix* A square matrix A is called nonsingular if $|A| \neq 0$. If $|A| = 0$, then we call it a singular matrix.

The rows and the columns of a nonsingular matrix are linearly independent. Since for any two square matrices A, B, $|AB| = |A|\,|B|$, we conclude that the product of two nonsingular matrices is a nonsingular matrix. However, the sum of two nonsingular matrices is not necessarily a nonsingular matrix. One such trivial case is $A = -B$ where both A and B are nonsingular matrices.

DEFINITION 1.2.8 *Inverse matrix* The inverse of a nonsingular matrix A of dimension $p \times p$ is the unique matrix A^{-1} such that $A^{-1}A = AA^{-1} = I$.

Let A_{ij} be the cofactor of the element a_{ij} of A and

$$C = \begin{pmatrix} \dfrac{A_{11}}{|A|} & \cdots & \dfrac{A_{1p}}{|A|} \\ \vdots & & \vdots \\ \dfrac{A_{p1}}{|A|} & \cdots & \dfrac{A_{pp}}{|A|} \end{pmatrix}. \tag{1.7}$$

From (1.6) and (1.7) we get $AC' = I$. Hence $A^{-1} = C'$. The inverse matrix is defined only for the nonsingular matrix and A^{-1} is symmetric if A is symmetric. Furthermore $|A^{-1}| = (|A|)^{-1}$, $(A')^{-1} = (A^{-1})'$, and $(AB)^{-1} = B^{-1}A^{-1}$.

1.3 RANK AND TRACE OF A MATRIX

Let A be a matrix of dimension $p \times q$. Let $R(A)$ be the vector space spanned by the rows of A and let $C(A)$ be the vector space spanned by the columns of A. The space $R(A)$ is called the *row space* of A and its rank $r(A)$ is called the *row rank* of A. The space $C(A)$ is called the *column space* of A and its rank $c(A)$ is called the *column rank* of A. For any matrix A, $r(A) = c(A)$.

DEFINITION 1.3.1 *Rank of a matrix* The common value of the row rank and the column rank is called the rank of the matrix A and is denoted by $\rho(A)$.

For any matrix A of dimension $p \times q$, $q < p$, $\rho(A)$ may vary from 0 to q. If $\rho(A) = q$, then A is called the matrix of full rank. The rank of the null matrix $\mathbf{0}$ is 0.

For any two matrices A, B for which AB is defined, the columns of AB are linear combinations of the columns of A. Thus the number of linearly independent columns of AB cannot exceed the number of linearly independent columns of A. Hence $\rho(AB) \le \rho(A)$. Similarly, considering the rows of AB we can argue that $\rho(AB) \le \rho(B)$. Hence $\rho(AB) \le \min(\rho(A), \rho(B))$.

THEOREM 1.3.1 If A, B, C are matrices of dimensions $p \times q$, $p \times p$, $q \times q$, respectively, then $\rho(A) = \rho(AC) = \rho(BA) = \rho(BAC)$.

DEFINITION 1.3.2 *Trace of a matrix* The trace of a square matrix $A = (a_{ij})$ of dimension $p \times p$ is defined by the sum of its diagonal elements and is denoted by $\operatorname{tr} A = \sum_1^p a_{ii}$.

Obviously $\operatorname{tr} A = \operatorname{tr} A'$, $\operatorname{tr}(A + B) = \operatorname{tr}(A) + \operatorname{tr}(B)$. Furthermore, $\operatorname{tr} AB = \operatorname{tr} BA$, provided both AB and BA are defined. Hence for any orthogonal matrix θ, $\operatorname{tr} \theta' A \theta = \operatorname{tr} A\theta\theta' = \operatorname{tr} A$.

1.4 QUADRATIC FORMS AND POSITIVE DEFINITE MATRIX

A quadratic form in the real variables $x_1, \ldots, x_p$ is an expression of the form $Q = \sum_{i=1}^p \sum_{j=1}^p a_{ij}x_i x_j$, where a_{ij} are real constants.

Writing $\mathbf{x} = (x_1, \ldots, x_p)'$, $A = (a_{ij})$ we can write $Q = \mathbf{x}'A\mathbf{x}$. Without any loss of generality we can take the matrix A in the quadratic form Q to be a symmetric one. Since Q is a scalar quantity,

$$Q = Q' = \mathbf{x}'A'\mathbf{x} = \tfrac{1}{2}(Q + Q') = \mathbf{x}'((A + A')/2)\mathbf{x}$$

and $\frac{1}{2}(A + A')$ is a symmetric matrix.

DEFINITION 1.4.1 *Positive definite matrix* A square matrix A or the associated quadratic form $\mathbf{x}'A\mathbf{x}$ is called positive definite if $\mathbf{x}'A\mathbf{x} > 0$ for all $\mathbf{x} \ne \mathbf{0}$ and is called positive semidefinite if $\mathbf{x}'A\mathbf{x} \ge 0$ for all $\mathbf{x}$.

The matrix A or the associated quadratic form $\mathbf{x}'A\mathbf{x}$ is *negative definite* or *negative semidefinite* if $-\mathbf{x}'A\mathbf{x}$ is positive definite or positive semidefinite, respectively.

EXAMPLE 1.4.1

$$(x_1, x_2)\begin{pmatrix} 2 & 1 \\ 1 & 3 \end{pmatrix}(x_1, x_2)' = 2x_1^2 + 2x_1x_2 + 3x_2^2 = 2(x_1 + \tfrac{1}{2}x_2)^2 + \tfrac{5}{2}x_2^2 > 0$$

for all $x_1 \ne 0$, $x_2 \ne 0$. Hence the matrix $\left(\begin{smallmatrix} 2 & 1 \\ 1 & 3 \end{smallmatrix}\right)$ is positive definite.

1.5 CHARACTERISTIC ROOTS AND VECTORS

The characteristic roots of a square matrix $A = (a_{ij})$ of dimension $p \times p$ are given by the roots of the characteristic equation

$$|A - \lambda I| = 0 \tag{1.8}$$

where λ is real. Obviously this is an equation of degree p in λ and thus has exactly p roots. If A is a diagonal matrix, then the diagonal elements are themselves the characteristic roots of A. In general we can write (1.8) as

$$(-\lambda)^p + (-\lambda)^{p-1}S_1 + (-\lambda)^{p-2}S_2 + \cdots + (-\lambda)S_{p-1} + |A| = 0 \tag{1.9}$$

where S_i is the sum of all principal minors of order i of A. In particular, $S_1 = \operatorname{tr} A$. Thus the product of the characteristic roots of A is equal to $|A|$ and the sum of the characteristic roots of A is equal to tr A. The vector $\mathbf{x} = (x_1, \ldots, x_p)'$, not identically zero, satisfying

$$(A - \lambda I)\mathbf{x} = \mathbf{0}, \tag{1.10}$$

is called the *characteristic vector* of the matrix A, corresponding to its characteristic root λ. Clearly, if $\mathbf{x}$ is a characteristic vector of the matrix A corresponding to its characteristic root λ, then any scalar multiple $c\mathbf{x}$, $c \neq 0$, is also a characteristic vector of A corresponding to λ. Since, for any orthogonal matrix θ of dimension $p \times p$,

$$|\theta A\theta' - \lambda I| = |\theta A\theta' - \lambda\theta\theta'| = |A - \lambda I|,$$

the characteristic roots of the matrix A remain invariant (unchanged) with respect to the transformation $A \to \theta A\theta'$.

THEOREM 1.5.1 If A is a real symmetric matrix (of order $p \times p$), then all its characteristic roots are real.

Proof Let λ be a complex characteristic root of A and let $\mathbf{x} + i\mathbf{y}$, $\mathbf{x} = (x_1, \ldots, x_p)'$, $\mathbf{y} = (y_1, \ldots, y_p)'$, be the characteristic vector (complex) corresponding to λ. Then from (1.10)

$$A(\mathbf{x} + i\mathbf{y}) = \lambda(\mathbf{x} + i\mathbf{y}), \qquad (\mathbf{x} - i\mathbf{y})'A(\mathbf{x} + i\mathbf{y}) = \lambda(\mathbf{x}'\mathbf{x} + \mathbf{y}'\mathbf{y}).$$

But

$$(\mathbf{x} - i\mathbf{y})'A(\mathbf{x} + i\mathbf{y}) = \mathbf{x}'A\mathbf{x} + \mathbf{y}'A\mathbf{y}.$$

Hence we conclude that λ must be real. Q.E.D.

Note The characteristic vector $\mathbf{z}$ corresponding to a complex characteristic root λ must be complex. Otherwise $A\mathbf{z} = \lambda\mathbf{z}$ will imply that a real vector is equal to a complex vector.

THEOREM 1.5.2 The characteristic vectors corresponding to distinct characteristic roots of a symmetric matrix are orthogonal.

Proof Let λ_1, λ_2 be two distinct characteristic roots of a symmetric (real) matrix A and let $\mathbf{x} = (x_1, \ldots, x_p)'$, $\mathbf{y} = (y_1, \ldots, y_p)'$ be the characteristic vectors corresponding to λ_1, λ_2, respectively. Then

$$A\mathbf{x} = \lambda_1\mathbf{x}, \qquad A\mathbf{y} = \lambda_2\mathbf{y}.$$

So

$$\mathbf{y}'A\mathbf{x} = \lambda_1\mathbf{y}'\mathbf{x}, \qquad \mathbf{x}'A\mathbf{y} = \lambda_2\mathbf{x}'\mathbf{y}.$$

Thus

$$\lambda_1\mathbf{x}'\mathbf{y} = \lambda_2\mathbf{x}'\mathbf{y}.$$

Since $\lambda_1 \neq \lambda_2$ we conclude that $\mathbf{x}'\mathbf{y} = 0$. Q.E.D.

Let λ be a characteristic root of a symmetric positive definite matrix A and let $\mathbf{x}$ be the corresponding characteristic vector. Then

$$\mathbf{x}'A\mathbf{x} = \lambda\mathbf{x}'\mathbf{x} > 0.$$

Hence we get the following theorem.

THEOREM 1.5.3 The characteristic roots of a symmetric positive definite matrix are all positive.

THEOREM 1.5.4 For every real symmetric matrix A, there exists an orthogonal matrix θ such that $\theta A\theta'$ is a diagonal matrix whose diagonal elements are the characteristic roots of A.

Proof Let $\lambda_1 \geq \lambda_2 \geq \cdots \geq \lambda_p$ denote the characteristic roots of A including multiplicities and let $\mathbf{x}_i$ be the characteristic vector of A, corresponding to the characteristic root λ_i, $i = 1, \ldots, p$. Write

$$\mathbf{y}_i = \mathbf{x}_i/\|\mathbf{x}_i\|, \qquad i = 1, \ldots, p;$$

obviously $\mathbf{y}_1, \ldots, \mathbf{y}_p$ are the normalized characteristic vectors of A.

Suppose there exists s $(\leq p)$ orthonormal vectors $\mathbf{y}_1, \ldots, \mathbf{y}_s$ such that $(A - \lambda_i I)\mathbf{y}_i = \mathbf{0}$, $i = 1, \ldots, s$. Denoting by A^r the product of r matrices each equal to A we get

$$A^r\mathbf{y}_i = \lambda_i A^{r-1}\mathbf{y}_i = \cdots = \lambda_i^r\mathbf{y}_i, \qquad i = 1, \ldots, s.$$

Let $\mathbf{x}$ be orthogonal to the vector space spanned by $\mathbf{y}_1, \ldots, \mathbf{y}_s$. Then

$$(A^r\mathbf{x})'\mathbf{y}_i = \mathbf{x}'A^r\mathbf{y}_i = \lambda_i^r\mathbf{x}'\mathbf{y}_i = \mathbf{0}$$

for all r including zero and $i = 1, \ldots, s$. Hence any vector belonging to the vector space spanned by the vectors $\mathbf{x}$, $A\mathbf{x}$, $A^2\mathbf{x}, \ldots$ is orthogonal to any vector spanned by $\mathbf{y}_1, \ldots, \mathbf{y}_s$. Obviously not all vectors $\mathbf{x}$, $A\mathbf{x}$, $A^2\mathbf{x}, \ldots$ are linearly independent. Let k be the smallest value of r such that for real constants $c_1, \ldots, c_k$

$$A^k\mathbf{x} + c_1 A^{k-1}\mathbf{x} + \cdots + c_k\mathbf{x} = \mathbf{0}.$$

Factorizing the left-hand side of this expression we can, for constants $u_1, \ldots, u_k$, write it as

$$\prod_{i=1}^{k} (A - u_i I)\mathbf{x} = \mathbf{0}.$$

Let

$$\mathbf{y}_{s+1} = \prod_{i=2}^{k} (A - u_i I)\mathbf{x}.$$

Then $(A - u_1 I)\mathbf{y}_{s+1} = \mathbf{0}$.

In other words there exists a normalized vector $\mathbf{y}_{s+1}$ in the space spanned by $(\mathbf{x}, A\mathbf{x}, A^2\mathbf{x}, \ldots)$ which is a characteristic vector of A corresponding to its root $u_1 = \lambda_{s+1}$ (say) and $\mathbf{y}_{s+1}$ is orthogonal to $\mathbf{y}_1, \ldots, \mathbf{y}_s$. Since $\mathbf{y}_1$ can be chosen corresponding to any characteristic root to start with, we have proved the existence of p orthonormal vectors $\mathbf{y}_1, \ldots, \mathbf{y}_p$ satisfying $A\mathbf{y}_i = \lambda_i \mathbf{y}_i$, $i = 1, \ldots, p$. Let θ be an orthogonal matrix of dimension $p \times p$ with $\mathbf{y}_i$ as its rows. Obviously then $\theta A \theta'$ is a diagonal matrix with diagonal elements $\lambda_1, \ldots, \lambda_p$. Q.E.D.

From this theorem it follows that any positive definite quadratic form $\mathbf{x}'A\mathbf{x}$ can be transformed into a diagonal form $\sum_{i=1}^{p} \lambda_i y_i^2$ where $\mathbf{y} = (y_1, \ldots, y_p)' = \theta\mathbf{x}$, and the orthogonal matrix θ is such that $\theta A \theta'$ is a diagonal matrix with diagonal elements $\lambda_1, \ldots, \lambda_p$ (characteristic roots of A). Note that $\mathbf{x}'A\mathbf{x} = (\theta\mathbf{x})'(\theta A\theta')(\theta\mathbf{x})$.

Since the characteristic roots of a positive definite matrix A are all positive, $|A| = |\theta A \theta'| = \prod_{i=1}^{p} \lambda_i > 0$.

THEOREM 1.5.5 For every positive definite matrix A there exists a nonsingular matrix C such that $A = C'C$.

Proof From Theorem 1.5.4, there exists an orthogonal matrix θ such that $\theta A \theta'$ is a diagonal matrix D with diagonal elements $\lambda_1, \ldots, \lambda_p$, the characteristic roots of A. Let $D^{1/2}$ be a diagonal matrix with diagonal elements $\lambda_1^{1/2}, \ldots, \lambda_p^{1/2}$ and let $D^{1/2}\theta = C$. Then $A = \theta' D \theta = C'C$ and obviously C is a nonsingular matrix. Q.E.D.

Any positive definite quadratic form $\mathbf{x}'A\mathbf{x}$ can be transformed to a diagonal form $\mathbf{y}'\mathbf{y}$ where $\mathbf{y} = C\mathbf{x}$ and C is a nonsingular matrix such that $A = C'C$.

Furthermore, given any positive definite matrix A there exists a nonsingular matrix B such that $B'AB = I$ $(B = C^{-1})$.

THEOREM 1.5.6 If A is a positive definite matrix, then A^{-1} is also positive definite.

Proof Let $A = C'C$ where C is a nonsingular matrix. Then

$$\mathbf{x}'A^{-1}\mathbf{x} = ((C')^{-1}\mathbf{x})'((C')^{-1}\mathbf{x}) > 0 \qquad \text{for all} \quad \mathbf{x} \neq \mathbf{0}. \qquad \text{Q.E.D.}$$

THEOREM 1.5.7 Let A be a symmetric and at least positive semidefinite matrix of dimension $p \times p$ and of rank $r \leq p$. Then A has exactly r positive characteristic roots and the remaining $p - r$ characteristic roots of A are zero.

The proof is left to the reader.

THEOREM 1.5.8 Let A be a symmetric nonsingular matrix of dimension $p \times p$. Then there exists a nonsingular matrix C such that

$$CAC' = \begin{pmatrix} I & 0 \\ 0 & -I \end{pmatrix}$$

where the order of I is the number of positive characteristic roots of A and that of $-I$ is the number of negative characteristic roots of A.

Proof From Theorem 1.5.4 there exists an orthogonal matrix θ such that $\theta A\theta'$ is a diagonal matrix with diagonal elements $\lambda_1, \ldots, \lambda_p$, the characteristic roots of A. Without any loss of generality let us assume that $\lambda_1 \geq \cdots \geq \lambda_q > 0 > \lambda_{q+1} \geq \cdots \geq \lambda_p$. Let D be a diagonal matrix with diagonal elements $(\lambda_1)^{-1/2}, \ldots (\lambda_q)^{-1/2}, (-\lambda_{q+1})^{-1/2}, \ldots, (-\lambda_p)^{-1/2}$, respectively. Then

$$D\theta A\theta' D' = \begin{pmatrix} I & 0 \\ 0 & -I \end{pmatrix}. \qquad \text{Q.E.D.}$$

THEOREM 1.5.9 Let A, B be two matrices of dimensions $p \times q$, $q \times p$, respectively. Then every nonzero characteristic root of AB is also a nonzero characteristic root of BA.

Proof Let λ be a nonzero characteristic root of AB. Then $|AB - \lambda I| = 0$. This implies

$$\begin{vmatrix} \lambda I & A \\ B & I \end{vmatrix} = 0.$$

But this we can obviously write as

$$\begin{vmatrix} \lambda I & B \\ A & I \end{vmatrix} = |BA - \lambda I| = 0. \qquad \text{Q.E.D.}$$

Thus it follows from Theorem 1.5.7 that a positive semidefinite quadratic form $\mathbf{x}'A\mathbf{x}$ of rank $r \leq p$ can be reduced to the diagonal form $\sum_1^r \lambda_i y_i^2$ where $\lambda_1, \ldots, \lambda_r$ are the positive characteristic roots of A and $y_1, \ldots, y_r$ are linear combinations of the components $x_1, \ldots, x_p$ of $\mathbf{x}$.

THEOREM 1.5.10 If A is positive definite and B is positive semidefinite of the same dimension $p \times p$, then there exists a nonsingular matrix C such that $CAC' = I$ and CBC' is a diagonal matrix with diagonal elements $\lambda_1, \ldots, \lambda_p$, the roots of the equation $|B - \lambda A| = 0$.

Proof Since A is positive definite, there exists a nonsingular matrix D such that $DAD' = I$. Let $DBD' = B^*$. Since B^* is a real symmetric matrix there exists an orthogonal matrix θ such that $\theta DBD'\theta'$ is a diagonal matrix. Write $\theta D = C$, where C is a nonsingular matrix. Obviously $CAC' = I$ and CBC' is a diagonal matrix whose diagonal elements are the characteristic roots of B^*, which are, in turn, the roots of $|B - \lambda A| = 0$. Q.E.D.

THEOREM 1.5.11 Let A be a matrix of dimension $p \times q$, $p < q$. Then AA' is symmetric and positive semidefinite if the rank of $A < p$ and positive definite if the rank of $A = p$.

Proof Obviously AA' is symmetric and the rank of AA' is equal to the rank of A. Let the rank of AA' be r $(\leq p)$. Since AA' is symmetric there exists an orthogonal $p \times p$ matrix θ such that $\theta AA'\theta'$ is a diagonal matrix with nonzero diagonal elements $\lambda_1, \ldots, \lambda_r$. Let $\mathbf{x} = (x_1, \ldots, x_p)'$, $\mathbf{y} = \theta\mathbf{x}$. Then

$$\mathbf{x}'AA'\mathbf{x} = \sum_1^r \lambda_i y_i^2 \geq 0 \qquad \text{for all} \quad \mathbf{x}.$$

If $r = p$, then

$$\mathbf{x}'AA'\mathbf{x} = \sum_1^p \lambda_i y_i^2 > 0 \qquad \text{for all} \quad \mathbf{x} \neq \mathbf{0}. \qquad \text{Q.E.D.}$$

THEOREM 1.5.12 Let A be a symmetric positive definite matrix of dimension $p \times p$ and let B be a $q \times p$ matrix. Then BAB' is symmetric and at least positive semidefinite of the same rank as B.

Proof Since A is positive definite there exists a nonsingular matrix C such that $A = CC'$. Hence $BAB' = (BC)(BC)'$. Proceeding exactly in the same way as in Theorem 1.5.11 we get the result. Q.E.D.

THEOREM 1.5.13 Let A be a symmetric positive definite matrix and let B be a symmetric positive semidefinite matrix of the same dimension $p \times p$ and of rank $r \leq p$. Then

(i) all roots of the equation $|B - \lambda A| = 0$ are zero if and only if $B = 0$;
(ii) all roots of $|B - \lambda A| = 0$ are unity if and only if $B = A$.

Proof Since A is positive definite there exists a nonsingular matrix C such that $CAC' = I$ and CBC' is a diagonal matrix whose diagonal elements are the roots of the equation $|CBC' - \lambda I| = 0$ (see Theorem 1.5.10). Since the rank of $CBC' =$ rank B, by Theorem 1.5.7, and the fact that $|CBC' - \lambda I| = 0$ implies $|B - \lambda A| = 0$ we conclude that all roots of $|B - \lambda A| = 0$ are zero if and only if the rank of B is zero, i.e., $B = 0$.

Let $\lambda = 1 - u$. Then $|B - \lambda A| = |B - A + uA|$. By part (i) all roots u of $|B - A + uA| = 0$ are zero if and only if $B - A = 0$. Q.E.D.

1.6 PARTITIONED MATRIX

A matrix $A = (a_{ij})$ of dimension $p \times q$ is said to be partitioned into submatrices A_{ij}, $i, j = 1, 2$, if A can be written as

$$A = \begin{pmatrix} A_{11} & A_{12} \\ A_{21} & A_{22} \end{pmatrix}$$

where $A_{11} = (a_{ij})$ $(i = 1, \ldots, m; j = 1, \ldots, n)$; $A_{12} = (a_{ij})$ $(i = 1, \ldots, m; j = n + 1, \ldots, q)$; $A_{21} = (a_{ij})(i = m + 1, \ldots, p; j = 1, \ldots, n)$; $A_{22} = (a_{ij})$ $(i = m + 1, \ldots, p; j = n + 1, \ldots, q)$. If two matrices A, B of the same dimension are similarly partitioned, then

$$A + B = \begin{pmatrix} A_{11} + B_{11} & A_{12} + B_{12} \\ A_{21} + B_{21} & A_{22} + B_{22} \end{pmatrix}.$$

Let the matrix A of dimension $p \times q$ be partitioned as above and let the matrix C of dimension $q \times r$ be partitioned into submatrices C_{ij} where C_{11}, C_{12} have n rows. Then

$$AC = \begin{pmatrix} A_{11}C_{11} + A_{12}C_{21} & A_{11}C_{12} + A_{12}C_{22} \\ A_{21}C_{11} + A_{22}C_{21} & A_{21}C_{12} + A_{22}C_{22} \end{pmatrix}.$$

THEOREM 1.6.1 For any square matrix

$$A = \begin{pmatrix} A_{11} & A_{12} \\ A_{21} & A_{22} \end{pmatrix}$$

where A_{11}, A_{22} are square submatrices and A_{22} is nonsingular, $|A| = |A_{22}|\,|A_{11} - A_{12}A_{22}^{-1}A_{21}|$.

Proof

$$\begin{vmatrix} A_{11} & A_{12} \\ A_{21} & A_{22} \end{vmatrix} = \begin{vmatrix} A_{11} & A_{12} \\ A_{21} & A_{22} \end{vmatrix} \begin{vmatrix} I & 0 \\ -A_{22}^{-1}A_{21} & I \end{vmatrix} = \begin{vmatrix} A_{11} - A_{12}A_{22}^{-1}A_{21} & A_{12} \\ 0 & A_{22} \end{vmatrix}$$

$$= |A_{22}|\,|A_{11} - A_{12}A_{22}^{-1}A_{21}|. \qquad \text{Q.E.D.}$$

THEOREM 1.6.2 Let the symmetric matrix A of dimension $p \times p$ be partitioned as

$$A = \begin{pmatrix} A_{11} & A_{12} \\ A_{21} & A_{22} \end{pmatrix}$$

where A_{11}, A_{22} are square submatrices of dimensions $q \times q$, $(p - q) \times (p - q)$, respectively, and let A_{22} be nonsingular. Then $A_{11} - A_{12}A_{22}^{-1}A_{21}$ is a symmetric matrix of rank $r - (p - q)$ where r is the rank of A.

Proof Since A is symmetric, $A_{11} - A_{12}A_{22}^{-1}A_{21}$ is obviously symmetric. Now

$$\begin{aligned} \operatorname{rank} A &= \operatorname{rank}\left[\begin{pmatrix} I & -A_{12}A_{22}^{-1} \\ 0 & I \end{pmatrix}\begin{pmatrix} A_{11} & A_{12} \\ A_{21} & A_{22} \end{pmatrix}\begin{pmatrix} I & 0 \\ -A_{22}^{-1}A_{21} & I \end{pmatrix}\right] \\ &= \operatorname{rank}\begin{pmatrix} A_{11} - A_{12}A_{22}^{-1}A_{21} & 0 \\ 0 & A_{22} \end{pmatrix}. \end{aligned}$$

But A_{22} is nonsingular of rank $p - q$. Hence the rank of $A_{11} - A_{12}A_{22}^{-1}A_{21}$ is $r - (p - q)$. Q.E.D.

THEOREM 1.6.3 A symmetric matrix

$$A = \begin{pmatrix} A_{11} & A_{12} \\ A_{21} & A_{22} \end{pmatrix}$$

of dimension $p \times p$ (A_{11} is of dimension $q \times q$) is positive definite if and only if A_{11}, $A_{22} - A_{21}A_{11}^{-1}A_{12}$ are positive definite.

Proof Let $\mathbf{x} = (\mathbf{x}'_{(1)}, \mathbf{x}'_{(2)})$ where $\mathbf{x}'_{(1)} = (x_1, \ldots, x_q)$, $\mathbf{x}'_{(2)} = (x_{q+1}, \ldots, x_p)$. Then

$$\begin{aligned} \mathbf{x}'A\mathbf{x} = {} & (\mathbf{x}_{(1)} + A_{11}^{-1}A_{12}\mathbf{x}_{(2)})'A_{11}(\mathbf{x}_{(1)} + A_{11}^{-1}A_{12}\mathbf{x}_{(2)}) \\ & + \mathbf{x}'_{(2)}(A_{22} - A_{21}A_{11}^{-1}A_{12})\mathbf{x}_{(2)}. \end{aligned} \tag{1.11}$$

Furthermore, if A is positive definite, then obviously A_{11} and A_{22} are both positive definite. Now from (1.11) if A_{11}, $A_{22} - A_{21}A_{11}^{-1}A_{12}$ are positive definite, then A is positive definite. Conversely, if A and consequently A_{11} are positive definite, then by taking $\mathbf{x}$ ($\neq \mathbf{0}$) such that $\mathbf{x}_{(1)} + A_{11}^{-1}A_{12}\mathbf{x}_{(2)} = \mathbf{0}$ we conclude that $A_{22} - A_{21}A_{11}^{-1}A_{12}$ is positive definite. Q.E.D.

THEOREM 1.6.4 Let a positive definite matrix A be partitioned into submatrices A_{ij}, $i, j = 1, 2$, where A_{11} is a square submatrix, and let the inverse matrix $A^{-1} = B$ be similarly partitioned into submatrices B_{ij}, $i, j = 1, 2$. Then

$$A_{11}^{-1} = B_{11} - B_{12}B_{22}^{-1}B_{21}, \qquad A_{22}^{-1} = B_{22} - B_{21}B_{11}^{-1}B_{12}.$$

Proof Since $AB = I$, we get

$$A_{11}B_{11} + A_{12}B_{21} = I, \qquad A_{11}B_{12} + A_{12}B_{22} = 0,$$
$$A_{21}B_{11} + A_{22}B_{21} = 0, \qquad A_{21}B_{12} + A_{22}B_{22} = I.$$

Solving these matrix equations we obtain

$$A_{11}B_{11} - A_{11}B_{12}B_{22}^{-1}B_{21} = I, \qquad A_{22}B_{22} - A_{22}B_{21}B_{11}^{-1}B_{12} = I,$$

or, equivalently,

$$A_{11}^{-1} = B_{11} - B_{12}B_{22}^{-1}B_{21}, \qquad A_{22}^{-1} = B_{22} - B_{21}B_{11}^{-1}B_{12}. \qquad \text{Q.E.D.}$$

From this it follows that $A_{11}^{-1}A_{12} = -B_{12}B_{22}^{-1}$.

THEOREM 1.6.5 A symmetric positive definite quadratic form $\mathbf{x}'A\mathbf{x}$, where $A = (a_{ij})$, can be transformed to $(T\mathbf{x})'(T\mathbf{x})$ where T is the unique upper triangular matrix with positive diagonal elements such that $A = T'T$.

Proof Let $Q_p(x_1, \ldots, x_p) = \mathbf{x}'A\mathbf{x}$. Then

$$\begin{aligned} Q_p(x_1, \ldots, x_p) &= \left((a_{11})^{1/2}x_1 + \sum_{j=2}^{p} \frac{a_{ij}}{(a_{11})^{1/2}} x_j\right)^2 \\ &\quad + \sum_{j,k=2}^{p} \left(\frac{a_{11}a_{jk} - a_{1j}a_{1k}}{a_{11}}\right) x_j x_k \\ &= \left((a_{11})^{1/2}x_1 + \sum_{j=2}^{p} \frac{a_{ij}}{(a_{11})^{1/2}} x_j\right)^2 \\ &\quad + Q_{p-1}(x_2, \ldots, x_p). \end{aligned} \tag{1.12}$$

Let

$$(a_{11})^{1/2}x_1 + \sum_{j=2}^{p} \frac{a_{ij}}{(a_{11})^{1/2}} x_j = \sum_{j=1}^{p} T_{ij}x_j.$$

Since Q_p is positive definite Q_{p-1} is also positive definite so that by continuing the procedure of completing the square, we can write

$$Q_p(x_1, \ldots, x_p) = \left(\sum_{j=1}^{p} T_{ij}x_j\right)^2 + \left(\sum_{j=2}^{p} T_{2j}x_j\right)^2 + \cdots + (T_{pp}x_p)^2 = (T\mathbf{x})'(T\mathbf{x})$$

where T is the unique upper triangular matrix

$$T = \begin{pmatrix} T_{11} & T_{12} & \cdots & T_{1p} \\ 0 & T_{22} & \cdots & T_{2p} \\ \vdots & \vdots & & \vdots \\ 0 & 0 & \cdots & T_{pp} \end{pmatrix}$$

with $T_{ii} > 0$, $i = 1, \ldots, p$. Q.E.D.

Thus a symmetric positive definite matrix A can be uniquely written as $A = T'T$ where T is the unique nonsingular upper triangular matrix with positive diagonal elements. From (1.12) it follows that

$$Q_p(x_1, \ldots, x_p) = \left((a_{pp})^{1/2} x_p + \sum_{j=p-1}^{1} \frac{a_{pi}}{(a_{pp})^{1/2}} x_j \right)^2 + Q_{p-1}(x_1, \ldots, x_{p-1})$$

so that we can write

$$Q_p(x_1, \ldots, x_p) = \left(\sum_{j=1}^{p} T_{pj} x_j \right)^2 + \left(\sum_{j=1}^{p-1} T_{p-1} x_j \right)^2 + \cdots + (T_{11} x_1)^2.$$

Hence, given any symmetric positive definite matrix A there exists a unique nonsingular lower triangular matrix T with positive diagonal elements, such that $A = T'T$. Let θ be an orthogonal matrix in the diagonal form. For any upper (lower) triangular matrix T, θT is also an upper (lower) triangular matrix and $T'T = (\theta T)'(\theta T)$. Thus given any symmetric positive definite matrix A, there exists a nonsingular lower triangular matrix T, not necessarily with positive diagonal elements, such that $A = T'T$. Obviously such decomposition is not unique.

1.7 SOME SPECIAL THEOREMS

Let $\mathbf{x} = (x_1, \ldots, x_p)'$ and let the partial derivative operator vector $\partial/\partial\mathbf{x}$ be defined by

$$\frac{\partial}{\partial \mathbf{x}} = \left(\frac{\partial}{\partial x_1}, \ldots, \frac{\partial}{\partial x_p} \right)'.$$

For any scalar function $f(\mathbf{x})$ of the vector $\mathbf{x}$

$$\frac{\partial f}{\partial \mathbf{x}} = \left(\frac{\partial f}{\partial x_1}, \ldots, \frac{\partial f}{\partial x_p} \right)'.$$

Let

$$f(\mathbf{x}) = \mathbf{x}' A \mathbf{x}$$

where $A = (a_{ij})$. Since

$$\mathbf{x}' A \mathbf{x} = a_{ii} x_i^2 + 2x_i \sum_{j \neq i} a_{ij} x_j + \sum_{\substack{k \neq i, \\ l \neq i}} a_{kl} x_k x_l$$

we obtain

$$\frac{\partial f(\mathbf{x})}{\partial x_i} = 2 \sum_{j=1}^{p} a_{ij} x_j.$$

Hence

$$\frac{\partial f(\mathbf{x})}{\partial \mathbf{x}} = 2A\mathbf{x}.$$

Let $A = (a_{ij})$ be a matrix of dimension $p \times p$. Denoting by A_{ij} the cofactor of a_{ij} we obtain $|A| = \sum_{j=1}^{p} a_{ij}A_{ij}$. Thus

$$\frac{\partial |A|}{\partial a_{ii}} = A_{ii}, \qquad \frac{\partial |A|}{\partial a_{ij}} = A_{ij}.$$

If the matrix A is symmetric, then

$$\frac{\partial |A|}{\partial a_{ii}} = A_{ii}, \qquad \frac{\partial |A|}{\partial a_{ij}} = 2A_{ij}.$$

THEOREM 1.7.1 Let A be a symmetric and at least positive semidefinite matrix of dimension $p \times p$. The largest and the smallest values of $\mathbf{x}'A\mathbf{x}/\mathbf{x}'\mathbf{x}$ for all $\mathbf{x} \neq \mathbf{0}$ are the largest and the smallest characteristic roots of A, respectively.

Proof Let $\mathbf{x}'A\mathbf{x}/\mathbf{x}'\mathbf{x} = \lambda$. Differentiating λ with respect to the components of $\mathbf{x}$ the stationary values of λ are given by the characteristic equation $(A - \lambda I)\mathbf{x} = \mathbf{0}$. Eliminating $\mathbf{x}$ we get $|A - \lambda I| = 0$. Thus the values of λ are the characteristic roots of the matrix A and consequently the largest value of λ corresponds to the largest characteristic root of A, and the smallest value of λ corresponds to the smallest characteristic root of A. Q.E.D.

From this theorem it follows that if $g_1 \leq \mathbf{x}'A\mathbf{x}/\mathbf{x}'\mathbf{x} \leq g_2$ for all $\mathbf{x} \neq \mathbf{0}$, then $g_1 \leq \lambda_1 \leq \lambda_p \leq g_2$ where λ_1, λ_p are the smallest and the largest characteristic roots of A, respectively.

THEOREM 1.7.2 Let A be a symmetric and at least positive semidefinite matrix of dimension $p \times p$ and let B be a symmetric and positive definite matrix of the same dimension. The largest and the smallest values of $\mathbf{x}'A\mathbf{x}/\mathbf{x}'B\mathbf{x}$ for all $\mathbf{x} \neq \mathbf{0}$ are the largest and the smallest roots respectively of the characteristic equation $|A - \lambda B| = 0$.

Proof Let $\mathbf{x}'A\mathbf{x}/\mathbf{x}'B\mathbf{x} = \lambda$. Differentiating λ with respect to the components of $\mathbf{x}$, the stationary values of λ are given by the characteristic equation $(A - \lambda B)\mathbf{x} = \mathbf{0}$; hence by eliminating $\mathbf{x}$ we conclude that the smallest and the largest values of λ are given by the smallest and the largest roots of the characteristic equation $|A - \lambda B| = 0$. Q.E.D.

If $g_1 \leq \mathbf{x}'A\mathbf{x}/\mathbf{x}'B\mathbf{x} \leq g_2$ for all $\mathbf{x} \neq \mathbf{0}$, then $g_1 \leq \lambda_1 \leq \lambda_p \leq g_2$ where λ_1, λ_p are the smallest and the largest roots of the characteristic equation $|A - \lambda B| = 0$.

1.8 COMPLEX MATRICES

In this section we shall briefly discuss complex matrices, matrices with complex elements, and state some theorems without proof concerning these matrices which are useful for the study of complex Gaussian distributions. For a proof the reader is referred to MacDuffee (1946). The adjoint operator (conjugate transpose) will be denoted by an asterisk (*). The adjoint A^* of a complex matrix $A = (a_{ij})$ of dimension $p \times q$ is the $q \times p$ matrix $A^* = (\bar{a}_{ij})'$, where the overbar ($\bar{\ }$) denotes the conjugate and the prime (′) denotes the transpose. Clearly for any two complex matrices A, B, $(A^*)^* = A$, $(AB)^* = B^*A^*$, provided AB is defined. A square complex matrix A is called *unitary* if $AA^* = I$ (real identity matrix) and it is called *Hermitian* if $A = A^*$. A square complex matrix is called normal if $AA^* = A^*A$. An Hermitian matrix A of dimension $p \times p$ is called positive definite (semidefinite) if for all complex nonnull p-vectors ξ, $\xi^*A\xi > 0\ (\geq 0)$. Since

$$(\xi^*A\xi)^* = \xi^*A\xi$$

for any Hermitian matrix A, the Hermitian quadratic form $\xi^*A\xi$ assumes only real values.

THEOREM 1.8.1 If A is an Hermitian matrix of dimension $p \times p$, there exists a unitary matrix U of dimension $p \times p$ such that U^*AU is a diagonal matrix whose diagonal elements $\lambda_1, \ldots, \lambda_p$ are the characteristic roots of A.

Since $(U^*AU)^* = U^*AU$, it follows that all characteristic roots of an Hermitian matrix are real.

THEOREM 1.8.2 An Hermitian matrix A is positive definite if all its characteristic roots are positive.

THEOREM 1.8.3 Every Hermitian positive definite (semidefinite) matrix A is uniquely expressible as $A = BB^*$ where B is Hermitian positive definite (semidefinite).

THEOREM 1.8.4 For every Hermitian positive definite matrix there exists a complex nonsingular matrix B such that $BAB^* = I$.

EXERCISES

1. Prove Theorem 1.1.2.

2. Show that for any basis $\boldsymbol{\alpha}_1, \ldots, \boldsymbol{\alpha}_k$ of V^k of rank k there exists an orthonormal basis $\boldsymbol{\gamma}_1, \ldots, \boldsymbol{\gamma}_k$ of V^k.

3. If $\boldsymbol{\alpha}_1, \ldots, \boldsymbol{\alpha}_k$ is a basis of a vector space V, show that no set of $k + 1$ vectors in V is linearly independent.

4. Find the orthogonal projection of the vector (1, 2, 3, 4) on the vector (1, 0, 1, 1).

5. Find the number of linearly independent vectors in the set $(a, c, \ldots, c)'$, $(c, a, c, \ldots, c)', \ldots, (c, \ldots, c, a)'$ such that the sum of components of each vector is zero.

6. Let V be a set of vectors of dimension p and let V^+ be the set of all vectors orthogonal to V. Show that $(V^+)^+ = V$ if V is a linear subspace of V.

7. Let V_1 and V_2 be two linear subspaces containing the null vector $\mathbf{0}$ and let V_i^+ denote the set of all vectors orthogonal to V_i, $i = 1, 2$. Show that $(V_1 \cup V_2)^+ = V_1^+ \cap V_2^+$.

8. Let $(\gamma_1, \ldots, \gamma_k)$ be an orthonormal basis of the subspace V^k of a vector space V^p. Show that it can be extended to an orthonormal basis $(\gamma_1, \ldots, \gamma_k, \gamma_{k+1}, \ldots, \gamma_p)$ of V^p.

9. Show that for any three vectors $\mathbf{x}, \mathbf{y}, \mathbf{z}$ in V^p, the function d, defined by

$$d(\mathbf{x}, \mathbf{y}) = \max_{1 \leq i \leq p} |x_i - y_i|,$$

satisfies

(a) $d(\mathbf{x}, \mathbf{y}) = d(\mathbf{y}, \mathbf{x}) \geq 0$ (symmetry),
(b) $d(\mathbf{x}, \mathbf{z}) \leq d(\mathbf{x}, \mathbf{y}) + d(\mathbf{y}, \mathbf{z})$ (triangular inequality).

10. Let W be a vector subspace of the vector space V. Show that the rank of $W \leq$ rank of V.

11. (*Cauchy–Schwarz inequality*) Show that for any two vectors $\mathbf{x}, \mathbf{y}$ in V^p,

$$(\mathbf{x}'\mathbf{y}) \leq \|\mathbf{x}\| \, \|\mathbf{y}\|.$$

12. (*Triangle inequality*) Show that for any two vectors $\mathbf{x}, \mathbf{y}$ in V^p

$$\|\mathbf{x} + \mathbf{y}\| \leq \|\mathbf{x}\| + \|\mathbf{y}\|.$$

13. Let A, B be two positive definite matrices of the same dimension. Show that for $0 \leq \alpha \leq 1$,

$$|\alpha A + (1 - \alpha)B| \geq |A|^{\alpha}|B|^{1-\alpha}.$$

14. (*Skew matrix*) A matrix A is skew if $A = -A$. Show that

(a) for any matrix A, AA is symmetric if A is skew,
(b) the determinant of a skew matrix with an odd number of rows is zero,
(c) the determinant of a skew symmetric matrix is nonnegative.

15. Show that for any square matrix A there exists an orthogonal matrix O such that AO is an upper triangular matrix.

16. (*Idempotent matrix*) A square matrix A is idempotent if $AA = A$. Show the following:

(a) if A is idempotent and nonsingular, then $A = I$;

(b) the characteristic roots of an idempotent matrix are either unity or zero;

(c) if A is idempotent of rank r, then $\operatorname{tr} A = r$;

(d) let $A_1, \ldots, A_k$ be symmetric matrices of the same dimension; if $A_i A_j = 0$ $(i \neq j)$ and if $\sum_{i=1}^k A_i$ is idempotent, then show that A_i for each i is an idempotent matrix and $\operatorname{rank}(\sum_{i=1}^k A_i) = \sum_{i=1}^k \operatorname{rank}(A_i)$.

17. Show that for any lower triangular matrix A the diagonal elements are its characteristic roots.

18. Show that any orthogonal transformation may be regarded as the change of axes about a fixed origin.

19. Show that for any nonsingular matrix A of dimension $p \times p$ and nonnull p-vector $\mathbf{x}$,

$$\mathbf{x}'(A + \mathbf{x}\mathbf{x}')^{-1}\mathbf{x} = \frac{\mathbf{x}'A^{-1}\mathbf{x}}{1 + \mathbf{x}'A^{-1}\mathbf{x}}.$$

20. Let A be a nonsingular matrix of dimension $p \times p$ and let $\mathbf{x}$, $\mathbf{y}$ be two nonnull p-vectors. Show that

$$(A + \mathbf{x}\mathbf{y}')^{-1} = A^{-1} - \frac{(A^{-1}\mathbf{x})(\mathbf{y}'A^{-1})}{1 + \mathbf{y}'A\mathbf{x}}.$$

21. Let X be a $p \times q$ matrix and let S be a $p \times p$ nonsingular matrix. Then show that $|XX' + S| = |S|\,|I + X'S^{-1}X|$.

22. Let X be a $p \times p$ matrix. Show that the nonzero characteristic roots of $X'X$ are the same as those of XX'.

23. Let A, X be two matrices of dimension $q \times p$. Show that

(a) $(\partial/\partial X)(\operatorname{tr} A'X) = A$

(b) $(\partial/\partial X)(\operatorname{tr} AX') = A$

where $\partial/\partial X = (\partial/\partial x_{ij})$, $X = (x_{ij})$.

24. For any square symmetric matrix A show that

$$\frac{\partial}{\partial A}(\operatorname{tr} AA) = 2A.$$

REFERENCES

Giri, N. (1974). "Introduction to Probability and Statistics," Part 1, Probability. Dekker, New York.

MacDuffee, C. (1946). "The Theory of Matrices." Chelsea, New York.

Mac Lane, S., and Birkoff, G. (1967). "Algebra." Macmillan, New York.

Markus, M., and Mine, H. (1967). "Introduction to Linear Algebra." Macmillan, New York.

Perlis, S. (1952). "Theory of Matrices." Addison Wesley, Reading, Massachusetts.

CHAPTER **II**

Groups and Jacobian of Some Transformations

2.0 INTRODUCTION

In multivariate analysis the most frequently used test procedures are often invariant with respect to a group of transformations, leaving the testing problems invariant. In such situations an application of group theory results leads us in a straightforward way to the desired test procedures (see Stein, 1959). In this chapter we shall describe the basic concepts and some basic results of group theory. Results on the Jacobian of some specific transformations which are very useful in deriving the distributions of multivariate test statistics are also discussed.

2.1 GROUPS

DEFINITION 2.1.1 *Group* A group is a nonempty set G of elements with an operation τ satisfying the following axioms:

O_1 For any $a, b \in G$, $a\tau b \in G$.

O_2 There exists a unit element $e \in G$ such that for all $a \in G$, $a\tau e \in G$.

O_3 For any $a, b, c \in G$, $(a\tau b)\tau c = a\tau(b\tau c)$.

O_4 For each $a \in G$, there exists $a^{-1} \in G$ such that $a\tau a^{-1} = e$.

The following properties follow directly from axioms O_1–O_4 ($a, b \in G$):

(1) $a\tau a^{-1} = a^{-1}\tau a$,
(2) $a\tau e = e\tau a$,
(3) $a\tau x = b$ has the unique solution $x = a^{-1}\tau b$.

Note For convenience we shall write $a\tau b$ as ab. The reader is cautioned not to confuse this with multiplication.

DEFINITION 2.1.2 *Abelian group* A group G is called Abelian if $ab = ba$ for $a, b \in G$.

DEFINITION 2.1.3 *Subgroup* If the restriction of the operation τ to a nonempty subset H of G satisfies the group axioms O_1–O_4, then H is called a subgroup of G.

The following lemma facilitates verifying whether a subset of a group is a subgroup.

LEMMA 2.1.1 Let G be a group and $H \subset G$. Then H is a subgroup of G (i) if and only if $H \neq \varnothing$ (nonempty), (ii) if $a, b \in H$, then $ab^{-1} \in H$.

Proof If H satisfies (i) and (ii), then H is a group. For if $a \in H$, then by (ii) $aa^{-1} = e \in H$. Also if $b \in H$, then $b^{-1} = eb^{-1} \in H$. Hence $a, b \in H$ implies $a(b^{-1})^{-1} = ab \in H$. Axiom O_3 is true in H as it is true in G. Hence H is a group. Conversely, if H is a group, then clearly H satisfies (i) and (ii). Q.E.D.

2.2 SOME EXAMPLES OF GROUPS

EXAMPLE 2.2.1 The additive group of real numbers is the set of all reals with the group operation $ab = a + b$.

EXAMPLE 2.2.2 The multiplicative group of nonzero real numbers is the set of all nonzero reals with the group operation $ab = a$ multiplied by b.

EXAMPLE 2.2.3 *Permutation group* Let X be a nonempty set and let G be the set of all one-to-one functions of X onto X. Define the group operation τ as follows: for $g_1, g_2 \in G$, $x \in X$, $(g_1\tau g_2)(x) = g_1(g_2(x))$. Then G is a group and is called the permutation group.

EXAMPLE 2.2.4 Let X be a linear space. Then under the operation of addition X is an Abelian group.

EXAMPLE 2.2.5 *Translation group* Let X be a linear space of dimension n and let $x_0 \in X$. Define $g_{x_0}(x) = x + x_0$, $x \in X$. The collection of all g_{x_0} forms an additive Abelian group.

EXAMPLE 2.2.6 *Full linear group* Let X be a linear space of dimension n. Let $G_l(n)$ denote the set of all nonsingular linear transformations of X onto X. $G_l(n)$ is obviously a group with matrix multiplication as the group operation and it is called the full linear group.

EXAMPLE 2.2.7 *Affine group* Let X be a linear space of dimension n and let $G_l(n)$ be the linear group. The affine group is the set of pairs (g, x), $g \in G_l(n)$, $x \in X$, with the following operation: $(g_1, x_1)(g_2, x_2) = (g_1 g_2, g_1 x_2 + x_1)$. For the affine group the unit element is $(I, 0)$, where I is the identity matrix and $(g, x)^{-1} = (g^{-1}, -g^{-1}x)$.

EXAMPLE 2.2.8 *Unimodular group* The unimodular group is the subgroup of $G_l(n)$ such that g is in this group if and only if the determinant of g is ± 1.

EXAMPLE 2.2.9 The set of all nonsingular lower (upper) triangular matrices of dimension n forms a group with the usual matrix multiplication as the group operation. Obviously the product of two nonsingular lower (upper) triangular matrices is a lower (upper) triangular matrix and the inverse of a nonsingular lower (upper) matrix is a nonsingular lower (upper) triangular matrix. The unit element for this group is the identity matrix.

EXAMPLE 2.2.10 The set of all orthogonal matrices of dimension n forms a group.

2.3 NORMAL SUBGROUPS, QUOTIENT GROUP, HOMOMORPHISM, ISOMORPHISM, DIRECT PRODUCT

DEFINITION 2.3.1 *Normal subgroup* A subgroup H of G is a normal subgroup if for all $h \in H$ and $g \in G$, $ghg^{-1} \in H$ or, equivalently, $gHg^{-1} = H$.

DEFINITION 2.3.2 *Quotient group* Let G be a group and let H be a normal subgroup of G. The set G/H is defined to be the set of elements of the form $g_1 H = \{g_1 h | h \in H\}$, $g_1 \in G$. For $g_1, g_2 \in G$ we define $(g_1 H)(g_2 H)$ as the set of all elements obtained by multiplying all elements of $g_1 H$ by all elements of $g_2 H$. With this operation defined on the elements of G/H, it is a group. We verify this as follows:

(1) $g_1 H = g_2 H \Leftrightarrow g_2^{-1} g_1 H = H \Leftrightarrow g_2^{-1} g_1 \in H \Leftrightarrow g_1 \in g_2 H$.

(2) Since H is a normal subgroup, we have for $g_1, g_2 \in G$, $g_2 H = H g_2$ and $(g_1 H)(g_2 H) = g_1(H g_2 H) = g_1(g_2 H)H = g_1 g_2 H \in G/H$.

(3) H is the identity element in G/H ($gHH = gH$). The group G/H is the quotient group of G (mod H).

EXAMPLE 2.3.1 The affine group (I, X), where X is a linear space of dimension n and I is the $n \times n$ identity matrix, is the normal subgroup of $(G_l(n), X)$. For $g \in G_l(n)$, $x \in X$,

$$(g, x)(I, x)(g, x)^{-1}$$
$$= (g, x)(I, x)(g^{-1}, -g^{-1}x) = (g, x)(g^{-1}, -g^{-1}x + x) = (I, gx) \in (I, X).$$

DEFINITION 2.3.3 *Homomorphism* Let G and H be two groups. Then a mapping f of G into H is called a homomorphism if it preserves the group operation; i.e., for $g_1, g_2 \in G$, $f(g_1 g_2) = f(g_1)f(g_2)$. This implies that if e is the identity element of G, then $f(e)$ is the identity element of H and $f(g_1^{-1}) = [f(g_1)]^{-1}$. For

(i) $f(g_1) = f(g_1 e) = f(g_1)f(e)$,
(ii) $f(e) = f(g_1 g_1^{-1}) = f(g_1)f(g_1^{-1})$.

If, in addition, f is a one-to-one mapping, it is called an *isomorphism*.

DEFINITION 2.3.4 *Direct products* Let G and H be groups and let $G \times H$ be the Cartesian product of G and H. With the operation $(g_1, h_1)(g_2, h_2) = (g_1 g_2, h_1 h_2)$, where $g_1, g_2 \in G$, $h_1, h_2 \in H$, and $g_1 g_2$, $h_1 h_2$ are the products in the groups G and H, respectively, $G \times H$ is a group and is known as the direct product of G and H.

DEFINITION 2.3.5 The group G operates on the space X from the left if there exists a function on $G \times X$ to X whose value at $x \in X$ is denoted by gx such that

(1) $ex = x$ for all $x \in X$ and e is the unit element of G;
(2) for $g_1, g_2 \in G$ and $x \in X$, $g_1(g_2 x) = g_1 g_2(x)$.

Note (1) and (2) imply that $g \in G$ is one-to-one on X to X. To see this, suppose $gx_1 = gx_2 = y$. Then $g^{-1}(gx_1) = g^{-1}(gx_2) = g^{-1}y$. Using (1) and (2) we then have $x_1 = x_2$.

DEFINITION 2.3.6 Let the group G operate from the left on the space X. G operates transitively on X if for every $x_1, x_2 \in X$ there exists a $g \in G$ such that $gx_1 = x_2$.

EXAMPLE 2.3.2 Let X be the space of all $n \times n$ nonsingular matrices and let $G = G_l(n)$. Given any two points $x_1, x_2 \in X$, there exists a nonsingular matrix $g \in G$ such that $x_1 = gx_2$. In other words, G acts transitively on X.

EXAMPLE 2.3.3 Let X be a linear space. $G_l(n)$ acts transitively on $X - \{0\}$.

2.4 JACOBIAN OF SOME TRANSFORMATIONS

Let $X_1, \ldots, X_n$ be a sequence of n continuous random variables with a joint probability density function $f_{X_1,\ldots,X_n}(x_1, \ldots, x_n)$. Let $Y_i = g_i(X_1, \ldots, X_n)$ be a set of continuous one-to-one transformations of the random variables $X_1, \ldots, X_n$. Let us assume that the functions $g_1, \ldots, g_n$ have continuous partial derivatives with respect to $x_1, \ldots, x_n$. Let the inverse function be denoted by $X_i = h_i(Y_1, \ldots, Y_n)$, $i = 1, \ldots, n$. Denote by J the determinant of the $n \times n$ square matrix

$$\begin{pmatrix} \partial x_1/\partial y_1 & \cdots & \partial x_1/\partial y_n \\ \vdots & & \vdots \\ \partial x_n/\partial y_1 & \cdots & \partial x_n/\partial y_n \end{pmatrix}.$$

Then J is called the Jacobian of the inverse transformation. We shall assume that there exists a region R of points $(x_1, \ldots, x_n)$ on which J is different from zero. Let S be the image of R under the transformations. Then

$$\underbrace{\int \cdots \int_R}_{n \text{ integrals}} f_{X_1,\ldots,X_n}(x_1, \ldots, x_n)\, dx_1, \ldots, dx_n$$
$$= \int \underset{S}{\cdots} \int f_{X_1,\ldots,X_n}(h_1(y_1, \ldots, y_n), \ldots, h_n(y_1, \ldots, y_n))|J|\, dy_1, \ldots, dy_n.$$

From this it follows that the joint probability density function of the random variables $Y_1, \ldots, Y_n$ is given by

$$f_{Y_1,\ldots,Y_n}(y_1, \ldots, y_n) = \begin{cases} f_{X_1,\ldots,X_n}(h_1(y_1, \ldots, y_n), \ldots, h_n(y_1, \ldots, y_n))|J| \\ \qquad\qquad \text{if} \quad (y_1, \ldots, y_n) \in S, \\ 0 \qquad \text{otherwise.} \end{cases}$$

We shall now state some theorems on the Jacobian (J^{-1}), but will not give all the proofs. For further results on the Jacobian the reader is referred to Olkin (1962), Rao (1965), Roy (1957), and Nachbin (1965).

THEOREM 2.4.1 Let V be a vector space of dimension p. For $\mathbf{x}, \mathbf{y} \in V$, the Jacobian of the linear transformation $\mathbf{x} \to \mathbf{y} = A\mathbf{x}$, where A is a nonsingular matrix of dimension $p \times p$, is given by $|A|$.

THEOREM 2.4.2 Let the $p \times n$ matrix X be transformed to the $p \times n$ matrix $Y = AX$ where A is a nonsingular matrix of dimension $p \times p$. The Jacobian of this transformation is given by $|A|^n$.

THEOREM 2.4.3 Let a $p \times q$ matrix X be transformed to the $p \times q$ matrix $Y = AXB$ where A and B are nonsingular matrices of dimensions $p \times p$ and $q \times q$, respectively. Then the Jacobian of this transformation is given by $|A|^q|B|^p$.

THEOREM 2.4.4 Let G_T be the multiplicative group of nonsingular lower triangular matrices of dimension $p \times p$. For $g = (g_{ij})$, $h = (h_{ij}) \in G_T$, the Jacobian of the transformation $g \to hg$ is $\prod_{i=1}^{p} (h_{ii})^i$.

Proof Let $hg = c = (c_{ij})$. Obviously, $c \in G_T$ and $c_{ij} = \sum_{k=1}^{p} h_{ik} g_{kj}$ with $h_{ij} = 0$, $g_{ij} = 0$ if $i < j$. The Jacobian of this transformation is given by the determinant of the $\frac{1}{2}p(p+1) \times \frac{1}{2}p(p+1)$ matrix

$$\begin{pmatrix} \partial c_{11}/\partial g_{11} & \partial c_{11}/\partial g_{21} & \cdots & \partial c_{11}/\partial g_{pp} \\ \partial c_{21}/\partial g_{11} & \partial c_{21}/\partial g_{21} & \cdots & \partial c_{21}/\partial g_{pp} \\ \vdots & \vdots & & \vdots \\ \partial c_{pp}/\partial g_{11} & \partial c_{pp}/\partial g_{21} & \cdots & \partial c_{pp}/\partial g_{pp} \end{pmatrix}.$$

It is easy to see that this matrix is a lower triangular matrix with diagonal element

$$\frac{\partial c_{ij}}{\partial g_{ij}} = \begin{cases} h_{ii} & \text{if } i \geq j, \\ 0 & \text{otherwise.} \end{cases}$$

Thus among the diagonal elements h_{ii} is repeated i times. Hence the Jacobian is given by $\prod_{i=1}^{p} (h_{ii})^i$. Q.E.D.

COROLLARY 2.4.1 The Jacobian of the transformation $g \to gh$ is $\prod_{i=1}^{p} (h_{ii})^{p+1-i}$.

Proof Let $gh = c = (c_{ij})$. Obviously c is a lower triangular matrix. Since

$$\frac{\partial c_{ij}}{\partial g_{ij}} = \begin{cases} h_{jj} & i \geq j \\ 0 & \text{otherwise,} \end{cases}$$

following the same argument as in Theorem 2.4.4 we conclude that the Jacobian is the determinant of a triangular matrix where h_{ii} is repeated $p + 1 - i$ times among its diagonal elements. Hence the result. Q.E.D.

THEOREM 2.4.5 Let G_{UT} be the group of $p \times p$ nonsingular upper triangular matrices. For $g = (g_{ij})$, $h = (h_{ij}) \in G_{UT}$, the Jacobian of the transformation $g \to hg$ is $\prod_{i=1}^{p} (h_{ii})^{p+1-i}$.

Proof Let $hg = c = (c_{ij})$. Obviously c is an upper triangular matrix and $c_{ij} = \sum_{k=1}^{p} h_{ik} g_{kj}$ with $h_{ij} = 0$, $g_{ij} = 0$ if $i > j$. The Jacobian of this transformation is given by the determinant of the matrix of dimension $\frac{1}{2}p(p+1) \times \frac{1}{2}p(p+1)$:

$$\begin{pmatrix} \partial c_{11}/\partial g_{11} & \partial c_{11}/\partial g_{12} & \cdots & \partial c_{11}/\partial g_{pp} \\ \partial c_{12}/\partial g_{11} & \partial c_{12}/\partial g_{12} & \cdots & \partial c_{12}/\partial g_{pp} \\ \vdots & \vdots & & \vdots \\ \partial c_{pp}/\partial g_{11} & \partial c_{pp}/\partial g_{12} & \cdots & \partial c_{pp}/\partial g_{pp} \end{pmatrix}.$$

Since

$$\frac{\partial c_{ij}}{\partial g_{ij}} = \begin{cases} h_{ii} & i \le j \\ 0 & \text{otherwise,} \end{cases}$$

the preceding matrix is an upper triangular matrix such that among its diagonal elements h_{ii} is repeated $p + 1 - i$ times. Hence the Jacobian of this transformation is $\prod_{i=1}^{p}(h_{ii})^{p+1-i}$. Q.E.D.

COROLLARY 2.4.2 The Jacobian of the transformation $g \to gh$ is $\prod_1^p (h_{ii})^i$.

The proof follows from an argument similar to that of the theorem.

THEOREM 2.4.6 Let S be a symmetric positive definite matrix of dimension $p \times p$. The Jacobian of the transformation $S \to B$, where B is the unique lower triangular matrix with positive diagonal elements such that $S = BB'$, is $\prod_{i=1}^{p}(b_{ii})^{p+1-i}$.

THEOREM 2.4.7 Let G_{BT} be the group of $p \times p$ lower triangular nonsingular matrices in block form, i.e., $g \in G_{BT}$,

$$g = \begin{pmatrix} g_{(11)} & 0 & 0 & \cdots & 0 \\ g_{(21)} & g_{(22)} & 0 & \cdots & 0 \\ \vdots & \vdots & & & \vdots \\ g_{(k1)} & g_{(k2)} & g_{(k3)} & \cdots & g_{(kk)} \end{pmatrix},$$

where $g_{(ii)}$ are submatrices of g of dimension $d_i \times d_i$ such that $\sum_1^k d_i = p$. The Jacobian of the transformation $g \to hg$, g, $h \in G_{BT}$, is $\prod_{i=1}^{k}|h_{(ii)}|^{\sigma_i}$ where $\sigma_i = \sum_{j=1}^{i} d_j$, $\sigma_0 = 0$. The Jacobian of the transformation $g \to gh$ is $\prod_{i=1}^{k}|h_{(ii)}|^{p-\sigma_{i-1}}$.

THEOREM 2.4.8 Let G_{BUT} be the group of nonsingular upper triangular $p \times p$ matrices in block form, i.e., $g \in G_{BUT}$,

$$g = \begin{pmatrix} g_{(11)} & g_{(12)} & \cdots & g_{(1k)} \\ 0 & g_{(22)} & \cdots & g_{(2k)} \\ \vdots & \vdots & & \vdots \\ 0 & 0 & 0 & g_{(kk)} \end{pmatrix},$$

where $g_{(ii)}$ are submatrices of dimension $d_i \times d_i$ and $\sum_{j=1}^{k} d_j = p$. For g, $h \in G_{BUT}$ the Jacobian of the transformation $g \to gh$ is $\prod_{i=1}^{k}|h_{(ii)}|^{\sigma_i}$ and that of $g \to hg$ is $\prod_{i=1}^{k}|h_{(ii)}|^{p-\sigma_{i-1}}$.

THEOREM 2.4.9 Let $S = (s_{ij})$ be a symmetric matrix of dimension $p \times p$. The Jacobian of the transformation $S \to CSC'$, where C is any nonsingular matrix of dimension $p \times p$, is $|C|^{p+1}$.

Proof To prove this theorem it is sufficient to show that it holds for the elementary $p \times p$ matrices $E(ij)$, $M_i(c)$, and $A(ij)$ where $E(ij)$ is the matrix obtained from the $p \times p$ identity matrix by interchanging the ith and the jth row; $M_i(c)$ is the matrix obtained from the $p \times p$ identity matrix by multiplying its ith row by the nonzero constant c; and $A(ij)$ is the matrix obtained from the $p \times p$ identity matrix by adding the jth row to the ith row. The fact that the theorem is valid for these matrices can be easily verified by the reader. For example, $M_i(c)SM_i(c)$ is obtained from S by multiplying s_{ii} by c^2 and s_{ij} by c $(i \neq j)$ so that the Jacobian is $c^{2+(p-1)} = c^{p+1}$. Q.E.D.

THEOREM 2.4.10 Let $S = (s_{ij})$ be a symmetric positive definite matrix of dimension $p \times p$. The Jacobian of the transformation $S \to gSg'$, $g \in G_T$, is $|g|^{p+1}$.

Proof Let $g = (g_{ij})$ with $g_{ij} = 0$ for $i < j$ and let $A = (a_{ij}) = gSg'$. Then $a_{ij} = \sum_{l,k} g_{il}s_{lk}g_{jk}$. Since

$$\frac{\partial a_{ij}}{\partial s_{kk}} = g_{ik}g_{jk}, \qquad \frac{\partial a_{ij}}{\partial s_{lk}} = g_{il}g_{jk} + g_{ik}g_{jl},$$

the Jacobian, which is the determinant of the $\frac{1}{2}p(p+1) \times \frac{1}{2}p(p+1)$ lower triangular matrix

$$\begin{pmatrix} \partial a_{11}/\partial s_{11} & \partial a_{11}/\partial s_{12} & \cdots & \partial a_{11}/\partial s_{pp} \\ \vdots & \vdots & & \vdots \\ \partial a_{pp}/\partial s_{11} & \partial a_{pp}/\partial s_{12} & \cdots & \partial a_{pp}/\partial s_{pp} \end{pmatrix},$$

is equal to $\prod_{i=1}^{p}(g_{ii})^{p+1} = |g|^{p+1}$. Q.E.D.

FURTHER READING

Nachbin, L. (1965). "The Haar Integral." Van Nostrand–Reinhold, Princeton, New Jersey.

Olkin, I. (1962). Note on the Jacobian of certain matrix transformations useful in multivariate analysis, *Biometrika* **40**, 43–46.

Rao, C. R. (1965). "Linear Statistical Inference and its Applications." Wiley, New York.

Roy, S. N. (1957). "Some Aspects of Multivariate Analysis." Wiley, New York.

Stein, C. (1959). Lecture notes on Multivariate Analysis. Dept. of Statistics, Stanford Univ., California.

CHAPTER **III**

Notions of Multivariate Distributions and Invariance in Statistical Inference

3.0 INTRODUCTION

In this chapter we shall discuss the distribution of vector random variables and its properties. Most of the commonly used test criteria in multivariate analysis are invariant test procedures with respect to a certain group of transformations leaving the problem in question invariant. Thus to study the basic properties of such test criteria we will outline here the "principle of invariance" in some details. For further details the reader is referred to Giri (1975), Lehmann (1959), and Ferguson (1969).

3.1 MULTIVARIATE DISTRIBUTIONS

By a multivariate distribution we mean the distribution of a random vector $\mathbf{X} = (X_1, \ldots, X_p)'$, where p (≥ 2) is arbitrary, whose elements X_i are univariate random variables with distribution function $F_{X_i}(x_i)$. Let $\mathbf{x} = (x_1, \ldots, x_p)'$. The distribution function of $\mathbf{X}$ is defined by

$$F_{\mathbf{X}}(\mathbf{x}) = \text{prob}(X_1 \leq x_1, \ldots, X_p \leq x_p),$$

which is also written as

$$F_{X_1,X_2,\ldots,X_p}(x_1, x_2, \ldots, x_p)$$

to indicate the fact that it is the joint distribution of $X_1, \ldots, X_p$. If each X_i is a discrete random variable, then $\mathbf{X}$ is called a discrete random vector and its probability mass function is given by

$$p_{\mathbf{X}}(\mathbf{x}) = p_{X_1,\ldots,X_p}(x_1, \ldots, x_p) = \operatorname{prob}(X_1 = x_1, \ldots, X_p = x_p).$$

It is also called the joint probability mass function of $X_1, \ldots, X_p$. If $F_{\mathbf{X}}(\mathbf{x})$ is continuous in $x_1, \ldots, x_p$, $-\infty < x_i < \infty$ for all i, and if there exists a nonnegative function $f_{X_1,\ldots,X_p}(x_1, \ldots, x_p)$ such that

$$F_{\mathbf{X}}(\mathbf{x}) = \int_{-\infty}^{x_1} \cdots \int_{-\infty}^{x_p} f_{\mathbf{X}}(\mathbf{y})\, dy_1, \ldots, dy_p \tag{3.1}$$

where $\mathbf{y} = (y_1, \ldots, y_p)'$, then $f_{\mathbf{X}}(\mathbf{x})$ is called the probability density function of the continuous random vector $\mathbf{X}$. [For clarity of exposition we have used $f_{\mathbf{X}}(\mathbf{y})$ instead of $f_{\mathbf{X}}(\mathbf{x})$ in (3.1).] If the components $X_1, \ldots, X_p$ are independent (statistically), then

$$F_{\mathbf{X}}(\mathbf{x}) = \prod_{i=1}^{p} F_{X_i}(x_i),$$

or, equivalently,

$$f_{\mathbf{X}}(\mathbf{x}) = \prod_{i=1}^{p} f_{X_i}(x_i), \qquad p_{\mathbf{X}}(\mathbf{x}) = \prod_{i=1}^{p} p_{X_i}(x_i).$$

Given $f_{\mathbf{X}}(\mathbf{x})$, the marginal probability density function of any subset of $\mathbf{X}$ is obtained by integrating $f_{\mathbf{X}}(\mathbf{x})$ over the domain of the variables not in the subset. For $q < p$

$$f_{X_1,\ldots,X_q}(x_1, \ldots, x_q) = \int \cdots \int f_{\mathbf{X}}(\mathbf{x})\, dx_{q+1}, \ldots, dx_p. \tag{3.2}$$

In the case of discrete $p_{\mathbf{X}}(\mathbf{x})$, the marginal probability mass function of $X_1, \ldots, X_q$ is obtained from $p_{\mathbf{X}}(\mathbf{x})$ by summing it over the domain of $X_{q+1}, \ldots, X_p$.

It is well known that

(i) $\lim_{x_p \to \infty} F_{\mathbf{X}}(\mathbf{x}) = F_{X_1,\ldots,X_{p-1}}(x_1, \ldots, x_{p-1})$;
(ii) for each i, $1 \le i \le p$, $\lim_{x_i \to -\infty} F_{\mathbf{X}}(\mathbf{x}) = 0$;
(iii) $F_{\mathbf{X}}(\mathbf{x})$ is continuous from above in each argument.

The notion of conditional probability of events can be used to obtain the conditional probability density function of a subset of components of $\mathbf{X}$ given that the variates of another subset of components of $\mathbf{X}$ have assumed constant specified values or have been constrained to lie in some subregion of the space described by their variate values. For a general discussion of this the reader is referred to Kolmogorov (1950). Given that $\mathbf{X}$ has a probability density function $f_{\mathbf{X}}(\mathbf{x})$, the conditional probability density function of $X_1, \ldots, X_q$ where $X_{q+1} = x_{q+1}, \ldots, X_p = x_p$ is given by

$$f_{X_1,\ldots,X_q|X_{q+1},\ldots,X_p}(x_1, \ldots, x_q|x_{q+1}, \ldots, x_p) = \frac{f_{\mathbf{X}}(\mathbf{x})}{f_{X_1,\ldots,X_q}(x_1, \ldots, x_q)}, \quad (3.3)$$

provided the marginal probability density function $f_{X_1,\ldots,X_q}(x_1, \ldots, x_q)$ of $X_1, \ldots, X_q$ is not zero. For discrete random variables, the conditional probability mass function $p_{X_1,\ldots,X_q|X_{q+1},\ldots,X_p}(x_1, \ldots, x_q|x_{q+1}, \ldots, x_q)$ of $X_1, \ldots, X_q$ given that $X_{q+1} = x_{q+1}, \ldots, X_p = x_p$ can be obtained from (3.3) by replacing the probability density functions by the corresponding mass functions.

The mathematical expectation of a random matrix X

$$X = \begin{pmatrix} X_{11} & \cdots & X_{q1} \\ \vdots & & \vdots \\ X_{1p} & \cdots & X_{qp} \end{pmatrix}$$

of dimension $p \times q$ (the components X_{ij} are random variables) is defined by

$$E(X) = \begin{pmatrix} E(X_{11}) & \cdots & E(X_{q1}) \\ \vdots & & \vdots \\ E(X_{1p}) & \cdots & E(X_{qp}) \end{pmatrix}. \quad (3.4)$$

Since a random vector $\mathbf{X} = (X_1, \ldots, X_p)'$ is a random matrix of dimension $p \times 1$, its mathematical expectation is given by

$$E(\mathbf{X}) = (E(X_1), \ldots, E(X_p))'. \quad (3.5)$$

Thus it follows that for any matrices A, B, C of real constants and for any random matrix X

$$E(AXB + C) = AE(X)B + C. \quad (3.6)$$

DEFINITION 3.1.1 For any random vector $\mathbf{X}$, $\boldsymbol{\mu} = E(\mathbf{X})$ and $\Sigma = E(\mathbf{X} - \boldsymbol{\mu})(\mathbf{X} - \boldsymbol{\mu})'$ are called, respectively, the mean and the covariance matrix of $\mathbf{X}$.

DEFINITION 3.1.2 For every real $\mathbf{t} = (t_1, \ldots, t_p)'$, the characteristic function of any random vector $\mathbf{X}$ is defined by $\phi_{\mathbf{X}}(\mathbf{t}) = E(e^{i\mathbf{t}'\mathbf{X}})$ where $i = (-1)^{1/2}$.

Since $E|e^{i\mathbf{t}'\mathbf{X}}| = 1$, $\phi_{\mathbf{X}}(\mathbf{t})$ always exists.

3.2 INVARIANCE IN STATISTICAL TESTING OF HYPOTHESES

Invariance is a mathematical term for symmetry and in practice many statistical testing problems exhibit symmetries. The notion of invariance in statistical tests is of old origin. The unpublished work of Hunt and Stein (see Lehmann, 1959) toward the end of World War II has given this principle strong support as to its applicability and meaningfulness in the framework of the general class of all statistical tests. It is now established as a very powerful tool for proving the admissibility and minimax property of many statistical tests. It is a generally accepted principle that if a problem with a unique solution is invariant under a certain transformation, then the solution should be invariant under that transformation. The main reason for the strong intuitive appeal of an invariant decision procedure is the feeling that there should be or exists a unique best way of analyzing a collection of statistical information. Nevertheless in cases in which the use of an invariant procedure conflicts violently with the desire to make a correct decision with high probability or to have a small expected loss, the procedure must be abandoned.

Let $\mathscr{X}$ be the sample space, let $\mathscr{A}$ be the σ-algebra of subsets of $\mathscr{X}$ (a class of subsets of $\mathscr{X}$ which contains $\mathscr{X}$ and is closed under complementation and countable unions), and let $\Omega = \{\theta\}$ be the parametric space. Denote by P the family of probability distributions P_θ on $\mathscr{A}$. We are concerned here with the problem of testing the null hypothesis $H_0 : \theta \in \Omega_{H_0}$ against the alternatives $H_1 : \theta \in \Omega_{H_1}$. The principle of invariance for testing problems involves transformations mainly on two spaces: the sample space $\mathscr{X}$ and the parametric space Ω. Between the two, the most basic is the transformation g on $\mathscr{X}$. The transformation on Ω is transformation $\bar{g}$, induced by g on Ω. All transformations g, considered in the context of invariance, will be assumed to be

(i) one-to-one from $\mathscr{X}$ onto $\mathscr{X}$; i.e., for every $x_1 \in \mathscr{X}$ there exists $x_2 \in \mathscr{X}$ such that $x_2 = g(x_1)$ and $g(x_1) = g(x_2)$ implies $x_1 = x_2$.

(ii) bimeasurable, to ensure that whenever X is a random variable with values in $\mathscr{X}$, $g(X)$ (usually written as gX) is also a random variable with values in $\mathscr{X}$ and for any set $A \in \mathscr{A}$, gA and $g^{-1}A$ (the image and the transformed set) both belong to $\mathscr{A}$.

The induced transformation $\bar{g}$ corresponding to g on $\mathscr{X}$ is defined as follows:

If the random variable X with values in $\mathscr{X}$ has probability distribution P_θ, gX is also a random variable with values in $\mathscr{X}$, and has probability distribution $P_{\theta'}$, where $\theta' = \bar{g}\theta \in \Omega$. An equivalent way of stating this fact is

(g^{-1} being the inverse transformation corresponding to g)

$$P_\theta(g^{-1}A) = P_{\bar{g}\theta}(A) \tag{3.7}$$

or

$$P_\theta(A) = P_{\bar{g}\theta}(gA) \tag{3.8}$$

for all $A \in \mathscr{A}$. In terms of mathematical expectation this is also equivalent to saying that for any integrable real-valued function ϕ

$$E_\theta(\phi(g^{-1}X)) = E_{\bar{g}\theta}(\phi(X)), \tag{3.9}$$

where E_θ refers to expectation when X has distribution P_θ.

If, in addition, all P_θ, $\theta \in \Omega$, are distinct, i.e., if $\theta_1 \neq \theta_2$, $\theta_1, \theta_2 \in \Omega$, implies $P_{\theta_1} \neq P_{\theta_2}$, then g determines $\bar{g}$ uniquely and the correspondence between g and $\bar{g}$ is a homomorphism.

The condition (3.7) or its equivalent is known as the condition of invariance of probability distributions with respect to the transformation g on $\mathscr{X}$.

DEFINITION 3.2.1 *Invariance of the parametric space* Ω The parametric space Ω remains invariant under a one-to-one transformation $g:\mathscr{X}$ onto $\mathscr{X}$ if the induced transformation $\bar{g}$ on Ω satisfies (i) $\bar{g}\theta \in \Omega$ for $\theta \in \Omega$, and (ii) for any $\theta' \in \Omega$ there exists a $\theta \in \Omega$ such that $\theta' = \bar{g}\theta$.

An equivalent way of writing (i) and (ii) is

$$\bar{g}\Omega = \Omega. \tag{3.10}$$

If the P_θ for different values of θ are distinct, then $\bar{g}$ is also one-to-one.

Given a set of transformations, each leaving Ω invariant, the following theorem will assert that we can always extend this set to a group G of transformations whose members also leave Ω invariant.

THEOREM 3.2.1. Let g_1 and g_2 be two transformations which leave Ω invariant. The transformations $g_2 g_1$ and g_1^{-1} defined by $g_2 g_1(x) = g_2(g_1(x))$, $g_1^{-1} g_1(x) = x$ for all $x \in \mathscr{X}$ leave Ω invariant and $\overline{g_2 g_1} = \bar{g}_2 \bar{g}_1$, $\overline{g_1^{-1}} = \bar{g}^{-1}$.

Proof If the random variable X with values in $\mathscr{X}$ has probability distribution P_θ, then for any transformation g, gX has probability distribution $P_{\bar{g}\theta}$ with $\bar{g}\theta \in \Omega$. Since $\bar{g}_1\theta \in \Omega$ and g_2 leaves Ω invariant, the probability distribution of $g_2 g_1(X) = g_2(g_1(X))$ is $P_{\bar{g}_2\bar{g}_1\theta}$, $\bar{g}_2\bar{g}_1\theta \in \Omega$. Thus $g_2 g_1$ leaves Ω invariant and obviously $\overline{g_2 g_1} = \bar{g}_2\bar{g}_1$. The reader may find it instructive to verify the other assertion. Q.E.D.

Very often in statistical problems there exists a measure λ on $\mathscr{X}$ such that P_θ is absolutely continuous with respect to λ, so that we can write for every

$A \in \mathscr{A}$,

$$P_\theta(A) = \int_A p_\theta(x)\, d\lambda(x). \tag{3.11}$$

It is possible to choose the measure λ such that it is left invariant under G, i.e.,

$$\lambda(A) = \lambda(gA) \tag{3.12}$$

for all $A \in \mathscr{A}$ and all $g \in G$. Then the condition of invariance of distribution reduces to

$$p_{\bar{g}\theta}(x) = p_\theta(g^{-1}x)$$

for all $x \in \mathscr{X}$ and $g \in G$.

EXAMPLE 3.2.1. Let $\mathscr{X}$ be the Euclidean space and G be the group of translations defined by

$$g_{x_1}(x) = x + x_1, \qquad x_1 \in \mathscr{X}, \quad g \in G.$$

Here G acts transitively on $\mathscr{X}$. The n-dimensional Lebesgue measure λ is invariant under G and it is unique up to a positive multiplicative constant.

Let us now consider the problem of testing $H_0 : \theta \in \Omega_{H_0}$ against the alternatives $H_1 : \theta \in \Omega_{H_1}$, where Ω_{H_0} and Ω_{H_1} are disjoint subsets of Ω. Let G be a group of transformations which operates from the left on $\mathscr{X}$, satisfying conditions (3.7) and (3.10).

DEFINITION 3.2.2 *Invariance of statistical problems* The problem of testing $H_0 : \theta \in \Omega_{H_0}$ against $H_1 : \theta \in \Omega_{H_1}$ remains invariant with respect to G if

(i) for $g \in G$, $A \in \mathscr{A}$, $P_{\bar{g}\theta}(gA) = P_\theta(A)$, and
(ii) $\Omega_{H_0} = \bar{g}\Omega_{H_0}, \qquad \Omega_{H_1} = \bar{g}\Omega_{H_1}$. (3.13)

EXAMPLE 3.2.2 Let $X_1, \ldots, X_n$ be a random sample of size n from a normal distribution with mean μ and variance σ^2 and let $x_1, \ldots, x_n$ be sample observations. Denote by $\mathscr{X}$ the space of all values $(x_1, \ldots, x_n)$. Let

$$\bar{x} = \frac{1}{n}\sum_{i=1}^n x_i, \qquad s^2 = \frac{1}{n}\sum_{i=1}^n (x_i - \bar{x})^2,$$

$$\bar{X} = \frac{1}{n}\sum_{i=1}^n X_i, \qquad S^2 = \frac{1}{n}\sum_{i=1}^n (X_i - \bar{X})^2, \qquad \theta = (\mu, \sigma^2).$$

The parametric space Ω is given by

$$\Omega = \{\theta = (\mu, \sigma^2), -\infty < \mu < \infty, \sigma^2 > 0\},$$

and let

$$\Omega_{H_0} = \{(0, \sigma^2) : \sigma^2 > 0\}, \qquad \Omega_{H_1} = \{(\mu, \sigma^2) : \mu \neq 0, \sigma^2 > 0\}.$$

The group of transformations G which leaves the problem invariant is the group of scale changes

$$X_i \to aX_i, \qquad i = 1, \ldots, n$$

with $a \neq 0$ and $\bar{g}\theta = (a\mu, a^2\sigma^2)$. Obviously $\bar{g}\Omega = \Omega$ for all $g \in G$. Since $x = (x_1, \ldots, x_n) \in g^{-1}A$ implies $gx \in A$, we have, with $y_i = ax_i, i = 1, \ldots, n$,

$$\begin{aligned} P_\theta(g^{-1}A) &= \int_{g^{-1}A} \frac{1}{(2\pi)^{n/2}(\sigma^2)^{n/2}} \exp\left[-\frac{1}{2}\sum_{i=1}^{n} \frac{(x_i - \mu)^2}{\sigma^2}\right] dx_1, \ldots, x_n \\ &= \int_A \frac{1}{(2\pi)^{n/2}(a^2\sigma^2)^{n/2}} \exp\left[-\frac{1}{2a^2\sigma^2}\sum_{i=1}^{n} (y_i - a\mu)^2\right] dy_1, \ldots, dy_n \\ &= P_{\bar{g}\theta}(A). \end{aligned}$$

Furthermore $\bar{g}\Omega_{H_0} = \Omega_{H_0}, \bar{g}\Omega_{H_1} = \Omega_{H_1}$.

If a statistical problem remains invariant under a group of transformations G operating on the sample space $\mathscr{X}$, it is then natural to restrict attention to statistical tests ϕ which are also invariant under G, i.e.,

$$\phi(x) = \phi(gx), \qquad x \in \mathscr{X}, \quad g \in G.$$

DEFINITION 3.2.3 *Invariant function* A function $T(x)$ defined on $\mathscr{X}$ is invariant under the group of transformations G if $T(x) = T(gx)$ for all $x \in \mathscr{X}, g \in G$.

DEFINITION 3.2.4 *Maximal invariant* A function $T(x)$ defined on $\mathscr{X}$ is a maximal invariant under G if (i) $T(x) = T(gx)$, $x \in \mathscr{X}$, $g \in G$, and (ii) $T(x) = T(y)$ for x, $y \in \mathscr{X}$ implies that there exists a $g \in G$ such that $y = gx$.

The reader is referred to Lehmann (1959) for the interpretation of invariant function and maximal invariant in terms of partition of the sample space.

Let $\mathscr{Y}$ be a space and let B be the σ-algebra of subsets of $\mathscr{Y}$. Suppose $T(x)$ is a measurable mapping from $\mathscr{X}$ into $\mathscr{Y}$. Let h be a one-to-one function on $\mathscr{Y}$ to $\mathscr{Z}$. If $T(x)$ with values in $\mathscr{Y}$ is a maximal invariant on $\mathscr{X}$, then $T \circ h$ is a maximal invariant on $\mathscr{X}$ with values in $\mathscr{Z}$. This fact is often used to write a maximal invariant in a convenient form.

Let $\phi(x)$ be a statistical test (probability of rejecting H_0 when x is observed). For a nonrandomized test $\phi(x)$ takes values 0 or 1. Suppose $\phi(x)$, $x \in \mathscr{X}$ is invariant under a group of transformations G, operating from the left on $\mathscr{X}$. A useful characterization of $\phi(x)$ in terms of the maximal invariant $T(x)$ (under G) on $\mathscr{X}$ is given by the following theorem.

THEOREM 3.2.2 A test $\phi(x)$ is invariant under G if and only if there exists a function h such that $\phi(x) = h(T(x))$.

Proof Let $\phi(x) = h(T(x))$. Obviously

$$\phi(x) = h(T(x)) = h(T(gx)) = \phi(gx)$$

for $x \in \mathscr{X}$, $g \in G$. Conversely, if $\phi(x)$ is invariant under G and $T(x) = T(y)$, $x, y \subset \mathscr{X}$, then there exists a $g \in G$ such that $y = gx$ and therefore $\phi(x) = \phi(y)$. Q.E.D.

In general h may not be a Borel measurable function. However, if the range of T is Euclidean and T is Borel measurable, then h is Borel measurable. See, for example, Blackwell (1956).

Let $\bar{G}$ be the group of induced (induced by G) transformations on Ω. We define on Ω, as on $\mathscr{X}$, a maximal invariant with respect to $\bar{G}$. Let $v(\theta)$ be a maximal invariant on Ω with respect to $\bar{G}$.

THEOREM 3.2.3 The distribution of $T(X)$ with values in the space of $\mathscr{Y}$, where X is a random variable with values in $\mathscr{X}$, depends on Ω only through $v(\theta)$.

Proof Suppose $v(\theta_1) = v(\theta_2), \theta_1, \theta_2 \in \Omega$. Since $v(\theta)$ is a maximal invariant on Ω under $\bar{G}$, there exists a $\bar{g} \in G$ such that $\theta_2 = \bar{g}\theta_1$. Now for any measurable set C in B [by (3.7)]

$$P_{\theta_1}(T(X) \in C) = P_{\theta_1}(T(gX) \in C) = P_{\bar{g}\theta_1}(T(X) \in C) = P_{\theta_2}(T(X) \in C). \qquad \text{Q.E.D.}$$

EXAMPLE 3.2.3 Consider Example 3.2.2. Here

$$T(x) = \sqrt{n}\bar{x}\left[\sum_{i=1}^{n} \frac{(x_i - \bar{x})^2}{n-1}\right]^{-1/2}, \qquad v(\theta) = (\sqrt{n}\mu)/\sigma = \lambda$$

where λ is an arbitrary designation. The probability density function of T is given by (see Giri, 1974)

$$f_T(t) = \frac{(n-1)^{(n-1)/2}}{\Gamma((n-1)/2)} \frac{\exp(-\lambda^2/2)}{(n-1+t^2)^{n/2}} \sum_{j=0}^{\infty} \Gamma\left(\frac{n+j}{2}\right) \frac{\lambda^j}{j!} \left(\frac{2t^2}{n-1+t^2}\right)^{j/2}.$$

3.2.1 Almost Invariance and Invariance

To study the relative performances of different test criteria we need to compare their power functions. Thus it is of interest to study the implication of the invariance of power functions of the tests rather than the tests themselves. Since the power function of invariant tests depends only on the maximal invariant on Ω, any invariant test has invariant power functions. The converse that if the power function of a test ϕ is invariant under the induced group $\bar{G}$, i.e.,

$$E_\theta \phi(g^{-1}X) = E_{\bar{g}\theta}\phi(X), \qquad (3.14)$$

then the test ϕ is invariant under G, does not always hold well. To investigate this further we need to define the notions of almost invariance and equivalence to an invariant test.

DEFINITION 3.2.5 *Equivalence to an invariant test* Let G be a group of transformations satisfying (3.8) and (3.10). A test $\psi(x)$, $x \in \mathscr{X}$, is equivalent to an invariant test $\phi(x)$, $x \in \mathscr{X}$, with respect to the group of transformations G if

$$\phi(x) = \psi(x) \qquad \text{for all} \quad x \in \mathscr{X} - N$$

where $P_\theta(N) = 0$ for $\theta \in \Omega$.

DEFINITION 3.2.6 *Almost invariance* Let G be a group of transformations on $\mathscr{X}$ satisfying (3.8) and (3.10). A test $\phi(x)$ is said to be almost invariant with respect to G if for $g \in G$, $\phi(x) = \phi(gx)$ for all $x \in \mathscr{X} - Ng$ where $P_\theta(Ng) = 0$, $\theta \in \Omega$.

It is tempting to conjecture that an almost invariant test is equivalent to an invariant test. If any test ψ is equivalent to an invariant test ϕ, then it is almost invariant. For, take

$$Ng = N \bigcup (g^{-1}N).$$

Obviously $x \in \mathscr{X} - Ng$ implies $x \in \mathscr{X} - N$ and $gx \in \mathscr{X} - N$. Hence for $x \in \mathscr{X} - Ng$, $\psi(x) = \phi(x) = \phi(gx) = \psi(gx)$. Since $P_\theta(g^{-1}N) = P_{\bar{g}\theta}(N) = 0$, $P_\theta(Ng) = 0$. Conversely, if the group G is countable, for any almost invariant test ψ, take

$$N = \bigcup_{g \in G} Ng,$$

where $\psi(x) = \psi(gx)$, $x \in \mathscr{X} - Ng$, $g \in G$, so that $P_\theta(N) = 0$. Then $\psi(x) = \psi(gx)$, $x \in \mathscr{X} - N$. Now define $\phi(x)$ such that

$$\phi(x) = \begin{cases} 1 & \text{if} \quad x \in N \\ \psi(x) & \text{if} \quad x \in \mathscr{X} - N. \end{cases}$$

Obviously $\phi(x)$ is an invariant function and $\psi(x)$ is equivalent to an invariant test. (Note that $gN = N$, $g \in G$.) If the group G is uncountable, such a result does not, in general, hold well.

Let $\mathscr{X}$ be the sample space and let $\mathscr{A}$ be the σ-field of subsets of $\mathscr{X}$. Suppose that G is a group of transformations operating on $\mathscr{X}$ and that B is a σ-field of subsets of G. Let for any $A \in \mathscr{A}$, the set of pairs (x, g), such that $gx \in A$, belong to $\mathscr{A} \times B$. Suppose further that there exists a σ-finite measure μ (i.e., for $B_1, B_2, \ldots$ in B such that $\bigcup B_i = G$ and $\mu(B_i) < \infty$ for all i) on G

such that

$$\mu(B) = 0 \qquad \text{implies} \quad \mu(Bg) = 0 \tag{3.15}$$

for all $g \in G$. Then any almost invariant function on $\mathscr{X}$ with respect to G is equivalent to an invariant function with respect to G. For a proof of this result the reader is referred to Lehmann (1959, p. 225). This requirement is satisfied in particular when

$$\mu(Bg) = \mu(B), \qquad B \in \mathscr{B}, \quad g \in G.$$

In other words, μ is a σ-finite right invariant measure. Such a right invariant measure μ exists for a large number of groups.

EXAMPLE 3.2.4 Let $G = E^p$ (Euclidean p-space) where the group operation is addition. The Lebesgue measure in the space of E^p is the right invariant measure. Since G is Abelian, the right invariant measure is also left invariant, i.e., $\mu(gB) = \mu(B)$ for $g \in G$.

EXAMPLE 3.2.5 Let G be the positive half of the real line with multiplication as the group operation. The right invariant measure μ is given by ($B \in \mathscr{B}$)

$$\mu(B) = \int_B \frac{dg}{g}.$$

EXAMPLE 3.2.6 Let G be the multiplicative group of $p \times p$ nonsingular real matrices $g = (g_{ij})$. Write

$$dg = \prod_{i,j} dg_{ij}.$$

The right invariant measure μ on G is given by

$$\mu(B) = \int_B \frac{dg}{|\det g|^p}.$$

This follows from the fact that the Jacobian of the transformation

$$g \to gh, \qquad g, h \in G,$$

is $(\det(h))^p$. Furthermore it is also left invariant.

EXAMPLE 3.2.7 Let G be the group of affine transformations of the real line R onto itself, i.e., $g \in G$ has the form (a, b) such that for $x \in R$

$$(a, b)x = ax + b.$$

Here the group operation is defined by

$$g_1 g_2 = (a_1 a_2, a_1 b_2 + b_1)$$

where $g_1 = (a_1, b_1)$, $g_2 = (a_2, b_2)$.

The Jacobian of the transformation $g_1 \to g_1 g_2$ is a_2. So the right invariant measure μ on G is given by

$$\mu(B) = \int_B \frac{da\, db}{a}.$$

Note The left invariant measure in this case is given by

$$\mu(B) = \int_B \frac{da\, db}{a^2}.$$

EXAMPLE 3.2.8 Let G be the group of affine transformations of V^p (a real vector space of dimension p) onto itself; i.e., for $x \in V^p$, $g = (c, b) \in G$ where $c \in G_l(p)$, the group of $p \times p$ nonsingular real matrices, and b is a p-vector,

$$gx = cx + b.$$

The group operation in G is defined by

$$g_1 g_2 = (c_1 c_2, c_1 b_2 + b_1)$$

where $g_1 = (c_1, b_1)$, $g_2 = (c_2, b_2)$, $b_1, b_2 \in V^p$, and $c_1, c_2 \in G_l(p)$. The right invariant measure μ on G is defined by

$$\mu(B) = \int_B \frac{dc\, db}{|\det c|^p}.$$

The left invariant measure in this case is given by

$$\mu(B) = \int_B \frac{dc\, db}{|\det c|^{p+1}}.$$

EXAMPLE 3.2.9 Let G_T be the multiplicative group of $p \times p$ nonsingular lower triangular matrices g, given by

$$g = \begin{pmatrix} g_{11} & 0 & 0 & \cdots & 0 \\ g_{21} & g_{22} & 0 & \cdots & 0 \\ \vdots & \vdots & & & \vdots \\ g_{k1} & g_{k2} & g_{k3} & \cdots & g_{kk} \end{pmatrix}$$

where g_{ii} is a submatrix of g of dimension $d_i \times d_i$ such that $\sum_1^k d_i = p$. The right invariant measure μ on G_T is given by

$$\mu(B) = \int_B \frac{dg}{\prod_{i=1}^k |\det g_{ii}|^{p-\sigma_i - 1}}$$

where $\sigma_i = \sum_{j=1}^{i} d_j$ with $\sigma_0 = 0$. The left invariant measure on G is given by

$$\mu(B) = \int_B \frac{dg}{\prod_{i=1}^{k} |\det g_{ii}|^{\sigma_i}}.$$

For further results on invariant measure the reader is referred to Nachbin (1965).

It is now evident that any almost invariant test function with respect to a group of transformations G on $\mathscr{X}$ has an invariant power function with respect to the induced group $\bar{G}$ on Ω. The converse of this is not true in general. However, in cases in which prior to the application of invariance the problem can be reduced to one based on a sufficient statistic on the sample space whose distributions constitute a boundedly complete family, the converse is true.

Let T be sufficient for $\{P_\theta, \theta \in \Omega\}$ and let the distribution $\{P_\theta{}^T, \theta \in \Omega\}$ of T be boundedly complete; i.e., for any bounded function $g(T)$ of T, if

$$E_\theta g(T) \equiv 0 \tag{3.15}$$

for all $\theta \in \Omega$, then $g(T) = 0$ almost everywhere with respect to the probability measure $P_\theta{}^T$. For any almost invariant test function $\psi(T)$ with respect to the group of transformations G on the space of the sufficient statistic T we have, for $g \in G$,

$$E_\theta \psi(T) = E_\theta \psi(gT) = E_{\bar{g}\theta} \psi(T).$$

Conversely, if

$$E_\theta \psi(T) = E_{\bar{g}\theta} \psi(T),$$

then for $g \in G$ [note $gT(x) = T(gx)$],

$$E_\theta \psi(T) = E_\theta \psi(gT)$$

or, equivalently,

$$E_\theta(\psi(gT) - \psi(T)) \equiv 0$$

for all $\theta \in \Omega$. Since the distribution of T is boundedly complete, we obtain

$$\psi(t) = \psi(gt)$$

almost everywhere with respect to the probability measure $P_\theta{}^T$.

Since for any test $\phi(x)$ on the original sample space $\mathscr{X}$,

$$\psi(t) = E(\phi(X)|T = t)$$

is also a test based on the sufficient statistic T with the same power function as that of $\phi(x)$, we can conclude that if there exists a uniformly most powerful almost invariant test among all tests based on the sufficient statistic T, then that test is uniformly most powerful among all tests based on the original observations x and its power function depends only on the maximal invariant on the parametric space Ω.

EXAMPLE 3.2.10 Consider Example 3.2.3. Let us first show that the sufficient statistic $(\bar{X}, S^2)$ for (μ, σ^2) is boundedly complete. The joint probability density function of $(\bar{X}, S^2)$ is given by (see Giri, 1974)

$$f_{\bar{X},S^2}(\bar{x}, s^2) = \frac{K}{(\sigma^2)^{n/2}} \exp\{-(1/2\sigma^2)(ns^2 + n(\bar{x} - \mu)^2)\} (ns^2)^{(n-3)/2}$$

where

$$K = \sqrt{n}/[(2\pi)^{1/2} 2^{(n-1)/2} \Gamma((n-1)/2)].$$

For any bounded function $g(\bar{x}, s^2)$

$$E(g(\bar{X}, S^2)) = K \int \frac{g(\bar{x}, s^2)}{(\sigma^2)^{n/2}} \exp\left\{-\frac{1}{2\sigma^2}(ns^2 + n(\bar{x} - \mu)^2)\right\} (ns^2)^{(n-3)/2}\, ds^2\, d\bar{x}.$$

Let $1/\sigma^2 = 1 - 2\theta$ and $(1 - 2\theta)^{-1} t = \mu$. Then

$$E(g(\bar{X}, S^2)) = K \int g(\bar{x}, s^2)(1 - 2\theta)^{n/2}(ns^2)^{(n-3)/2}$$
$$\times \exp\{-\tfrac{1}{2}[(1 - 2\theta)(ns^2 + n\bar{x}^2) - 2nt\bar{x} + nt^2/(1 - 2\theta)]\}\, ds^2\, d\bar{x}. \tag{3.16}$$

If

$$E(g(\bar{X}, S^2)) \equiv 0$$

for all (μ, σ^2), then from (3.16) we obtain

$$K \int g(\bar{x}, ns^2 + n\bar{x}^2 - n\bar{x}^2)(ns^2)^{(n-3)/2}$$
$$+ \exp\{-\tfrac{1}{2}(ns^2 + n\bar{x}^2) + \theta(ns^2 + n\bar{x}^2) + nt\bar{x}\}\, ds^2\, d\bar{x} \equiv 0.$$

This is the Laplace transform of

$$g(\bar{x}, ns^2 + n\bar{x}^2 - n\bar{x}^2) K (ns^2)^{(n-3)/2} \exp\{-\tfrac{1}{2}(n\bar{x}^2 + ns^2)\}$$

with respect to the variables $n\bar{x}$, $ns^2 + n\bar{x}^2$. Since this is zero for all (μ, σ^2), we obtain

$$g(\bar{x}, s^2) = 0$$

except for a set of $(\bar{x}, s^2)$ with probability measure 0. So the distribution of $(\bar{X}, S^2)$ is boundedly complete.

Second, from Example 3.2.3 we can conclude that for testing $H_0 : \mu = 0$, the test which rejects H_0 whenever $|t| \geq t_{1-\alpha/2}$, where $t_{1-\alpha/2}$ is the upper

$1 - \alpha/2$ percent point of the central t-distribution with $n - 1$ degrees of freedom, is uniformly most powerful among all tests whose power function depends only on $\sqrt{n}\mu/\sigma$.

3.3 SUFFICIENCY AND INVARIANCE

It is well known that some simplification is introduced in a testing problem by characterizing the statistical tests as a function of the sufficient statistic and thus reducing the dimension of the sample space to the dimension of the space of the sufficient statistic. On the other hand, invariance by reducing the dimension of the sample space to that of the space of the maximal invariant also shrinks the parametric space. Thus a question naturally arises: Is it possible to use both principles simultaneously and if so in what order, i.e., first sufficiency and then invariance, or first invariance and then sufficiency. Under certain conditions this reduction can be done by using both principles, and the order in which the reduction is made is immaterial in such cases. The reader is referred to Hall *et al.* (1965) for these conditions and some related results.

One can also avoid the task of verifying these conditions by replacing the sample space by the space of the sufficient statistic before looking for the group of transformations which leave the problem invariant and then look for the group of transformations on the space of the sufficient statistic that leave the problem invariant.

3.4 UNBIASEDNESS AND INVARIANCE

The discussions presented in this and the following section are sketchy. For further study and details relevant references are given.

In testing statistical hypotheses, the principle of unbiasedness plays an important role in deriving a suitable test statistic in complex situations involving composite hypotheses. A size α test ϕ is said to be unbiased for testing $H_0{:}\theta \in \Omega_{H_0}$ against $H_1{:}\theta \in \Omega_{H_1}$ if $E_\theta\phi(X) \geq \alpha$ for $\theta \in \Omega_{H_1}$. In many such problems the principle of unbiasedness and the principle of invariance seem to complement each other in the sense that each is successful in the cases in which the other is not. For example, it is well known that a uniformly most powerful unbiased test exists for testing the hypothesis $H_0{:}\sigma^2 = \sigma_0^{\,2}$ (specified) against the alternatives $H_1{:}\sigma^2 \neq \sigma_0^{\,2}$ in a normal distribution with mean μ whereas the principle of invariance does not reduce the problem sufficiently far to ensure the existence of a uniformly most powerful invariant test. On the other hand, for problems involving general linear hypotheses there exists a uniformly most powerful invariant test (F-test) but no uniformly most powerful unbiased test exists if the null hypothesis has more than one degree of freedom. However, if both principles can be applied successfully, then they lead to the same (almost everywhere) optimum test. Consider the problem of testing $H_0:\theta \in \Omega_{H_0}$ against the alternatives $H_1:\theta \in \Omega_{H_1}$. Let us assume that it is invariant under the group of transformations G. Let C_α be the class of unbiased tests of size α $(0 < \alpha < 1)$. For any test $\phi(x)$ define the test function ϕg by

$$\phi g(x) = \phi(gx), \qquad x \in \mathscr{X}, \quad g \in G.$$

Obviously $\phi \in C_\alpha$ if and only if $\phi g \in C_\alpha$. Thus if the test ϕ^* is a unique (up to measure 0) uniformly most powerful unbiased test for this problem, then

$$E_\theta(\phi^* g(X)) = E_{\bar{g}\theta}(\phi^*(X)) = \sup_{\phi \in C_\alpha} E_{\bar{g}\theta}(\phi(X)) = \sup_{\phi g \in C_\alpha} E_\theta(\phi(g(X)))$$

$$= \sup_{\phi g \in C_\alpha} E_\theta(\phi g(X)) = E_\theta \phi^*(X).$$

Thus ϕ^* and $\phi^* g$ have the same power function. Hence under the assumption of completeness of the sufficient statistic, ϕ^* is almost invariant. Therefore if there exists a uniformly most powerful almost invariant test ϕ^{**}, we have

$$E_\theta \phi^{**}(X) \geq E_\theta \phi^*(X) \tag{3.17}$$

for $\theta \in \Omega_{H_1}$. Comparing this with the trivial level α invariant test $\phi(x) = \alpha$, we conclude that ϕ^{**} is also unbiased, and hence

$$E_\theta \phi^{**}(X) \leq E_\theta \phi^*(X) \tag{3.18}$$

for $\theta \in \Omega_{H_1}$. Thus from (3.17) and (3.18) it follows that ϕ^* and ϕ^{**} have the same power function. Since ϕ^* is unique $\phi^* = \phi^{**}$ almost everywhere.

Thus for a testing problem which is invariant under G, if there exists a unique uniformly most powerful unbiased test ϕ^* and if there exists a uniformly most powerful almost invariant test ϕ^{**}, then $\phi^* = \phi^{**}$ almost everywhere.

3.5 INVARIANCE AND OPTIMUM TESTS

Apart from the important fact that the performance of an invariant test is independent of the nuisance parameters, a powerful support of the principle comes from the famous unpublished Hunt–Stein theorem which asserts that under certain conditions on the group G there exists an invariant test which is minimax among all size α tests. It is well known that given any test function on the sample space we can always replace it by a test which depends only on the sufficient statistic such that both have the same power function. Such a result is too strong to expect from the maximal invariant statistic on the sample space. The appropriate weakening of this property and the conditions under which it holds constitute the Hunt–Stein theorem which asserts that for testing $H_0:\theta \in \Omega_{H_0}$ against $H_1:\theta \in \Omega_{H_1}$ (which is invariant under G), under certain conditions on the group G, given any test function ϕ on the sample space $\mathscr{X}$, there exists an invariant test ψ such that

$$\sup_{\theta \in \Omega_{H_0}} E_\theta \phi \geq \sup_{\theta \in \Omega_{H_0}} E_\theta \psi, \qquad \inf_{\theta \in \Omega_{H_1}} E_\theta \phi \leq \inf_{\theta \in \Omega_{H_1}} E_\theta \psi. \tag{3.19}$$

In other words, ψ performs at least as well as ϕ in the worst possible cases. For the exact statement of this theorem the reader is referred to Lehmann (1959, p. 335).

This method has been successfully used by Giri *et al.* (1963), Giri and Kiefer (1964a), Linnik *et al.* (1966), and Salaevskii (1968) to solve the long time open problem of the minimax character of Hotelling's T^2-test, and by Giri and Kiefer (1964b) to prove the minimax character of the R^2-test in some special cases.

It may be remarked here that the conditions of Hunt–Stein's theorem, whether algebraic or topological, are almost entirely on the group and are nonstatistical in nature. For verifying the admissibility of statistical tests through invariance the situation is more complicated. Aside from the trivial case of compact groups only the one-dimensional translation parameter case has been studied by Lehmann and Stein (1953). If G is a finite or a compact group, the most powerful invariant test is admissible. For other groups statistical structure plays an important role.

For further relevant results in this context the reader is referred to Kiefer (1957, 1966), Ghosh (1967), and Pitman (1939).

3.6 MOST STRINGENT TESTS AND INVARIANCE

Consider the problem of testing $H_0:\theta \in \Omega_{H_0}$ against the alternatives H_1: $\theta \in \Omega_{H_1}$ where $\Omega_{H_0} \cap \Omega_{H_1}$ is a null set. Let Q_α denote the class of all level α tests of H_0 and let

$$\beta_\alpha^*(\theta) = \sup_{\phi \in Q_\alpha} E_\theta \phi, \qquad \theta \in \Omega_{H_1}.$$

$\beta_\alpha^*(\theta)$ is called the envelope power function and it is the maximum power that can be obtained at level α against the alternative θ.

DEFINITION 3.6.1 *Most stringent test* A test ϕ that minimizes $\sup_{\theta \in \Omega_{H_1}}(\beta_\alpha^*(\theta) - E_\theta(\phi))$ is said to be most stringent. In other words, it minimizes the maximum shortcomings.

If the testing problem is invariant under a group of transformations G and if there exists a uniformly most powerful almost invariant test ϕ^* with respect to G such that the group satisfies the conditions of the Hunt–Stein theorem (see Lehmann, 1959, p. 336), then ϕ^* is most stringent. For details and further reading in this context the reader is referred to Lehmann (1959, 1959a), Kiefer (1958), and Giri and Kiefer (1964a).

EXERCISES

1. Let $\{P_\theta, \theta \in \Omega\}$, the family of distributions on $(\mathscr{X}, \mathscr{A})$, be such that each P_θ is absolutely continuous with respect to a σ-finite measure μ; i.e., if $\mu(A) = 0$ for $A \in \mathscr{A}$, then $P_\theta(A) = 0$. Let $p_\theta = \partial P_\theta / \partial \mu$ and define the measure μg^{-1} for $g \in G$, the group of transformations on $\mathscr{X}$, by

$$\mu g^{-1}(A) = \mu(g^{-1}A).$$

Suppose that

(a) μ is absolutely continuous with respect to μg^{-1} for all $g \in G$;
(b) $p_\theta(x)$ is absolutely continuous in θ for all x;
(c) Ω is separable;
(d) the subspaces Ω_{H_0} and Ω_{H_1} are invariant with respect to G. Then show that

$$\sup_{\Omega_{H_1}} p_\theta(x) \Big/ \sup_{\Omega_{H_0}} p_\theta(x)$$

is almost invariant under G.

2. Let $X_1, \ldots, X_n$ be a random sample of size n from a normal population with unknown mean μ and variance σ^2. Find the uniformly most powerful invariant test of $H_0: \sigma^2 < \sigma_0^2$ (specified) against the alternatives $\sigma^2 > \sigma_0^2$ with respect to the group of transformations which transform $X_i \to X_i + c$, $-\infty < c < \infty$, $i = 1, \ldots, n$.

3. Let $X_1, \ldots, X_{n_1}$ be a random sample of size n_1 from a normal population with mean μ and variance σ_1^2; and let $Y_1, \ldots, Y_{n_2}$ be a random sample of size n_2 from another normal population with mean v and variance σ_2^2. Let $(X_1, \ldots, X_{n_1})$ be independent of $(Y_1, \ldots, Y_{n_2})$. Write

$$\bar{X} = \frac{1}{n_1} \sum_{i=1}^{n_1} X_i, \qquad S_1^2 = \sum_{i=1}^{n_1} (X_i - \bar{X})^2,$$

$$\bar{Y} = \frac{1}{n_2} \sum_{i=1}^{n_2} Y_i, \qquad S_2^2 = \sum_{i=1}^{n_2} (Y_i - \bar{Y})^2.$$

The problem of testing $H_0:\sigma_1^2/\sigma_2^2 \le \lambda_0$ (specified) against the alternatives $H_1:\sigma_1^2/\sigma_2^2 > \lambda_0$ remains invariant under the group of transformations

$$\bar{X} \to \bar{X} + c_1, \qquad \bar{Y} \to \bar{Y} + c_2, \qquad S_1^2 \to S_1^2, \qquad S_2^2 \to S_2^2,$$

where $-\infty < c_1, c_2 < \infty$ and also under the group of common scale changes

$$\bar{X} \to a\bar{X}, \qquad \bar{Y} \to a\bar{Y}, \qquad S_1^2 \to a^2 S_1^2, \qquad S_2^2 \to a^2 S_2^2,$$

where $a > 0$. A maximal invariant under these two groups of transformations is

$$F = \frac{S_1^2}{n_1 - 1} \bigg/ \frac{S_2^2}{n_2 - 1}.$$

Show that for testing H_0 against H_1 the test which rejects H_0 whenever $F \ge C_\alpha$ where C_α is a constant such that $P(F \ge C_\alpha) = \alpha$ when H_0 is true, is the uniformly most powerful invariant. Is it uniformly most powerful unbiased for testing H_0 against H_1?

4. In Exercise 3 assume that $\sigma_1^2 = \sigma_2^2$. Let $S^2 = S_1^2 + S_2^2$.

(a) The problem of testing $H_0: v - \mu \le 0$ against the alternatives $H_1: v - \mu > 0$ is invariant under the group of transformations

$$\bar{X} \to \bar{X} + c, \qquad \bar{Y} \to \bar{Y} + c, \qquad S^2 \to S^2,$$

where $-\infty < c < \infty$, and also under the group of transformations

$$\bar{X} \to a\bar{X}, \qquad \bar{Y} \to a\bar{Y}, \qquad S^2 \to a^2 S^2,$$

$0 < a < \infty$. Find the uniformly most powerful invariant test with respect to these transformations.

(b) The problem of testing $H_0: v - \mu = 0$ against the alternatives $H_1: v - \mu \ne 0$ is invariant under the group of affine transformations

$$X_i \to aX_i + b, \qquad Y_j = aY_j + b,$$

$a \ne 0$, $-\infty < b < \infty$, $i = 1, \ldots, n_1$, $j = 1, \ldots, n_2$. Find the uniformly most powerful test of H_0 against H_1 with respect to this group of transmations.

5. (*Linear hypotheses*) Let $Y_1, \ldots, Y_n$ be independently distributed normal random variables with a common variance σ^2 and with means

$$E(Y_i) = \begin{cases} \mu_i, & i = 1, \ldots, s \qquad (s < n) \\ 0, & i = s + 1, \ldots, n \end{cases}$$

and let $\delta^2 = \sum_{i=1}^r \mu_i^2/\sigma^2$. Show that the test which rejects $H_0: \mu_1 = \cdots = \mu_r = 0$, $r < s$, whenever

$$W = \frac{\sum_{i=1}^r Y_i^2}{r} \bigg/ \frac{\sum_{i=s+1}^n Y_i^2}{n - s} \ge k,$$

where the constant k is determined so that the probability of rejection is α whenever H_0 is true, is uniformly most powerful among all tests whose power function depends only on δ^2.

6. (*General linear hypotheses*) Let $X_1, \ldots, X_n$ be n independently distributed normal random variables with mean ξ_i, $i = 1, \ldots, n$, and common variance σ^2. Assume that $\xi = (\xi_1, \ldots, \xi_n)$ lies in a linear subspace of Π_Ω of dimension $s < n$. Show that the problem of testing $H_0: \xi \in \Pi_\omega \subset \Pi_\Omega$ can be reduced to Exercise 5 by means of an orthogonal transformation. Find the test statistic W (of Exercise 5) in terms of $X_1, \ldots, X_n$.

7. (*Analysis of variance, one-way classification*) Let Y_{ij}, $j = 1, \ldots, n_i$, $i = 1, \ldots, k$, be independently distributed normal random variables with means $E(Y_{ij}) = \mu_i$ and common variance σ^2. Let $H_0: \mu_1 = \cdots = \mu_k$. Identify this as a problem of general linear hypotheses. Find the uniformly most powerful invariant test with respect to a suitable group of transformations.

8. In Example 3.2.3 show that for testing $H_0: \mu = 0$ against $H_1: \delta^2 > 0$, student's test is minimax. Is it stringent for H_0 against H_1?

REFERENCES

Blackwell, D. (1956). On a class of probability spaces, *Proc. Berkeley Symp. Math. Statist. Probability. 3rd.* Univ. of California Press, Berkeley, California.

Ferguson, T. S. (1969). "Mathematical Statistics." Academic Press, New York.

Ghosh, J. K. (1967). Invariance in Testing and Estimation. Indian Statist. Inst., Calcutta, Publ. No. SM67/2.

Giri, N. (1974). "Introduction to Probability and Statistics," Part I, Probability. Dekker, New York.

Giri, N. (1975). "Invariance and Minimax Statistical Tests." The Univ. Press of Canada and Hindusthan Publ. Corp., India.

Giri, N., and Kiefer, J. (1964a). Local and asymptotic minimax properties of multivariate tests, *Ann. Math. Statist.* **35**, 21–35.

Giri, N., and Kiefer, J. (1964b). Minimax character of R^2-test in the simplest case, *Ann. Math. Statist.* **35**, 1475–1490.

Giri, N., Kiefer, J., and Stein, C. (1963). Minimax character of Hotelling's T^2-test in the simplest case, *Ann. Math. Statist.* **34**, 1524–1535.

Hall, W. J., Wijsman, R. A., and Ghosh, J. K. (1965). The relationship between sufficiency and invariance with application in sequential analysis, *Ann. Math. Statist.* **36**, 575–614.

Halmos, P. R. (1958). "Measure Theory." Van Nostrand–Reinhold, Princeton, New Jersey.

Kiefer, J. (1957). Invariance sequential estimation and continuous time processes, *Ann. Math. Statist.* **28**, 675–699.

Kiefer, J. (1958). On the nonrandomized optimality and randomized nonoptimality of symmetrical designs, *Ann. Math. Statist.* **29**, 675–699.

Kiefer, J. (1966). Multivariate optimality results, *In* "Multivariate Analysis" (P. R. Krishnaiah, ed.). Academic Press, New York.

Kolmogorov, A. N. (1950). "Foundations of the Theory of Probability." Chelsea, New York.

Lehmann, E. L. (1959). "Testing Statistical Hypotheses." Wiley, New York.

Lehmann, E. L. (1959a). Optimum invariant tests, *Ann. Math. Statist.* **30**, 881–884.
Lehmann, E. L., and Stein, C. (1953). The admissibility of certain invariant statistical tests involving a translation parameter, *Ann. Math. Statist.* **24**, 473–479.
Linnik, Ju, V., Pliss, V. A., and Salaevskii, O. V. (1966). *Sov. Math. Dokl.* **7**, 719.
Nachbin, L. (1965). "The Haar Integral." Van Nostrand–Reinhold, Princeton, New Jersey.
Pitman, E. J. G. (1939). Tests of hypotheses concerning location and scale parameters, *Biometrika* **31**, 200–215.
Salaevskii, O. V. (1968). Minimax character of Hotelling's T^2-test, *Sov. Math. Dokl.* **9**, 733–735.

CHAPTER IV

Multivariate Normal Distribution, Its Properties and Characterization

4.0 INTRODUCTION

We will first define the multivariate normal distribution in the classical way by means of its probability density function and study some of its basic properties. This definition does not include the cases in which the covariance matrix is singular and also the cases in which the dimension of the random vector is countable or uncountable. We will then define the multivariate normal distribution in a general way to include such cases. A number of characterizations of the multivariate normal distribution will also be given in order to enable the reader to study this distribution in Hilbert and Banach spaces. The complex multivariate normal distribution plays an important role in describing the statistical variability of estimators of the spectral density function and of functions of the elements of a multiple stationary Gaussian time series. We will include it for completeness. For further discussion the reader is referred to Goodman (1962).

4.1 MULTIVARIATE NORMAL DISTRIBUTION (CLASSICAL APPROACH)

DEFINITION 4.1.1 *Multivariate normal distribution* A random vector $\mathbf{X} = (X_1, \ldots, X_p)'$ taking values $\mathbf{x} = (x_1, \ldots, x_p)'$ in E^p (Euclidean space of dimension p) is said to have a p-variate normal distribution if its probability density function can be written as

$$f_{\mathbf{X}}(\mathbf{x}) = \frac{1}{(2\pi)^{p/2}(\det \Sigma)^{1/2}} \exp\{-\tfrac{1}{2}(\mathbf{x} - \boldsymbol{\mu})'\Sigma^{-1}(\mathbf{x} - \boldsymbol{\mu})\}, \tag{4.1}$$

where $\boldsymbol{\mu} = (\mu_1, \ldots, \mu_p)' \in E^p$ and Σ is a positive definite symmetric matrix of dimension $p \times p$.

In what follows a random vector will always imply a real vector unless it is specifically stated otherwise.

Since Σ is positive definite, $(\mathbf{x} - \boldsymbol{\mu})'\Sigma^{-1}(\mathbf{x} - \boldsymbol{\mu}) \geq 0$ for all $\mathbf{x} \in E^p$ and $\det(\Sigma) > 0$. Hence $f_{\mathbf{X}}(\mathbf{x}) \geq 0$ for all $\mathbf{x} \in E^p$. Now to show that it is an honest probability density function we proceed as follows. Since Σ is positive definite, there exists a nonsingular matrix C such that $\Sigma = CC'$. Let $\mathbf{y} = C^{-1}\mathbf{x}$. The Jacobian of the transformation $\mathbf{x} \to \mathbf{y} = C^{-1}\mathbf{x}$ is det C.

Writing $\mathbf{v} = (v_1, \ldots, v_p)' = C^{-1}\boldsymbol{\mu}$, we obtain

$$\int_{E_p} \frac{1}{(2\pi)^{p/2}(\det \Sigma)^{1/2}} \exp\{-\tfrac{1}{2}(\mathbf{x} - \boldsymbol{\mu})'\Sigma^{-1}(\mathbf{x} - \boldsymbol{\mu})\}\, d\mathbf{x}$$

$$= \int_{E_p} \frac{1}{(2\pi)^{p/2}} \exp\{-\tfrac{1}{2}(\mathbf{y} - \mathbf{v})'(\mathbf{y} - \mathbf{v})\}\, d\mathbf{y}$$

$$= \prod_{i=1}^{p} \int \frac{1}{(2\pi)^{1/2}} \exp\{-\tfrac{1}{2}(y_i - v_i)^2\}\, dy_i = 1.$$

We would now like to establish that if the random vector $\mathbf{X}$ has multivariate normal distribution with probability density function $f_{\mathbf{X}}(\mathbf{x})$, then the parameters $\boldsymbol{\mu}$ and Σ are given by

$$E(\mathbf{X}) = \boldsymbol{\mu}, \qquad E(\mathbf{X} - \boldsymbol{\mu})(\mathbf{X} - \boldsymbol{\mu})' = \Sigma.$$

From these calculations it is obvious that $Y = C^{-1}X$, with values $\mathbf{y} = (y_1, \ldots, y_p)' \in E^p$, has probability density function

$$f_{\mathbf{Y}}(\mathbf{y}) = \prod_{i=1}^{p} \frac{1}{(2\pi)^{1/2}} \exp\{-\tfrac{1}{2}(y_i - v_i)^2\}.$$

Thus

$$E(\mathbf{Y}) = (E(Y_1), \ldots, E(Y_p))' = \mathbf{v} = C^{-1}\boldsymbol{\mu}, \qquad E(\mathbf{Y} - \mathbf{v})(\mathbf{Y} - \mathbf{v})' = I. \tag{4.2}$$

From (4.2)

$$E(C^{-1}\mathbf{X}) = C^{-1}E(\mathbf{X}) = C^{-1}\boldsymbol{\mu}$$
$$E(C^{-1}\mathbf{X} - C^{-1}\boldsymbol{\mu})(C^{-1}\mathbf{X} - C^{-1}\boldsymbol{\mu})' = C^{-1}E(\mathbf{X} - \boldsymbol{\mu})(\mathbf{X} - \boldsymbol{\mu})'C^{-1'} = I.$$

Hence

$$E(\mathbf{X}) = \boldsymbol{\mu}, \qquad E(\mathbf{X} - \boldsymbol{\mu})(\mathbf{X} - \boldsymbol{\mu})' = \Sigma.$$

We will frequently write Σ as

$$\Sigma = \begin{pmatrix} \sigma_1{}^2 & \sigma_{12} & \cdots & \sigma_{1p} \\ \sigma_{21} & \sigma_2{}^2 & \cdots & \sigma_{2p} \\ \vdots & \vdots & & \vdots \\ \sigma_{p1} & \sigma_{p2} & \cdots & \sigma_p{}^2 \end{pmatrix} \quad \text{with} \quad \sigma_{ij} = \sigma_{ji}.$$

We will now prove some properties of normal distributions in the following theorems.

THEOREM 4.1.1 If the covariance matrix Σ of a normal random vector $\mathbf{X} = (X_1, \ldots, X_p)'$ is a diagonal matrix, then the components of $\mathbf{X}$ are independently distributed normal random variables.

Proof Let

$$\Sigma = \begin{pmatrix} \sigma_1{}^2 & 0 & \cdots & 0 \\ 0 & \sigma_2{}^2 & \cdots & 0 \\ \vdots & \vdots & & \vdots \\ 0 & 0 & \cdots & \sigma_p{}^2 \end{pmatrix}.$$

Then

$$(\mathbf{x} - \boldsymbol{\mu})'\Sigma^{-1}(\mathbf{x} - \boldsymbol{\mu}) = \sum_{i=1}^{p} \left(\frac{x_i - \mu_i}{\sigma_i}\right)^2, \qquad \det \Sigma = \prod_{i=1}^{p} \sigma_i{}^2.$$

Hence

$$f_{\mathbf{X}}(\mathbf{x}) = \prod_{i=1}^{p} \frac{1}{(2\pi)^{1/2}} \exp\left\{-\frac{1}{2}\left(\frac{x_i - \mu_i}{\sigma_i}\right)^2\right\},$$

which implies that the components are independently distributed normal random variables with means μ_i and variance $\sigma_i{}^2$. Q.E.D.

THEOREM 4.1.2 Let $\mathbf{X} = (\mathbf{X}_{(1)}, \mathbf{X}_{(2)})'$, $\mathbf{X}_{(1)} = (X_1, \ldots, X_q)'$, $\mathbf{X}_{(2)} = (X_{q+1}, \ldots, X_p)'$, let $\boldsymbol{\mu}$ be similarly partitioned as $\boldsymbol{\mu} = (\boldsymbol{\mu}_{(1)}, \boldsymbol{\mu}_{(2)})'$, and let Σ be partitioned as

$$\Sigma = \begin{pmatrix} \Sigma_{11} & \Sigma_{12} \\ \Sigma_{21} & \Sigma_{22} \end{pmatrix}$$

where Σ_{11} is the upper left-hand corner submatrix of Σ of dimension $q \times q$. If $\mathbf{X}$ has normal distribution with mean $\boldsymbol{\mu}$ and covariance matrix Σ (postive definite) and $\Sigma_{12} = \Sigma_{21} = 0$, then $\mathbf{X}_{(1)}, \mathbf{X}_{(2)}$ are independently normally distributed with means $\boldsymbol{\mu}_{(1)}, \boldsymbol{\mu}_{(2)}$ and covariance matrices Σ_{11}, Σ_{22}, respectively.

Proof Under the assumption that $\Sigma_{12} = \Sigma_{21} = 0$, we obtain

$$\begin{aligned}(\mathbf{x} - \boldsymbol{\mu})'\Sigma^{-1}(\mathbf{x} - \boldsymbol{\mu}) &= (\mathbf{x}_{(1)} - \boldsymbol{\mu}_{(1)})'\Sigma_{11}^{-1}(\mathbf{x}_{(1)} - \boldsymbol{\mu}_{(1)}) \\ &\quad + (\mathbf{x}_{(2)} - \boldsymbol{\mu}_{(2)})'\Sigma_{22}^{-1}(\mathbf{x}_{(2)} - \boldsymbol{\mu}_{(2)}), \\ \det \Sigma &= (\det \Sigma_{11})(\det \Sigma_{22}).\end{aligned}$$

Hence

$$\begin{aligned}f_{\mathbf{X}}(\mathbf{x}) &= \frac{1}{(2\pi)^{q/2}(\det \Sigma_{11})^{1/2}} \exp\{-\tfrac{1}{2}(\mathbf{x}_{(1)} - \boldsymbol{\mu}_{(1)})'\Sigma_{11}^{-1}(\mathbf{x}_{(1)} - \boldsymbol{\mu}_{(1)})\} \\ &\quad \times \frac{1}{(2\pi)^{(p-q)/2}(\det \Sigma_{22})^{1/2}} \exp\{-\tfrac{1}{2}(\mathbf{x}_{(2)} - \boldsymbol{\mu}_{(2)})'\Sigma_{22}^{-1}(\mathbf{x}_{(2)} - \boldsymbol{\mu}_{(2)})\}\end{aligned}$$

and the result follows. Q.E.D.

THEOREM 4.1.3 Let $\mathbf{X} = (X_1, \ldots, X_p)'$ with values $\mathbf{x}$ in E^p be normally distributed with mean $\boldsymbol{\mu}$ and positive definite covariance matrix Σ. Then the random vector $\mathbf{Y} = C\mathbf{X}$ with values $\mathbf{y}$ in E^p where C is a nonsingular matrix of dimension $p \times p$ has p-variate normal distribution with mean $C\boldsymbol{\mu}$ and covariance matrix $C\Sigma C'$.

Proof The Jacobian of the transformation $\mathbf{x} \to \mathbf{y} = C\mathbf{x}$ is $(\det C)^{-1}$. Hence the probability density function of $\mathbf{Y}$ is given by

$$f_{\mathbf{Y}}(\mathbf{y}) = \frac{1}{(2\pi)^{p/2}(\det C\Sigma C')^{1/2}} \exp\{-\tfrac{1}{2}(\mathbf{y} - C\boldsymbol{\mu})'(C\Sigma C')^{-1}(\mathbf{y} - C\boldsymbol{\mu})\}.$$

Thus $\mathbf{Y}$ has p-variate normal distribution with mean $C\boldsymbol{\mu}$ and positive definite covariance matrix $C\Sigma C'$. Q.E.D.

COROLLARY 4.1.1 (a) $\mathbf{X}_{(1)}, \mathbf{X}_{(2)} - \Sigma_{21}\Sigma_{11}^{-1}\mathbf{X}_{(1)}$ are independently normally distributed with means $\boldsymbol{\mu}_{(1)}, \boldsymbol{\mu}_{(2)} - \Sigma_{21}\Sigma_{11}^{-1}\boldsymbol{\mu}_{(1)}$, and covariance matrices (positive definite) $\Sigma_{11}, \Sigma_{22.1} = \Sigma_{22} - \Sigma_{21}\Sigma_{11}^{-1}\Sigma_{12}$.

(b) The marginal distribution of $\mathbf{X}_{(1)}$ is q-variate normal with mean $\boldsymbol{\mu}_{(1)}$ and covariance matrix Σ_{11}.

(c) The conditional distribution of $\mathbf{X}_{(2)}$ given $\mathbf{X}_{(1)} = \mathbf{x}_{(1)}$ is normal with mean $\boldsymbol{\mu}_{(2)} + \Sigma_{21}\Sigma_{11}^{-1}(\mathbf{x}_{(1)} - \boldsymbol{\mu}_{(1)})$ and covariance matrix $\Sigma_{22.1}$.

Proof (a) Let

$$C = \begin{pmatrix} I_1 & 0 \\ -\Sigma_{21}\Sigma_{11}^{-1} & I_2 \end{pmatrix} \tag{4.3}$$

where I_1 and I_2 are identity matrices of dimension $q \times q$ and $(p-q) \times (p-q)$, respectively. Obviously C is a nonsingular matrix. Hence

$$C\mathbf{X} = \begin{pmatrix} \mathbf{X}_{(1)} \\ \mathbf{X}_{(2)} - \Sigma_{21}\Sigma_{11}^{-1}\mathbf{X}_{(1)} \end{pmatrix}$$

has p-variate normal distribution with mean

$$C\boldsymbol{\mu} = \begin{pmatrix} \boldsymbol{\mu}_{(1)} \\ \boldsymbol{\mu}_{(2)} - \Sigma_{21}\Sigma_{11}^{-1}\boldsymbol{\mu}_{(1)} \end{pmatrix}$$

and covariance matrix

$$C\Sigma C' = \begin{pmatrix} \Sigma_{11} & 0 \\ 0 & \Sigma_{22.1} \end{pmatrix}.$$

Hence by Theorem 4.1.2 we get the result.

(b) It follows trivially from part (a).

(c) The Jacobian of the inverse transformation $Y = CX$ with C given by (4.3) is unity. From (a) the probability density function of $\mathbf{X}$ can be written as

$$f_{\mathbf{X}}(\mathbf{x}) = \frac{\exp\{-\frac{1}{2}(\mathbf{x}_{(1)} - \boldsymbol{\mu}_{(1)})'\Sigma_{11}^{-1}(\mathbf{x}_{(1)} - \boldsymbol{\mu}_{(1)})\}}{(2\pi)^{q/2}(\det \Sigma_{11})^{1/2}}$$

$$\times \frac{\begin{array}{c}\exp\{-\frac{1}{2}(\mathbf{x}_{(2)} - \boldsymbol{\mu}_{(2)} - \Sigma_{21}\Sigma_{11}^{-1}(\mathbf{x}_{(1)} - \boldsymbol{\mu}_{(1)}))'(\Sigma_{22.1})^{-1} \\ (\mathbf{x}_{(2)} - \boldsymbol{\mu}_{(2)} - \Sigma_{21}\Sigma_{11}^{-1}(\mathbf{x}_{(1)} - \boldsymbol{\mu}_{(1)}))\}\end{array}}{(2\pi)^{p-q/2}(\det \Sigma_{22.1})^{1/2}}. \tag{4.4}$$

Hence the result. Q.E.D.

THEOREM 4.1.4 Let $\mathbf{X} = (X_1, \ldots, X_p)'$ be normally distributed with mean $\boldsymbol{\mu}$ and positive definite covariance matrix Σ. The characteristic function of the random vector $\mathbf{X}$ is given by

$$E(e^{i\mathbf{t}'\mathbf{X}}) = \exp\{i\mathbf{t}'\boldsymbol{\mu} - \tfrac{1}{2}\mathbf{t}'\Sigma\mathbf{t}\} \tag{4.5}$$

where $\mathbf{t} = (t_1, \ldots, t_p)' \in E^p$, $i = (-1)^{1/2}$.

Proof Since Σ is positive definite there exists a nonsingular matrix C such that $\Sigma = CC'$. Write $\mathbf{y} = C^{-1}\mathbf{x}$, $\boldsymbol{\alpha} = (\alpha_1, \ldots, \alpha_p)' = C'\mathbf{t}$, $\mathbf{v} = C^{-1}\boldsymbol{\mu} =$

$(v_1, \ldots, v_p)'$. Then

$$E(e^{i\mathbf{t}'\mathbf{X}}) = \int_{E^p} \frac{1}{(2\pi)^{p/2}} \exp\{i\boldsymbol{\alpha}'\mathbf{y} - \tfrac{1}{2}(\mathbf{y} - \mathbf{v})'(\mathbf{y} - \mathbf{v})\}\, d\mathbf{y}$$

$$= \prod_{j=1}^{p} \int_{-\infty}^{\infty} \frac{1}{(2\pi)^{1/2}} \exp\{i\alpha_j y_j - \tfrac{1}{2}(y_j - v_j)^2\}\, dy_j$$

$$= \prod_{j=1}^{p} \exp\{i\alpha_j v_j - \tfrac{1}{2}\alpha_j^2\} = \exp\{i\boldsymbol{\alpha}'\mathbf{v} - \tfrac{1}{2}\boldsymbol{\alpha}'\boldsymbol{\alpha}\} = \exp\{i\mathbf{t}'\boldsymbol{\mu} - \tfrac{1}{2}\mathbf{t}'\Sigma\mathbf{t}\} \tag{4.6}$$

as the characteristic function of a univariate normal random variable with mean μ and variance σ^2 is $\exp\{it\mu - \frac{1}{2}t^2\sigma^2\}$. Q.E.D.

Since the characteristic function determines uniquely the distribution function it follows from (4.5) that the p-variate normal distribution is completely specified by its mean vector $\boldsymbol{\mu}$ and covariance matrix Σ. A standard notation for a p-variate normal distribution with mean $\boldsymbol{\mu}$ and covariance matrix Σ is $N_p(\boldsymbol{\mu}, \Sigma)$.

THEOREM 4.1.5 Let $\mathbf{X} = (X_1, \ldots, X_p)'$ be distributed as $N_p(\boldsymbol{\mu}, \Sigma)$ and let $\mathbf{Y} = A\mathbf{X}$ where A is a matrix of dimension $q \times p$ ($q < p$) of rank q. Then $\mathbf{Y}$ is distributed as $N_q(A\boldsymbol{\mu}, A\Sigma A')$.

Proof Let C be a nonsingular matrix of dimension $p \times p$ such that

$$C = \begin{pmatrix} A \\ B \end{pmatrix}$$

where B is a matrix of dimension $(p - q) \times p$ and is of rank $p - q$, and let $\mathbf{Z} = B\mathbf{X}$. Then by Theorem 4.1.3 $\binom{\mathbf{Y}}{\mathbf{Z}}$ has p-variate normal distribution with mean

$$C\boldsymbol{\mu} = \begin{pmatrix} A\boldsymbol{\mu} \\ B\boldsymbol{\mu} \end{pmatrix}$$

and covariance matrix

$$CAC' = \begin{pmatrix} A\Sigma A' & A\Sigma B' \\ B\Sigma A' & B\Sigma B' \end{pmatrix}.$$

By Corollary 4.1.1(b) we get the result. Q.E.D.

This theorem tells us that if $\mathbf{X}$ is distributed as $N_p(\boldsymbol{\mu}, \Sigma)$, then every linear combination of $\mathbf{X}$ has a univariate normal distribution. We will now show that if for a random vector $\mathbf{X}$ with mean $\boldsymbol{\mu}$ and covariance matrix Σ every linear combination of the components of $\mathbf{X}$ has a univariate normal distribution, then $\mathbf{X}$ has a multivariate normal distribution. For any nonnull fixed real vector $\mathbf{L}$, let $\mathbf{L}'\mathbf{X}$ have a univariate normal distribution with mean $\mathbf{L}'\boldsymbol{\mu}$

and variance $\mathbf{L}'\Sigma\mathbf{L}$. Then for any real t the characteristic function of $\mathbf{L}'\mathbf{X}$ is

$$\phi(t, \mathbf{L}) = E(e^{it\mathbf{L}'\mathbf{X}}) = \exp\{it\mathbf{L}'\boldsymbol{\mu} - \tfrac{1}{2}t^2\mathbf{L}'\Sigma\mathbf{L}\}. \tag{4.7}$$

Hence

$$\phi(1, \mathbf{L}) = \exp\{i\mathbf{L}'\mu - \tfrac{1}{2}\mathbf{L}'\Sigma\mathbf{L}\},$$

which as a function of $\mathbf{L}$ is the characteristic function of $\mathbf{X}$. By the inversion theorem of the characteristic function (see Giri, 1974) the probability density function of $\mathbf{X}$ is given by (4.1).

Motivated by this result we define multivariate normal distribution as follows.

DEFINITION 4.1.2 *Multivariate normal distribution* A p-variate random vector $\mathbf{X}$ with values in E^p is said to have a normal distribution if and only if every linear combination of the components of $\mathbf{X}$ has a univariate normal distribution.

When Σ is nonsingular, this definition is equivalent to that of the multivariate normal distribution given in Definition 4.1.1.

If $\mathbf{X}$ has a multivariate normal distribution according to Definition 4.1.2, then each component X_i of $\mathbf{X}$ is distributed as univariate normal so that $E(X_i) < \infty$, $\text{var}(X_i) < \infty$, and hence $\text{cov}(X_i, X_j)$ exists. Let $E(X_i) = \mu_i$, $\text{var}(X_i) = \sigma_i^2$, $\text{cov}(X_i, X_j) = \sigma_{ij}$. Then $E(\mathbf{X})$, $\text{cov}(\mathbf{X})$ exist and we denote them by $\boldsymbol{\mu}$, Σ respectively. In Definition 4.1.2 it is not necessary that Σ be positive definite; it can be semipositive definite also.

Definition 4.1.2 can be extended to the definition of a normal probability measure on Hilbert and Banach spaces by demanding that the induced distribution of every linear functional be univariate normal. The reader is referred to Frechet (1951) for further details. One big advantage of Definition 4.1.2 over Definition 4.1.1 is that certain results of univariate normal distribution can be immediately generalized to the multivariate case. Readers may find it instructive to prove Theorems 4.1.1–4.1.5 by using Definition 4.1.2. As an illustration let us first prove Theorem 4.1.2 and then Theorem 4.1.5.

For any nonzero real p-vector $\mathbf{L} = (l_1, \ldots, l_p)'$ the characteristic function of $\mathbf{L}'\mathbf{X}$ is

$$\phi(t, \mathbf{L}) = \exp\{it\mathbf{L}'\boldsymbol{\mu} - \tfrac{1}{2}t^2\mathbf{L}'\Sigma\mathbf{L}\}.$$

Write $\mathbf{L} = (\mathbf{L}_{(1)}, \mathbf{L}_{(2)})'$, where $\mathbf{L}_{(1)} = (l_1, \ldots, l_q)'$. Then

$$\mathbf{L}'\boldsymbol{\mu} = \mathbf{L}'_{(1)}\boldsymbol{\mu}_{(1)} + \mathbf{L}'_{(2)}\boldsymbol{\mu}_{(2)}, \qquad \mathbf{L}'\Sigma\mathbf{L} = \mathbf{L}'_{(1)}\Sigma_{11}\mathbf{L}_{(1)} + \mathbf{L}'_{(2)}\Sigma_{22}\mathbf{L}_{(2)}.$$

Hence

$$\phi(t, \mathbf{L}) = \exp\{it\mathbf{L}'_{(1)}\boldsymbol{\mu}_{(1)} - \tfrac{1}{2}t^2\mathbf{L}'_{(1)}\Sigma_{11}\mathbf{L}_{(1)}\} \exp\{it\mathbf{L}'_{(2)}\boldsymbol{\mu}_{(2)} - \tfrac{1}{2}t^2\mathbf{L}'_{(2)}\Sigma_{22}\mathbf{L}_{(2)}\}.$$

In other words the characteristic function of $\mathbf{X}$ is the product of the characteristic functions of $\mathbf{X}_{(1)}$ and $\mathbf{X}_{(2)}$ and each one is the characteristic function of a multivariate normal distribution. Hence Theorem 4.1.2 is proved.

To prove Theorem 4.1.5 let $\mathbf{Y} = A\mathbf{X}$. For any fixed nonnull vector $\mathbf{L}$,

$$\mathbf{L}'\mathbf{Y} = (\mathbf{L}'A)\mathbf{X}.$$

By Definition 4.1.2 $\mathbf{L}'A\mathbf{X}$ has univariate normal distribution with mean $\mathbf{L}'A\boldsymbol{\mu}$ and variance $\mathbf{L}'A\Sigma A'\mathbf{L}$. Since $\mathbf{L}$ is arbitrary, this implies that $\mathbf{Y}$ has q-variate normal distribution with mean $A\boldsymbol{\mu}$ and covariance matrix $A\Sigma A'$.

Using the definition of normal distribution 4.1.2 we need to establish the existence of the probability density function of the multivariate normal distribution. We would now examine the following question: Does Definition 4.1.2 always guarantee the existence of the probability density function? If not, under what conditions can we determine explicitly the probability density function?

Evidently Definition 4.1.2 does not restrict the covariance matrix to be positive definite. If Σ is nonnegative definite, of rank q, then for any real nonnull vector $\mathbf{L}$, $\mathbf{L}'\Sigma\mathbf{L}$ can be written as

$$\mathbf{L}'\Sigma\mathbf{L} = (\boldsymbol{\alpha}_1'L)^2 + \cdots + (\boldsymbol{\alpha}_q'L)^2 \tag{4.8}$$

where $\boldsymbol{\alpha}_i = (\alpha_{i1}, \ldots, \alpha_{ip})'$, $i = 1, \ldots, q$, are linearly independent p-vectors. Hence the characteristic function of $\mathbf{X}$ can be written as

$$\exp\left\{i\mathbf{L}'\boldsymbol{\mu} - \frac{1}{2}\sum_{j=1}^{q} (\boldsymbol{\alpha}_j'\mathbf{L})^2\right\}. \tag{4.9}$$

Now $\exp\{i\mathbf{L}'\boldsymbol{\mu}\}$ is the characteristic function of a p-dimensional random variable $\mathbf{Z}_0$ which assumes value $\boldsymbol{\mu}$ with probability 1 and $\exp\{-\frac{1}{2}(\boldsymbol{\alpha}_i'L)^2\}$ is the characteristic function of the p-dimensional random variable

$$\mathbf{Z}_i = (\alpha_{i1}U_i, \ldots, \alpha_{ip}U_i)'$$

where $U_1, \ldots, U_q$ are independently, identically distributed normal random variables with mean zero and variance unity.

THEOREM 4.1.6 The random vector $\mathbf{X} = (X_1, \ldots, X_p)'$ has p-variate normal distribution with mean $\boldsymbol{\mu}$ and with covariance matrix Σ of rank q $(q \le p)$ if and only if

$$\mathbf{X} = \boldsymbol{\mu} + \alpha\mathbf{U}, \qquad \alpha\alpha' = \Sigma$$

where α is a $p \times q$ matrix of rank q and $\mathbf{U} = (U_1, \ldots, U_q)'$ has q-variate normal distribution with mean $\mathbf{0}$ and covariance matrix I (identity matrix).

Proof Let $\mathbf{X} = \boldsymbol{\mu} + \alpha\mathbf{U}$, $\alpha\alpha' = \Sigma$, and $\mathbf{U}$ be normally distributed with mean $\mathbf{0}$ and covariance matrix I. For any nonnull fixed real p-vector $\mathbf{L}$

$$\mathbf{L}'\mathbf{X} = \mathbf{L}'\boldsymbol{\mu} + (\mathbf{L}'\alpha)\mathbf{U}.$$

But $(\mathbf{L}'\alpha)\mathbf{U}$ has univariate normal distribution with mean zero and variance $\mathbf{L}'\alpha\alpha'\mathbf{L}$. Hence $\mathbf{L}'\mathbf{X}$ has univariate normal distribution with mean $\mathbf{L}'\boldsymbol{\mu}$ and variance $\mathbf{L}'\alpha\alpha'\mathbf{L}$. Since $\mathbf{L}$ is arbitrary, by Definition 4.1.2, $\mathbf{X}$ has p-variate normal distribution with mean $\boldsymbol{\mu}$ and covariance matrix $\Sigma = \alpha\alpha'$ of rank q.

Conversely, if the rank of Σ is q and $\mathbf{X}$ has a p-variate normal distribution with mean $\boldsymbol{\mu}$ and covariance matrix Σ, then from (4.9) we can write

$$\mathbf{X} = \mathbf{Z}_0 + \mathbf{Z}_1 + \cdots + \mathbf{Z}_q = \boldsymbol{\mu} + \alpha\mathbf{U},$$

thus satisfying the conditions of the theorem. Q.E.D.

4.2 SOME CHARACTERIZATIONS OF THE NORMAL DISTRIBUTION

Before we begin to discuss characterization results we need to state the following result due to Cramer (1937) regarding univariate random variables.

If the sum of two independent random variables X, Y is normally distributed, then each one is normally distributed.

For a proof of this the reader is referred to Cramer (1937). The following results are due to Basu (1955).

THEOREM 4.2.1 If $\mathbf{X}$, $\mathbf{Y}$ are two independent p-vectors and if $\mathbf{X} + \mathbf{Y}$ has a p-variate normal distribution, then both $\mathbf{X}$ and $\mathbf{Y}$ have p-variate normal distribution.

Proof Since $\mathbf{X} + \mathbf{Y}$ has a normal distribution, for any nonnull p-vector $\mathbf{L}$, $\mathbf{L}'(\mathbf{X} + \mathbf{Y}) = \mathbf{L}'\mathbf{X} + \mathbf{L}'\mathbf{Y}$ has univariate normal distribution. Since $\mathbf{L}'\mathbf{X}$, $\mathbf{L}'\mathbf{Y}$ are independent, by Cramer's result, $\mathbf{L}'\mathbf{X}$, $\mathbf{L}'\mathbf{Y}$ are both univariate normal random variables. This, by Definition 4.1.2, implies that both $\mathbf{X}$ and $\mathbf{Y}$ have p-variate normal distribution. Q.E.D.

THEOREM 4.2.2 Let $\mathbf{X}_1, \ldots, \mathbf{X}_n$ be a set of mutually independent p-vectors and let

$$\mathbf{X} = \sum_{i=1}^{n} a_i\mathbf{X}_i, \qquad \mathbf{Y} = \sum_{i=1}^{n} b_i\mathbf{X}_i$$

where $a_1, \ldots, a_n; b_1, \ldots, b_n$ are two sets of real constants.

(a) If $\mathbf{X}_1, \ldots, \mathbf{X}_n$ are identically normally distributed p-vectors and if $\sum_{i=1}^{n} a_ib_i = 0$, then $\mathbf{X}$ and $\mathbf{Y}$ are independent.

(b) If $\mathbf{X}$ and $\mathbf{Y}$ are independently distributed, then each $\mathbf{X}_i$ for which $a_ib_i \neq 0$ has p-variate normal distribution.

Note Part (b) of this theorem is a generalization of the Darmois–Skitovic theorem which states that if $X_1, \ldots, X_n$ are independently distributed random variables, then the independence of $\sum_{i=1}^{n} a_i X_i$, $\sum_{i=1}^{n} b_i X_i$ implies that each X_i is normally distributed provided $a_i b_i \neq 0$. (See Darmois, 1953; Skitovic, 1954; or Basu, 1951.)

Proof (a) For any nonnull p-vector $\mathbf{L}$

$$\mathbf{L}'\mathbf{X} = a_1(\mathbf{L}'\mathbf{X}_1) + \cdots + a_n(\mathbf{L}'\mathbf{X}_n).$$

If $\mathbf{X}_1, \ldots, \mathbf{X}_n$ are independent and identically distributed normal random vectors, then $\mathbf{L}'\mathbf{X}_1, \ldots, \mathbf{L}'\mathbf{X}_n$ are independently and identically distributed normal random variables and hence $\mathbf{L}'\mathbf{X}$ has univariate normal distribution for all $\mathbf{L}$. This implies that $\mathbf{X}$ has a p-variate normal distribution. Similarly $\mathbf{Y}$ has a p-variate normal distribution. Furthermore, the joint distribution of $\mathbf{X}$, $\mathbf{Y}$ is a $2p$-variate normal. Now

$$\operatorname{cov}(\mathbf{X}, \mathbf{Y}) = \sum_{i=1}^{n} a_i b_i \operatorname{cov}(\mathbf{X}_i) = \Sigma \cdot 0 = 0.$$

Thus $\mathbf{X}$ and $\mathbf{Y}$ are independent.

(b) For any nonnull real p-vector $\mathbf{L}$

$$\mathbf{L}'\mathbf{X} = \sum_{i=1}^{n} a_i(\mathbf{L}'\mathbf{X}_i), \qquad \mathbf{L}'\mathbf{Y} = \sum_{i=1}^{n} b_i(\mathbf{L}'\mathbf{Y}_i).$$

Since $\mathbf{L}'\mathbf{X}_i$ are independent random variables, independence of $\mathbf{L}'\mathbf{X}$, $\mathbf{L}'\mathbf{Y}$ and $a_i b_i \neq 0$ implies $\mathbf{L}'\mathbf{X}_i$ has a univariate normal distribution. Since $\mathbf{L}$ is arbitrary, $\mathbf{X}_i$ has a p-variate normal distribution. Q.E.D.

4.3 COMPLEX MULTIVARIATE NORMAL DISTRIBUTION

A complex normal random variable

$$Z = X + iY,$$

where $i = (-1)^{1/2}$, is a complex random variable whose real and imaginary parts have a joint bivariate normal distribution. A p-variate complex normal random vector

$$\mathbf{Z} = (Z_1, \ldots, Z_p)',$$

with $Z_j = X_j + iY_j$, is a p-tuple of complex normal random variables $Z_1, \ldots, Z_p$ such that the real random $2p$-vector, consisting of the real and the imaginary parts of $Z_1, \ldots, Z_p$, has a $2p$-variate normal distribution. Let $\mathbf{Z}$ be a complex random vector with mean $E(\mathbf{Z}) = \boldsymbol{\alpha}$ and Hermitian positive definite complex covariance matrix

$$\Sigma = E(\mathbf{Z} - \boldsymbol{\alpha})(\mathbf{Z} - \boldsymbol{\alpha})^*$$

where $(\mathbf{Z} - \boldsymbol{\alpha})^*$ is the adjoint of $(\mathbf{Z} - \boldsymbol{\alpha})$. Its probability density function is given by

$$f_{\mathbf{Z}}(\mathbf{z}) = \frac{1}{(\pi)^p(\det \Sigma)} \exp\{-(\mathbf{z} - \boldsymbol{\alpha})^*\Sigma^{-1}(\mathbf{z} - \boldsymbol{\alpha})\}.$$

The special case with the added restriction

$$E(\mathbf{Z} - \boldsymbol{\alpha})(\mathbf{Z} - \boldsymbol{\alpha})' = 0 \tag{4.10}$$

is of considerable interest in the literature. This condition implies that the real and imaginary parts of different components are pairwise independent, and the real and the imaginary parts of the same component are independent with the same variance. With this probability density function one can obtain results analogous to Theorems 4.1.1–4.1.5. In particular, if $\mathbf{Z}$ has a complex p-variate normal distribution with mean $\boldsymbol{\alpha}$ and Hermitian positive definite covariance matrix Σ, then $C\mathbf{Z}$, where C is a complex nonsingular matrix of dimension $p \times p$, has a complex p-variate normal distribution with mean $C\boldsymbol{\alpha}$ and Hermitian positive definite covariance matrix $C\Sigma C^*$, and any subset of components of $\mathbf{Z}$ has a complex multivariate normal distribution.

Furthermore if the covariance between any two subsets of components of $\mathbf{Z}$ is zero, then these subsets are statistically independent.

4.4 CONCENTRATION ELLIPSOID AND AXES

It may be observed that the probability density function [given in Eq. (4.1)] of a p-variate normal distribution is constant on the ellipsoid

$$(\mathbf{x} - \boldsymbol{\mu})'\Sigma^{-1}(\mathbf{x} - \boldsymbol{\mu}) = c \tag{4.11}$$

in E^p for every positive value of c. The family of ellipsoids obtained by varying c ($c > 0$) has the same center $\boldsymbol{\mu}$, their shapes and orientation are determined by Σ, and their sizes (for given Σ) are determined by c. In particular,

$$(\mathbf{x} - \boldsymbol{\mu})'\Sigma^{-1}(\mathbf{x} - \boldsymbol{\mu}) = p + 2$$

is called the concentration ellipsoid of $\mathbf{X}$ (see Cramer, 1946). It may be verified that the probability density function defined by the uniform distribution

$$f_{\mathbf{X}}(\mathbf{x}) = \begin{cases} \dfrac{\Gamma(\frac{1}{2}p + 1)}{(\det \Sigma)^{1/2}((p + 2)\pi)^{p/2}} & \text{if } (\mathbf{x} - \boldsymbol{\mu})'\Sigma^{-1}(\mathbf{x} - \mu) \le p + 2, \\ 0 & \text{otherwise,} \end{cases}$$

has the same mean $E(\mathbf{X}) = \boldsymbol{\mu}$ and the same covariance matrix

$$E(\mathbf{X} - \boldsymbol{\mu})(\mathbf{X} - \boldsymbol{\mu})' = \Sigma$$

of the p-variate normal distribution. Representing any line through the center $\boldsymbol{\mu}$ to the surface of the ellipsoid

$$(\mathbf{x} - \boldsymbol{\mu})'\Sigma^{-1}(\mathbf{x} - \boldsymbol{\mu}) = c$$

by its coordinates on the surface, the principal axis of the ellipsoid

$$(\mathbf{x} - \boldsymbol{\mu})'\Sigma^{-1}(\mathbf{x} - \boldsymbol{\mu}) = c$$

will have its coordinates $\mathbf{x}$ which maximize its squared half-length

$$(\mathbf{x} - \boldsymbol{\mu})'(\mathbf{x} - \boldsymbol{\mu})$$

subject to the restriction that

$$(\mathbf{x} - \boldsymbol{\mu})'\Sigma^{-1}(\mathbf{x} - \boldsymbol{\mu}) = c.$$

Using the Lagrange multiplier λ we can conclude that $\mathbf{x}$ must maximize the expression (given Σ, $\boldsymbol{\mu}$)

$$(\mathbf{x} - \boldsymbol{\mu})'(\mathbf{x} - \boldsymbol{\mu}) - \lambda(\mathbf{x} - \boldsymbol{\mu})'\Sigma^{-1}(\mathbf{x} - \boldsymbol{\mu}).$$

Differentiating this expression with respect to the coordinates of $\mathbf{x}$ and equating these partial derivatives to zero one can conclude that the coordinates $\mathbf{x}$ of the first (longest) principal axis must satisfy

$$(I - \lambda\Sigma^{-1})(\mathbf{x} - \boldsymbol{\mu}) = 0 \tag{4.12}$$

or, equivalently,

$$(\Sigma - \lambda I)(\mathbf{x} - \boldsymbol{\mu}) = 0. \tag{4.13}$$

From (4.12) the squared length of the first principal axis of the ellipsoid

$$(\mathbf{x} - \boldsymbol{\mu})'\Sigma^{-1}(\mathbf{x} - \boldsymbol{\mu}) = c$$

for fixed c, is equal to

$$4(\mathbf{x} - \boldsymbol{\mu})'(\mathbf{x} - \boldsymbol{\mu}) = 4\lambda_1(\mathbf{x} - \boldsymbol{\mu})'\Sigma^{-1}(\mathbf{x} - \boldsymbol{\mu}) = 4\lambda_1 c$$

where λ_1 is the largest characteristic root of Σ. The coordinates of $\mathbf{x}$, specifying the first principal axis, are proportional to the characteristic vector corresponding to λ_1. Thus the position of the first principal axis of the ellipsoid

$$(\mathbf{x} - \boldsymbol{\mu})'\Sigma^{-1}(\mathbf{x} - \boldsymbol{\mu}) = c$$

is specified by the direction cosines which are the elements of the normalized characteristic vector corresponding to the largest characteristic root of Σ.

The second (longest) axis has the orientation given by the characteristic vector corresponding to the second largest characteristic root of Σ. In Chapter I we observed that if the characteristic roots of Σ are all different, then the corresponding characteristic vectors are all orthogonal and hence in this case the positions of the axes are uniquely specified by p mutually perpendicular axes. But if any two successive roots of Σ (in descending order of magnitude) are equal, the ellipsoid is circular through the plane generated by the corresponding characteristic vectors. However, two perpendicular axes can be constructed for the common root, though their position through the circle is hardly unique. If λ_i is a characteristic root of Σ of multiplicity r_i, then the ellipsoid has a hyperspherical shape in the r_i-dimensional subspace.

Thus for any p-variate normal random vector $\mathbf{X}$ we can define a new p-variate normal random vector $\mathbf{Y} = (Y_1, \ldots, Y_p)'$ whose elements Y_i have values on the ellipsoid by means of the transformations (called the principal axis transformations)

$$\mathbf{Y} = A'(\mathbf{X} - \boldsymbol{\mu}),$$

where the columns of A are the normalized characteristic vectors $\boldsymbol{\alpha}_i$ of Σ. If the characteristic roots of Σ are all different or if the characteristic vectors corresponding to the multiple characteristic roots of Σ have been constructed to be orthogonal, then the covariance matrix of the principal axis variates $\mathbf{Y}$ is a diagonal matrix whose diagonal elements are the characteristic roots of Σ. Thus the principal axis transformation of the p-variate normal vector $\mathbf{X}$ results in uncorrelated variates whose variances are proportional to axis length of any specified ellipsoid.

4.5 SOME EXAMPLES

EXAMPLE 4.5.1 Let

$$\Sigma = \begin{pmatrix} \sigma_1^2 & \rho\sigma_1\sigma_2 \\ \rho\sigma_1\sigma_2 & \sigma_2^2 \end{pmatrix}$$

with $\sigma_1^2 > 0, \sigma_2^2 > 0, -1 < \rho < 1$. Since $\det \Sigma = \sigma_1^2\sigma_2^2(1 - \rho^2) > 0$, Σ^{-1} exists and is given by

$$\Sigma^{-1} = \begin{pmatrix} \dfrac{1}{\sigma_1^2} & \dfrac{-\rho}{\sigma_1\sigma_2} \\ \dfrac{-\rho}{\sigma_1\sigma_2} & \dfrac{1}{\sigma_2^2} \end{pmatrix} \cdot \frac{1}{1 - \rho^2}.$$

Furthermore, for $\mathbf{x} = (x_1, x_2)' \neq \mathbf{0}$

$$\mathbf{x}'\Sigma\mathbf{x} = (\sigma_1 x_1 + \rho\sigma_2 x_2)^2 + (1 - \rho^2)\sigma_2{}^2 x_2{}^2 > 0.$$

Hence Σ is positive definite. With $\boldsymbol{\mu} = (\mu_1, \mu_2)'$

$$(\mathbf{x} - \boldsymbol{\mu})'\Sigma^{-1}(\mathbf{x} - \boldsymbol{\mu})$$

$$= \frac{1}{1 - \sigma^2}\left(\left(\frac{x_1 - \mu_1}{\sigma_1}\right)^2 + \left(\frac{x_2 - \mu_2}{\sigma_2}\right)^2 - 2\rho\left(\frac{x_1 - \mu_1}{\sigma_1}\right)\left(\frac{x_2 - \mu_2}{\sigma_2}\right)\right).$$

The probability density function of a bivariate normal random variable with values in E^2 is

$$\frac{1}{2\pi\sigma_1\sigma_2(1 - \rho^2)^{1/2}} \exp\left\{-\frac{1}{2(1 - \rho^2)}\left[\left(\frac{x_1 - \mu_1}{\sigma_1}\right)^2 + \left(\frac{x_2 - \mu_2}{\sigma_2}\right)^2 - 2\rho\left(\frac{x_1 - \mu_1}{\sigma_1}\right)\left(\frac{x_2 - \mu_2}{\sigma_2}\right)\right]\right\}.$$

The coefficient of correlation between X_1 and X_2 is

$$\frac{\operatorname{cov}(X_1, X_2)}{(\operatorname{var}(X_1)\operatorname{var}(X_2))^{1/2}} = \rho. \tag{4.14}$$

If $\rho = 0$, X_1, X_2 are independently normally distributed with means μ_1, μ_2 and variances $\sigma_1{}^2, \sigma_2{}^2$, respectively. If $\rho > 0$, then X_1, X_2 are positively related; and if $\rho < 0$, then X_1, X_2 are negatively related.

The marginal distributions of X_1 and X_2 are both normal with means μ_1 and μ_2, and with variances $\sigma_1{}^2$ and $\sigma_2{}^2$, respectively. The conditional probability density function of X_2, given $X_1 = x_1$, is normal with

$$E(X_2|X_1 = x_1) = \mu_2 + \rho\left(\frac{\sigma_2}{\sigma_1}\right)(x_1 - \mu_1), \qquad \operatorname{var}(X_2|X_1 = x_1) = \sigma_2{}^2(1 - \rho^2).$$

EXAMPLE 4.5.2 Let

$$\Sigma = \begin{pmatrix} 2 & 1 \\ 1 & 2 \end{pmatrix}.$$

Since, for $\mathbf{x} = (x_1, x_2)' \neq \mathbf{0}$,

$$\mathbf{x}'\Sigma\mathbf{x} = 2(x_1 + x_2/2)^2 + \tfrac{3}{2}x_2{}^2 > 0,$$

Σ is positive definite. Hence

$$f_{\mathbf{X}}(\mathbf{x}) = \frac{1}{2\pi\sqrt{3}} \exp\left\{-\frac{2}{3}\left[\frac{(x_1 - 2)^2}{2} + \frac{(x_2 - 3)^2}{2} - \frac{(x_1 - 2)(x_2 - 3)}{2}\right]\right\}$$

is the probability density function of a bivariate normal random variable $\mathbf{X} = (X_1, X_2)'$ with mean $(2, 3)'$ and covariance Σ. Here $\rho = \frac{1}{2}$.

EXAMPLE 4.5.3 Consider Example 4.5.1 with $\boldsymbol{\mu} = \mathbf{0}$ and $\sigma_1{}^2 = \sigma_2{}^2 = 1$. The characteristic roots of Σ are $\lambda_1 = 1 + \rho$, $\lambda_2 = 1 - \rho$ and the corresponding characteristic vectors are

$$(1/\sqrt{2}, 1/\sqrt{2}), \qquad (1/\sqrt{2}, -1/\sqrt{2}).$$

If $\rho > 0$, the first principal axis (major axis) is $y_2 = y_1$ and the second axis (minor axis) is $y_2 = -y_1$. For $\rho < 0$ the first principal axis is $y_2 = -y_1$ and the second axis is $y_2 = y_1$.

4.6 REGRESSION, MULTIPLE AND PARTIAL CORRELATION

We observed in Corollary 4.1.1(c) that the conditional probability density function of $\mathbf{X}_{(2)}$, given that $\mathbf{X}_{(1)} = \mathbf{x}_{(1)}$, is a $(p - q)$-variate normal with mean $\boldsymbol{\mu}_{(2)} + \Sigma_{21}\Sigma_{11}^{-1}(\mathbf{x}_{(1)} - \boldsymbol{\mu}_{(1)})$ and covariance matrix $\Sigma_{22.1} = \Sigma_{22} - \Sigma_{21}\Sigma_{11}^{-1}\Sigma_{12}$. The matrix $\Sigma_{21}\Sigma_{11}^{-1}$ is called the matrix of regression coefficients of $\mathbf{X}_{(2)}$ on $\mathbf{X}_{(1)} = \mathbf{x}_{(1)}$. The quantity

$$E(\mathbf{X}_{(2)}|\mathbf{X}_{(1)} = \mathbf{x}_{(1)}) = \boldsymbol{\mu}_{(2)} + \Sigma_{21}\Sigma_{11}^{-1}(\mathbf{x}_{(1)} - \boldsymbol{\mu}_{(1)}) \tag{4.15}$$

is called the regression surface of $\mathbf{X}_{(2)}$ on $\mathbf{X}_{(1)}$. This is used to predict $\mathbf{X}_{(2)}$ from the observed value $\mathbf{x}_{(1)}$ of $\mathbf{X}_{(1)}$. It will be shown in Theorem 4.6.1 that among all linear combinations $\boldsymbol{\alpha}\mathbf{X}_{(1)}$, the one that minimizes

$$\operatorname{var}(X_{q+i} - \boldsymbol{\alpha}\mathbf{X}_{(1)})$$

is the linear combination $\boldsymbol{\beta}_{(i)}\Sigma_{11}^{-1}\mathbf{X}_{(1)}$ where $\boldsymbol{\alpha}$ is a row vector and $\boldsymbol{\beta}_{(i)}$ denotes the ith row of the matrix Σ_{21}. This regression terminology is due to Galton (1889) who first introduced it in his studies of the correlation between the heights of fathers and sons. He observed that the heights of sons of either unusually short or tall fathers tend more closely to the average height than their deviant father's values did to the mean for their generation. Galton called this phenomenon "regression to mediocrity" and the parameters of the linear relationship as regression parameters.

From (4.15) it follows that

$$E(X_{q+i}|\mathbf{X}_{(1)} = \mathbf{x}_{(1)}) = \mu_{q+i} + \boldsymbol{\beta}_{(i)}\Sigma_{11}^{-1}(\mathbf{x}_{(1)} - \boldsymbol{\mu}_{(1)})$$

where $\boldsymbol{\beta}_{(i)}$ denotes the ith row of the matrix Σ_{21} of dimension $(p - q) \times p$. Furthermore the covariance between X_{q+i} and $\boldsymbol{\beta}_{(i)}\Sigma_{11}^{-1}\mathbf{X}_{(1)}$ is given by

$$\begin{aligned} &E((X_{q+i} - \mu_{q+i})(\boldsymbol{\beta}_{(i)}\Sigma_{11}^{-1}[\mathbf{X}_{(1)} - \boldsymbol{\mu}_{(1)}])') \\ &\quad = E((X_{q+i} - \mu_{q+i})(\mathbf{X}_{(1)} - \boldsymbol{\mu}_{(1)})'\Sigma_{11}^{-1}\boldsymbol{\beta}'_{(i)}) = \boldsymbol{\beta}_{(i)}\Sigma_{11}^{-1}\boldsymbol{\beta}'_{(i)} \end{aligned}$$

and

$$\begin{aligned}
\operatorname{var}(X_{q+i}) &= \sigma^2_{q+i},\\
\operatorname{var}(\boldsymbol{\beta}_{(i)}\Sigma_{11}^{-1}\mathbf{X}_{(1)}) &= E(\boldsymbol{\beta}_{(i)}\Sigma_{11}^{-1}(\mathbf{X}_{(1)} - \boldsymbol{\mu}_{(1)})(\mathbf{X}_{(1)} - \boldsymbol{\mu}_{(1)})'\Sigma_{11}^{-1}\boldsymbol{\beta}'_{(i)})\\
&= \boldsymbol{\beta}_{(i)}\Sigma_{11}^{-1}\boldsymbol{\beta}'_{(i)}.
\end{aligned}$$

The coefficient of correlation between X_{q+i} and $\boldsymbol{\beta}_{(i)}\Sigma_{11}^{-1}\mathbf{X}_{(1)}$ is

$$\rho = \frac{(\boldsymbol{\beta}_{(i)}\Sigma_{11}^{-1}\boldsymbol{\beta}'_{(i)})^{1/2}}{\sigma_{q+i}}. \tag{4.16}$$

DEFINITION 4.6.1 *Multiple correlation* The term ρ, as defined above, is called the multiple correlation between the component X_{q+i} of $\mathbf{X}_{(2)}$ and the linear function $\boldsymbol{\beta}_{(i)}\Sigma_{11}^{-1}\mathbf{X}_{(1)}$.

THEOREM 4.6.1 Of all linear combinations $\boldsymbol{\alpha}\mathbf{X}_{(1)}$ of $\mathbf{X}_{(1)}$, the one that minimizes the variance of $(X_{q+i} - \boldsymbol{\alpha}\mathbf{X}_{(1)})$ and maximizes the correlation between X_{q+i} and $\boldsymbol{\alpha}\mathbf{X}_{(1)}$ is the linear function $\boldsymbol{\beta}_{(i)}\Sigma_{11}^{-1}\mathbf{X}_{(1)}$.

Proof Let $\boldsymbol{\beta} = \boldsymbol{\beta}_{(i)}\Sigma_{11}^{-1}$. Then

$$\begin{aligned}
\operatorname{var}(X_{q+i} - \boldsymbol{\alpha}\mathbf{X}_{(1)}) &= E(X_{q+i} - \mu_{q+i} - \boldsymbol{\alpha}(\mathbf{X}_{(1)} - \boldsymbol{\mu}_{(1)}))^2\\
&= E(X_{q+i} - \mu_{q+i} - \boldsymbol{\beta}(\mathbf{X}_{(1)} - \boldsymbol{\mu}_{(1)}))^2 + E((\boldsymbol{\beta} - \boldsymbol{\alpha})(\mathbf{X}_{(1)} - \boldsymbol{\mu}_{(1)}))^2\\
&\quad + 2E(X_{q+i} - \mu_{q+i} - \boldsymbol{\beta}(\mathbf{X}_{(1)} - \boldsymbol{\mu}_{(1)}))((\boldsymbol{\beta} - \boldsymbol{\alpha})(\mathbf{X}_{(1)} - \boldsymbol{\mu}_{(1)}))'.
\end{aligned}$$

But

$$\begin{aligned}
E(\boldsymbol{\beta}(\mathbf{X}_{(1)} - \boldsymbol{\mu}_{(1)})(X_{q+i} - \mu_{q+i})) &= \boldsymbol{\beta}\boldsymbol{\beta}'_{(i)} = \boldsymbol{\beta}_{(i)}\Sigma_{11}^{-1}\boldsymbol{\beta}'_{(i)}\\
E((X_{q+i} - \mu_{q+i} - \boldsymbol{\beta}(\mathbf{X}_{(1)} - \boldsymbol{\mu}_{(1)}))(\mathbf{X}_{(1)} - \boldsymbol{\mu}_{(1)})') &= \boldsymbol{\beta}_{(i)} - \boldsymbol{\beta}_{(i)} = \mathbf{0};\\
E((\boldsymbol{\beta} - \boldsymbol{\alpha})(\mathbf{X}_{(1)} - \boldsymbol{\mu}_{(1)}))^2 &= (\boldsymbol{\beta} - \boldsymbol{\alpha})E((\mathbf{X}_{(1)} - \boldsymbol{\mu}_{(1)})(\mathbf{X}_{(1)} - \boldsymbol{\mu}_{(1)})')(\boldsymbol{\beta} - \boldsymbol{\alpha})'\\
&= (\boldsymbol{\beta} - \boldsymbol{\alpha})\Sigma_{11}(\boldsymbol{\beta} - \boldsymbol{\alpha})';\\
E(X_{q+i} - \mu_{q+i} - \boldsymbol{\beta}(\mathbf{X}_{(1)} - \boldsymbol{\mu}_{(1)}))^2 &= \sigma^2_{q+i} - 2\boldsymbol{\beta}_{(i)}\Sigma_{11}^{-1}\boldsymbol{\beta}_{(i)} + \boldsymbol{\beta}_{(i)}\Sigma_{11}^{-1}\boldsymbol{\beta}'_{(i)}\\
&= \sigma^2_{q+i} - \boldsymbol{\beta}_{(i)}\Sigma_{11}^{-1}\boldsymbol{\beta}'_{(i)}.
\end{aligned}$$

Hence

$$\operatorname{var}(X_{q+i} - \boldsymbol{\alpha}\mathbf{X}_{(1)}) = \sigma^2_{q+i} - \boldsymbol{\beta}_{(i)}\Sigma_{11}^{-1}\boldsymbol{\beta}'_{(i)} + (\boldsymbol{\beta} - \boldsymbol{\alpha})\Sigma_{11}(\boldsymbol{\beta} - \boldsymbol{\alpha})'.$$

Since Σ_{11} is positive definite, $(\boldsymbol{\beta} - \boldsymbol{\alpha})\Sigma_{11}(\boldsymbol{\beta} - \boldsymbol{\alpha})' \geq 0$ and is equal to zero if $\boldsymbol{\beta} = \boldsymbol{\alpha}$. Thus $\boldsymbol{\beta}\mathbf{X}_{(1)}$ is the linear function such that $X_{q+i} - \boldsymbol{\beta}\mathbf{X}_{(1)}$ has the minimum variance.

We now consider the correlation between X_{q+i} and $\boldsymbol{\alpha}\mathbf{X}_{(1)}$ and show that this correlation is maximum when $\boldsymbol{\alpha} = \boldsymbol{\beta}$. For any nonzero scalar c, $c\boldsymbol{\alpha}\mathbf{X}_{(1)}$

is a linear function of $\mathbf{X}_{(1)}$. Hence

$$E(X_{q+i}-\mu_{q+i}-\boldsymbol{\beta}(\mathbf{X}_{(1)}-\boldsymbol{\mu}_{(1)}))^2 \leq E(X_{q+i}-\mu_{q+i}-c\boldsymbol{\alpha}(\mathbf{X}_{(1)}-\boldsymbol{\mu}_{(1)}))^2. \tag{4.17}$$

Dividing both sides of (4.17) by $\sigma_{q+i}[E(\boldsymbol{\beta}(\mathbf{X}_{(1)} - \boldsymbol{\mu}_{(1)}))^2]^{1/2}$ and choosing

$$c = \left[\frac{E(\boldsymbol{\beta}(\mathbf{X}_{(1)} - \boldsymbol{\mu}_{(1)}))^2}{E(\boldsymbol{\alpha}(\mathbf{X}_{(1)} - \boldsymbol{\mu}_{(1)}))^2}\right]^{1/2}$$

we get from (4.17)

$$\frac{E(X_{q+i} - \mu_{q+i})(\boldsymbol{\beta}(\mathbf{X}_{(1)} - \boldsymbol{\mu}_{(1)}))}{\sigma_{q+i}[E(\boldsymbol{\beta}(\mathbf{X}_{(1)} - \boldsymbol{\mu}_{(1)}))^2]^{1/2}} \geq \frac{E(X_{q+i} - \mu_{q+i})(\boldsymbol{\alpha}(\mathbf{X}_{(1)} - \boldsymbol{\mu}_{(1)}))}{\sigma_{q+i}[E(\boldsymbol{\alpha}(\mathbf{X}_{(1)} - \boldsymbol{\mu}_{(1)}))^2]^{1/2}}. \qquad \text{Q.E.D.}$$

DEFINITION 4.6.2 *Partial correlation coefficient* Let $\sigma_{ij.1,\ldots,q}$ be the (i, j)th element of the matrix $\Sigma_{22} - \Sigma_{21}\Sigma_{11}^{-1}\Sigma_{12}$ of dimension $(p - q) \times (p - q)$. Then

$$\rho_{ij.1,\ldots,q} = \frac{\sigma_{ij.1,\ldots,q}}{(\sigma_{ii.1,\ldots,q}\sigma_{jj.1,\ldots,q})^{1/2}} \tag{4.18}$$

is called the partial correlation coefficient (of order q) between the components X_{q+i} and X_{q+j} of $\mathbf{X}_{(2)}$ when $X_1, \ldots, X_q$ are held fixed.

We would now like to find a recursive relation to compute the partial correlations of order k (say) from the partial correlations of order $k - 1$.

Let $\mathbf{X} = (X_1, \ldots, X_p)'$ be normally distributed with mean $\boldsymbol{\mu}$ and positive definite covariance matrix Σ. Write

$$\mathbf{X} = (\mathbf{X}_{(1)}, \mathbf{X}_{(2)}, \mathbf{X}_{(3)})'$$

where

$$\mathbf{X}_{(1)} = (X_1, \ldots, X_{p_1})', \qquad \mathbf{X}_{(2)} = (X_{p_1+1}, \ldots, X_{p_1+p_2})',$$
$$\mathbf{X}_{(3)} = (X_{p_1+p_2+1}, \ldots, X_p)',$$

and

$$\Sigma = \begin{pmatrix} \Sigma_{11} & \Sigma_{12} & \Sigma_{13} \\ \Sigma_{21} & \Sigma_{22} & \Sigma_{23} \\ \Sigma_{31} & \Sigma_{32} & \Sigma_{33} \end{pmatrix},$$

where Σ_{ii} are submatrices of Σ of dimension $p_i \times p_i$, $i = 1, 2, 3$. From Corollary 4.1.1(c),

$$\operatorname{cov}\left(\begin{pmatrix} \mathbf{X}_{(2)} \\ \mathbf{X}_{(3)} \end{pmatrix} \middle| \mathbf{X}_{(1)} = \mathbf{x}_{(1)}\right) = \begin{pmatrix} \Sigma_{22} & \Sigma_{23} \\ \Sigma_{32} & \Sigma_{33} \end{pmatrix}^{-1} - \begin{pmatrix} \Sigma_{21} \\ \Sigma_{31} \end{pmatrix} \Sigma_{11}^{-1}(\Sigma_{12} \quad \Sigma_{13}).$$

Following the same argument we can deduce that

$$
\begin{aligned}
&\operatorname{cov}(\mathbf{X}_{(3)} | \mathbf{X}_{(2)} = \mathbf{x}_{(2)}, \mathbf{X}_{(1)} = \mathbf{x}_{(1)}) \\
&\quad = \Sigma_{33} - (\Sigma_{31} \quad \Sigma_{32}) \begin{pmatrix} \Sigma_{11} & \Sigma_{12} \\ \Sigma_{21} & \Sigma_{22} \end{pmatrix}^{-1} \begin{pmatrix} \Sigma_{13} \\ \Sigma_{23} \end{pmatrix} \\
&\quad = (\Sigma_{33} - \Sigma_{31}\Sigma_{11}^{-1}\Sigma_{13}) \\
&\qquad - (\Sigma_{32} - \Sigma_{31}\Sigma_{11}^{-1}\Sigma_{12})(\Sigma_{22} - \Sigma_{21}\Sigma_{11}^{-1}\Sigma_{12})^{-1}(\Sigma_{23} - \Sigma_{21}\Sigma_{11}^{-1}\Sigma_{13}).
\end{aligned}
$$

Now taking $p_1 = q - 1$, $p_2 = 1$, $p_3 = p - q$ we get for the (i, j)th element, $i, j = q + 1, \ldots, p$,

$$\sigma_{ij.1,\ldots,q} = \sigma_{ij.1,\ldots,q-1} - \frac{\sigma_{iq.1,\ldots,q-1}\sigma_{jq.1,\ldots,q-1}}{\sigma_{qq.1,\ldots,q-1}}. \tag{4.19}$$

If $j = i$, we obtain

$$\sigma_{ii.1,\ldots,q} = \sigma_{ii.1,\ldots,q-1}(1 - \rho_{iq.1,\ldots,q-1}^2).$$

Hence from (4.19) we obtain

$$\rho_{ij.1,\ldots,q} = \frac{\rho_{ij.1,\ldots,q-1} - \rho_{iq.1,\ldots,q-1}\rho_{jq.1,\ldots,q-1}}{(1 - \rho_{iq.1,\ldots,q-1}^2)^{1/2}(1 - \rho_{jq.1,\ldots,q-1}^2)^{1/2}}. \tag{4.20}$$

In particular,

$$\rho_{34.12} = \frac{\rho_{34.1} - \rho_{32.1}\rho_{42.1}}{[(1 - \rho_{32.1}^2)(1 - \rho_{42.1}^2)]^{1/2}}$$

and

$$\rho_{23.1} = \frac{\rho_{23} - \rho_{21}\rho_{31}}{[(1 - \rho_{21}^2)(1 - \rho_{31}^2)]^{1/2}}$$

where $\rho_{ij} = \operatorname{cor}(X_i, X_j)$. Thus if all partial correlations of certain order are zero, then all higher order partial correlations must be zero.

EXERCISES

1. Find the mean and the covariance matrix of the random vector $\mathbf{X} = (X_1, X_2)'$ with probability density function

$$f_{\mathbf{X}}(\mathbf{x}) = (1/2\pi)\exp\{-\tfrac{1}{2}(2x_1{}^2 + x_2{}^2 + 2x_1x_2 - 22x_1 - 14x_2 + 65)\}$$

and $\mathbf{x} \in E^2$.

2. Show that if the sum of two independent random variables is normally distributed, then each one is normally distributed.

3. Let $\mathbf{X} = (X_1, X_2)'$ be a random vector with the moment generating function

$$E(\exp(t_1X_1 + t_2X_2)) = a(\exp(t_1 + t_2) + 1) + b(\exp(t_1) + \exp(t_2)),$$

where a, b are positive constants satisfying $a + b = \frac{1}{2}$. Find the covariance matrix of $\mathbf{X}$.

4. (*The best linear predictor*) Let $\mathbf{X} = (X_1, \ldots, X_p)'$ be a random vector with $E(\mathbf{X}) = \mathbf{0}$ and covariance matrix Σ. Show that among all functions g of $X_2, \ldots, X_p$

$$E(X_1 - g(X_2, \ldots, X_p))^2$$

is minimum when

$$g(x_2, \ldots, x_p) = E(X_1 | X_2 = x_2, \ldots, X_p = x_p).$$

5. Let $\mathbf{X} = (X_2, X_2, X_3)'$ be a random vector whose first and second moments are assumed known. Show that among all linear functions $a + bX_2 + cX_3$ the linear function that minimizes

$$E(X_1 - a - bX_2 - cX_3)^2$$

is given by

$$E(X_1) + \beta(X_2 - E(X_2)) + \gamma(X_3 - E(X_3))$$

where

$$\begin{aligned}
\beta &= \operatorname{cov}(X_1, X_2)\sigma_{11} + \operatorname{cov}(X_1, X_3)\sigma_{12}, \\
\gamma &= \operatorname{cov}(X_1, X_2)\sigma_{21} + \operatorname{cov}(X_1, X_3)\sigma_{22} \\
\sigma_{11} &= \operatorname{var}(X_3)/\Delta, \qquad \sigma_{22} = \operatorname{var}(X_2)/\Delta, \\
\sigma_{12} &= \sigma_{21} = -\operatorname{cov}(X_2, X_3)/\Delta, \\
\Delta &= \operatorname{var}(X_2)\operatorname{var}(X_3)(1 - \rho^2(X_2, X_3)),
\end{aligned}$$

and $\rho(X_2, X_3)$ is the coefficient of correlation between X_2 and X_3.

6. (*Residual variates*) Let $\mathbf{X} = (\mathbf{X}_{(1)}, \mathbf{X}_{(2)})'$, $\mathbf{X}_{(1)} = (X_1, \ldots, X_q)'$, be a p-dimensional normal random vector with mean $\boldsymbol{\mu} = (\boldsymbol{\mu}_{(1)}, \boldsymbol{\mu}_{(2)})'$ and covariance matrix

$$\Sigma = \begin{pmatrix} \Sigma_{11} & \Sigma_{12} \\ \Sigma_{21} & \Sigma_{22} \end{pmatrix},$$

where $\Sigma_{11} = \operatorname{cov}(\mathbf{X}_{(1)})$. The random vector

$$\mathbf{X}_{1.2} = \mathbf{X}_{(1)} - \boldsymbol{\mu}_{(1)} - \Sigma_{12}\Sigma_{22}^{-1}(\mathbf{X}_{(2)} - \boldsymbol{\mu}_{(2)})$$

is called the set of residual variates since it represents the discrepancies of the elements of $\mathbf{X}_{(1)}$ from their values as predicted from the mean vector of the conditional distribution of $\mathbf{X}_{(1)}$ given $\mathbf{X}_{(2)} = \mathbf{x}_{(2)}$. Show that

(a) $E(\mathbf{X}_{(1)} - \boldsymbol{\mu}_{(1)})\mathbf{X}'_{1.2} = \Sigma_{11} - \Sigma_{12}\Sigma_{22}^{-1}\Sigma_{21}$,
(b) $E(\mathbf{X}_{(2)} - \boldsymbol{\mu}_{(2)})\mathbf{X}'_{1.2} = \mathbf{0}$.

7. Show that the multiple correlation coefficient $\rho_{1(2,\ldots,p)}$ of X_1 on $X_2, \ldots, X_p$ of the normal vector $\mathbf{X} = (X_1, \ldots, X_p)'$ satisfies

$$1 - \rho^2_{1(2,\ldots,p)} = (1 - \rho^2_{12})(1 - \rho^2_{13.2}) \cdots (1 - \rho^2_{1p.2,3,\ldots,p-1}).$$

8. In Corollary 4.1.1(c) show that the conditional distribution of $\mathbf{X}_{(1)}$ given $\mathbf{X}_{(2)} = \mathbf{x}_{(2)}$ is a multivariate normal with mean $\boldsymbol{\mu}_{(1)} + \Sigma_{12}\Sigma_{22}^{-1}(\mathbf{x}_{(2)} - \boldsymbol{\mu}_{(2)})$ and positive definite covariance matrix

$$\Sigma_{11} - \Sigma_{12}\Sigma_{22}^{-1}\Sigma_{21}.$$

9. Show that the multiple correlations $\rho_{1(2,\ldots,j)}$ between X_1 and $(X_2, \ldots, X_j)$, $j = 2, \ldots, p$, satisfy

$$\rho^2_{1(2)} \leq \rho^2_{1(23)} \leq \cdots \leq \rho^2_{1(2,3,\ldots,p)}.$$

In other words, the multiple correlation cannot be reduced by adding to the set of variables on which the dependence of X_1 has to be measured.

10. Let the covariance matrix of a four-dimensional normal random vector $\mathbf{X} = (X_1, \ldots, X_4)'$ be given by

$$\Sigma = \sigma^2 \begin{pmatrix} 1 & \rho & \rho^2 & \rho^3 \\ \rho & 1 & \rho & \rho^2 \\ \rho^2 & \rho & 1 & \rho \\ \rho^3 & \rho^2 & \rho & 1 \end{pmatrix}.$$

Find the partial correlation coefficient between the $(i-1)$th and $(i+1)$th components of $\mathbf{X}$ when the ith component is held fixed.

11. Let $\mathbf{X} = (X_1, X_2, X_3)'$ be normally distributed with mean $\mathbf{0}$ and covariance matrix

$$\Sigma = \begin{pmatrix} 3 & 1 & 1 \\ 1 & 3 & 1 \\ 1 & 1 & 3 \end{pmatrix}.$$

Show that the first principal axis of its concentration ellipsoid passes through the point (1, 1, 1).

12. (*Multinomial distribution*) Let $\mathbf{X} = (X_1, \ldots, X_p)'$ be a discrete p-dimensional random vector with probability mass function

$$p_{X_1,\ldots,X_p}(x_1, \ldots, x_p) = \begin{cases} \dfrac{n!}{x_1! \cdots x_p!} \displaystyle\prod_{i=1}^{p} p_i^{x_i} & \text{if } 0 \leq x_i \leq n \text{ for all } n, \\ & \displaystyle\sum_1^p x_i = n. \\ 0 & \text{otherwise,} \end{cases}$$

where $p_i \geq 0$, $\sum_i^p p_i = 1$.

(a) Show that

$$E(X_i) = np_i, \qquad V(X_i) = np_i(1 - p_i)$$
$$\operatorname{cov}(X_i, X_j) = -np_ip_j \qquad (i \neq j).$$

(b) Find the characteristic function of $\mathbf{X}$.

(c) Show that the marginal probability mass function of $X_{(1)} = (X_1, \ldots, X_q)'$, $q \leq p$, is given by

$$p_{X_1,\ldots,X_q}(x_1, \ldots, x_q) = \frac{n!}{x_1! \cdots x_q!(n - n_0)!} \prod_{i=1}^{q} p_i^{x_i}(1 - p_1 - \cdots - p_q)^{n-n_0}$$

if $\sum_1^q x_i = n_0$.

(d) Find the conditional distribution of X_1 given $X_3 = x_3, \ldots, X_q = x_q$.

(e) Show that the partial correlation coefficient is

$$\rho_{12.3,\ldots,q} = -\left[\frac{p_1p_2}{(1 - p_2 - p_3 - \cdots - p_q)(1 - p_1 - p_3 - \cdots - p_q)}\right]^{1/2}.$$

(f) Show that the squared multiple correlation between X_1 and $(X_2, \ldots, X_q)$ is

$$\rho^2 = \frac{p_1(p_2 + \cdots + p_q)}{(1 - p_1)(1 - p_2 - \cdots - p_q)}.$$

(g) Let $Y_i = (X_i - np_i)/\sqrt{n}$. Show that as $n \to \infty$ the distribution of $Y = (Y_1, \ldots, Y_{p-1})$ tends to a multivariate normal distribution. Find its mean and its covariance matrix.

13. (*The multivariate log-normal distribution*) Let $\mathbf{X} = (X_1, \ldots, X_p)'$ be normally distributed with mean vector $\boldsymbol{\mu}$ and positive definite (symmetric) covariance matrix $\Sigma = (\sigma_{ij})$. For any random vector $\mathbf{Y} = (Y_1, \ldots, Y_p)'$ let us define

$$\log \mathbf{Y} = (\log Y_1, \ldots, \log Y_p)'$$

and let $\log Y_i = X_i$, $i = 1, \ldots, p$. Then $\mathbf{Y}$ is said to have a p-variate log-normal distribution with probability density function

$$f_{\mathbf{Y}}(\mathbf{y}) = (2\pi)^{-p/2}(\det \Sigma)^{1/2} \prod_{i=1}^{p} y_i^{-1} \exp\{-\tfrac{1}{2}(\log \mathbf{y} - \boldsymbol{\mu})'\Sigma^{-1}(\log \mathbf{y} - \boldsymbol{\mu})\}$$

when $y_i > 0$, $i = 1, \ldots, p$, and is zero otherwise.

(a) Show that for any positive integer r

$$E(Y_i^r) = \exp\{r\mu_i + \tfrac{1}{2}r^2\sigma_{ii}\}, \qquad V(Y_i) = \exp\{2\mu_i + 2\sigma_{ii}\} - \exp\{2\mu_i + \sigma_{ii}\}$$
$$\operatorname{cov}(Y_i Y_j) = \exp\{\mu_i + \mu_j + \tfrac{1}{2}(\sigma_{ii} + \sigma_{jj}) + \sigma_{ij}\} - \exp\{\mu_i + \mu_j + \tfrac{1}{2}(\sigma_{ii} + \sigma_{jj})\}.$$

(b) Find the marginal probability density function of $(Y_1, \ldots, Y_q)'$, $q < p$.

14. (*The multivariate beta (Dirichlet) distribution*) Let $\mathbf{X} = (X_1, \ldots, X_p)'$ be a p-variate random vector with values in the simplex

$$S = \left\{\mathbf{x} = (x_1, \ldots, x_p)' : x_i \geq 0 \text{ for all } i, \sum_1^p x_i \leq 1\right\}.$$

$\mathbf{X}$ has a multivariate beta distribution with parameters $v_1, \ldots, v_{p+1}$, $v_i > 0$, if its probability density function is given by

$$f_{\mathbf{X}}(\mathbf{x}) = \begin{cases} \dfrac{\Gamma(v_1 \cdots v_{p+1})}{\Gamma(v_1) \cdots \Gamma(v_{p+1})} \displaystyle\prod_{i=1}^{p} x_i^{v_i - 1}\left(1 - \sum_1^p x_i\right)^{v_{p+1}-1} & \text{if } \mathbf{x} \in S, \\ 0 & \text{otherwise.} \end{cases}$$

(a) Show that

$$E(X_i) = \frac{v_i}{v_1 + \cdots + v_{p+1}}, \qquad i = 1, \ldots, p$$

$$\operatorname{var}(X_i) = \frac{v_i(v_1 + \cdots + v_{p+1} - v_i)}{(v_1 + \cdots + v_{p+1})^2(1 + v_1 + \cdots + v_{p+1})}$$

$$\operatorname{cov}(X_i X_j) = \frac{-v_i v_j}{(v_1 + \cdots + v_{p+1})^2(1 + v_1 + \cdots + v_{p+1})} \qquad (i \neq j).$$

(b) Show that the marginal probability density function of $X_1, \ldots, X_q$ is multivariate beta with parameters $v_1, \ldots, v_q, v_{q+1} + \cdots + v_{p+1}$.

15. (*Multivariate Student t-distribution*) The random vector $\mathbf{X} = (X_1, \ldots, X_p)'$ has p-variate Student t-distribution if the probability density function of $\mathbf{X}$ can be written as

$$f_{\mathbf{X}}(\mathbf{x}) = C(\det \Sigma)^{-1/2}(N + (\mathbf{x} - \boldsymbol{\mu})'\Sigma^{-1}(\mathbf{x} - \boldsymbol{\mu}))^{-(N+p)/2}$$

where $\mathbf{x} \in E^p$, $\boldsymbol{\mu} = (\mu_1, \ldots, \mu_p)'$, Σ is a symmetric positive definite matrix of dimension $p \times p$, and

$$C = N^{N/2}\Gamma\left(\frac{N+p}{2}\right)\Big/\pi^{p/2}\Gamma\left(\frac{N}{2}\right).$$

(a) Show that

$$E(\mathbf{X}) = \boldsymbol{\mu} \quad \text{if} \quad N > 1; \qquad \operatorname{cov}(\mathbf{X}) = [N/(N-2)]\Sigma, \quad N > 2.$$

(b) Show that the marginal probability density function of $(X_1, \ldots, X_q)'$, $q < p$, is distributed as a q-variate Student t.

(c) Prove that $A\mathbf{X} + \mathbf{b}$, where A is a $q \times p$ matrix of rank q and $\mathbf{b}$ is a $q \times 1$ vector, has a q-variate Student t-distribution.

REFERENCES

Basu, D. (1951). On the independence of linear functions of independence chance variables, *Bull. Int. Statist. Inst.* **33**, 83–86.

Basu, D. (1955). A note on the multivariate extension of some theorems related to the univariate normal distribution, *Sankhya* **17**, 221–224.

Cramér, H. (1937). "Random Variables and Probability Distributions" (Cambridge Tracts, no. 36). Cambridge Univ. Press, London and New York.

Cramér, H. (1946). "Mathematical Methods of Statistics." Princeton Univ. Press, Princeton, New Jersey.

Darmois, G. (1953). Analyse générale des liaisons stochastiques, étude particulière de l'analyse factorielle linéaire, *Rev. Inst. Int. Statist.* **21**, 2–8.

Frechet, M. (1951). Généralisation de la loi de probabilité de Laplace, *Ann. Inst. Henri Poincaré* **12**, Fasc. L.

Galton, F. (1889). "Natural Inheritance." MacMillan, New York.

Giri, N. (1974). "An Introduction to Probability and Statistics," Part I, Probability. Dekker, New York.

Goodman, N. R. (1962). Statistical analysis based on a certain multivariate complex Gaussian distribution (an introduction), *Ann. Math. Statist.* **33**, 152–176.

Skitovic, V. P. (1954). Linear combinations of independent random variables and the normal distribution law, *Izv. Akad. Nauk. SSSR. Ser. Mat.* **18**, 185–200.

CHAPTER V

Estimators of Parameters and Their Functions in a Multivariate Normal Distribution

5.0 INTRODUCTION

We observed in Chapter IV that the probability density function (when it exists) of a multivariate normal distribution is completely specified by its mean vector $\boldsymbol{\mu}$ and the positive definite covariance matrix Σ. In this chapter we will estimate these parameters and some of their functions, namely, multiple correlation coefficient, partial correlation coefficients of different orders, and regression coefficients on the basis of information contained in a random sample of size N from this distribution.

The method of maximum likelihood (Fisher, 1925) has been very successful in finding suitable estimators of parameters in many problems. Under certain regularity conditions on the probability density function, the maximum likelihood estimator is strongly consistent in large samples (see Wald, 1943; Wolfowitz, 1949; Le Cam, 1953; Bahadur, 1960). Under such conditions, if the dimension p of the random vector is not large, it seems likely that the sample size N occurring in practice would usually be large enough for this optimum result to hold. However, if p is large, then it may be that the sample size N needs to be extremely large for this result to apply; for example, there are cases where N/p^3 must be large. The fact that the maximum likelihood estimator is not universally good has been demonstrated by Basu (1955), Neyman and Scott (1948), and Kiefer and Wolfowitz (1956), among others.

5.1 MAXIMUM LIKELIHOOD ESTIMATORS OF $\boldsymbol{\mu}, \Sigma$

Let $\mathbf{x}^\alpha = (x_{\alpha 1}, \ldots, x_{\alpha p})'$, $\alpha = 1, \ldots, N$, be a sample of size N from a normal distribution with mean $\boldsymbol{\mu}$ and positive definite covariance matrix Σ, and let

$$\bar{\mathbf{x}} = \sum_{\alpha=1}^{N} \mathbf{x}^\alpha / N, \qquad s = \sum_{\alpha=1}^{N} (\mathbf{x}^\alpha - \bar{\mathbf{x}})(\mathbf{x}^\alpha - \bar{\mathbf{x}})'.$$

We are interested here in finding the maximum likelihood estimates $(\hat{\boldsymbol{\mu}}, \hat{\Sigma})$ of $(\boldsymbol{\mu}, \Sigma)$. The likelihood of the sample observations $\mathbf{x}^\alpha$, $\alpha = 1, \ldots, N$, is given by

$$\begin{aligned} L(\mathbf{x}^1, \ldots, \mathbf{x}^N | \boldsymbol{\mu}, \Sigma) &= (2\pi)^{-Np/2} (\det \Sigma)^{-N/2} \exp\left\{ -\frac{1}{2} \sum_{\alpha=1}^{N} (\mathbf{x}^\alpha - \boldsymbol{\mu})' \Sigma^{-1} (\mathbf{x}_\alpha - \boldsymbol{\mu}) \right\} \\ &= (2\pi)^{-Np/2} (\det \Sigma)^{-N/2} \exp\left\{ -\frac{1}{2} \operatorname{tr} \Sigma^{-1} \sum_{\alpha=1}^{N} (\mathbf{x}^\alpha - \boldsymbol{\mu})(\mathbf{x}^\alpha - \boldsymbol{\mu})' \right\} \\ &= (2\pi)^{-Np/2} (\det \Sigma)^{-N/2} \exp\{ -\tfrac{1}{2} \operatorname{tr} \Sigma^{-1} (s + N(\bar{\mathbf{x}} - \boldsymbol{\mu})(\bar{\mathbf{x}} - \boldsymbol{\mu})' \} \end{aligned}$$

as

$$\begin{aligned} \sum_{\alpha=1}^{N} (\mathbf{x}^\alpha - \boldsymbol{\mu})(\mathbf{x}^\alpha - \boldsymbol{\mu})' &= \sum_{\alpha=1}^{N} (\mathbf{x}^\alpha - \bar{\mathbf{x}})(\mathbf{x}^\alpha - \bar{\mathbf{x}})' + N(\bar{\mathbf{x}} - \boldsymbol{\mu})(\bar{\mathbf{x}} - \boldsymbol{\mu})' + 2 \sum_{\alpha=1}^{N} (\mathbf{x}^\alpha - \bar{\mathbf{x}})(\bar{\mathbf{x}} - \boldsymbol{\mu})' \\ &= \sum_{\alpha=1}^{N} (\mathbf{x}^\alpha - \bar{\mathbf{x}})(\mathbf{x}^\alpha - \bar{\mathbf{x}})' + N(\bar{\mathbf{x}} - \boldsymbol{\mu})(\bar{\mathbf{x}} - \boldsymbol{\mu})'. \end{aligned}$$

Since Σ is positive definite

$$N(\bar{\mathbf{x}} - \boldsymbol{\mu})' \Sigma^{-1} (\bar{\mathbf{x}} - \boldsymbol{\mu}) \geq 0$$

for all $\bar{\mathbf{x}} - \boldsymbol{\mu}$ and is zero if and only if $\bar{\mathbf{x}} = \boldsymbol{\mu}$. Hence $\boldsymbol{\mu} = \bar{\mathbf{x}}$ maximizes the likelihood function. Obviously $\hat{\boldsymbol{\mu}} = \bar{\mathbf{x}}$. We will assume throughout that $N > p$; the reason for such an assumption will be evident from Lemma 5.1.2. Given $\mathbf{x}^\alpha$, $\alpha = 1, \ldots, N$, L is a function of $\boldsymbol{\mu}$ and Σ only and we will denote it simply by $L(\boldsymbol{\mu}, \Sigma)$. Hence

$$L(\hat{\boldsymbol{\mu}}, \Sigma) = (2\pi)^{-Np/2}(\det \Sigma)^{-N/2} \exp\{-\tfrac{1}{2} \operatorname{tr} \Sigma^{-1}s\}. \tag{5.1}$$

We now prove three lemmas which are useful in the sequel and for subsequent presentations.

LEMMA 5.1.1 Let A be any symmetric positive definite matrix and let

$$f(A) = c(\det A)^{N/2} \exp\{-\tfrac{1}{2} \operatorname{tr} A\}$$

where c is a constant. Then $f(A)$ has a maximum in the space of all positive definite matrices when $A = NI$, where I is the identity matrix of dimension $p \times p$.

Proof Clearly

$$f(A) = c \prod_{i=1}^{p} (\theta_i^{N/2} \exp\{-\theta_i/2\})$$

where $\theta_1, \ldots, \theta_p$ are the characteristic roots of the matrix A. But this is maximum when $\theta_1 = \cdots = \theta_p = N$, which holds if and only if $A = NI$. Hence $f(A)$ is maximum if $A = NI$. Q.E.D.

LEMMA 5.1.2 Let $\mathbf{X}^\alpha = (X_{\alpha 1}, \ldots, X_{\alpha p})'$, $\alpha = 1, \ldots, N$, be independently distributed normal random vectors with the same mean vector $\boldsymbol{\mu}$ and the same positive definite covariance matrix Σ, and let

$$S = \sum_{\alpha=1}^{N} (\mathbf{X}^\alpha - \bar{\mathbf{X}})(\mathbf{X}^\alpha - \bar{\mathbf{X}})' \qquad \text{where} \quad \bar{\mathbf{X}} = \frac{1}{N}\sum_{1}^{N} \mathbf{X}^\alpha.$$

Then

(a) $\bar{\mathbf{X}}$, S are independent, $\sqrt{N}\,\bar{\mathbf{X}}$ has p-dimensional normal distribution with mean $\sqrt{N}\,\boldsymbol{\mu}$ and covariance matrix Σ, and S is distributed as

$$\sum_{\alpha=1}^{N-1} \mathbf{Z}^\alpha \mathbf{Z}^{\alpha\prime}$$

where $\mathbf{Z}^\alpha$, $\alpha = 1, \ldots, N-1$, are independently distributed normal p-vectors with the same mean $\mathbf{0}$ and the same covariance matrix Σ:

(b) S is positive definite with probability 1 if and only if $N > p$.

Proof (a) Let O be an orthogonal matrix of dimension $N \times N$ of the form

$$O = \begin{pmatrix} O_{11} & \cdots & O_{1N} \\ \vdots & & \vdots \\ O_{N-1\,1} & \cdots & O_{N-1\,N} \\ 1/\sqrt{N} & \cdots & 1/\sqrt{N} \end{pmatrix}.$$

The last row of O is the equiangular vector of unit length. Since $\mathbf{X}^\alpha$, $\alpha = 1, \ldots, N$, are independent,

$$E(\mathbf{X}^\alpha - \boldsymbol{\mu})(\mathbf{X}^\beta - \boldsymbol{\mu})' = \begin{cases} 0 & \text{if} \quad \alpha \neq \beta \\ \Sigma & \text{if} \quad \alpha = \beta \end{cases}.$$

Let

$$\mathbf{Z}^\alpha = \sum_{\beta=1}^{N} O_{\alpha\beta}\mathbf{X}^\beta, \qquad \alpha = 1, \ldots, N.$$

The set of vectors $\mathbf{Z}^\alpha$, $\alpha = 1, \ldots, N$, has a joint normal distribution because the entire set of components is a set of linear combinations of the components of the set of vectors $\mathbf{X}^\alpha$, $\alpha = 1, \ldots, N$, which has a joint normal distribution. Now

$$E(\mathbf{Z}^N) = \sqrt{N}\,\boldsymbol{\mu},$$

$$E(\mathbf{Z}^\alpha) = \sum_{\beta=1}^{N} O_{\alpha\beta}\boldsymbol{\mu} = \sqrt{N}\,\boldsymbol{\mu} \sum_{\beta=1}^{N} O_{\alpha\beta}\frac{1}{\sqrt{N}} = 0, \qquad \alpha < N,$$

$$\operatorname{cov}(\mathbf{Z}^\alpha, \mathbf{Z}^\gamma) = \sum_{\beta=1}^{N} O_{\alpha\beta}O_{\gamma\beta}E(\mathbf{X}^\beta - \boldsymbol{\mu})(\mathbf{X}^\beta - \boldsymbol{\mu})' = \begin{cases} 0 & \text{if} \quad \alpha \neq \gamma \\ \Sigma & \text{if} \quad \alpha = \gamma \end{cases}.$$

Furthermore,

$$\sum_{\alpha=1}^{N} \mathbf{Z}^\alpha\mathbf{Z}^{\alpha'} = \sum_{\alpha=1}^{N}\sum_{\beta=1}^{N} O_{\alpha\beta}\mathbf{X}^\beta \sum_{\gamma=1}^{N} O_{\alpha\gamma}\mathbf{X}^{\gamma'}$$

$$= \sum_{\beta=1}^{N}\sum_{\gamma=1}^{N}\left(\sum_{\alpha=1}^{N} O_{\alpha\beta}O_{\alpha\gamma}\right)\mathbf{X}^\beta\mathbf{X}^{\gamma'} = \sum_{\beta=1}^{N} \mathbf{X}^\beta\mathbf{X}^{\beta'}.$$

Thus it is evident that $\mathbf{Z}^\alpha$, $\alpha = 1, \ldots, N$, are independent and $\mathbf{Z}^\alpha$, $\alpha = 1, \ldots, N-1$, are normally distributed with mean $\mathbf{0}$ and covariance matrix Σ. Since $\mathbf{Z}^N = \sqrt{N}\,\bar{\mathbf{X}}$ and $S = \sum_{\alpha=1}^{N} \mathbf{X}^\alpha\mathbf{X}^{\alpha'} - \mathbf{Z}^N\mathbf{Z}^{N'} = \sum_{\alpha=1}^{N-1} \mathbf{Z}^\alpha\mathbf{Z}^{\alpha'}$, we conclude that $\bar{\mathbf{X}}$, S are independent, $\mathbf{Z}^N$ has p-variate normal distribution with mean $\sqrt{N}\,\boldsymbol{\mu}$ and covariance matrix Σ, and S is distributed as $\sum_{\alpha=1}^{N-1} \mathbf{Z}^\alpha\mathbf{Z}^{\alpha'}$.

(b) Let $B = (\mathbf{Z}^1, \ldots, \mathbf{Z}^{N-1})$. Then $S = BB'$ where B is a matrix of dimension $p \times (N-1)$. This part will be proved if we can show that B has rank p with probability 1 if and only if $N > p$. Obviously by adding more columns to B we cannot diminish its rank and if $N \leq p$, then the rank of B is less than p. Thus it will suffice to show that B has rank p with probability 1 when $N - 1 = p$.

For any set of $(p-1)$ p-vectors $(\boldsymbol{\alpha}^1, \ldots, \boldsymbol{\alpha}^{p-1})$ in R^p let $S(\boldsymbol{\alpha}^1, \ldots, \boldsymbol{\alpha}^{p-1})$ be the subspace spanned by $\boldsymbol{\alpha}^1, \ldots, \boldsymbol{\alpha}^{p-1}$. Since Σ is nonsingular, for any given $\boldsymbol{\alpha}^1, \ldots, \boldsymbol{\alpha}^{p-1}$,

$$P\{\mathbf{Z}^i \in S(\boldsymbol{\alpha}^1, \ldots, \boldsymbol{\alpha}^{p-1})\} = 0.$$

Now, as $\mathbf{Z}^1, \ldots, \mathbf{Z}^p$ are independent and identically distributed random p-vectors,

$$\begin{aligned}
P\{\mathbf{Z}^1, \ldots, \mathbf{Z}^p \text{ are linearly dependent}\} \\
&\leq \sum_{i=1}^{p} P\{\mathbf{Z}^i \in S(\mathbf{Z}^1, \ldots, \mathbf{Z}^{i-1}, \mathbf{Z}^{i+1}, \ldots, \mathbf{Z}^p)\} \\
&= pP\{\mathbf{Z}^1 \in S(\mathbf{Z}^2, \ldots, \mathbf{Z}^p)\} \\
&= pE[P\{\mathbf{Z}^1 \in S(\mathbf{Z}^2, \ldots, \mathbf{Z}^p) | \mathbf{Z}^2 = \mathbf{z}^2, \ldots, \mathbf{Z}^p = \mathbf{z}^p\}] \\
&= pE(0) = 0. \qquad \text{Q.E.D.}
\end{aligned}$$

This lemma is due to Dykstra (1970). A similar proof also appears in the lecture notes of Stein (1969). This result depends heavily on the normal distribution of $\mathbf{Z}^1, \ldots, \mathbf{Z}^p$. Subsequently Eaton and Pearlman (1973) have given conditions in the case of a random matrix whose columns are independent but not necessarily normal or identically distributed.

Note The distribution of S is called the Wishart distribution with parameter Σ and degrees of freedom $N - 1$. We will show in Chapter VI that its probability density function is given by

$$\frac{(\det s)^{(N-p-2)/2} \exp\{-\frac{1}{2} \operatorname{tr} \Sigma^{-1} s\}}{2^{(N-1)p/2} \pi^{p(p-1)/4} (\det \Sigma)^{(N-1)/2} \prod_{i=1}^{p} \Gamma((N-i)/2)}. \tag{5.2}$$

The following lemma gives an important property, usually called the invariance property of the method of maximum likelihood in statistical estimation. Briefly stated, if $\hat{\theta}$ is a maximum likelihood estimator of $\theta \in \Omega$, then $f(\hat{\theta})$ is a maximum likelihood estimator of $f(\theta)$ where $f(\theta)$ is some function of θ. As Zehna (1966) observed, some textbooks dealing with this topic do not explicitly mention the properties which the function f must possess in order to satisfy this invariance property. However, in the proof it is assumed that f is a one-to-one function defining a unique inverse, for example, see Anderson (1958). It will be clear from the following lemma that the restriction to a one-to-one function is not necessary.

LEMMA 5.1.3 Let $\theta \in \Omega$ (an interval in a K-dimensional Euclidean space) and let $L(\theta)$ denote the likelihood function—a mapping from Ω to the real line R. Assume that the maximum likelihood estimator $\hat{\theta}$ of θ exists so that $\hat{\theta} \in \Omega$ and $L(\hat{\theta}) \geq L(\theta)$ for all $\theta \in \Omega$. Let f be any arbitrary transformation mapping Ω to Ω^* (an interval in an r-dimensional Euclidean space, $1 \leq r \leq k$). Then $f(\hat{\theta})$ is a maximum induced likelihood estimator of $f(\theta)$.

Proof Let $\omega = f(\theta)$. Since f is a function, $f(\hat{\theta})$ is a unique number $\hat{\omega}$ of Ω^*. For each $\omega \in \Omega^*$ let

$$F(\omega) = \{\theta, \theta \in \Omega \text{ such that } f(\theta) = \omega\}, \qquad M(\omega) = \sup_{\theta \in F(\omega)} L(\theta).$$

The function $M(\omega)$ on Ω^* is the induced likelihood function of $f(\theta)$. Clearly

$$\{F(\omega):\omega \in \Omega^*\}$$

is a partition of Ω and $\hat{\theta}$ belongs to one and only one set of this partition; let us denote this set by $F(\hat{\omega})$. Moreover

$$L(\hat{\theta}) = \sup_{\theta \in F(\hat{\omega})} L(\theta) = M(\hat{\omega}) \leq \sup_{\omega \in \Omega^*} M(\omega) = \sup_{\theta \in \Omega} L(\theta) = L(\hat{\theta}),$$

and $M(\hat{\omega}) = \sup_{\Omega^*} M(\omega)$. Hence $\hat{\omega}$ is a maximum likelihood estimator of $f(\theta)$. Since $\hat{\theta} \in F(\hat{\omega})$ we get $f(\hat{\theta}) = \hat{\omega}$. Q.E.D.

From this it follows that if $\hat{\boldsymbol{\theta}} = (\hat{\theta}_1, \ldots, \hat{\theta}_k)$ is a maximum likelihood estimator of $\boldsymbol{\theta} = (\theta_1, \ldots, \theta_k)$ and if the transformation

$$\boldsymbol{\theta} \to (f_1(\boldsymbol{\theta}), \ldots, f_k(\boldsymbol{\theta}))$$

is one-to-one, then $f_1(\hat{\boldsymbol{\theta}}), \ldots, f_k(\hat{\boldsymbol{\theta}})$ are the maximum likelihood estimators of $f_1(\boldsymbol{\theta}), \ldots, f_k(\boldsymbol{\theta})$, respectively. Furthermore, if $\hat{\theta}_1, \ldots, \hat{\theta}_k$ are unique, then $f_1(\hat{\boldsymbol{\theta}}), \ldots, f_k(\hat{\boldsymbol{\theta}})$ are also unique.

Since $N > p$ by assumption, from Lemma 5.1.2 we conclude that s is positive definite. Hence we can write $s = \alpha\alpha'$ where α is a nonsingular matrix of dimension $p \times p$. From (5.1) we can write

$$\begin{aligned} L(\hat{\mu}, \Sigma) &= (2\pi)^{-Np/2}(\det \Sigma^{-1})^{N/2} \exp\{-\tfrac{1}{2} \operatorname{tr} \Sigma^{-1}s\} \\ &= (2\pi)^{-Np/2}(\det s)^{-N/2}(\det(\alpha'\Sigma^{-1}\alpha))^{N/2} \exp\{-\tfrac{1}{2} \operatorname{tr}(\alpha'\Sigma^{-1}\alpha)\}. \end{aligned}$$

Using Lemmas 5.1.1 and 5.1.3 we conclude that $\alpha'(\hat{\Sigma})^{-1}\alpha = NI$ or, equivalently, $\hat{\Sigma} = s/N$. Hence the maximum likelihood estimator of $\boldsymbol{\mu}$ is $\bar{X}$ and that of Σ is S/N.

5.1.1 Maximum Likelihood Estimator of Regression, Multiple and Partial Correlation Coefficients

Let the covariance matrix Σ of the random vector $\mathbf{X} = (X_1, \ldots, X_p)'$ be denoted by

$$\Sigma = (\sigma_{ij})$$

with $\sigma_{ii} = \sigma_i^2$. Then

$$\rho_{ij} = \sigma_{ij}/\sigma_i\sigma_j \tag{5.3}$$

is called the Pearson correlation coefficient between the ith and jth components of the random vector $\mathbf{X}$. (Karl Pearson, in 1896, gave the first justification for the estimate of ρ_{ij}.)

Write $s = (s_{ij})$. The maximum likelihood estimate of σ_{ij}, on the basis of observations $\mathbf{x}^\alpha = (x_{\alpha 1}, \ldots, x_{\alpha p})'$, $\alpha = 1, \ldots, N$, is $(1/N)s_{ij}$. Since $\mu_i = \mu_i$ and

$\sigma_i^2 = \sigma_{ii}$, $\rho_{ij} = \sigma_{ij}/(\sigma_{ii}\sigma_{ij})^{1/2}$ is a function of the σ_{ij}, the maximum likelihood estimates of μ_i, σ_i^2, and ρ_{ij} are

$$\hat{\mu}_i - \bar{x}_i, \qquad \hat{\sigma}_i^2 - \frac{1}{N}\sum_{\alpha=1}^{N}(x_{\alpha i} - \bar{x}_i)^2 - \frac{s_i^2}{N},$$

$$\hat{\rho}_{ij} = \frac{s_{ij}}{(s_i^2 s_j^2)^{1/2}} = \frac{\sum_{\alpha=1}^{N}(x_{\alpha i} - \bar{x}_i)(x_{\alpha j} - \bar{x}_j)}{(\sum_{\alpha=1}^{N}(x_{\alpha i} - \bar{x}_i)^2)^{1/2}(\sum_{\alpha=1}^{N}(x_{\alpha j} - \bar{x}_j)^2)^{1/2}}$$

$$= \frac{\sum_{\alpha=1}^{N}(x_{\alpha i} - \bar{x}_i)x_{\alpha j}}{(\sum_{\alpha=1}^{N}(x_{\alpha i} - \bar{x}_i)^2)^{1/2}(\sum_{\alpha=1}^{N}(x_{\alpha j} - \bar{x}_j)^2)^{1/2}} = r_{ij} \qquad \text{(say)}. \tag{5.4}$$

Let $\mathbf{X} = (X_1, \ldots, X_p)'$ be normally distributed with mean vector $\boldsymbol{\mu} = (\mu_1, \ldots, \mu_p)'$ and positive definite covariance matrix Σ. We observed in Chapter IV that the regression surface of $\mathbf{X}_{(2)} = (X_{q+1}, \ldots, X_p)'$ on $\mathbf{X}_{(1)} = (X_1, \ldots, X_q)' = \mathbf{x}_{(1)} = (x_1, \ldots, x_q)'$ is given by

$$E(\mathbf{X}_{(2)}|\mathbf{X}_{(1)} = \mathbf{x}_{(1)}) = \boldsymbol{\mu}_{(2)} + \beta(\mathbf{x}_{(1)} - \boldsymbol{\mu}_{(1)})$$

where $\beta = \Sigma_{21}\Sigma_{11}^{-1}$ is the matrix of regression coefficients of $\mathbf{X}_{(2)}$ on $\mathbf{X}_{(1)} = \mathbf{x}_{(1)}$ and Σ, $\boldsymbol{\mu}$ are partitioned as

$$\Sigma = \begin{pmatrix} \Sigma_{11} & \Sigma_{12} \\ \Sigma_{21} & \Sigma_{22} \end{pmatrix}, \qquad \boldsymbol{\mu} = (\boldsymbol{\mu}'_{(1)}\boldsymbol{\mu}'_{(2)}), \qquad \boldsymbol{\mu}_{(1)} = (\mu_1, \ldots, \mu_q)'$$

with Σ_{11} a $q \times q$ submatrix of Σ. Let

$$s = \sum_{\alpha=1}^{N}(\mathbf{x}_\alpha - \bar{\mathbf{x}})(\mathbf{x}_\alpha - \bar{\mathbf{x}})'$$

be similarly partitioned as

$$s = \begin{pmatrix} s_{11} & s_{12} \\ s_{21} & s_{22} \end{pmatrix}$$

where s_{11} is the upper left-hand corner submatrix of s of dimension $q \times q$. From Lemma 5.1.3 we obtain the following theorem.

THEOREM 5.1.1 On the basis of observations $\mathbf{x}^\alpha = (x_{\alpha 1}, \ldots, x_{\alpha p})'$, $\alpha = 1, \ldots, N$, from the p-dimensional normal distribution with mean $\boldsymbol{\mu}$ and positive definite covariance matrix Σ, the maximum likelihood estimates of β, $\Sigma_{22.1}$, and Σ_{11} are given by

$$\hat{\beta} = s_{21}s_{11}^{-1}, \qquad \hat{\Sigma}_{22.1} = (1/N)(s_{22} - s_{21}s_{11}^{-1}s_{12}), \qquad \hat{\Sigma}_{11} = s_{11}/N.$$

Let $s_{ij.1,\ldots,q}$ be the (i, j)th element of the matrix $s_{22} - s_{21}s_{11}^{-1}s_{12}$ of dimension $(p - q) \times (p - q)$. From Theorem 5.1.1 the maximum likelihood estimate of the partial correlation coefficient between the components X_i and X_j $(i \neq j)$, $i, j = q + 1, \ldots, p$, of $\mathbf{X}$ when $\mathbf{X}_{(1)} = (X_1, \ldots, X_q)'$ is held fixed is given by

$$\hat{\rho}_{ij.1,\ldots,q} = \frac{s_{ij.1,\ldots,q}}{(s_{ii.1,\ldots,q})^{1/2}(s_{jj.1,\ldots,q})^{1/2}} = r_{ij.1,\ldots,q} \tag{5.5}$$

where $r_{ij.1,\ldots,q}$ is an arbitrary designation.

In Chapter IV we defined the multiple correlation coefficient between the ith component X_{q+i} of $\mathbf{X}_{(2)} = (X_{q+1}, \ldots, X_p)'$ and $\mathbf{X}_{(1)}$ as

$$\rho = \left(\frac{\boldsymbol{\beta}'_{(i)}\Sigma_{11}^{-1}\boldsymbol{\beta}_{(i)}}{\sigma_{q+i}^2}\right)^{1/2}$$

where $\boldsymbol{\beta}_{(i)}$ is the ith row of the submatrix Σ_{21} of dimension $(p-q) \times q$ of Σ. If $q = p - 1$, then the multiple correlation coefficient between X_p and $(X_1, \ldots, X_{p-1})'$ is

$$\rho = \left(\frac{\Sigma_{21}\Sigma_{11}^{-1}\Sigma_{12}}{\Sigma_{22}}\right)^{1/2}.$$

Since $(\rho, \Sigma_{21}, \Sigma_{11})$ is a one-to-one function of Σ the maximum likelihood estimate of ρ is given by

$$\hat{\rho} = \left(\frac{s_{21}s_{11}^{-1}s_{12}}{s_{22}}\right)^{1/2} = r \tag{5.6}$$

where r is an arbitrary designation. Since S is positive definite, $R^2 \geq 0$. Furthermore,

$$1 - R^2 = \frac{S_{22} - S_{21}S_{11}^{-1}S_{12}}{S_{22}} = \frac{\det S}{S_{22}(\det S_{11})}.$$

In the general case the maximum likelihood estimate of ρ is obtained by replacing the parameters by their maximum likelihood estimates.

5.2 PROPERTIES OF MAXIMUM LIKELIHOOD ESTIMATORS OF $\boldsymbol{\mu}$ AND Σ

5.2.1 Unbiasedness

Let $\mathbf{X}^\alpha = (X_{\alpha 1}, \ldots, X_{\alpha p})'$, $\alpha = 1, 2, \ldots, N$, be independently and identically distributed normal p-vectors with the same mean vector $\boldsymbol{\mu}$ and the same positive definite covariance matrix Σ and let $N > p$. The maximum likelihood estimator of $\boldsymbol{\mu}$ is

$$\bar{\mathbf{X}} = \frac{1}{N}\sum_{\alpha=1}^{N} \mathbf{X}^\alpha$$

and that of Σ is

$$\frac{S}{N} = \frac{1}{N} \sum_{\alpha=1}^{N} (\mathbf{X}^\alpha - \bar{\mathbf{X}})(\mathbf{X}^\alpha - \bar{\mathbf{X}})'.$$

Furthermore we have observed that S is distributed independently of $\bar{\mathbf{X}}$ as

$$S = \sum_{\alpha=1}^{N-1} \mathbf{Z}^\alpha \mathbf{Z}^{\alpha\prime}$$

where $\mathbf{Z}^\alpha = (Z_{\alpha 1}, \ldots, Z_{\alpha p})'$, $\alpha = 1, \ldots, N-1$, are independently and identically distributed normal p-vectors with the same mean $\mathbf{0}$ and the same positive definite covariance matrix Σ. Since

$$E(\bar{\mathbf{X}}) = \boldsymbol{\mu}, \qquad E\left(\frac{S}{N-1}\right) = \frac{1}{N-1} \sum_{\alpha=1}^{N-1} E(\mathbf{Z}^\alpha \mathbf{Z}^{\alpha\prime}) = \Sigma,$$

we conclude that $\bar{\mathbf{X}}$ is an unbiased estimator of $\boldsymbol{\mu}$ and $S/(N-1)$ is an unbiased estimator of Σ.

5.2.2 Sufficiency

A statistic $T(\mathbf{X}^1, \ldots, \mathbf{X}^N)$, which is a function of the random sample $\mathbf{X}^\alpha$, $\alpha = 1, \ldots, N$, only, is said to be minimal sufficient for $(\boldsymbol{\mu}, \Sigma)$ if the sample space of $\mathbf{X}^\alpha$, $\alpha = 1, \ldots, N$, cannot be reduced beyond that of $T(\mathbf{X}^1, \ldots, \mathbf{X}^N)$ without losing sufficiency. Explicit procedures for obtaining minimal sufficient statistics are given by Lehman and Scheffé (1950) and Bahadur (1955). It has been established that the sufficient statistic obtained through the following Fisher–Neyman factorization theorem is minimal sufficient.

FISHER–NEYMAN FACTORIZATION THEOREM Let

$$\mathbf{X}^\alpha = (X_{\alpha 1}, \ldots, X_{\alpha p})',$$

$\alpha = 1, \ldots, N$, be a random sample of size N from a distribution with probability density function $f_{\mathbf{X}}(\mathbf{x}|\theta)$, $\theta \in \Omega$. The statistic $T(X^1, \ldots, X^N)$ is sufficient for θ if and only if we can find two nonnegative functions $g_T(t|\theta)$ (not necessarily a probability density function) and $K(\mathbf{X}^1, \ldots, \mathbf{X}^N)$ such that

$$\prod_{\alpha=1}^{N} f_{\mathbf{X}^\alpha}(\mathbf{x}^\alpha) = g_T(t|\theta) \cdot K(\mathbf{x}^1, \ldots, \mathbf{x}^N)$$

where $g_T(t|\theta)$ depends on $\mathbf{x}^1, \ldots, \mathbf{x}^N$ only through $T(\mathbf{x}^1, \ldots, \mathbf{x}^N)$ and depends on θ, and K is independent of θ.

For a proof of this theorem the reader is referred to Giri (1975), or to Halmos and Savage (1949) for a general proof involving some deeper theorems of measure theory.

If $\mathbf{X}^\alpha$, $\alpha = 1, \ldots, N$, is a random sample of size N from the p-dimensional normal distribution with mean $\boldsymbol{\mu}$ and positive definite covariance matrix Σ, the joint probability density function of $\mathbf{X}^\alpha$, $\alpha = 1, \ldots, N$, is given by

$$(2\pi)^{-Np/2}(\det \Sigma)^{-N/2} \exp\{-\tfrac{1}{2} \operatorname{tr}(\Sigma^{-1}s + N\Sigma^{-1}(\bar{\mathbf{x}} - \boldsymbol{\mu})(\bar{\mathbf{x}} - \boldsymbol{\mu})')\}.$$

Using the Fisher–Neyman factorization theorem we conclude that $(\bar{\mathbf{X}}, S)$ is a minimal sufficient statistic for $(\boldsymbol{\mu}, \Sigma)$. In the sequel we will use sufficiency to indicate minimal sufficiency.

5.2.3 Consistency

A real valued estimator T_N (function of a random sample of size N) is said to be weakly consistent for a parametric function $g(\theta)$, $\theta \in \Omega$, if T_N converges to $g(\theta)$ in probability, i.e., for every $\varepsilon > 0$

$$\operatorname*{Limit}_{N \to \infty} P\{|T_N - g(\theta)| < \varepsilon\} = 1$$

and is strongly consistent if

$$P\left\{\operatorname*{Limit}_{N \to \infty} T_N = g(\theta)\right\} = 1.$$

In the case of a normal univariate random variable with mean μ and variance σ^2, the sample mean $\bar{X}$ of a random sample $X_1, \ldots, X_N$ of size N is both weakly and strongly consistent (see Giri, 1975). When the estimator T_N is a random matrix there are various ways of defining the stochastic convergence $T_N \to g(\theta)$. Let

$$T_N = (T_{ij}(N)), \qquad g(\theta) = (g_{ij}(\theta))$$

be matrices of dimension $p \times q$. For any matrix $A = (a_{ij})$ let us define two different norms

$$N_1(A) = \operatorname{tr} AA', \qquad N_2(A) = \max_{ij} |a_{ij}|$$

where $|\ |$ is the absolute value symbol.

Some alternative ways of defining the convergence of T_N to $g(\theta)$ are

$$\begin{aligned}
&(1)\quad T_{ij}(N) \text{ converges stochastically to } g_{ij}(\theta) \text{ for all } i, j.\\
&(2)\quad N_1(T_N - g(\theta)) \text{ converges stochastically to zero.}\\
&(3)\quad N_2(T_N - g(\theta)) \text{ converges stochastically to zero.}
\end{aligned} \tag{5.7}$$

It may be verified that these three different ways of defining stochastic convergence are equivalent. We shall establish stochastic convergence by using the first criterion.

To show that $\bar{\mathbf{X}}$ converges stochastically to $\boldsymbol{\mu}$ and S/N converges stochastically to $\Sigma = (\sigma_{ij})$, $\sigma_{ii} = \sigma_i^2$, we need to show that $\bar{X}_i$ converges stochastically to μ_i and $S_{ij}/(N-1)$ converges stochastically to σ_{ij} where $S = (S_{ij})$ for all i, j. Since

$$\bar{X}_i = \frac{1}{N} \sum_{\alpha=1}^{N} X_{\alpha i}$$

where $X_{\alpha i}$, $\alpha = 1, \ldots, N$, are independently and identically distributed normal random variables with mean μ_i and variance σ_i^2, using the Chebychev inequality and the Kolmogorov theorem (see Giri, 1975), we conclude that $\bar{X}_i$ is both weakly and strongly consistent for μ_i, $i = 1, \ldots, p$. Thus $\bar{\mathbf{X}}$ is a consistent estimator for $\boldsymbol{\mu}$.

From Lemma 5.1.2 S can be written as

$$S = \sum_{\alpha=1}^{N-1} \mathbf{Z}^\alpha \mathbf{Z}^{\alpha\prime}$$

where $\mathbf{Z}^\alpha$, $\alpha = 1, \ldots, N$, are independently and identically distributed normal p-vectors with mean $\mathbf{0}$ and positive definite covariance matrix Σ. Hence

$$\frac{S_{ij}}{N-1} = \frac{1}{N-1} \sum_{\alpha=1}^{N} Z_{\alpha i} Z_{\alpha j} = \frac{1}{N-1} \sum_{\alpha=1}^{N} Z_\alpha(i, j)$$

where $Z_\alpha(i, j) = Z_{\alpha i} Z_{\alpha j}$. Obviously $Z_\alpha(i, j)$, $\alpha = 1, \ldots, N-1$, are independently and identically distributed random variables with

$$\begin{aligned} E(Z_\alpha(i, j)) &= \sigma_{ij} \\ V(Z_\alpha(i, j)) &= E(Z_{\alpha i}^2 Z_{\alpha j}^2) - E^2(Z_{\alpha i} Z_{\alpha j}) \\ &\leq (E(Z_{\alpha i}^4) E(Z_{\alpha j}^4))^{1/2} - \sigma_{ij}^2 \leq \sigma_i^2 \sigma_j^2 (3 - \rho_{ij}^2) < \infty \end{aligned}$$

where ρ_{ij} is the coefficient of correlation between the ith and the jth component of $\mathbf{Z}_\alpha$. Now applying the Chebychev inequality and the Kolmogorov theorem we conclude that $S_{ij}/(N-1)$ is weakly and strongly consistent for σ_{ij} for all i, j.

5.2.4 Completeness

Let T be a continuous random variable (univariate or multivariate) with probability density function $f_T(t|\theta)$, $\theta \in \Omega$—the parametric space. The family of probability density functions $\{f_T(t|\theta), \theta \in \Omega\}$ is said to be complete if for any real valued function $g(T)$

$$E_\theta(g(T)) = 0 \tag{5.8}$$

for every $\theta \in \Omega$ implies that $g(T) = 0$ for all values of T for which $f_T(t|\theta)$ is greater than zero for some $\theta \in \Omega$. If the family of probability density functions of a sufficient statistic is complete, we call it a complete sufficient statistic.

We would like to show that $(\bar{\mathbf{X}}, S)$ is a complete sufficient statistic for $(\boldsymbol{\mu}, \Sigma)$. From (5.2) the joint probability density function of $\bar{\mathbf{X}}, S$ is given by

$$c(\det \Sigma)^{-N/2}(\det s)^{(N-p-2)/2} \exp\{-\tfrac{1}{2}\operatorname{tr}(\Sigma^{-1}s + N(\bar{\mathbf{x}} - \boldsymbol{\mu})'\Sigma^{-1}(\bar{\mathbf{x}} - \boldsymbol{\mu}))\} \quad (5.9)$$

where

$$c^{-1} = 2^{Np/2}\Pi^{p(p+1)/4}N^{-p/2}\prod_{i=1}^{p}\Gamma\left(\frac{N-i}{2}\right).$$

For any real valued function $g(\bar{\mathbf{X}}, S)$ of $(\bar{\mathbf{X}}, S)$

$$Eg(\bar{\mathbf{X}}, S) = c\int g(\bar{\mathbf{x}}, s)(\det \Sigma)^{-N/2}(\det s)^{(N-p-2)/2}$$
$$\times \exp\{-\tfrac{1}{2}\operatorname{tr}(\Sigma^{-1}s + N(\bar{\mathbf{x}} - \boldsymbol{\mu})'\Sigma^{-1}(\bar{\mathbf{x}} - \boldsymbol{\mu}))\}\, d\bar{\mathbf{x}}\, ds \quad (5.10)$$

where $d\bar{\mathbf{x}} = \Pi_i\, d\bar{x}_i$, $ds = \Pi_{ij}\, ds_{ij}$. Write $\Sigma^{-1} = I - 2\theta$ where I is the identity matrix of dimension $p \times p$ and θ is symmetric. Let

$$\boldsymbol{\mu} = (I - 2\theta)^{-1}\boldsymbol{\alpha}.$$

If $Eg(\bar{\mathbf{X}}, S) = 0$ for all $(\boldsymbol{\mu}, \Sigma)$, then from (5.10) we get

$$c\int g(\bar{\mathbf{x}}, s)(\det(I - 2\theta))^{N/2}(\det s)^{(N-p-2)/2}$$
$$\times \exp\{-\tfrac{1}{2}[\operatorname{tr}(I - 2\theta)(s + N\bar{\mathbf{x}}\bar{\mathbf{x}}') - 2N\boldsymbol{\alpha}'\bar{\mathbf{x}} + N\boldsymbol{\alpha}'(I - 2\theta)^{-1}\boldsymbol{\alpha}]\}\, d\bar{\mathbf{x}}\, ds$$
$$= 0,$$

or

$$c\int g(\bar{\mathbf{x}}, s + N\bar{\mathbf{x}}\bar{\mathbf{x}}' - N\bar{\mathbf{x}}\bar{\mathbf{x}}')(\det s)^{(N-p-2)/2}$$
$$\times \exp\{-\tfrac{1}{2}\operatorname{tr}(s + N\bar{\mathbf{x}}\bar{\mathbf{x}}') + \operatorname{tr}\theta(s + N\bar{\mathbf{x}}\bar{\mathbf{x}}') + N\boldsymbol{\alpha}'\bar{\mathbf{x}}\}\, d\bar{\mathbf{x}}\, ds$$
$$= 0 \quad (5.11)$$

identically in θ and $\boldsymbol{\alpha}$. We now identify (5.11) as the Laplace transform of

$$cg(\bar{\mathbf{x}}, s + N\bar{\mathbf{x}}\bar{\mathbf{x}}' - N\bar{\mathbf{x}}\bar{\mathbf{x}}')(\det s)^{(N-p-2)/2}\exp\{-\tfrac{1}{2}\operatorname{tr}(s + N\bar{\mathbf{x}}\bar{\mathbf{x}}')\}$$

with respect to variables $N\bar{\mathbf{x}}$, $s + N\bar{\mathbf{x}}\bar{\mathbf{x}}'$. Since this is identically equal to zero for all $\boldsymbol{\alpha}$ and θ we conclude that $g(\bar{\mathbf{x}}, s) = 0$ except possibly for a set of values of $(\bar{\mathbf{X}}, S)$ with probability measure 0. In other words, $(\bar{\mathbf{X}}, S)$ is a complete sufficient statistic for $(\boldsymbol{\mu}, \Sigma)$.

5.2.5 Efficiency

Let $X^\alpha = (X_{\alpha 1}, \ldots, X_{\alpha p})'$, $\alpha = 1, \ldots, N$, be a random sample of size N from a distribution with probability density function $f_{\mathbf{X}}(\mathbf{x}|\boldsymbol{\theta})$, $\boldsymbol{\theta} \in \Omega$. Assume that $\boldsymbol{\theta} = (\theta_1, \ldots, \theta_k)'$ and Ω is E^k (Euclidean k-space) or an interval in E^k.

Consider the problem of estimating parametric functions

$$\mathbf{g}(\boldsymbol{\theta}) = (g_1(\boldsymbol{\theta}), \ldots, g_r(\boldsymbol{\theta}))'.$$

We shall denote an estimator

$$\mathbf{T}(\mathbf{X}^1, \ldots, \mathbf{X}^N) = (T_1(\mathbf{X}^1, \ldots, \mathbf{X}^N), \ldots, T_r(\mathbf{X}^1, \ldots, \mathbf{X}^N))'$$

simply by $\mathbf{T} = (T_1, \ldots, T_r)'$.

An unbiased estimator $\mathbf{T}$ of $\mathbf{g}(\boldsymbol{\theta})$ is said to be an efficient estimator of $\mathbf{g}(\boldsymbol{\theta})$ if for any other unbiased estimator $\mathbf{U}$ of $\mathbf{g}(\boldsymbol{\theta})$

$$\operatorname{cov}(\mathbf{T}) \leq \operatorname{cov}(\mathbf{U}) \qquad \text{for all} \quad \boldsymbol{\theta} \in \Omega \tag{5.12}$$

in the sense that $\operatorname{cov}(\mathbf{U}) - \operatorname{cov}(\mathbf{T})$ is nonnegative definite for all $\boldsymbol{\theta} \in \Omega$.

The efficient unbiased estimator of $\mathbf{g}(\boldsymbol{\theta})$ can be obtained by the following two methods.

Generalized Rao–Cramer Inequality for a Vector Parameter

Let

$$L(\boldsymbol{\theta}) = L(\bar{\mathbf{x}}^1, \ldots, \bar{\mathbf{x}}^N|\boldsymbol{\theta}) = \prod_{\alpha=1}^{N} f_{\mathbf{X}^\alpha}(\bar{\mathbf{x}}^\alpha|\boldsymbol{\theta}),$$

$$P_{ij} = -\frac{\partial^2 \log L(\boldsymbol{\theta})}{\partial\theta_i\, \partial\theta_j}, \qquad I_{ij} = E(P_{ij}).$$

The $k \times k$ matrix

$$I = (I_{ij}) \tag{5.13}$$

is called the Fisher information measure on $\boldsymbol{\theta}$ or simply the information matrix (provided the P_{ij} exist).

For any unbiased estimator $\mathbf{T}^{**}$ of $\mathbf{g}(\boldsymbol{\theta})$ let us assume that

$$\frac{\partial}{\partial\theta_j} \int T_i^{**} L(\mathbf{x}^1, \ldots, \mathbf{x}^N|\boldsymbol{\theta})\, d\mathbf{x}^1, \ldots, d\mathbf{x}^N = \int T_i^{**} \frac{\partial L(\boldsymbol{\theta})}{\partial\theta_j}\, d\mathbf{x}^1, \ldots, d\mathbf{x}^N$$

$$= \frac{\partial g_i(\boldsymbol{\theta})}{\partial\theta_j}, \qquad i = 1, \ldots, r, \quad j = 1, \ldots, k,$$

and let

$$\Delta = \left(\frac{\partial g_i(\boldsymbol{\theta})}{\partial\theta_j}\right) \tag{5.14}$$

be a matrix of dimension $r \times k$. Then it can be verified that (see, e.g., Rao, 1965)

$$\operatorname{cov}(\mathbf{T}^{**}) - \Delta I^{-1} \Delta' \tag{5.15}$$

is nonnegative definite. Since $\Delta I^{-1}\Delta'$ is defined independently of any estimation procedure it follows that for any unbiased estimator $\mathbf{T}^{**}$ of $\mathbf{g}(\boldsymbol{\theta})$

$$\operatorname{var}(T_i^{**}) \geq \sum_{m=1}^{k} \sum_{n=1}^{k} I^{mn} \frac{\partial g_i}{\partial \theta_m} \frac{\partial g_i}{\partial \theta_n}, \qquad i = 1, \ldots, r, \tag{5.16}$$

where $I^{-1} = (I^{mn})$. Hence the efficient unbiased estimator of $\mathbf{g}(\boldsymbol{\theta})$ is an estimator $\mathbf{T}$ (if it exists) such that

$$\operatorname{cov}(\mathbf{T}) = \Delta I^{-1}\Delta'. \tag{5.17}$$

If $\mathbf{g}(\boldsymbol{\theta}) = \boldsymbol{\theta}$, then Δ is the identity matrix and the covariance of the efficient unbiased estimator is I^{-1}. From (5.15) it follows that if for any unbiased estimator $T = (T_1, \ldots, T_r)'$ of $\mathbf{g}(\boldsymbol{\theta})$

$$\operatorname{var}(T_i) = \sum_{m=1}^{k} \sum_{n=1}^{k} I^{mn} \frac{\partial g_i}{\partial \theta_m} \frac{\partial g_j}{\partial \theta_n}, \qquad i = 1, \ldots, r, \tag{5.18}$$

then

$$\operatorname{cov}(T_i, T_j) = \sum_{m=1}^{k} \sum_{n=1}^{k} I^{mn} \frac{\partial g_i}{\partial \theta_m} \frac{\partial g_j}{\partial \theta_n} \qquad \text{for all} \quad i \neq j. \tag{5.19}$$

Thus (5.18) implies that

$$\operatorname{cov}(\mathbf{T}) = \Delta I^{-1}\Delta'. \tag{5.20}$$

Thus any unbiased estimator of $\mathbf{g}(\boldsymbol{\theta})$ is efficient if (5.18) holds. Now we would like to establish that (5.18) holds well if

$$T_i = g_i(\boldsymbol{\theta}) + \sum_{j=1}^{k} \xi_{ij} \frac{1}{L(\boldsymbol{\theta})} \frac{\partial L(\boldsymbol{\theta})}{\partial \theta_j}, \qquad i = 1, \ldots, r, \tag{5.21}$$

where $\boldsymbol{\xi}_i = (\xi_{i1}, \ldots, \xi_{ik})' = \text{const} \times I^{-1}\boldsymbol{\beta}_i$ with

$$\boldsymbol{\beta}_i = \left(\frac{\partial g_i(\boldsymbol{\theta})}{\partial \theta_1}, \ldots, \frac{\partial g_i(\boldsymbol{\theta})}{\partial \theta_k}\right)'.$$

To do that let

$$U = T_i - g_i(\boldsymbol{\theta}), \qquad W = \sum_{j=1}^{k} \xi_{ij} \frac{1}{L(\boldsymbol{\theta})} \frac{\partial L(\boldsymbol{\theta})}{\partial \theta_j}$$

where $\boldsymbol{\xi}_i = (\xi_{i1}, \ldots, \xi_{ik})'$ is a constant nonnull vector which is independent of $\mathbf{x}^\alpha$, $\alpha = 1, \ldots, N$, but possibly dependent on $\boldsymbol{\theta}$. Since

$$\int \frac{\partial}{\partial \theta_i} L(\mathbf{x}^1, \ldots, \mathbf{x}^N | \boldsymbol{\theta})\, d\mathbf{x}^1 \cdots d\mathbf{x}^N = \frac{\partial}{\partial \theta_i} \int L(\mathbf{x}^1, \ldots, \mathbf{x}^N | \boldsymbol{\theta})\, d\mathbf{x}^1 \cdots d\mathbf{x}^N$$

$$= 0 \qquad \text{for all} \quad i,$$

we get $E(W) = 0$. Also $E(U) = 0$. The variances and covariance of U, W are given by

$$\begin{aligned}
\operatorname{var}(U) &= \operatorname{var}(T_i), \\
\operatorname{var}(W) &= \operatorname{var}\left(\sum_{j=1}^{k} \xi_{ij} \frac{1}{L(\boldsymbol{\theta})} \frac{\partial L(\boldsymbol{\theta})}{\partial \theta_j}\right) \\
&= \sum_{j=1}^{k} \sum_{j'=1}^{k} \xi_{ij}\xi_{ij'} \operatorname{cov}\left(\frac{1}{L(\boldsymbol{\theta})} \frac{\partial L(\boldsymbol{\theta})}{\partial \theta_j}, \frac{1}{L(\boldsymbol{\theta})} \frac{\partial L(\boldsymbol{\theta})}{\partial \theta_{j'}}\right) \\
&= \sum_{j=1}^{k} \sum_{j'=1}^{k} \xi_{ij}\xi_{ij'} I_{jj'} = \boldsymbol{\xi}_i' I \boldsymbol{\xi}_i, \\
\operatorname{cov}(U, W) &= \sum_{j=1}^{k} \xi_{ij} E\left((T_i - g_i(\boldsymbol{\theta}))\frac{1}{L(\boldsymbol{\theta})} \frac{\partial L(\boldsymbol{\theta})}{\partial \theta_j}\right) \\
&= \sum_{j=1}^{k} \xi_{ij} E\left(T_i \frac{1}{L(\boldsymbol{\theta})} \frac{\partial L(\boldsymbol{\theta})}{\partial \theta_j}\right) \\
&= \sum_{j=1}^{k} \xi_{ij} \frac{\partial g_i(\boldsymbol{\theta})}{\partial \theta_j} = \boldsymbol{\xi}_i' \boldsymbol{\beta}_i
\end{aligned}$$

where $\boldsymbol{\beta}_i = (\partial g_i(\boldsymbol{\theta})/\partial \theta_1, \ldots, \partial g_i(\boldsymbol{\theta})/\partial \theta_k)'$. Applying the Cauchy–Schwarz inequality, we obtain

$$(\boldsymbol{\xi}_i' \boldsymbol{\beta}_i)^2 \leq \operatorname{var}(T_i)(\boldsymbol{\xi}_i' I \boldsymbol{\xi}_i),$$

which implies that

$$\operatorname{var}(T_i) \geq \frac{(\boldsymbol{\xi}_i' \boldsymbol{\beta}_i)^2}{(\boldsymbol{\xi}_i' I \boldsymbol{\xi}_i)}.$$

Since $\boldsymbol{\xi}_i$ is arbitrary (nonnull), this implies

$$\operatorname{var}(T_i) \geq \sup_{\boldsymbol{\xi}_i \neq \mathbf{0}} \frac{(\boldsymbol{\xi}_i' \boldsymbol{\beta}_i)^2}{(\boldsymbol{\xi}_i' I \boldsymbol{\xi}_i)} = \boldsymbol{\beta}_i' I^{-1} \boldsymbol{\beta}_i \tag{5.22}$$

and the supremum is attained when

$$\boldsymbol{\xi}_i = c I^{-1} \boldsymbol{\beta}_i = \boldsymbol{\xi}_i^0,$$

where c is a constant and $\boldsymbol{\xi}_i^0$ is an arbitrary designation.

The equality in (5.22) holds if and only if

$$U = \text{const} \times W = \text{const} \times \sum_{j=1}^{k} \xi_{ij}^0 \frac{1}{L(\boldsymbol{\theta})} \frac{\partial L(\boldsymbol{\theta})}{\partial \theta_j}$$

with probability 1, i.e.,

$$T_i = g_i(\boldsymbol{\theta}) + \sum_{j=1}^{k} \xi_{ij} \frac{1}{L(\boldsymbol{\theta})} \frac{\partial L(\boldsymbol{\theta})}{\partial \theta_j}$$

with probability 1 where $\boldsymbol{\xi}_i = \text{const} \times I^{-1}\boldsymbol{\beta}_i$.

To prove that the sample mean $\bar{\mathbf{X}}$ is efficient for $\boldsymbol{\mu}$ we first observe that $\bar{\mathbf{X}}$ is unbiased for $\boldsymbol{\mu}$. Let

$$\Sigma^{-1} = (\sigma^{ij}), \qquad \theta = (\mu_1, \ldots, \mu_p, \sigma^{11}, \ldots, \sigma^{pp})'$$

where $\boldsymbol{\theta}$ is a vector of dimension $p(3p + 1)/2$. Let

$$\mathbf{g}(\boldsymbol{\theta}) = (g_1(\boldsymbol{\theta}), \ldots, g_p(\boldsymbol{\theta}))' = (\mu_1, \ldots, \mu_p)'.$$

Take $T_i = \bar{X}_i$, $g_i(\boldsymbol{\theta}) = \mu_i$. The likelihood of $\mathbf{x}^1, \ldots, \mathbf{x}^N$ is

$$\begin{aligned} L(\mathbf{x}^1, \ldots, \mathbf{x}^N | \boldsymbol{\theta}) &= L(\boldsymbol{\theta}) \\ &= (2\pi)^{-Np/2} (\det \Sigma^{-1})^{N/2} \\ &\quad \times \exp\{-\tfrac{1}{2} \operatorname{tr}(\Sigma^{-1} s + N\Sigma^{-1}(\bar{\mathbf{x}} - \boldsymbol{\mu})(\bar{\mathbf{x}} - \boldsymbol{\mu})')\}. \end{aligned}$$

Hence

$$\frac{\partial \log L}{\partial \mu_i} = N\sigma^{ii}(\bar{x}_i - \mu_i) + N \sum_{j(\neq i)} \sigma^{ij}(\bar{x}_j - \mu_j),$$

$$\frac{\partial g_i(\boldsymbol{\theta})}{\partial \mu_j} = \begin{cases} 1 & \text{if } j = i, \\ 0 & \text{if } j \neq i, \end{cases}$$

$$\frac{\partial g_i(\boldsymbol{\theta})}{\partial \sigma^{i'j'}} = 0 \qquad \text{for all} \quad i', j', i.$$

Hence

$$\boldsymbol{\beta}_i = (0, \ldots, 0, 1, 0, \ldots, 0)', \qquad i = 1, \ldots, p,$$

which is a unit vector with unity as its ith coordinate. Since

$$\frac{\partial^2 \log L(\boldsymbol{\theta})}{\partial \mu_i^2} = -N\sigma^{ii}, \qquad \frac{\partial^2 \log L(\boldsymbol{\theta})}{\partial \mu_i \, \partial \mu_j} = -N\sigma^{ij},$$

we get, for $i \neq j$, $l', l = 1, \ldots, p$,

$$E\left(-\frac{\partial^2 \log L(\boldsymbol{\theta})}{\partial \mu_i^2}\right) = N\sigma^{ii}, \qquad E\left(-\frac{\partial^2 \log L(\boldsymbol{\theta})}{\partial \mu_i \, \partial \mu_j}\right) = N\sigma^{ij},$$

$$E\left(-\frac{\partial^2 \log L(\boldsymbol{\theta})}{\partial \mu_i \, \partial \sigma^{ll'}}\right) = 0.$$

Thus, the information matrix I is given by

$$I = \begin{pmatrix} N\Sigma^{-1} & 0 \\ 0 & A \end{pmatrix}$$

where A is a nonsingular matrix of dimension $\frac{1}{2}p(p+1) \times \frac{1}{2}p(p+1)$. (It is not necessary, in this context, to evaluate A specifically.) So

$$I^{-1}\boldsymbol{\beta}_i = (1/N)(\sigma_{1i}, \ldots, \sigma_{pi}, 0, \ldots, 0)'.$$

Choosing $\boldsymbol{\xi}_{(i)} = I^{-1}\boldsymbol{\beta}_i$, we obtain

$$\begin{aligned} \sum_{j=1}^{k} \xi_{ij} \frac{1}{L(\boldsymbol{\theta})} \frac{\partial L(\boldsymbol{\theta})}{\partial \theta_j} &= (\bar{X}_i - \mu_i)(\sigma^{1i}\sigma_{1i} + \cdots + \sigma^{pi}\sigma_{pi}) \\ &\quad + \sum_{j(\neq i)} (X_j - \mu_j)(\sigma^{1j}\sigma_{1i} + \cdots + \sigma^{pj}\sigma_{pi}) \\ &= \bar{X}_i - \mu_i \end{aligned}$$

since $\Sigma^{-1}\Sigma$ is the identity matrix. Hence we conclude that $\bar{\mathbf{X}}$ is efficient for $\boldsymbol{\mu}$.

Second Method

Let $\mathbf{T}^* = (T_1^*, \ldots, T_k^*)'$ be a sufficient (minimal) estimator of $\boldsymbol{\theta}$ and let the distribution of $\mathbf{T}^*$ be complete. Given any unbiased estimator $\mathbf{T}^{**} = (T_1^{**}, \ldots, T_r^{**})'$ of $\mathbf{g}(\boldsymbol{\theta})$, the estimator

$$\mathbf{T} = E(\mathbf{T}^{**}|\mathbf{T}^*) = (E(T_1^{**}|\mathbf{T}^*), \ldots, E(T_r^{**}|\mathbf{T}^*))'$$

is at least as good as $\mathbf{T}^{**}$ for $\mathbf{g}(\boldsymbol{\theta})$, in the sense that $\text{cov}(\mathbf{T}^{**}) - \text{cov}(\mathbf{T})$ is nonnegative definite for all $\boldsymbol{\theta} \in \Omega$.

This follows from the fact that for any nonnull vector $\mathbf{L}$, $\mathbf{L}'\mathbf{T}^{**}$ is an unbiased estimator of the parametric function $\mathbf{L}'\mathbf{g}(\boldsymbol{\theta})$ and by the Rao–Blackwell theorem (see Giri, 1975), the estimator

$$\mathbf{L}'\mathbf{T} = E(\mathbf{L}'\mathbf{T}^{**}|T^*)$$

is at least as good as $\mathbf{L}'\mathbf{T}^{**}$ for all $\boldsymbol{\theta}$. Since this holds well for all $\mathbf{L} \neq \mathbf{0}$ it follows that $\text{cov}(\mathbf{T}^{**}) - \text{cov}(\mathbf{T})$ is nonnegative definite for all $\boldsymbol{\theta}$. Thus given any unbiased estimator $\mathbf{T}^{**}$ of $g(\boldsymbol{\theta})$ which is not a function of $\mathbf{T}^*$, the estimator $\mathbf{T}$ is better than $\mathbf{T}^{**}$. Hence in our search for efficient unbiased estimators we can restrict attention to unbiased estimators which are functions of $\mathbf{T}^*$ alone. Furthermore, if $f(\mathbf{T}^*)$ and $g(\mathbf{T}^*)$ are two unbiased estimators of $\mathbf{g}(\boldsymbol{\theta})$, then

$$E_\theta(f(\mathbf{T}^*) - g(\mathbf{T}^*)) \equiv 0 \tag{5.23}$$

for all $\boldsymbol{\theta} \in \Omega$. Since the distribution of $\mathbf{T}^*$ is complete (5.23) will imply $f(\mathbf{T}^*) - g(\mathbf{T}^*) = 0$ almost everywhere. Thus we conclude that there exists a unique unbiased efficient estimator of $\mathbf{g}(\boldsymbol{\theta})$ and this is obtained by exhibiting a function of $\mathbf{T}^*$ which is unbiased for $\mathbf{g}(\boldsymbol{\theta})$.

We established earlier that $(\bar{\mathbf{X}}, S)$ is a complete sufficient statistic for $(\boldsymbol{\mu}, \Sigma)$ of the p-variate normal distribution. Since $E(\bar{\mathbf{X}}) = \boldsymbol{\mu}$ and $E(S/(N-1)) = \Sigma$, it follows that $\bar{\mathbf{X}}$ and $S/(N-1)$ are unbiased efficient estimators of $\boldsymbol{\mu}$ and Σ, respectively.

5.3 BAYES, MINIMAX, AND ADMISSIBLE CHARACTERS OF THE MAXIMUM LIKELIHOOD ESTIMATOR OF $(\boldsymbol{\mu}, \Sigma)$

Let $\mathcal{X}$ be the sample space and let $\mathcal{A}$ be the σ-algebra of subsets of $\mathcal{X}$, and let $P_{\boldsymbol{\theta}}$, $\boldsymbol{\theta} \in \Omega$, be the probability on $(\mathcal{X}, \mathcal{A})$, where Ω is an interval in E^p. Let D be the set of all possible estimators of $\boldsymbol{\theta}$. A function

$$L(\boldsymbol{\theta}, d), \qquad \boldsymbol{\theta} \in \Omega, \quad d \in D,$$

defined on $\Omega \times D$, represents the loss of erroneously estimating $\boldsymbol{\theta}$ by d. (It may be remarked that d is a vector quantity.) Let

$$R(\boldsymbol{\theta}, d) = E_{\boldsymbol{\theta}}(L(\boldsymbol{\theta}, d)) = \int L(\boldsymbol{\theta}, d(x)) f_X(x|\boldsymbol{\theta})\, d\mathbf{x} \tag{5.24}$$

where $f_X(x|\boldsymbol{\theta})$ denotes the probability density function of X with values $x \in \mathcal{X}$, corresponding to $P_{\boldsymbol{\theta}}$ with respect to the Lebesgue measure $d\mathbf{x}$. $R(\boldsymbol{\theta}, d)$ is called the risk function of the estimator $d \in D$ for the parameter $\boldsymbol{\theta} \in \Omega$. Let $h(\boldsymbol{\theta})$, $\boldsymbol{\theta} \in \Omega$, denote the prior probability density on Ω. The posterior probability density function of $\boldsymbol{\theta}$ given that $X = x$ is given by

$$h(\boldsymbol{\theta}|x) = \frac{f_X(x|\boldsymbol{\theta}) h(\boldsymbol{\theta})}{\int f_X(x|\boldsymbol{\theta}) h(\boldsymbol{\theta})\, d\boldsymbol{\theta}}. \tag{5.25}$$

The prior risk [Bayes risk of d with respect to the prior $h(\boldsymbol{\theta})$] is given by

$$R(h, d) = \int R(\boldsymbol{\theta}, d) h(\boldsymbol{\theta})\, d\boldsymbol{\theta}. \tag{5.26}$$

If $R(\boldsymbol{\theta}, d)$ is bounded, we can interchange the order of integration in $R(h, d)$ and obtain

$$\begin{aligned} R(h, d) &= \int \left\{ \int L(\boldsymbol{\theta}, d(x)) f_X(x|\boldsymbol{\theta})\, dx \right\} h(\boldsymbol{\theta})\, d\boldsymbol{\theta} \\ &= \int \bar{f}(x) \left\{ \int L(\boldsymbol{\theta}, d(x)) h(\boldsymbol{\theta}|x)\, d\boldsymbol{\theta} \right\} dx \end{aligned} \tag{5.27}$$

where

$$\tilde{f}(x) = \int f_X(x|\boldsymbol{\theta})h(\boldsymbol{\theta})\, d\boldsymbol{\theta}. \tag{5.28}$$

The quantity

$$\int L(\boldsymbol{\theta}, d(x))h(\boldsymbol{\theta}|x)\, d\boldsymbol{\theta} \tag{5.29}$$

is called the posterior risk of d given $X = x$ (the posterior conditional expected loss).

DEFINITION 5.3.1 *Bayes estimator* A Bayes estimator of $\boldsymbol{\theta}$ with respect to the prior density $h(\boldsymbol{\theta})$ is the estimator $d_0 \in D$ which takes the value $d_0(x)$ for $X = x$ and minimizes the posterior risk given $X = x$. In other words, for every $x \in \mathscr{X}$, $d_0(x)$ is defined as

$$\int L(\boldsymbol{\theta}, d_0(x))h(\boldsymbol{\theta}|x)\, d\boldsymbol{\theta} = \inf_{d \in D} \int L(\boldsymbol{\theta}, d)h(\boldsymbol{\theta}|x)\, d\boldsymbol{\theta}. \tag{5.30}$$

Note (i) It is easy to check that the Bayes estimator d_0 also minimizes the prior risk.

(ii) The Bayes estimator is not necessarily unique. However, if $L(\boldsymbol{\theta}, d)$ is strictly convex in d for given $\boldsymbol{\theta}$, then d_0 is essentially unique.

For a thorough discussion of this the reader is referred to Zacks (1971) or Ferguson (1967). Raiffa and Schlaifer (1961) have discussed in considerable detail the problem of choosing prior distributions for various models.

Let $\mathbf{x}^\alpha = (x_{\alpha 1}, \ldots, x_{\alpha p})'$, $\alpha = 1, \ldots, N$, be a sample of size N from a p-dimensional normal distribution with mean $\boldsymbol{\mu}$ and positive definite covariance matrix Σ. Let

$$X = (\mathbf{X}^1, \ldots, \mathbf{X}^N), \qquad x = (\mathbf{x}^1, \ldots, \mathbf{x}^N).$$

Then

$$f_X(x) = (2\pi)^{-Np/2}(\det \Sigma)^{-N/2} \times \exp\{-\tfrac{1}{2}\operatorname{tr}(\Sigma^{-1}s + N\Sigma^{-1}(\bar{\mathbf{x}} - \boldsymbol{\mu})(\bar{\mathbf{x}} - \boldsymbol{\mu})')\}. \tag{5.31}$$

Let

$$L(\boldsymbol{\theta}, \mathbf{d}) = (\boldsymbol{\mu} - \mathbf{d})'(\boldsymbol{\mu} - \mathbf{d}). \tag{5.32}$$

The posterior risk

$$E((\boldsymbol{\mu} - \mathbf{d})'(\boldsymbol{\mu} - \mathbf{d})|X = x) = E(\boldsymbol{\mu}'\boldsymbol{\mu}|X = x) - 2\mathbf{d}'E(\boldsymbol{\mu}|X = x) + \mathbf{d}'\mathbf{d}$$

is a minimum when

$$\mathbf{d}(x) = E(\boldsymbol{\mu}|X = x).$$

In other words, the Bayes estimator is the mean of the marginal posterior density function of $\boldsymbol{\mu}$. Since

$$\frac{\partial^2 E((\boldsymbol{\mu} - \mathbf{d})'(\boldsymbol{\mu} - \mathbf{d})|X = x)}{\partial \mathbf{d}' \, \partial \mathbf{d}} = 2I,$$

$E(\boldsymbol{\mu}|X = x)$ actually corresponds to the minimum value.

Let us take the prior as

$$\begin{aligned} h(\boldsymbol{\theta}) = h(\boldsymbol{\mu}, \Sigma) = {} & K(\det \Sigma)^{-(v+1)/2} \\ & \times \exp\{-\tfrac{1}{2}[(\boldsymbol{\mu} - \mathbf{a})'\Sigma^{-1}(\boldsymbol{\mu} - \mathbf{a})b + \operatorname{tr} \Sigma^{-1}H]\} \end{aligned} \tag{5.33}$$

where $b > 0$, $v > 2p$, H is a positive definite matrix, and K is the normalizing constant. From (5.31) and (5.33) we get

$$\begin{aligned} h(\boldsymbol{\theta}|X = x) = {} & K'(\det \Sigma)^{-(N+v+1)/2} \exp\Bigg\{-\tfrac{1}{2} \operatorname{tr} \Sigma^{-1}\Bigg[s + H + (N + b) \\ & \times \left(\boldsymbol{\mu} - \frac{N\bar{\mathbf{x}} + \mathbf{a}b}{N + b}\right)\left(\boldsymbol{\mu} - \frac{N\bar{\mathbf{x}} + \mathbf{a}b}{N + b}\right)' \\ & + \frac{Nb}{N + b}(\bar{\mathbf{x}} - \mathbf{a})(\bar{\mathbf{x}} - \mathbf{a})'\Bigg]\Bigg\} \end{aligned} \tag{5.34}$$

where K' is a constant. Using (5.2), we get from (5.34)

$$\begin{aligned} h(\boldsymbol{\mu}|X = x) &= C\Bigg[\det\Bigg(s + H + \frac{Nb}{N+b}(\bar{\mathbf{x}} - \mathbf{a})(\bar{\mathbf{x}} - \mathbf{a})' + (N+b)\left(\boldsymbol{\mu} - \frac{N\mathbf{x} + \mathbf{a}b}{N+b}\right) \\ & \qquad \times \left(\boldsymbol{\mu} - \frac{N\bar{\mathbf{x}} + \mathbf{a}b}{N+b}\right)'\Bigg)\Bigg]^{-(N+v-p)/2} \\ &= C\frac{\left[\det\left(s + H + \dfrac{Nb}{N+b}(\bar{\mathbf{x}} - \mathbf{a})(\bar{\mathbf{x}} - \mathbf{a})'\right)\right]^{-(N+v-p)/2}}{\left[1 + (N+b)\left(\boldsymbol{\mu} - \dfrac{N\bar{\mathbf{x}} + \mathbf{a}b}{N+b}\right)' \left(s + H + \dfrac{Nb}{N+b}(\bar{\mathbf{x}} - \mathbf{a})(\bar{\mathbf{x}} - \mathbf{a})'\right)^{-1}\left(\boldsymbol{\mu} - \dfrac{N\bar{\mathbf{x}} + \mathbf{a}b}{N+b}\right)\right]^{(N+v-p)/2}} \end{aligned} \tag{5.35}$$

where C is a constant.

From Exercise 4.15 it is easy to calculate that

$$E(\boldsymbol{\mu}|X = x) = (N\bar{\mathbf{x}} + \mathbf{a}b)/(N + b),$$

which is the Bayes estimate of $\boldsymbol{\mu}$ for the prior (5.33).

For estimating Σ by an estimator d let us consider the loss function

$$L(\boldsymbol{\theta}, d) = \operatorname{tr}(\Sigma - d)(\Sigma - d). \tag{5.36}$$

The posterior risk with respect to this loss function is given by

$$E(\operatorname{tr} \Sigma\Sigma | X = x) - 2E(\operatorname{tr} d\Sigma | X = x) + \operatorname{tr} dd. \tag{5.37}$$

The posterior risk is minimized (see Exercise 1.14) when

$$d = E(\Sigma | X = x). \tag{5.38}$$

From (5.34), integrating out $\boldsymbol{\mu}$, the marginal posterior probability density function of Σ is given by

$$h(\Sigma | X = x) = K(\det \Sigma^{-1})^{(N+\nu)/2}$$
$$\times \exp\left\{-\tfrac{1}{2} \operatorname{tr} \Sigma^{-1}\left(s + H + \frac{Nb}{N+b}(\bar{\mathbf{x}} - \mathbf{a})(\bar{\mathbf{x}} - \mathbf{a})'\right)\right\} \tag{5.39}$$

where K is the normalizing constant independent of Σ. Identifying the marginal distribution of Σ as an inverted Wishart distribution,

$$W^{-1}\left(s + H + \frac{Nb}{N+b}(\bar{\mathbf{x}} - \mathbf{a})(\bar{\mathbf{x}} - \mathbf{a})', p, N + \nu\right)$$

we get from Exercise 5.8

$$E(\Sigma | X = x)$$
$$= \frac{s + H + [Nb/(N+b)](\bar{\mathbf{x}} - \mathbf{a})(\bar{\mathbf{x}} - \mathbf{a})'}{N + \nu - 2p - 2}, \qquad N + \nu - 2p > 2 \tag{5.40}$$

as the Bayes estimate of Σ for the prior (5.33).

Note If we work with the diffuse prior $h(\boldsymbol{\theta}) \propto (\det \Sigma)^{-(p+1)/2}$ which is obtained from (5.33) by putting $b = 0$, $H = 0$, $\nu = p$, and which ceases to be a probability density on Ω, we get $\bar{\mathbf{x}}$ and $s/(N - p - 2)$, $(N > p + 2)$, as the Bayes estimates of $\boldsymbol{\mu}$ and Σ, respectively. Such estimates are called generalized Bayes estimates. Thus for the multivariate normal distribution, the maximum likelihood estimates of $\boldsymbol{\mu}$ and Σ are not exactly Bayes estimates.

DEFINITION 5.3.2 *Extended Bayes estimator* An estimator $d_0 \in D$ is an extended Bayes estimator for $\boldsymbol{\theta} \in \Omega$ if it is ε-Bayes for every $\varepsilon > 0$; i.e., given any $\varepsilon > 0$, there exists a prior $h_\varepsilon(\boldsymbol{\theta})$ on Ω such that

$$E_{h_\varepsilon(\boldsymbol{\theta})}(R(\boldsymbol{\theta}, d_0)) \leq \inf_{d \in D} E_{h_\varepsilon(\boldsymbol{\theta})}(R(\boldsymbol{\theta}, d)) + \varepsilon. \tag{5.41}$$

DEFINITION 5.3.3 *Minimax estimator* An estimator $d^* \in D$ is minimax for estimating $\boldsymbol{\theta} \in \Omega$ if

$$\sup_{\boldsymbol{\theta} \in \Omega} R(\boldsymbol{\theta}, d^*) = \inf_{d \in D} \sup_{\boldsymbol{\theta} \in \Omega} R(\boldsymbol{\theta}, d). \tag{5.42}$$

In other words, the minimax estimator protects against the largest possible risk when $\boldsymbol{\theta}$ varies over Ω.

To show that $\bar{\mathbf{X}}$ is minimax for the mean $\boldsymbol{\mu}$ of the normal distribution (with known covariance matrix Σ) with respect to the loss function

$$L(\boldsymbol{\mu}, d) = (\boldsymbol{\mu} - d)'\Sigma^{-1}(\boldsymbol{\mu} - d), \tag{5.43}$$

we need the following theorem.

THEOREM 5.3.1 An extended Bayes estimator with constant risk is minimax.

Proof Let $d_0 \in D$ be such that $R(\boldsymbol{\theta}, d_0) = C$, a constant for all $\boldsymbol{\theta} \in \Omega$, and let d_0 also be an extended Bayes estimator; i.e., given any $\varepsilon > 0$, there exists a prior density $h_\varepsilon(\boldsymbol{\theta})$ on Ω such that

$$E_{h_\varepsilon(\boldsymbol{\theta})}(R(\boldsymbol{\theta}, d)) \leq \inf_{d \in D} E_{h_\varepsilon(\boldsymbol{\theta})}\{R(\boldsymbol{\theta}, d)\} + \varepsilon. \tag{5.44}$$

Suppose d_0 is not minimax; then there exists an estimator $d^* \in D$ such that

$$\sup_{\boldsymbol{\theta} \in \Omega} R(\boldsymbol{\theta}, d^*) < \sup_{\boldsymbol{\theta} \in \Omega} R(\boldsymbol{\theta}, d_0) = C. \tag{5.45}$$

This implies that

$$\sup_{\boldsymbol{\theta} \in \Omega} R(\boldsymbol{\theta}, d^*) < C - \varepsilon_0 \qquad \text{for some} \quad \varepsilon_0 > 0,$$

or

$$R(\boldsymbol{\theta}, d^*) < C - \varepsilon_0 \qquad \text{for all} \quad \boldsymbol{\theta} \in \Omega,$$

which implies

$$E\{R(\boldsymbol{\theta}, d^*)\} \leq C - \varepsilon_0, \tag{5.46}$$

where the expectation is taken with respect to any prior distribution over Ω. From (5.44) and (5.46) we get for every $\varepsilon > 0$ and the corresponding prior density $h_\varepsilon(\boldsymbol{\theta})$ over Ω

$$C - \varepsilon \leq \inf_{d \in D} E_{h_\varepsilon(\boldsymbol{\theta})}(R(\boldsymbol{\theta}, d)) \leq E_{h_\varepsilon(\boldsymbol{\theta})}(R(\boldsymbol{\theta}, d^*)) \leq C - \varepsilon_0,$$

which is a contradiction for $0 < \varepsilon < \varepsilon_0$. Hence d_0 is minimax. Q.E.D.

We now show that $\bar{\mathbf{X}}$ is the minimax estimator for $\boldsymbol{\mu}$. Let $\mathbf{X}^\alpha = (X_{\alpha 1}, \ldots, X_{\alpha p})$, $\alpha = 1, \ldots, N$, be independently distributed normal vectors with mean $\boldsymbol{\mu}$ and with a known positive definite covariance matrix Σ. Let

$$X = (\mathbf{X}^1, \ldots, \mathbf{X}^N), \qquad \bar{\mathbf{X}} = \frac{1}{N}\sum_{\alpha=1}^{N} \mathbf{X}^\alpha, \qquad S = \sum_{\alpha=1}^{N} (\mathbf{X}^\alpha - \bar{\mathbf{X}})(\mathbf{X}^\alpha - \bar{\mathbf{X}})'.$$

Assume that the prior density $h(\boldsymbol{\mu})$ of $\boldsymbol{\mu}$ is a p-variate normal with mean $\mathbf{0}$ and covariance matrix $\sigma^2\Sigma$ with $\sigma^2 > 0$. The joint probability density function of X and $\boldsymbol{\mu}$ is given by

$$\begin{aligned} h(\boldsymbol{\mu}, x) &= (2\pi)^{-(N+1)p/2}(\det \Sigma)^{-(N+1)/2}(\sigma^2)^{-p/2} \\ &\quad \times \exp\left\{-\tfrac{1}{2}\operatorname{tr}\Sigma^{-1}s - \tfrac{1}{2}N(\bar{\mathbf{x}}-\boldsymbol{\mu})'\Sigma^{-1}(\bar{\mathbf{x}}-\boldsymbol{\mu}) - \frac{1}{2\sigma^2}\boldsymbol{\mu}'\Sigma^{-1}\boldsymbol{\mu}\right\} \\ &= (2\pi)^{-(N+1)p/2}(\det \Sigma)^{-(N+1)/2}(\sigma^2)^{-p/2} \\ &\quad \times \exp\{-\tfrac{1}{2}\operatorname{tr}\Sigma^{-1}s\}\exp\left\{-\frac{1}{2}\left(\frac{N}{N\sigma^2+1}\right)\bar{\mathbf{x}}'\Sigma^{-1}\bar{\mathbf{x}}\right\} \\ &\quad \times \exp\left\{-\frac{1}{2}\left(N+\frac{1}{\sigma^2}\right)\left(\boldsymbol{\mu}-\frac{N\bar{\mathbf{x}}}{N+1/\sigma^2}\right)'\Sigma^{-1}\left(\boldsymbol{\mu}-\frac{N\bar{\mathbf{x}}}{N+1/\sigma^2}\right)\right\}. \end{aligned} \tag{5.47}$$

From above the marginal probability density function of X is

$$(2\pi)^{-Np/2}(\det \Sigma)^{-N/2}(1 + N\sigma^2)^{-p/2}\exp\{-(N/2)\bar{\mathbf{x}}'\Sigma^{-1}\bar{\mathbf{x}}(N\sigma^2+1)^{-1}\}. \tag{5.48}$$

From (5.47) and (5.48) the posterior probability density function of $\boldsymbol{\mu}$, given $X = x$, is a p-variate normal with mean $N(N + 1/\sigma^2)^{-1}\bar{\mathbf{x}}$ and covariance matrix $(N + 1/\sigma^2)^{-1}\Sigma$. The Bayes risk of $N(N + 1/\sigma^2)^{-1}\bar{\mathbf{X}}$ with respect to the loss function given in (5.43) is

$$\begin{aligned} &E\{(\boldsymbol{\mu} - N(N + 1/\sigma^2)^{-1}\bar{\mathbf{x}})'\Sigma^{-1}(\boldsymbol{\mu} - N(N + 1/\sigma^2)^{-1}\bar{\mathbf{x}})|X = x\} \\ &\quad = E\{\operatorname{tr}\Sigma^{-1}(\boldsymbol{\mu} - N(N + 1/\sigma^2)^{-1}\bar{\mathbf{x}})(\boldsymbol{\mu} - N(N + 1/\sigma^2)^{-1}\bar{\mathbf{x}})'|X = x\} \\ &\quad = p(N + 1/\sigma^2)^{-1}. \end{aligned} \tag{5.49}$$

Thus, although $\bar{\mathbf{X}}$ is not a Bayes estimator of $\boldsymbol{\mu}$ with respect to the prior density $h(\boldsymbol{\mu})$, it is almost Bayes in the sense that the Bayes estimators $N(N + 1/\sigma^2)^{-1}\bar{\mathbf{X}}$, which are Bayes with respect to the prior density $h(\boldsymbol{\mu})$ [with the loss function as given in (5.43)], tend to $\bar{\mathbf{X}}$ as $\sigma^2 \to \infty$. Furthermore, since $N(N + 1/\sigma^2)^{-1}\bar{\mathbf{X}}$ is Bayes with respect to the prior density $h(\boldsymbol{\mu})$, we obtain

$$\inf_{d \in D} E_{h(\boldsymbol{\mu})}(R(\boldsymbol{\mu}, d)) = E_{h(\boldsymbol{\mu})}R(\boldsymbol{\mu}, N(N + 1/\sigma^2)^{-1}\bar{\mathbf{X}}) = p(N + 1/\sigma^2)^{-1}.$$

To show that $\bar{\mathbf{X}}$ is extended Bayes we first compute

$$E_{h(\boldsymbol{\mu})}(R(\boldsymbol{\mu}, \bar{\mathbf{X}})) = p/N.$$

Hence

$$E_{h(\boldsymbol{\mu})}R(\boldsymbol{\mu}, \bar{\mathbf{X}}) = \inf_{d \in D} E_{h(\boldsymbol{\mu})}(R(\boldsymbol{\mu}, d)) + \varepsilon$$

where $\varepsilon = p/N(N\sigma^2 + 1) > 0$. Thus $\bar{\mathbf{X}}$ is ε-Bayes for every $\varepsilon > 0$.

Also, $\bar{\mathbf{X}}$ has constant risk and hence, by Theorem 5.3.1, $\bar{\mathbf{X}}$ is minimax for estimating $\boldsymbol{\mu}$.

5.3.1 Admissibility of Estimators

An estimator $d_1 \in D$ is said to be as good as $d_2 \in D$ for estimating $\boldsymbol{\theta} \in \Omega$ if

$$R(\boldsymbol{\theta}, d_1) \leq R(\boldsymbol{\theta}, d_2)$$

for all $\boldsymbol{\theta} \in \Omega$.

An estimator $d_1 \in D$ is said to be better than or strictly dominates $d_2 \in D$ if

$$R(\boldsymbol{\theta}, d_1) \leq R(\boldsymbol{\theta}, d_2)$$

for all $\boldsymbol{\theta} \in \Omega$ with strict inequality for at least one $\boldsymbol{\theta} \in \Omega$.

DEFINITION 5.3.4 *Admissible estimator* An estimator $d^* \in D$ which is not dominated by any other estimator in D is called admissible.

For further study on minimax and admissible estimators the reader is referred to Zacks (1971) Stein and James (1961), Lehmann (1959) and Ferguson (1967). It is well known that if the dimension p of the normal random vector is unity and the variance is known, the sample mean is admissible and minimax for the population mean with the squared error loss function. As we have seen earlier for general p with known covariance matrix the sample mean $\bar{\mathbf{X}}$ is minimax for the population mean with the quadratic error loss function. We would then suspect that for general p the sample mean $\bar{\mathbf{X}}$ is an admissible estimator for the population mean $\boldsymbol{\mu}$ with the quadratic error loss function. However, Stein (1956) has shown that with the square error loss function and $\Sigma = I$ (identity matrix), $\bar{\mathbf{X}}$ is admissible for $p = 2$ and it becomes inadmissible for $p \geq 3$. An estimator that improves on $\bar{\mathbf{X}}$ for all values of $\boldsymbol{\mu}$ in this case is given by

$$\left(1 - \frac{p-2}{\bar{\mathbf{X}}'\bar{\mathbf{X}}}\right)\bar{\mathbf{X}}. \tag{5.50}$$

Stein and James (1961) have shown that even with one observation $\mathbf{x}$ on the random vector $\mathbf{X}$ having a p-variate normal distribution with mean $\boldsymbol{\mu}$

and covariance matrix I, the estimator

$$\left(1 - \frac{p-2}{\mathbf{X}'\mathbf{X}}\right)\mathbf{X} \tag{5.51}$$

dominates $\mathbf{X}$ for all $\boldsymbol{\mu}$ for $p \geq 3$. Actually this estimator is a special case of the more general estimator

$$\left(1 - \frac{p-2}{\mathbf{X}'\Sigma^{-1}\mathbf{X}}\right)\mathbf{X} \tag{5.52}$$

where $\mathbf{X}$ has normal distribution with mean $\boldsymbol{\mu}$ and known positive definite covariance matrix Σ. We now add Stein's proof of inadmissibility of one observation for the mean vector $\boldsymbol{\mu}$ from the p-dimensional normal distribution ($p \geq 3$) with mean $\boldsymbol{\mu}$ and known covariance matrix I under the squared error loss function.

Let

$$d_1 = \mathbf{X}, \qquad d_2 = \left(1 - \frac{p-2}{\mathbf{X}'\mathbf{X}}\right)\mathbf{X}.$$

Now

$$R(\boldsymbol{\mu}, d_1) - R(\boldsymbol{\mu}, d_2) = E_{\boldsymbol{\mu}}\left\{\left(2 - \frac{p-2}{\mathbf{X}'\mathbf{X}}\right)(p-2) - 2\frac{(p-2)\mathbf{X}'\boldsymbol{\mu}}{\mathbf{X}'\mathbf{X}}\right\}. \tag{5.53}$$

To achieve our goal we need the following lemma.

LEMMA 5.3.1 Let $\mathbf{X} = (X_1, \ldots, X_p)'$ be normally distributed with mean $\boldsymbol{\mu}$ and covariance matrix I. Then

$$\text{(a)} \quad E_{\boldsymbol{\mu}}\left(\frac{\boldsymbol{\mu}'\mathbf{X}}{\mathbf{X}'\mathbf{X}}\right) = E_{\delta^2}\left(\frac{2\lambda}{p-2+2\lambda}\right) \tag{5.54}$$

$$\text{(b)} \quad E_{\boldsymbol{\mu}}\left(\frac{p-2}{\mathbf{X}'\mathbf{X}}\right) = E_{\delta^2}\left(\frac{p-2}{p-2+2\lambda}\right) \tag{5.55}$$

where $\delta^2 = \boldsymbol{\mu}'\boldsymbol{\mu}$ and λ is a Poisson random variable with parameter $\frac{1}{2}\delta^2$.

Proof (a) Let $\mathbf{Y} = O\mathbf{X}$ where O is an orthogonal $p \times p$ matrix such that $Y_1 = \boldsymbol{\mu}'\mathbf{X}/\delta$. It follows that

$$E\left(\frac{\boldsymbol{\mu}'\mathbf{X}}{\mathbf{X}'\mathbf{X}}\right) = \delta E\left(\frac{Y_1}{\mathbf{Y}'\mathbf{Y}}\right), \tag{5.56}$$

where $\mathbf{Y} = (Y_1, \ldots, Y_p)'$ and $Y_1, \ldots, Y_p$ are independently distributed normal random variables with unit variance and $E(Y_1) = \delta$, $E(Y_i) = 0$, $i > 1$.

The conditional probability density function of Y_1, given $\mathbf{Y}'\mathbf{Y} = v$, is given by

$$f_{Y_1|\mathbf{Y}'\mathbf{Y}}(y_1|v) = \begin{cases} \dfrac{K\exp\{\delta y_1\}(v - y_1{}^2)^{(p-1)/2-1}}{\displaystyle\sum_{j=0}^{\infty} (\delta^2/2)^j \frac{v^{(p+2j-2)/2}}{j!2^{(p+2j)/2}\Gamma((p+2j)/2)}} & \text{if} \quad y_1{}^2 \le v, \\ 0 \qquad \text{otherwise,} \end{cases} \tag{5.57}$$

where K is the normalizing constant independent of $\boldsymbol{\mu}$. From (5.57) we get

$$\int_{y_1{}^2 \le v} \exp\{\delta y_1\}(v - y_1{}^2)^{(p-1)/2}\, dy_1 = \sum_{j=0}^{\infty} \frac{(\delta^2/2)^j v^{(p+2j)/2-1}}{j!2^{(p+2)/2}\Gamma((p+2j)/2)} \tag{5.58}$$

identically in $\boldsymbol{\mu} \in \Omega$. Differentiating (5.58) with respect to δ

$$E(Y_1|\mathbf{Y}'\mathbf{Y} = v) = \frac{\displaystyle\delta \sum_{j=0}^{\infty} \frac{(\delta^2/2)^{j-1} v^{(p+2j)/2-1}}{(j-1)!2^{(p+2j)/2}\Gamma((p+2j)/2)}}{\displaystyle\sum_{j=0}^{\infty} \frac{(\delta^2/2)^j v^{(p+2j)/2-1}}{j!2^{(p+2j)/2}\Gamma((p+2j)/2)}}. \tag{5.59}$$

The probability density function of $\mathbf{Y}'\mathbf{Y}$ is given by

$$f_{\mathbf{Y}'\mathbf{Y}}(v) = \exp\{-\tfrac{1}{2}\delta^2\} \sum_{j=0}^{\infty} \frac{(\delta^2/2)^j}{j!} \frac{e^{-v/2} v^{-(p+2j)/2-1}}{2^{(p+2j)/2}\Gamma((p+2j)/2)} \tag{5.60}$$

which is gamma $G(\frac{1}{2}, p/2 + \lambda)$ where λ is a Poisson random variable with parameter $\frac{1}{2}\delta^2$. From (5.59) and (5.60) we obtain

$$\begin{aligned} E\left(\frac{Y_1}{\mathbf{Y}'\mathbf{Y}}\right) &= E\left[\{E(Y_1|\mathbf{Y}'\mathbf{Y} = v)\}\frac{1}{v}\right] \\ &= \delta \exp\left(-\frac{1}{2}\delta^2\right) \sum_{j=0}^{\infty} \frac{(\delta^2/2)^j}{(j-1)!} \int_0^{\infty} \frac{e^{-v/2} v^{(p+2j)/2-1}\, dv}{2^{(p+2j)/2}\Gamma((p+2j)/2)} \\ &= \delta \exp\left\{-\frac{1}{2}\delta^2\right\} \sum_{j=0}^{\infty} \frac{(\delta^2/2)^{j-1}}{(j-1)!} \frac{1}{p+2j-2}. \end{aligned} \tag{5.61}$$

Hence

$$\begin{aligned} E_{\boldsymbol{\mu}}\left(\frac{\boldsymbol{\mu}'\mathbf{X}}{\mathbf{X}'\mathbf{X}}\right) &= \delta E_{\boldsymbol{\mu}}\left(\frac{Y_1}{\mathbf{Y}'\mathbf{Y}}\right) = \exp(-\tfrac{1}{2}\delta^2) \sum_{j=0}^{\infty} \frac{(\delta^2/2)^j}{j!} \frac{2j}{p-2+2j} \\ &= E_{\delta^2}\left(\frac{2\lambda}{p-2+2\lambda}\right). \end{aligned}$$

(b) Since $\mathbf{X}'\mathbf{X}$ is distributed as gamma $G(\frac{1}{2}, \frac{1}{2} + \lambda)$, where λ is a Poisson random variable with mean $\frac{1}{2}\delta^2$, we can easily show as in (a) that

$$E\left(\frac{p-2}{\mathbf{X}'\mathbf{X}}\right) = E_\delta\left(\frac{p-2}{p-2+2\lambda}\right). \qquad \text{Q.E.D.}$$

From (5.53) and Lemma 5.3.1 we get

$$R(\boldsymbol{\mu}, d_1) - R(\boldsymbol{\mu}, d_2) = (p-2)^2 E\left(\frac{1}{p-2+2\lambda}\right) > 0$$

if $p \geq 3$. In other words, $\bar{\mathbf{X}}$ is inadmissible for $\boldsymbol{\mu}$ for $p \geq 3$.

TABLE 5.1

Plant	X_1		X_2		X_3		X_4		X_5		X_6	
No.	1971	1972	1971	1972	1971	1972	1971	1972	1971	1972	1971	1972
1	82.85	74.35	150	162	8.97	9.76	12.6	12.2	261	337	11.8	13.7
2	79.10	66.05	163	145	10.19	10.10	13.1	12.5	320	351	14.3	13.9
3	86.95	80.30	181	156	9.63	10.71	13.5	13.8	339	424	15.4	17.7
4	83.31	77.60	205	148	9.47	10.75	13.8	13.0	287	379	12.7	17.3
5	88.90	80.45	187	142	9.59	9.56	13.3	12.4	308	327	14.3	13.8
6	83.10	81.00	182	200	9.19	10.48	12.8	13.9	314	378	13.9	15.7
7	89.50	85.05	152	163	9.60	10.90	13.5	13.3	311	367	13.5	16.3
8	86.50	80.75	188	170	9.30	10.65	12.5	13.0	281	372	12.9	15.1
9	87.30	80.95	170	165	9.00	10.57	12.7	13.8	264	357	11.8	14.6
10	88.75	64.40	193	142	9.78	10.21	13.4	12.2	293	352	14.0	14.8
11	84.60	75.90	188	157	10.43	10.79	14.2	13.6	346	357	16.7	13.8
12	83.60	69.00	164	170	9.58	8.61	13.9	9.8	290	258	13.0	9.4
13	86.60	82.25	193	156	10.43	11.06	15.3	13.8	336	404	15.7	17.5
14	84.55	80.75	200	156	9.07	11.14	12.6	14.7	237	412	10.4	17.1
15	87.95	82.25	202	164	9.31	10.30	14.4	13.3	287	390	11.7	17.2
16	85.50	79.55	225	174	10.32	10.75	12.9	13.4	355	400	16.5	16.9
17	86.30	81.90	184	163	9.50	10.75	12.7	13.4	300	355	13.4	16.5
18	86.10	83.55	198	182	9.73	11.43	13.7	14.3	295	406	12.8	15.0
19	81.80	65.45	203	147	10.41	9.55	13.9	11.3	314	300	13.1	12.0
20	75.20	68.00	185	156	10.10	9.88	13.4	11.7	320	330	15.9	13.2
21	78.60	66.85	174	194	9.77	9.56	12.6	11.9	310	304	13.8	9.3
22	85.20	81.45	159	192	9.92	11.12	13.5	14.2	286	384	12.1	17.8
23	81.05	75.65	189	191	9.74	10.93	15.0	13.7	307	380	13.6	13.0
24	86.65	77.30	198	170	10.22	11.09	14.0	14.1	324	404	13.4	16.3
25	89.30	81.35	212	186	9.90	10.41	13.5	13.3	323	340	14.4	12.5
26	84.50	79.45	173	165	9.86	10.79	13.0	13.6	282	384	12.2	14.4
27	88.30	81.35	212	198	10.08	10.53	13.6	13.4	328	310	15.0	13.8

For a random sample $\mathbf{X}^\alpha = (X_{\alpha 1}, \ldots, X_{\alpha p})'$, $\alpha = 1, \ldots, N$, from the p-dimensional normal distribution with mean $\boldsymbol{\mu}$ and positive definite covariance matrix Σ and with the loss function

$$L(\boldsymbol{\mu}, d) = (\boldsymbol{\mu} - d)'\Sigma^{-1}(\boldsymbol{\mu} - d)$$

the estimator

$$\left(1 - \frac{C}{\bar{\mathbf{X}}'S^{-1}\bar{\mathbf{X}}}\right)\bar{\mathbf{X}},$$

where $C = (p - 2)/(N - p - 2)$, will dominate $\bar{\mathbf{X}}$. Since it is possible to make

$$\bar{\mathbf{X}}'S^{-1}\bar{\mathbf{X}} \leq C,$$

even if all the components of $\boldsymbol{\mu}$ are positive, it is not unreasonable to try to improve upon

$$\left(1 - \frac{C}{\bar{\mathbf{X}}'S^{-1}\bar{\mathbf{X}}}\right)\bar{\mathbf{X}}$$

by using the estimator

$$\max\left(0, 1 - \frac{C}{\bar{\mathbf{X}}'S^{-1}\bar{\mathbf{X}}}\right)\bar{\mathbf{X}}$$

(see Stein, 1962).

The problem of the determination of the confidence region for $\boldsymbol{\mu}$ is discussed in Chapter VI.

EXAMPLE 5.3.1 Observations were made in the Indian Agricultural Research Institute, New Delhi, India, on six different characters:

X_1 plant height at harvesting (cm)
X_2 number of effective tillers
X_3 length of ear (cm)
X_4 number of fertile spikelets per 10 ears
X_5 number of grains per 10 ears
X_6 weight of grains per 10 ears

for 27 randomly selected plants of Sonalika, a late-sown variety of wheat in two consecutive years (1971, 1972). The observations are recorded in Table 5.1.

Assuming that each year's data constitute a sample from a six-variate normal distribution with mean $\boldsymbol{\mu}$ and covariance matrix Σ, we obtain the following maximum likelihood estimates.

For 1971

(i)

$$\hat{\boldsymbol{\mu}} = \begin{pmatrix} 84.8911 \\ 186.2963 \\ 9.7411 \\ 13.4593 \\ 304.3701 \\ 13.6259 \end{pmatrix}$$

(ii) $\hat{\Sigma} = s/27$

	X_1	X_2	X_3	X_4	X_5	X_6
X_1	705.10					
X_2	36.73	2.279				
X_3	−1.34	−0.510	11.79			
X_4	164.40	7.715	13.85	340.20		
X_5	8.775	0.446	−0.235	2.60	0.184	
X_6	7.333	0.301	0.202	3.16	0.149	0.500

(iii) The matrix of sample correlation coefficients $R = (r_{ij})$ is

	X_1	X_2	X_3	X_4	X_5	X_6
X_1	1.00000	0.91658	−0.01494	0.33562	0.77012	0.39081
X_2		1.00000	−0.09888	0.27697	0.68804	0.28188
X_3			1.00000	0.21875	−0.15991	0.08276
X_4				1.00000	0.32884	0.24215
X_5					1.00000	0.49069
X_6						1.00000

(iv) The maximum likelihood estimate of the regression of X_6 on $X_1 = x_1, \ldots, X_5 = x_5$ is

$$\hat{E}(X_6|X_1 = x_1, \ldots, X_5 = x_5) = 3.39768 - 0.03721x_1 - 0.00008x_2 - 0.14427x_3 - 0.15360x_4 + 0.05544x_5.$$

(v) The maximum likelihood estimate of the square of the multiple correlation coefficient of X_6 on $(X_1, \ldots, X_5)$ is

$$r^2 = 0.85358.$$

(vi) The maximum likelihood estimates of some partial correlation coefficients are

$r_{23.5} = 0.0156$ $\quad r_{23.45} = -0.0063$ $\quad r_{23.14} = -0.2000$

$r_{23.1} = -0.2130$ $\quad r_{23.46} = -0.1823$ $\quad r_{23.56} = 0.0328$

$r_{23.15} = -0.2363 \quad r_{23.145} = -0.2252 \quad r_{23.156} = -0.2074$

$r_{23.16} = -0.1982 \quad r_{23.146} = -0.1906 \quad r_{23.1456} = -0.1999$

$r_{23.456} = 0.0100$

For 1972

(i)

$$\hat{\boldsymbol{\mu}} = \begin{pmatrix} 77.1444 \\ 167.1852 \\ 10.4585 \\ 13.0963 \\ 361.5553 \\ 14.7630 \end{pmatrix}$$

(ii) $\hat{\Sigma} = s/27$

	X_1	X_2	X_3	X_4	X_5	X_6
X_1	1496					
X_2	73.83	5.122				
X_3	136.50	8.695	37.39			
X_4	−11.33	−6.438	31.65	288.7		
X_5	20.67	1.050	2.508	2.026	0.390	
X_6	34.26	1.714	4.720	4.817	0.609	1.097

(iii) The matrix of sample correlation coefficients $R = (r_{ij})$ is

	X_1	X_2	X_3	X_4	X_5	X_6
X_1	1.00000	0.84339	0.57699	−0.01729	0.85654	0.84636
X_2		1.00000	0.62820	−0.16753	0.74304	0.72303
X_3			1.00000	0.30469	0.65693	0.73694
X_4				1.00000	0.19090	0.27053
X_5					1.00000	0.93191
X_6						1.00000

(iv) The maximum likelihood estimate of the regression of X_6 on $X_1 = x_1, \ldots, X_5 = x_5$ is

$$\hat{E}(X_6 | X_1 = x_1, \ldots, X_5 = x_5) = -4.82662 + 0.12636x_1 - 0.03436x_2 + 0.61897x_3 - 0.28526x_4 + 0.03553x_5.$$

(v) The maximum likelihood estimate of the square of the multiple correlation coefficient of X_6 on $(X_1, \ldots, X_5)$ is

$$r^2 = 0.80141.$$

(vi) The maximum likelihood estimates of some of the partial correlation coefficients are

$$
\begin{array}{lll}
r_{23.4} = 0.7234 & r_{23.46} = 0.3539 & r_{23.456} = 0.3861 \\
r_{23.5} = 0.2776 & r_{23.14} = 0.4887 & r_{23.145} = 0.4578 \\
r_{23.6} = 0.2042 & r_{23.56} = 0.2494 & r_{23.146} = 0.4439 \\
r_{23.1} = 0.3226 & r_{23.15} = 0.3194 & r_{23.156} = 0.3795 \\
r_{23.45} = 0.4576 & r_{23.16} = 0.3709 & r_{23.1456} = 0.4532
\end{array}
$$

EXERCISES

1. The data in Table 5.2 were collected in an experiment on jute in a village of West Bengal, India, in which the weights of green jute plants (X_2) and their dry jute fibers (X_1) were recorded for 20 randomly selected individual plants. Assume that $\mathbf{X} = (X_1, X_2)'$ is normally distributed with mean $\boldsymbol{\mu} = (\mu_1, \mu_2)'$ and positive definite covariance matrix Σ.

TABLE 5.2

Plant No.	Weight (gm) X_1	Weight (gm) X_2	Plant No.	Weight (gm) X_1	Weight (gm) X_2
1	68	971	11	33	462
2	63	892	12	27	352
3	70	1125	13	21	305
4	6	82	14	5	84
5	65	931	15	14	229
6	9	112	16	27	332
7	10	162	17	17	185
8	12	321	18	53	703
9	20	315	19	62	872
10	30	375	20	65	740

(a) Find maximum likelihood estimates of $\boldsymbol{\mu}$, Σ.

(b) Find the maximum likelihood estimate of the coefficient of correlation ρ between the components.

(c) Find the maximum likelihood estimate of $E(X_1|X_2 = x_2)$.

2. The variability in the price of farmland per acre is to be studied in relation to three factors which are assumed to have major influence in determining the selling price. For 20 randomly selected farms, the price (in dollars) per acre (X_1), the depreciated cost (in dollars) of building per acre (X_2), and the distance to the nearest shopping center (in miles) (X_3) are recorded in Table 5.3. Assuming that $\mathbf{X} = (X_1, X_2, X_3)'$ has three-variate normal distribution, find the maximum likelihood estimates of the following:

(a) $E(X_1|X_2 = x_2, X_3 = x_3)$;

(b) the partial correlation coefficient between X_1 and X_3 when X_2 is kept fixed;

(c) the multiple correlation coefficient between X_1 and (X_2, X_3).

TABLE 5.3

Farm	X_1	X_2	X_3	Farm	X_1	X_2	X_3
1	75	15	6.0	11	135	13	0.5
2	156	6	2.5	12	175	12	2.5
3	145	60	0.5	13	240	7	2.0
4	175	24	3.0	14	175	27	4.0
5	70	5	2.0	15	197	16	6.0
6	179	8	1.5	16	125	6	5.0
7	165	14	4.0	17	227	13	5.0
8	134	13	4.0	18	172	13	11.0
9	137	7	1.5	19	170	34	2.0
10	175	19	2.5	20	172	19	6.5

3. Let $X^\alpha = (X_{\alpha 1}, \ldots, X_{\alpha p})'$, $\alpha = 1, \ldots, N$, be a random sample of size N from a p-variate normal distribution with mean $\boldsymbol{\mu}$ and positive definite covariance matrix Σ. Show that the distribution of $\bar{\mathbf{X}} = (1/N)\sum_{\alpha=1}^{N} \mathbf{X}^\alpha$ is complete for given Σ.

4. Prove the equivalence of the three criteria of stochastic convergence of a random matrix as given in (5.7).

5. Let $\mathbf{X}^\alpha = (X_{\alpha 1}, \ldots, X_{\alpha p})'$, $\alpha = 1, \ldots, N$, be a random sample of size N from a p-dimensional normal distribution with mean $\boldsymbol{\mu}$ positive definite covariance matrix Σ.

(a) Let $\boldsymbol{\mu} = (\mu, \ldots, \mu)'$,

$$\Sigma = \begin{pmatrix} 1 & \rho & \rho & \cdots & \rho \\ \rho & 1 & \rho & \cdots & \rho \\ \vdots & \vdots & \vdots & & \vdots \\ \rho & \rho & \rho & \cdots & 1 \end{pmatrix} \sigma^2,$$

with $-1/(p-1) < \rho < 1$. Find the maximum likelihood estimators of ρ, σ^2, and $\boldsymbol{\mu}$.

(b) Let $\boldsymbol{\mu} = (\mu_1, \ldots, \mu_p)'$,

$$\Sigma = \begin{pmatrix} \sigma_1^2 & \rho\sigma_1\sigma_2 & \rho\sigma_1\sigma_3 & \cdots & \rho\sigma_1\sigma_p \\ \rho\sigma_1\sigma_2 & \sigma_2^2 & \rho\sigma_2\sigma_3 & \cdots & \rho\sigma_2\sigma_p \\ \vdots & \vdots & \vdots & & \vdots \\ \rho\sigma_1\sigma_p & \rho\sigma_2\sigma_p & \rho\sigma_3\sigma_p & \cdots & \sigma_p^2 \end{pmatrix} \quad \text{with} \quad -\frac{1}{p-1} < \rho < 1.$$

Find the maximum likelihood estimators of $\boldsymbol{\mu}, \rho, \sigma_1^2, \ldots, \sigma_p^2$.

6. Find the maximum likelihood estimator of the parameters of the multivariate log-normal distribution and of the multivariate Student t-distribution as defined in Exercise 4.

7. Let $\mathbf{Y} = (Y_1, \ldots, Y_N)'$ be normally distributed with

$$E(\mathbf{Y}) = X\boldsymbol{\beta}, \qquad \operatorname{cov}(\mathbf{Y}) = \sigma^2 I$$

where $X = (x_{ij})$ is an $N \times p$ matrix of known constants x_{ij}, and $\boldsymbol{\beta} = (\beta_1, \ldots, \beta_p)'$, σ^2 are unknown constants.

(a) Let the rank of X be p. Find the maximum likelihood estimators $\hat{\boldsymbol{\beta}}$, $\hat{\sigma}^2$ of $\boldsymbol{\beta}$, σ^2. Show that $\hat{\boldsymbol{\beta}}$, $\hat{\sigma}^2$ are stochastically independent and $N\hat{\sigma}^2/\sigma^2$ is distributed as chi-square with $N - p$ degrees of freedom.

(b) A linear parametric function $\mathbf{L}'\boldsymbol{\beta}$, $\mathbf{L} = (l_1, \ldots, l_p)' \neq \mathbf{0}$, is called estimable if there exists a linear estimator $\mathbf{b}'\mathbf{Y}$, $\mathbf{b} = (b_1, \ldots, b_N)' \neq \mathbf{0}$, such that

$$E(\mathbf{b}'\mathbf{Y}) = \mathbf{L}'\boldsymbol{\beta}.$$

Let the rank of X be less than p and let the linear parametric function $\mathbf{L}'\boldsymbol{\beta}$ be estimable. Find the unique minimum variance linear unbiased estimator of $\mathbf{L}'\boldsymbol{\beta}$.

8. [*Inverted Wishart distribution*—$W_p^{-1}(A, N)$] A $p \times p$ symmetric random matrix V has an inverted Wishart distribution with parameter V (symmetric positive definite matrix) and with N degrees of freedom if its probability density function is given by

$$c(\det A)^{(N-p-1)/2}(\det V^{-1})^{N/2} \exp\{-\tfrac{1}{2} \operatorname{tr} V^{-1}A\}$$

where

$$c^{-1} = 2^{(N-p-1)p/2}\Pi^{p(p-1)/4} \prod_{i=1}^{p} (N - p - i)/2,$$

provided $2p < N$ and V is positive definite, and is zero otherwise.

(a) Show that if a $p \times p$ random matrix S has a Wishart distribution as given in (5.2), then S^{-1} has an inverted Wishart distribution with parameters Σ^{-1} and with $N + p$ degrees of freedom.

(b) Show that $E(S^{-1}) = \Sigma^{-1}/(N - p - 1)$.

9. Let $\mathbf{X}^\alpha = (X_{\alpha 1}, \ldots, X_{\alpha p})'$, $\alpha = 1, \ldots, N_1$, be a random sample of size N_1 from a p-dimensional normal distribution with mean $\boldsymbol{\mu} = (\mu_1, \ldots, \mu_p)'$ and positive definite covariance matrix Σ, and let $\mathbf{Y}^\alpha = (Y_{\alpha 1}, \ldots, Y_{\alpha p})'$, $\alpha = 1, \ldots, N_2$, be a random sample of size N_2 (independent of $\mathbf{X}^\alpha$, $\alpha = 1, \ldots, N_1$) from a normal distribution with mean $\mathbf{v} = (v_1, \ldots, v_p)'$ and the same covariance matrix Σ.

(a) Find the maximum likelihood estimators $\hat{\boldsymbol{\mu}}$, $\hat{\mathbf{v}}$, $\hat{\Sigma}$ of $\boldsymbol{\mu}$, $\mathbf{v}$, and Σ, respectively.

(b) Show that $\hat{\boldsymbol{\mu}}, \hat{\mathbf{v}}, \hat{\Sigma}$ are stochastically independent and that $(N_1 + N_2)\hat{\Sigma}$ is distributed as $\sum_{\alpha=1}^{N_1+N_2-2} \mathbf{Z}^\alpha \mathbf{Z}^{\alpha'}$, where $\mathbf{Z}^\alpha = (Z_{\alpha 1}, \ldots, Z_{\alpha p})'$, $\alpha = 1, \ldots, N_1 + N_2 - 2$, are independently distributed p-variate normal random variables with mean $\mathbf{0}$ and the same covariance matrix Σ.

10. [*Giri, 1965; Goodman, 1963*] Let $\boldsymbol{\xi}^\beta = (\xi_{\beta 1}, \ldots, \xi_{\beta p})'$, $\beta = 1, \ldots, N$, be N independent and identically distributed p-variate complex Gaussian random variables with the same mean $E(\boldsymbol{\xi}^\beta) = \boldsymbol{\alpha}$ and with the same Hermitian positive definite complex covariance matrix $\Sigma = E(\boldsymbol{\xi}^\beta - \boldsymbol{\alpha})(\boldsymbol{\xi}^\beta - \boldsymbol{\alpha})^*$, where $(\boldsymbol{\xi}^\beta - \boldsymbol{\alpha})^*$ is the adjoint of $(\boldsymbol{\xi}^\beta - \boldsymbol{\alpha})$.

(a) Show that the maximum likelihood estimators of $\boldsymbol{\alpha}$ and Σ are given by

$$\hat{\alpha} = \frac{1}{N} \sum_{\beta=1}^{N} \xi^\beta = \bar{\boldsymbol{\xi}}, \qquad \hat{\Sigma} = \frac{1}{N} \sum_{\beta=1}^{N} (\boldsymbol{\xi}^\beta - \bar{\boldsymbol{\xi}})(\boldsymbol{\xi}^\beta - \bar{\boldsymbol{\xi}})^*.$$

(b) (i) Show that $\hat{\boldsymbol{\alpha}}$ has a complex p-variate Gaussian distribution with mean $\boldsymbol{\alpha}$ and Hermitian positive definite covariance matrix Σ/N.

(ii) Show that $N\hat{\Sigma}$ is distributed as $\sum_{\beta=1}^{N-1} \boldsymbol{\eta}^\beta \boldsymbol{\eta}^{\beta *}$ where $\boldsymbol{\eta}^\beta = (\eta_{\beta 1}, \ldots, \eta_{\beta p})'$, $\beta = 1, \ldots, N-1$, are independent p-variate complex Gaussian random variables with the same mean $\mathbf{0}$ and Hermitian positive definite covariance matrix Σ.

(iii) Show that $\hat{\boldsymbol{\alpha}}, \hat{\Sigma}$ are stochastically independent.

(c) Show that $(\hat{\boldsymbol{\alpha}}, \hat{\Sigma})$ is sufficient for $(\boldsymbol{\alpha}, \Sigma)$.

REFERENCES

Anderson, T. W. (1958). "An Introduction to Multivariate Analysis." Wiley, New York.

Bahadur, R. R. (1955). Statistics and subfields, *Ann. Math. Statist.* **26**, 490–497.

Bahadur, R. R. (1960). On the asymptotic efficiency of tests and estimators, *Sankhya* **22**, 229–252.

Basu, D. (1955). An inconsistency of the method of maximum likelihood, *Ann. Math. Statist.* **26**, 144–145.

Dykstra, R. L. (1970). Establishing the positive definiteness of the sample covariance matrix, *Ann. Math. Statist.* **41**, 2153–2154.

Eaton, M., and Pearlman, M. (1973). The nonsingularity of generalized sample covariance matrix, *Ann. Statist.* **1**, 710–717.

Ferguson, T. S. (1967). "Mathematical Statistics, A Decision Theoretic Approach." Academic Press, New York.

Fisher, R. A. (1925). Theory of statistical estimation, *Proc. Cambridge Phil. Soc.* **22**, 700–715.

Giri, N. (1965). On the complex analogues of T^2- and R^2-tests, *Ann. Math. Statist.* **36**, 664–670.

Giri, N. (1975). "Introduction to Probability and Statistics," Part 2, Statistics. Dekker, New York.

Goodman, N. R. (1963). Statistical analysis based on a certain multivariate complex Gaussian distribution (an introduction), *Ann. Math. Statist.* **34**, 152–177.

Halmos, P. L., and Savage, L. J. (1949). Application of Radon-Nikodym Theorem of the theory of sufficient statistics, *Ann. Math. Statist.* **20**, 225–241.

Kiefer, J., and Wolfowitz, J. (1956). Consistency of the maximum likelihood estimator in the presence of infinitely many incidental parameters, *Ann. Math. Statist.* **27**, 887–906.

LeCam, L. (1953). On some asymptotic properties of the maximum likelihood estimates and related Bayes estimates, *Univ. California Publ. Statist.* **1**, 277–330.

Lehmann, E. L. (1959). "Testing Statistical Hypotheses." Wiley, New York.

Lehmann, E. L., and Scheffe, H. (1950). Completeness, similar regions and unbiased estimation, part I, *Sankhya* **10**, 305–340.

Neyman, J., and Scott, E. L. (1948). Consistent estimates based on partially consistent observations, *Econometrika* **16**, 1–32.

Pearson, Karl (1896). Mathematical contribution to the theory of evolution III, Regression, heredity and panmixia, *Phil. Trans. A.* **187**, 253–318.

Press, S. J. (1972). "Applied Multivariate Analysis." Holt, New York.

Raiffa, H., and Schlaifer, R. (1961). "Applied Statistical Decision Theory." Harvard Univ. Press, Cambridge, Massachusetts.

Rao, C. R. (1965). "Linear Statistical Inference and its Applications." Wiley, New York.

Stein, C. (1956). Inadmissibility of the usual estimator for the mean of a multivariate normal distribution, *Barkeley symp., Math. Statist. Probability, 3rd* **5**, 196–207.

Stein, C. (1962). Confidence sets for the mean of a multivariate normal distribution, *J. Roy. Statist. Soc. Ser. B* **24**, 265–285.

Stein, C. (1969). Multivariate analysis I (Notes recorded by M. L. Eaton), Tech. Rep. No. 42, Statist. Dept., Stanford Univ., California.

Stein, C., and James, W. (1961). Estimation with quadratic loss, *Barkeley Symp. Math. Statist. I, Probability 2, 4th* 361–379.

Wald, A. (1943). Tests of statistical hypotheses concerning several parameters when the number of observations is large, *Trans. Am. Math. Soc.* **54**, 426–482.

Wolfowitz, J. (1949). On Wald's proof of the consistency of the maximum likelihood estimate, *Ann. Math. Statist.* **20**, 601–602.

Zacks, S. (1971). "The Theory of Statistical Inference." Wiley, New York.

Zehna, P. W. (1966). Invariance of maximum likelihood estimation, *Ann. Math. Statist.* **37**, 755.

CHAPTER VI

Basic Multivariate Sampling Distributions

6.0 INTRODUCTION

This chapter deals with some basic distributions connected with multivariate normal distributions. The distributions of other multivariate test statistics needed for testing hypotheses concerning the parameters of multivariate normal populations will be derived where relevant. For better understanding and future reference we will also describe briefly the noncentral chi-square, noncentral Student's t-, and the noncentral F-distributions. For derivations of these noncentral distributions the reader is referred to Giri (1974).

6.1 NONCENTRAL CHI-SQUARE, STUDENT'S t-, F-DISTRIBUTIONS

6.1.1 Noncentral Chi-Square

Let $X_1, \ldots, X_N$ be independently distributed normal random variables with $E(X_i) = \mu_i$, $\operatorname{var}(X_i) = \sigma_i^2$, $i = 1, \ldots, N$. Then the random variable

$$Z = \sum_{i=1}^{N} \frac{X_i^2}{\sigma_i^2}$$

has the probability density function given by

$$f_Z(z|\delta^2) = \begin{cases} \dfrac{\exp\{-\frac{1}{2}(\delta^2 + z)\} z^{(N/2-1)}}{\sqrt{\pi}\, 2^{N/2}} \displaystyle\sum_{j=0}^{\infty} \frac{(\delta^2)^j z^j \Gamma(j + \frac{1}{2})}{(2j)!\Gamma(N/2 + j)}, & z \geq 0; \\ 0 \quad \text{otherwise,} & \end{cases} \tag{6.1}$$

where $\delta^2 = \sum_{i=1}^{N} (\mu_i^2/\sigma_i^2)$. This is called the noncentral chi-square distribution with N degrees of freedom and with the noncentrality parameter δ^2. The random variable Z is often written as $\chi_N^2(\delta^2)$.

The characteristic function of Z is (t real)

$$\phi_Z(t) = E(e^{itZ}) = (1 - 2it)^{-N/2} \exp\{it\delta^2/(1 - 2it)\} \tag{6.2}$$

with $i = (-1)^{1/2}$. From this it follows that if $Y_1, \ldots, Y_k$ are independently distributed noncentral chi-square random variables $\chi^2_{N_i}(\delta_i^2)$, $i = 1, \ldots, k$, then $\sum_1^k Y_i$ is distributed as $\chi^2_{\Sigma_1^k N_i}(\sum_1^k \delta_i^2)$. Furthermore,

$$E(\chi_N^2(\delta^2)) = N + \delta^2, \qquad \operatorname{var}(\chi_N^2(\delta^2)) = 2N + 4\delta^2. \tag{6.3}$$

Since for any integer k

$$\Gamma(2k + 1)\sqrt{\pi} = 2^{2k}\Gamma(k + \tfrac{1}{2})\Gamma(k + 1), \tag{6.4}$$

we can write $f_Z(z|\delta^2)$ as

$$f_Z(z|\delta^2) = \sum_{k=0}^{\infty} p_K(k) f_{\chi^2_{N+2k}}(z) \tag{6.5}$$

where $p_K(k)$ is the probability mass function of the Poisson random variable K with parameter $\frac{1}{2}\delta^2$ and $f_{\chi^2_{N+2k}}(z)$ is the probability density function of the central chi-square random variable with $N + 2k$ degrees of freedom.

6.1.2 Noncentral Student's *t*

Let the random variable X, distributed normally with mean μ and variance σ^2, and the random variable Y such that Y/σ^2 has a chi-square distribution with n degrees of freedom, be independent and let $t = \sqrt{n}\, X/\sqrt{Y}$. The probability density function of t is given by

$$f_t(t|\lambda) = \begin{cases} \dfrac{n^{n/2} \exp\{-\frac{1}{2}\lambda^2\}}{(n + t^2)^{(n+1)/2}} \displaystyle\sum_{j=0}^{\infty} \frac{\Gamma((n + j + 1)/2)\lambda^j}{j!} \left(\frac{2t^2}{n + t^2}\right)^{j/2}, \\ \qquad\qquad -\infty < t < \infty; \\ 0 \quad \text{otherwise,} \end{cases} \tag{6.6}$$

where $\lambda = \mu/\sigma$. The distribution of t is known as the noncentral t-distribution with n degrees of freedom and the noncentrality parameter λ.

6.1.3 Noncentral *F*-Distribution

Let the random variable X, distributed as $\chi_m^{\,2}(\delta^2)$, and the random variable Y, distributed as $\chi_n^{\,2}$, be independent and let

$$F = \frac{n}{m}\,\frac{\chi_m^{\,2}(\delta^2)}{\chi_n^{\,2}}.$$

The distribution F is known as the noncentral F-distribution and its probability density function is given by

$$f_F(z) = \begin{cases} \dfrac{m}{n}\exp(-\tfrac{1}{2}\delta^2) \displaystyle\sum_{j=0}^{\infty} \dfrac{(\delta^2/2)^j \Gamma((m+n)/2+j)((m/n)z)^{m/2+j-1}}{\Gamma(m/2+j)\Gamma(n/2)(1+(m/n)z)^{(m+n)/2+j}}, \\ \qquad\qquad\qquad\qquad\qquad\qquad z \geq 0; \\ 0 \qquad \text{otherwise.} \end{cases} \tag{6.7}$$

6.2 DISTRIBUTION OF QUADRATIC FORMS, COCHRAN'S THEOREM

THEOREM 6.2.1 Let $\mathbf{X} = (X_1, \ldots, X_p)'$ be normally distributed with mean $\boldsymbol{\mu}$ and symmetric positive definite covariance matrix Σ, and let

$$\mathbf{X}'\Sigma^{-1}\mathbf{X} = Q_1 + \cdots + Q_k, \tag{6.8}$$

where $Q_i = \mathbf{X}'A_i\mathbf{X}$ and the rank of A_i is p_i, $i = 1, \ldots, k$. Then the Q_i are independently distributed as noncentral chi-square $\chi_{p_i}^2(\boldsymbol{\mu}'A_i\boldsymbol{\mu})$ with p_i degrees of freedom and the noncentrality parameter $\boldsymbol{\mu}'A_i\boldsymbol{\mu}$ if and only if $\sum_1^k p_i = p$, in which case $\boldsymbol{\mu}'\Sigma^{-1}\boldsymbol{\mu} = \sum_1^k \boldsymbol{\mu}'A_i\boldsymbol{\mu}$.

Proof Since Σ is symmetric and positive definite there exists a nonsingular matrix C such that $\Sigma = CC'$. Let $\mathbf{Y} = C^{-1}\mathbf{X}$. Obviously $\mathbf{Y}$ has a p-variate normal distribution with mean $\mathbf{v} = C^{-1}\boldsymbol{\mu}$ and covariance matrix I (identity matrix). From (6.8) we get

$$\mathbf{Y}'\mathbf{Y} = \mathbf{Y}'B_1\mathbf{Y} + \cdots + \mathbf{Y}'B_k\mathbf{Y}, \tag{6.9}$$

where $B_i = C'A_iC$. Since C is nonsingular, $\text{rank}(A_i) = \text{rank}(B_i)$, $i = 1, \ldots, k$.

Obviously the theorem will be proved if we show that $\mathbf{Y}'B_i\mathbf{Y}$, $i = 1, \ldots, k$, are independently distributed noncentral chi-squares $\chi_{p_i}^2(\mathbf{v}'B_i\mathbf{v})$ if and only if $\sum_1^k p_i = p$, in which case $\mathbf{v}'\mathbf{v} = \sum_{i=1}^k \mathbf{v}'B_i\mathbf{v}$.

Let us suppose that $\mathbf{Y}'B_i\mathbf{Y}$, $i = 1, \ldots, k$, are independently distributed as $\chi_{p_i}^2(\mathbf{v}'B_i\mathbf{v})$. Then $\sum_{i=1}^k \mathbf{Y}'B_i\mathbf{Y}$ is distributed as noncentral chi-square $\chi_{\Sigma_1^k p_i}^2(\sum_{i=1}^k \mathbf{v}'B_i\mathbf{v})$. Since $\mathbf{Y}'\mathbf{Y}$ is distributed as $\chi_p^{\,2}(\mathbf{v}'\mathbf{v})$ and (6.9) holds, it follows from the uniqueness of the characteristic function that $\sum_1^k p_i = p$ and $\mathbf{v}'\mathbf{v} = \sum_{i=1}^k \mathbf{v}'B_i\mathbf{v}$, which proves the necessity part of the theorem.

To prove the sufficiency part of the theorem let us assume that $\sum_1^k p_i = p$. Since Q_i is a quadratic form in $\mathbf{Y}$ of rank p_i (rank of B_i) by Theorem 1.5.8, Q_i can be expressed as

$$Q_i = \sum_{j=1}^{p_i} \pm Z_{ij}^2 \tag{6.10}$$

where the Z_{ij} are linear functions of $Y_1, \ldots, Y_p$. Let

$$\mathbf{Z} = (Z_{11}, \ldots, Z_{1p_1}, \ldots, Z_{k1}, \ldots, Z_{kp_k})'$$

be a vector of dimension $\sum_1^k p_i = p$. Then

$$\mathbf{Y}'\mathbf{Y} = \sum_1^k Q_i = \mathbf{Z}'\Delta\mathbf{Z}, \tag{6.11}$$

where Δ is a diagonal matrix of dimension $p \times p$ with diagonal elements $+1$ or -1. Let $\mathbf{Z} = A\mathbf{Y}$ be the linear transformation which transforms the positive definite quadratic form $\mathbf{Y}'\mathbf{Y}$ to $\mathbf{Z}'\Delta\mathbf{Z}$. Since

$$\mathbf{Y}'\mathbf{Y} = \mathbf{Z}'\Delta\mathbf{Z} = \mathbf{Y}'A'\Delta A\mathbf{Y} \tag{6.12}$$

for all values of $\mathbf{Y}$ we conclude that $A'\Delta A = I$. In other words, A is nonsingular. Thus $\mathbf{Z}'\Delta\mathbf{Z}$ is positive definite and hence $\Delta = I$, $A'A = I$. Since A is orthogonal and $\mathbf{Y}$ has p-variate normal distribution with mean $\mathbf{v}$ and covariance matrix I, the components of $\mathbf{Z}$ are independently normally distributed with unit variance. So Q_i $(i = 1, \ldots, k)$ are independently distributed chi-square random variables with p_i degrees of freedom and noncentrality parameter $\mathbf{v}'B_i\mathbf{v}$, $i = 1, \ldots, k$ (see Exercise 6.1). But $\mathbf{Y}'\mathbf{Y}$ is distributed as $\chi_p^2(\mathbf{v}'\mathbf{v})$. Therefore

$$\mathbf{v}'\mathbf{v} = \sum_1^k \mathbf{v}'B_i\mathbf{v}. \qquad \text{Q.E.D.}$$

THEOREM 6.2.2 Let $\mathbf{X} = (X_1, \ldots, X_p)'$ be normally distributed with mean $\boldsymbol{\mu}$ and positive definite covariance matrix Σ. Then $\mathbf{X}'A\mathbf{X}$ is distributed as a noncentral chi-square with k degrees of freedom if and only if ΣA is an idempotent matrix of rank k, i.e., $A\Sigma A = A$.

Proof Since Σ is positive definite there exists a nonsingular matrix C such that $\Sigma = CC'$. Let $\mathbf{X} = C\mathbf{Y}$. Then $\mathbf{Y}$ has a p-variate normal distribution with mean $\mathbf{v} = C^{-1}\boldsymbol{\mu}$ and covariance matrix I, and

$$\mathbf{X}'A\mathbf{X} = \mathbf{Y}'B\mathbf{Y}, \tag{6.13}$$

where $B = C'AC$ and $\text{rank}(A) = \text{rank}(B)$. The theorem will now be proved if we show that $\mathbf{Y}'B\mathbf{Y}$ has a noncentral chi-square distribution $\chi_k^2(\mathbf{v}'B\mathbf{v})$ if and only if B is an idempotent matrix of rank k.

Let us assume that B is an idempotent matrix of rank k. Then there exists an orthogonal matrix θ such that $\theta B \theta'$ is a diagonal matrix

$$D = \begin{pmatrix} I & O \\ O & O \end{pmatrix}$$

where I is the identity matrix of dimension $k \times k$ (see Chapter I). Write $\mathbf{Z} = (Z_1, \ldots, Z_p)' = \theta \mathbf{Y}$. Then

$$\mathbf{Y}'B\mathbf{Y} = \mathbf{Z}'D\mathbf{Z} = \sum_{i=1}^{k} Z_i^2 \tag{6.14}$$

is distributed as chi-square $\chi_k^2(\mathbf{v}'B\mathbf{v})$ (see Exercise 6.1).

To prove the necessity of the condition let us assume that $\mathbf{Y}'B\mathbf{Y}$ is distributed as $\chi_k^2(\mathbf{v}'B\mathbf{v})$. If B is of rank m, there exists an orthogonal matrix θ such that $\theta B \theta'$ is a diagonal matrix with m nonzero diagonal elements $\lambda_1, \ldots, \lambda_m$, the characteristic roots of B (we can without any loss of generality assume that the first m diagonal elements are nonzero). Let $\mathbf{Z} = \theta \mathbf{Y}$. Then

$$\mathbf{Y}'B\mathbf{Y} = \sum_{i=1}^{m} \lambda_i Z_i^2. \tag{6.15}$$

Since the Z_i^2 are independently distributed each as noncentral chi-square with one degree of freedom and $\mathbf{Y}'B\mathbf{Y}$ is distributed as noncentral $\chi_k^2(\mathbf{v}'B\mathbf{v})$, it follows from the uniqueness of the characteristic function that $m = k$ and $\lambda_i = 1$, $i = 1, \ldots, k$. In other words, $\theta B \theta'$ is a diagonal matrix with k diagonal elements each equal to unity and the rest are zero. This implies that B is an idempotent matrix of rank k. Q.E.D.

From this theorem it follows trivially that

(a) $\mathbf{X}'\Sigma^{-1}\mathbf{X}$ is distributed as noncentral chi-square $\chi_p^2(\boldsymbol{\mu}'\Sigma^{-1}\boldsymbol{\mu})$;
(b) $(\mathbf{X} - \boldsymbol{\mu})'\Sigma^{-1}(\mathbf{X} - \boldsymbol{\mu})$ is distributed as χ_p^2;
(c) for any vector $\boldsymbol{\alpha} = (\alpha_1, \ldots, \alpha_p)'$, $(\mathbf{X} - \boldsymbol{\alpha})'\Sigma^{-1}(\mathbf{X} - \boldsymbol{\alpha})$ is distributed as $\chi_p^2((\boldsymbol{\mu} - \boldsymbol{\alpha})'\Sigma^{-1}(\boldsymbol{\mu} - \boldsymbol{\alpha}))$.

THEOREM 6.2.3 Let $\mathbf{X} = (X_1, \ldots, X_p)'$ be a normally distributed p-vector with mean $\boldsymbol{\mu}$ and positive definite covariance matrix Σ and let B be an $m \times p$ matrix of rank m ($<p$). Then the quadratic form $\mathbf{X}'A\mathbf{X}$ is distributed independently of the linear form $B\mathbf{X}$ if $B\Sigma A = O$.

Proof Since Σ is positive definite there exists a nonsingular matrix C such that $\Sigma = CC'$. Write $\mathbf{X} = C\mathbf{Y}$. Obviously $\mathbf{Y}$ is normally distributed with mean $\mathbf{v} = C^{-1}\boldsymbol{\mu}$ and covariance matrix I. Now

$$\mathbf{X}'A\mathbf{X} = \mathbf{Y}'D\mathbf{Y}, \qquad B\mathbf{X} = E\mathbf{Y} \tag{6.16}$$

where $D = C'AC$, $E = BC$. To prove the theorem we need to show that $\mathbf{Y}'D\mathbf{Y}$, $E\mathbf{Y}$ are independently distributed if $ED = 0$.

Assume that $ED = 0$ and that the rank of D is k ($<p$). There exists an orthogonal matrix θ such that $\theta D \theta'$ is a diagonal matrix

$$\begin{pmatrix} D_1 & 0 \\ 0 & 0 \end{pmatrix}$$

where D_1 is a diagonal matrix of dimension $k \times k$ with nonzero diagonal elements. Now

$$\mathbf{Y}'D\mathbf{Y} = \mathbf{Z}'_{(1)}D_1\mathbf{Z}_{(1)}, \qquad E\mathbf{Y} = E\theta'\theta\mathbf{Y} = E^*\mathbf{Z},$$

where $\mathbf{Z} = \theta\mathbf{Y} = (Z_1, \ldots, Z_p)'$, $Z_{(1)} = (Z_1, \ldots, Z_k)'$, and

$$E^* = E\theta' = \begin{pmatrix} E_{11}^* & E_{12}^* \\ E_{21}^* & E_{22}^* \end{pmatrix}$$

with E_{11}^* a $k \times k$ submatrix of E^*. Since $ED = O$ implies that $E^*\theta D\theta' = O$, we get $E_{11}^*D_1 = E_{21}^*D_1 = O$, and hence

$$E^* = \begin{pmatrix} O & E_{12}^* \\ O & E_{22}^* \end{pmatrix} = (O \quad E_2^*) \qquad \text{(say)},$$

and $E\mathbf{Y}$ is distributed as $E_2^*\mathbf{Z}_{(2)}$, where $\mathbf{Z}_{(2)} = (Z_{k+1}, \ldots, Z_p)'$. Since $Y_1, \ldots, Y_p$ are independently distributed normal random variables and θ is an orthogonal matrix we conclude that $\mathbf{Y}'D\mathbf{Y}$ is independent of $E\mathbf{Y}$. Q.E.D.

THEOREM 6.2.4 *Cochran's theorem* Let $\mathbf{X}^\alpha = (X_{\alpha 1}, \ldots, X_{\alpha p})'$, $\alpha = 1, \ldots, N$, be a random sample of size N from a p-variate normal distribution with mean $\mathbf{0}$ and positive definite covariance matrix Σ. Assume that

$$\sum_{\alpha=1}^{N} (\mathbf{X}^\alpha)\mathbf{X}^{\alpha\prime} = Q_1 + \cdots + Q_k, \tag{6.17}$$

where $Q_i = \sum_{\alpha,\beta=1}^{N} (\mathbf{X}^\alpha)' a_{\alpha\beta}^i \mathbf{X}^\beta$ with $A_i = (a_{\alpha\beta}^i)$ of rank N_i, $i = 1, \ldots, k$. Then the Q_i are independently distributed as

$$\sum_{\alpha=N_1+\cdots+N_{i-1}+1}^{N_1+\cdots+N_i} (\mathbf{Z}^\alpha)(\mathbf{Z}^\alpha)' \tag{6.18}$$

where $\mathbf{Z}^\alpha = (Z_{\alpha 1}, \ldots, Z_{\alpha p})'$, $\alpha = 1, \ldots, \sum_1^k N_i$, are independently distributed normal p-vectors with mean $\mathbf{0}$ and covariance matrix Σ if and only if $\sum_1^k N_i = N$.

Proof Suppose that the Q_i are independently distributed as in (6.18). Hence $\sum_1^k Q_i$ is distributed as

$$\sum_{\alpha=1}^{N_1+\cdots+N_k} (\mathbf{Z}^\alpha)(\mathbf{Z}^\alpha)'. \tag{6.19}$$

From (6.17) and (6.19) and the uniqueness of the characteristic function we conclude that $\sum_1^k N_i = N$.

To prove the sufficiency part of the theorem let us assume that $\sum_1^k N_i = N$. In the same way as in Theorem 6.2.1 we can assert that there exists an orthogonal matrix B

$$B = \begin{pmatrix} B_1 \\ \vdots \\ B_k \end{pmatrix} \qquad \text{with} \quad A_i = B_i B_i'.$$

Since $B = (b_{\alpha\beta})$ is orthogonal,

$$\mathbf{Z}^\alpha = \sum_{\beta=1}^{N} b_{\alpha\beta}\mathbf{X}^\beta, \qquad \alpha = 1, \ldots, N,$$

are independently distributed normal p-vectors with mean $\mathbf{0}$ and covariance matrix Σ. It easy to see that for $i = 1, \ldots, k$,

$$Q_i = \sum_{\alpha,\beta=1}^{N} (\mathbf{X}^\alpha) a_{\alpha\beta}^i (\mathbf{X}^\beta)' = \sum_{\alpha=N_1+\cdots+N_{i-1}+1}^{N_1+\cdots+N_i} (\mathbf{Z}^\alpha)(\mathbf{Z}^\alpha)'. \qquad \text{Q.E.D.}$$

This theorem is useful in generalizing the univariate analysis of variance results to multivariate analysis of variance problems. There is considerable literature on the distribution of quadratic forms and related results. The reader is referred to Cochran (1934), Hogg and Craig (1958), Ogawa (1949), Rao (1965), and Graybill (1961) for further references and details.

6.3 THE WISHART DISTRIBUTION

In Chapter V we remarked that a symmetric random matrix S of dimension $p \times p$ has a Wishart distribution with n degrees of freedom ($n \geq p$) and parameter Σ (a positive definite matrix) if S can be written as

$$S = \sum_{\alpha=1}^{n} \mathbf{X}^\alpha(\mathbf{X}^\alpha)'$$

where $\mathbf{X}^\alpha = (X_{\alpha 1}, \ldots, X_{\alpha p})'$, $\alpha = 1, \ldots, n$, are independently distributed normal p-vectors with mean $\mathbf{0}$ and covariance matrix Σ. In this section we shall derive the Wishart probability density function as given in (5.2). In the sequel we shall need the following lemma.

LEMMA 6.3.1 Suppose X with values in the sample space $\mathscr{X}$ is a random variable with probability density function $f(t(x))$ with respect to a σ-finite measure μ on $\mathscr{X}$ where $t: \mathscr{X} \to \mathscr{Y}$ is measurable. For any measurable subset $B \in \mathscr{Y}$ define the measure v by

$$v(B) = \mu(t^{-1}(B)). \tag{6.20}$$

Then the probability density function of $Y = t(X)$ with respect to the measure v is $f(y)$.

Proof It suffices to show that if $g:\mathscr{Y} \to R$ (real line), then

$$E(g(Y)) = \int_{\mathscr{Y}} g(y)f(y)\, dv(y).$$

From (6.20)

$$E(g(Y)) = Eg(t(X)) = \int_{\mathscr{X}} g(t(x))f(t(x))\, d\mu(x) = \int_{\mathscr{Y}} g(y)f(y)\, dv(y). \qquad \text{Q.E.D.}$$

We shall assume that $n \geq p$ so that S is positive definite with probability 1. The joint probability density function of $\mathbf{X}^\alpha$, $\alpha = 1, \ldots, n$, is given by

$$f(\mathbf{x}^1, \ldots, \mathbf{x}^n) = (2\pi)^{-np/2}(\det \Sigma^{-1})^{n/2} \exp\left\{-\tfrac{1}{2} \operatorname{tr} \Sigma^{-1} \sum_{\alpha=1}^{n} \mathbf{x}^\alpha(\mathbf{x}^\alpha)'\right\}. \qquad (6.21)$$

For any measurable set A in the space of S, the probability that S belongs to A depends on Σ and is given by

$$\begin{aligned} P_\Sigma(S \in A) &= (2\pi)^{-np/2} \int_{\Sigma_{\alpha=1}^n \mathbf{X}^\alpha(\mathbf{X}^\alpha)' = s \in A} (\det \Sigma)^{-n/2} \exp\{-\tfrac{1}{2} \operatorname{tr} \Sigma^{-1}s\} \prod_{\alpha=1}^{n} d\mathbf{x}^\alpha \\ &= (2\pi)^{-np/2} \int_{s \in A} (\det \Sigma)^{-n/2} \exp\{-\tfrac{1}{2} \operatorname{tr} \Sigma^{-1}s\}\, dm(s) \qquad (6.22) \end{aligned}$$

where m is the measure corresponding to the measure v of (6.20). Let us now define the measure m^* on the space of S by

$$dm^*(s) = \frac{dm(s)}{(\det s)^{n/2}}. \qquad (6.23)$$

Then

$$P_\Sigma(S \in A) = (2\pi)^{-np/2} \int_A (\det(\Sigma^{-1}s))^{n/2} \exp\{-\tfrac{1}{2} \operatorname{tr} \Sigma^{-1}s\}\, dm^*(s). \qquad (6.24)$$

Obviously to find the probability density function of S it is sufficient to find $dm^*(s)$. To do this let us first observe the following:

(i) Since Σ is positive definite there exists $C \in G_l(p)$, the multiplicative group of $p \times p$ nonsingular matrices, such that $\Sigma = CC'$.

(ii) Let

$$\tilde{S} = C^{-1}S(C^{-1})' = \sum_{\alpha=1}^{N} (C^{-1}\mathbf{X}^\alpha)(C^{-1}\mathbf{X}^\alpha)'. \qquad (6.25)$$

Since $C^{-1}\mathbf{X}^\alpha$ are independently normally distributed with mean $\mathbf{0}$ and covariance matrix I, $\tilde{S}$ is distributed as $W_p(n, I)$. Thus by (6.20)

$$P_{CC'}(S \in A) = P_I(C\tilde{S}C' \in A). \tag{6.26}$$

Now

$$P_{CC'}(S \in A) = (2\pi)^{-np/2} \int_A (\det((CC')^{-1}s))^{n/2} \times \exp\{-\tfrac{1}{2}\operatorname{tr}(CC')^{-1}s\}\, dm^*(s), \tag{6.27}$$

$$\begin{aligned} P_I(C\tilde{S}C' \in A) &= (2\pi)^{-np/2} \int_{C\tilde{s}C' \in A} (\det(\tilde{s}))^{n/2} \exp\{-\tfrac{1}{2}\operatorname{tr}\tilde{s}\}\, dm^*(\tilde{s}) \\ &= (2\pi)^{-np/2} \int_A (\det((CC')^{-1}s))^{n/2} \\ &\quad \times \exp\{-\tfrac{1}{2}\operatorname{tr}(CC')^{-1}s\}\, dm^*(C^{-1}sC'^{-1}). \end{aligned} \tag{6.28}$$

Since (6.26) holds for all measurable sets A in the space of S we must then have

$$dm^*(s) = dm^*(CsC') \tag{6.29}$$

for all $C \in G_l(p)$ and all s in the space of S. This implies that for some positive constant k

$$dm^*(s) = \frac{kds}{(\det s)^{(p+1)/2}} \tag{6.30}$$

where ds stands for the Lebesgue measure $\prod_{i \le j} ds_{ij}$ in the space of S. By Theorem 2.4.10 the Jacobian of the transformation $s \to CsC'$, $C \in G_l(p)$, is $(\det(CC'))^{(p+1)/2}$. Hence

$$dm^*(CsC') = \frac{kd(CsC')}{(\det(CsC'))^{(p+1)/2}} = \frac{kds}{(\det s)^{(p+1)/2}}. \tag{6.31}$$

In other words, $dm^*(s)$ is an invariant measure on the space of S under the action of the group of transformations defined by $s \to CsC'$, $C \in G_l(p)$. Now (6.30) follows from the uniqueness of invariant measures on homogeneous spaces (see Nachbin, 1965; or Eaton, 1972). From (6.24) and (6.30) the probability density function $W_p(n, \Sigma)$ of a Wishart random variable S with n degrees of freedom and parameter Σ is given by (with respect to the Lebesgue measure ds)

$$W_p(n, \Sigma) = \begin{cases} K(\det \Sigma)^{-n/2}(\det s)^{(n-p-1)/2} \exp\{-\tfrac{1}{2}\operatorname{tr}\Sigma^{-1}s\} & \text{if } s \text{ is positive definite,} \\ 0 \quad \text{otherwise,} & \end{cases} \tag{6.32}$$

where K is the normalizing constant independent of Σ. To specify the probability density function we need to evaluate the constant K. Since K is independent of Σ, we can in particular take $\Sigma = I$ for the evaluation of K. Since K is a function of n and p, we shall denote it by $C_{n,p}$. Let us partition $S = (S_{ij})$ as

$$S = \begin{pmatrix} S_{(11)} & S_{(12)} \\ S_{(21)} & S_{(22)} \end{pmatrix}$$

with $S_{(11)}$ a $(p-1) \times (p-1)$ submatrix of S, and let

$$Z = S_{(22)} - S_{(21)}S_{(11)}^{-1}S_{(12)}.$$

From (6.32)

$$\begin{aligned}
1 &= \int C_{n,p}(\det s_{(11)})^{(n-p-1)/2}(s_{(22)} - s_{(21)}s_{(11)}^{-1}s_{(12)})^{(n-p-1)/2} \\
&\quad \times \exp\{-\tfrac{1}{2}\operatorname{tr}(s_{(22)} + s_{(11)})\}\, ds_{(11)}\, ds_{(12)}\, ds_{(22)} \\
&= C_{n,p} \int \det(s_{(11)})^{(n-p-1)/2} \exp\{-\tfrac{1}{2}\operatorname{tr} s_{(11)}\} \\
&\quad \times \left(\int \exp(-\tfrac{1}{2}s_{(21)}s_{(11)}^{-1}s_{(12)})\, ds_{(12)} \right) ds_{(11)} \int_0^\infty z^{(n-p-1)/2} \exp\{-\tfrac{1}{2}z\}\, dz \\
&= C_{n,p}2^{(n-p+1)/2}\Gamma\left(\frac{n-p+1}{2}\right)(2\pi)^{(p-1)/2} \int (\det s_{(11)})^{(n-p)/2} \\
&\quad \times \exp\{-\tfrac{1}{2}\operatorname{tr} s_{(11)}\}\, ds_{(11)} \qquad (6.33)
\end{aligned}$$

as

$$\int_0^\infty z^{(n-p-1)/2} \exp\{-\tfrac{1}{2}z\}\, dz = 2^{(n-p+1)/2}\Gamma\left(\frac{n-p+1}{2}\right),$$

$$\int \exp\{-\tfrac{1}{2}s_{(21)}s_{(11)}^{-1}s_{(12)}\}\, ds_{(12)} = (2\pi)^{(p-1)/2}(\det(s_{(11)}))^{1/2}.$$

Since $W_p(n, I)$ is a probability density function with the constant $K = C_{n,p}$, we obtain

$$\int (\det s_{(11)})^{(n-p)/2} \exp\{-\tfrac{1}{2}\operatorname{tr} s_{(11)}\}\, ds_{(11)} = (C_{n,p-1})^{-1}. \qquad (6.34)$$

From (6.33) and (6.34) we get

$$\begin{aligned}
C_{n,p} &= \left[\Gamma\left(\frac{n-p+1}{2}\right)2^{n/2}\pi^{(p-1)/2}\right]^{-1} C_{n,p-1} \\
&= \left[\Gamma\left(\frac{n-p+1}{2}\right)2^{n/2}\pi^{(p-1)/2}\right]^{-1} \cdots \left[\Gamma\left(\frac{n-1}{2}\right)2^{n/2}\pi^{1/2}\right]^{-1} C_{n,1}. \qquad (6.35)
\end{aligned}$$

But $C_{n,1}$ is given by

$$C_{n,1} \int_0^\infty x^{(n-2)/2} \exp\{-\tfrac{1}{2}x\}\, dx = 1,$$

which implies

$$C_{n,1} = [\Gamma(n/2)2^{n/2}]^{-1}. \tag{6.36}$$

From (6.35) and (6.36) we get

$$(C_{n,p})^{-1} = \left(\prod_{i=0}^{p-1} \Gamma\left(\frac{n-i}{2}\right)\right) 2^{np/2} \pi^{p(p-1)/4} = K^{-1}. \tag{6.37}$$

The derivation of the Wishart distribution, which is very fundamental in multivariate analysis, was a major breakthrough for the development of multivariate analysis. Several derivations of the Wishart distribution are available in the literature. The derivation given here involves a property of invariant measure and is quite short and simple in nature.

Alternate Derivation Since the preceding derivation of the Wishart distribution involves some deep theoretical concepts, we will now give a straightforward derivation.

S is distributed as $\sum_{\alpha=1}^{N-1} \mathbf{X}^\alpha(\mathbf{X}^\alpha)'$, where $\mathbf{X}^\alpha$, $\alpha = 1, 2, \ldots, N-1$, are independently distributed normal p-vectors with mean $\mathbf{0}$ and positive definite covariance matrix Σ. Let

$$\Sigma = CC', \qquad \mathbf{Y}^\alpha = C^{-1}\mathbf{X}^\alpha, \qquad \alpha = 1, 2, \ldots, N-1,$$

where C is a nonsingular matrix. Let us first consider the distribution of

$$A = \sum_{\alpha=1}^{N-1} \mathbf{Y}^\alpha(\mathbf{Y}^\alpha)'.$$

Write $Y = (\mathbf{Y}^1, \ldots, \mathbf{Y}^{N-1})$. Then $A = YY'$. By the Gram–Schmidt orthogonalization process on the row vectors $\mathbf{Y}_1, \ldots, \mathbf{Y}_p$ of Y we obtain new row vectors $\mathbf{Z}_1, \ldots, \mathbf{Z}_p$ such that

$$ZZ' = I,$$

where

$$Z = \begin{pmatrix} \mathbf{Z}_1 \\ \vdots \\ \mathbf{Z}_p \end{pmatrix}.$$

Let the transformation involved in transforming Y to Z be given by $Z = B^{-1}Y$. Obviously $B = (b_{ij})$ is a random lower triangular nonsingular matrix. Now

$$A = YY' = BZZ'B' = BB',$$

where $B = (b_{ij})$ is a random lower triangular nonsingular matrix with positive diagonal elements satisfying $Y = BZ$. Thus we get

$$\mathbf{Y}_i = \sum_{j=1}^{i} b_{ij}\mathbf{Z}_j, \qquad i = 1, \ldots, p,$$

and $\mathbf{Z}_j\mathbf{Y}_i' = b_{ij}$. Hence with $A = (a_{ij})$,

$$a_{ii} = \mathbf{Y}_i\mathbf{Y}_i' = \sum_{j=1}^{i} b_{ij}^2, \qquad b_{ii}^2 = a_{ii} - \sum_{j=1}^{i-1} b_{ij}^2.$$

In other words,

$$\begin{pmatrix} b_{i1} \\ \vdots \\ b_{i,i-1} \end{pmatrix} = \begin{pmatrix} \mathbf{Z}_1 \\ \vdots \\ \mathbf{Z}_{i-1} \end{pmatrix} \mathbf{Y}_i'.$$

Since $ZZ' = I$, the components of Y^α, $\alpha = 1, \ldots, N - 1$, are independently distributed normal variables with mean 0 and variance 1. $\mathbf{Z}_1, \ldots, \mathbf{Z}_{i-1}$ are functions of $\mathbf{Y}_1, \ldots, \mathbf{Y}_{i-1}$, the conditional distributions of $b_{i1}, \ldots, b_{i,i-1}$, given that $\mathbf{Y}_1, \ldots, \mathbf{Y}_{i-1}$ are independent normal with mean 0 and variance 1,

$$b_{ii}^2 = \mathbf{Y}_i\mathbf{Y}_i' - \sum_{j=1}^{i-1} b_{ij}^2$$

is dristributed as χ_{N-i}^2, and all b_{ij}, $j = 1, \ldots, i$, are independent. Since the conditional distributions of b_{ij}, $j = 1, \ldots, i$, do not involve $\mathbf{Y}_1, \ldots, \mathbf{Y}_{i-1}$, these conditional distributions are also the unconditional distributions of $b_{ij}, j = 1, \ldots, i - 1$. Furthermore, $b_{i1}, \ldots, b_{ii}$ are distributed independently of $\mathbf{Y}_1, \ldots, \mathbf{Y}_{i-1}$ and hence of b_{rs}, $r, s = 1, \ldots, i - 1$ $(r \geq s)$, and $\mathbf{Z}_1, \ldots, \mathbf{Z}_{i-1}$, which are functions of $\mathbf{Y}_1, \ldots, \mathbf{Y}_{i-1}$ only. Hence b_{ij}, $i, j = 1, \ldots, p$ $(i > p)$, are independently distributed normal random variables with mean 0 and variance 1, and b_{ii}^2, $i = 1, \ldots, p$, are independently distributed (independently of the b_{ij}) as χ_{N-i}^2. From Theorem 2.4.6 the Jacobian of the transformations $B \to A = BB'$ is $2^p \prod_{i=1}^{p}(b_{ii})^{i-p-1}$. Hence the distribution of A is

$$f_A(a) = K(\det a)^{(N-p)/2} \exp\{-\tfrac{1}{2} \operatorname{tr} a\},$$

provided a is positive definite, where K is a constant depending on N and p. By Theorem 2.4.1 the probability density function of $S = CAC'$ is given by (6.32).

The Wishart distribution was first derived by Fisher (1915) for $p = 2$. Wishart (1928) gave a geometrical derivation of this distribution for general p. Ingham (1933) derived this distribution from its characteristic function. Elfving (1947), Mauldan (1955), and Olkin and Roy (1954) used matrix transformations to derive the Bartlett decomposition of the Wishart matrix from sample observations and then derived the distribution of the Wishart matrix. Khirsagar (1959) used random orthogonal transformations to derive the Wishart distribution and the distribution of Bartlett decomposition. Sverdrup (1947) derived this distribution by straightforward integration over the sample space. Narain (1948) and Ogawa (1953) used the regression approach, Ogawa's approach being more elegant. Rasch (1948) and Khatri (1963) also gave alternative derivations of this distribution.

6.4 TENSOR PRODUCT, PROPERTIES OF THE WISHART DISTRIBUTION

This section deals with some important properties of the Wishart distribution which are often used in multivariate analysis. Before we state and prove these we need to define the concept of tensor product.

Suppose that $\alpha = (\alpha_{ij})$ and $\beta = (\beta_{ij})$ are matrices of dimensions $n \times m$ and $l \times k$, respectively. The tensor product $\alpha \otimes \beta$ acts on matrices $\gamma = (\gamma_{ij})$ of dimension $m \times k$ such that

$$\alpha\gamma\beta' = \delta, \tag{6.38}$$

where $\delta = (\delta_{ij})$ is a matrix of dimension $n \times l$. Consider the vectors v_1 of dimension $mk \times 1$ and v_2 of dimension $nl \times 1$, defined by

$$v_1 = (\gamma_{11}, \ldots, \gamma_{m1}, \gamma_{12}, \ldots, \gamma_{m2}, \ldots, \gamma_{mk})',$$
$$v_2 = (\delta_{11}, \ldots, \delta_{n1}, \delta_{12}, \ldots, \delta_{n2}, \ldots, \delta_{nl})'.$$

The tensor product $\alpha \otimes \beta$ is a matrix of dimension $nl \times mk$ operating on the left of the column vector v_1 to give the column vector v_2, where δ is defined by

$$(\alpha \otimes \beta)\gamma = \delta. \tag{6.39}$$

Here the matrix $\alpha \otimes \beta$ is such that

$$(\alpha \otimes \beta)v_1 = v_2. \tag{6.40}$$

It is easy to see that $\alpha \otimes \beta$ is given by

$$\alpha \otimes \beta = \begin{pmatrix} \beta_{11}\alpha & \cdots & \beta_{1k}\alpha \\ \vdots & & \vdots \\ \beta_{l1}\alpha & \cdots & \beta_{lk}\alpha \end{pmatrix}, \tag{6.41}$$

which is an $nl \times mk$ matrix commonly called the Kronecker product of two matrices α and β. Let α, β be square matrices of dimensions $n \times n$ and

$m \times m$, respectively, and let $\lambda_1, \ldots, \lambda_n$ denote the characteristic roots of α and $\mu_1, \ldots, \mu_m$ denote the characteristic roots of β. Then the characteristic roots of $\alpha \otimes \beta$ consist of the set

$$\{\mu_i \lambda_j;\, i = 1, \ldots, m, j - 1, \ldots, n\}. \tag{6.42}$$

From this it follows that

$$\det(\alpha \otimes \beta) = (\det(\alpha))^m (\det(\beta))^n, \qquad \operatorname{tr}(\alpha \otimes \beta) = (\operatorname{tr} \alpha)(\operatorname{tr} \beta). \tag{6.43}$$

Furthermore, if α and β are nonsingular matrices, then

$$(\alpha \otimes \beta)^{-1} = \alpha^{-1} \otimes \beta^{-1}. \tag{6.44}$$

EXAMPLE 6.4.1 Let $\mathbf{X}^\alpha = (X_{\alpha 1}, \ldots, X_{\alpha p})'$, $\alpha = 1, \ldots, N$, be independently distributed normal random vectors with the same mean $\boldsymbol{\mu}$ and the same covariance matrix Σ and let

$$X = (\mathbf{X}^1, \ldots, \mathbf{X}^N), \qquad x = (\mathbf{x}^1, \ldots, \mathbf{x}^N).$$

Then the probability density function of X is

$$\begin{aligned} f_X(x) &= \prod_{\alpha=1}^{N} f_{\mathbf{X}^\alpha}(\mathbf{x}^\alpha) \\ &= (2\pi)^{-Np/2} (\det \Sigma)^{-N/2} \exp\left\{ -\tfrac{1}{2} \operatorname{tr} \Sigma^{-1} \left(\sum_{\alpha=1}^{N} (\mathbf{x}^\alpha - \boldsymbol{\mu})(\mathbf{x}^\alpha - \boldsymbol{\mu})' \right) \right\} \\ &= (2\pi)^{-Np/2} (\det(\Sigma \otimes I))^{-1/2} \exp\{ -\tfrac{1}{2} \operatorname{tr}(\Sigma \otimes I)^{-1} (x - \mathbf{e} \otimes \boldsymbol{\mu})(x - \mathbf{e} \otimes \boldsymbol{\mu})' \} \end{aligned}$$

where $\mathbf{e} = (1, \ldots, 1)'$ is an N vector with all components equal to unity and I is the identity matrix of dimension $N \times N$.

THEOREM 6.4.1 Let

$$S = \begin{pmatrix} S_{(11)} & S_{(12)} \\ S_{(21)} & S_{(22)} \end{pmatrix},$$

where $S_{(11)}$ is the $q \times q$ left-hand corner submatrix of S ($q < p$), be distributed as $W_p(n, \Sigma)$, and let Σ be similarly partitioned into

$$\Sigma = \begin{pmatrix} \Sigma_{(11)} & \Sigma_{(12)} \\ \Sigma_{(21)} & \Sigma_{(22)} \end{pmatrix}.$$

Then

(a) $S_{(11)} - S_{(12)} S_{(22)}^{-1} S_{(21)}$ is distributed as Wishart $W_q(n - (p - q), \Sigma_{(11)} - \Sigma_{(12)} \Sigma_{(22)}^{-1} \Sigma_{(21)})$;
(b) $S_{(22)}$ is distributed as Wishart $W_{p-q}(n, \Sigma_{(22)})$;

(c) the conditional distribution of $S_{(12)}S_{(22)}^{-1}$ given that $S_{(22)} = s_{(22)}$ is normal in the sense of Example 6.4.1 with mean $\Sigma_{(12)}\Sigma_{(22)}^{-1}$ and covariance matrix $(\Sigma_{(22)} - \Sigma_{(21)}\Sigma_{(11)}^{-1}\Sigma_{(12)}) \otimes s_{(22)}^{-1}$;

(d) $S_{(11)} - S_{(12)}S_{(22)}^{-1}S_{(21)}$ is independent of $(S_{(12)}, S_{(22)})$.

Proof Let $\Sigma^{-1} = \Lambda$ be partitioned into

$$\Lambda = \begin{pmatrix} \Lambda_{(11)} & \Lambda_{(12)} \\ \Lambda_{(21)} & \Lambda_{(22)} \end{pmatrix}$$

where $\Lambda_{(11)}$ is a $q \times q$ submatrix of Λ and let $(S_{(11)}, S_{(12)}, S_{(22)})$ be transformed to (W, U, V) where

$$W = S_{(11)} - S_{(12)}S_{(22)}^{-1}S_{(21)}, \qquad U = S_{(12)}S_{(22)}^{-1/2}, \qquad V = S_{(22)}, \tag{6.45}$$

or, equivalently,

$$S_{(11)} = W + UU', \qquad S_{(12)} = UV^{1/2}, \qquad S_{(22)} = V. \tag{6.46}$$

The Jacobian of this transformation is given by the absolute value of the determinant of the following matrix of partials:

$$\begin{matrix} \\ S_{(11)} \\ S_{(12)} \\ S_{(22)} \end{matrix} \begin{pmatrix} W & U & V \\ I & - & - \\ 0 & A & - \\ 0 & 0 & I \end{pmatrix},$$

where the dash indicates some matrix which need not be known and A is the matrix of partial derivatives of the transformation $S_{(12)} \to UV^{1/2}$ (V fixed). By a result analogous to Theorem 2.4.1, the Jacobian is

$$|\det(A)| = |\det(V)|^{q/2}. \tag{6.47}$$

Now

$$\begin{aligned}
\operatorname{tr} \Sigma^{-1}s &= \operatorname{tr} \Lambda s \\
&= \operatorname{tr}(\Lambda_{(11)}s_{(11)} + \Lambda_{(12)}s_{(21)}) + \operatorname{tr}(\Lambda_{(21)}s_{(12)} + \Lambda_{(22)}s_{(22)}) \\
&= \operatorname{tr}(\Lambda_{(11)}s_{(11)} + \Lambda_{(12)}s_{(21)} + s_{(12)}\Lambda_{(21)} + \Lambda_{(11)}^{-1}\Lambda_{(12)}s_{(22)}\Lambda_{(21)}) \\
&\quad + \operatorname{tr}(\Lambda_{(22)} - \Lambda_{(21)}\Lambda_{(11)}^{-1}\Lambda_{(12)})s_{(22)} \\
&= \operatorname{tr} \Lambda_{(11)}(w + uu') + 2 \operatorname{tr} \Lambda_{(12)}v^{1/2}u' + \operatorname{tr} \Lambda_{(21)}\Lambda_{(11)}^{-1}\Lambda_{(12)}v \\
&\quad + \operatorname{tr}(\Lambda_{(22)} - \Lambda_{(21)}\Lambda_{(11)}^{-1}\Lambda_{(12)})v \\
&= \operatorname{tr} \Lambda_{(11)}w + \operatorname{tr} \Lambda_{(11)}uu' + 2 \operatorname{tr} \Lambda_{(12)}v^{1/2}u' \\
&\quad + \operatorname{tr} \Lambda_{(12)}v\Lambda_{(21)}\Lambda_{(11)}^{-1} + \operatorname{tr}(\Lambda_{(22)} - \Lambda_{(21)}\Lambda_{(11)}^{-1}\Lambda_{(12)})v \\
&= \operatorname{tr} \Lambda_{(11)}w + \operatorname{tr} \Lambda_{(11)}(u + \Lambda_{(11)}^{-1}\Lambda_{(12)}v^{1/2})(u + \Lambda_{(11)}^{-1}\Lambda_{(12)}v^{1/2})' \\
&\quad + \operatorname{tr}(\Lambda_{(22)} - \Lambda_{(21)}\Lambda_{(11)}^{-1}\Lambda_{(12)})v.
\end{aligned} \tag{6.48}$$

Since

$$\begin{aligned}
\Lambda_{(11)} &= (\Sigma_{(11)} - \Sigma_{(12)}\Sigma_{(22)}^{-1}\Sigma_{(21)})^{-1}, \\
\Sigma_{(22)}^{-1} &= (\Lambda_{(22)} - \Lambda_{(21)}\Lambda_{(11)}^{-1}\Lambda_{(12)}), \\
\Lambda_{(11)}^{-1}\Lambda_{(12)} &= -\Sigma_{(12)}\Sigma_{(22)}^{1}, \\
\det(S) &= \det(S_{(22)}) \det(S_{(11)} - S_{(12)}S_{(22)}^{-1}S_{(21)}), \\
\det \Sigma &= \det(\Sigma_{(22)}) \det(\Sigma_{(11)} - \Sigma_{(12)}\Sigma_{(22)}^{-1}\Sigma_{(21)}),
\end{aligned}$$

from (6.47), (6.48), and (6.32), the joint probability density function of (W, U, V) can be written as

$$f_{W,U,V}(w, u, v) = f_W(w) f_{U|V}(u|v) f_V(v) \tag{6.49}$$

where

$$\begin{aligned}
f_W(w) &= k_1(\det(\Sigma_{(11)} - \Sigma_{(12)}\Sigma_{(22)}^{-1}\Sigma_{(21)}))^{-(n-(p-q))/2}(\det(w))^{(n-(p-q)-q-1)/2} \\
&\quad \times \exp\{-\tfrac{1}{2}\operatorname{tr}(\Sigma_{(11)} - \Sigma_{(12)}\Sigma_{(22)}^{-1}\Sigma_{(21)})^{-1}w\}, \\
f_{U|V}(u|v) &= k_2(\det((\Sigma_{(11)} - \Sigma_{(12)}\Sigma_{(22)}^{-1}\Sigma_{(21)}) \otimes I_{p-q}))^{-1/2} \\
&\quad \times \exp\{-\tfrac{1}{2}\operatorname{tr}((\Sigma_{(11)} - \Sigma_{(12)}\Sigma_{(22)}^{-1}\Sigma_{(21)}) \otimes I_{p-q})^{-1} \\
&\quad \times (u - \Sigma_{(12)}\Sigma_{(22)}^{-1}v^{1/2})(u - \Sigma_{(12)}\Sigma_{(22)}^{-1}v^{1/2})'\}, \\
f_V(v) &= k_3(\det \Sigma_{(22)})^{-n/2}(\det v)^{(n-(p-q)-1)/2} \exp\{-\tfrac{1}{2}\operatorname{tr} \Sigma_{(22)}^{-1}v\},
\end{aligned}$$

where k_1, k_2, k_3 are normalizing constants independent of Σ. Thus $S_{(11)} - S_{(12)}S_{(22)}^{-1}S_{(21)}$ is distributed as Wishart

$$W_q(n - (p - q), \Sigma_{(11)} - \Sigma_{(12)}\Sigma_{(22)}^{-1}\Sigma_{(21)})$$

and is independent of $(S_{(12)}, S_{(22)})$.

The conditional distribution of $S_{(12)}S_{(22)}^{-1/2}$, given $S_{(22)} = s_{(22)}$, is normal (in the sense of Example 6.4.1) with mean $\Sigma_{(12)}\Sigma_{(22)}^{-1}s_{(22)}^{1/2}$ and covariance matrix $(\Sigma_{(11)} - \Sigma_{(12)}\Sigma_{(22)}^{-1}\Sigma_{(21)}) \otimes I_{p-q}$.

Multiplying $S_{(12)}S_{(22)}^{-1/2}$ by $S_{(22)}^{-1/2}$, we conclude that the conditional distribution of $S_{(12)}S_{(22)}^{-1}$ given $S_{(22)} = s_{(22)}$ is normal in the sense of Example 6.4.1 with mean $\Sigma_{(12)}\Sigma_{(22)}^{-1}$ and covariance matrix $(\Sigma_{(11)} - \Sigma_{(12)}\Sigma_{(22)}^{-1}\Sigma_{(21)}) \otimes s_{(22)}^{-1}$. Finally, $S_{(22)}$ is distributed as Wishart $W_{(p-q)}(n, \Sigma_{(22)})$. Q.E.D.

THEOREM 6.4.2 If S is distributed as $W_p(n, \Sigma)$ and C is a nonsingular matrix of dimension $p \times p$, then CSC' is distributed as $W_p(n, C\Sigma C')$.

Proof Since S is distributed as $W_p(n, \Sigma)$, S can be written as

$$S = \sum_{\alpha=1}^{n} \mathbf{Y}^\alpha(\mathbf{Y}^\alpha)'$$

where $\mathbf{Y}^\alpha = (Y_{\alpha 1}, \ldots, Y_{\alpha p})'$, $\alpha = 1, \ldots, n$, are independently distributed normal p-vectors with mean $\mathbf{0}$ and the same covariance matrix Σ. Hence CSC' is distributed as

$$\sum_{\alpha=1}^{n} (C\mathbf{Y}^{\alpha})(C\mathbf{Y}^{\alpha})' = \sum_{\alpha=1}^{n} \mathbf{Z}^{\alpha}(\mathbf{Z}^{\alpha})',$$

where $\mathbf{Z}^{\alpha} = (Z_{\alpha 1}, \ldots, Z_{\alpha p})'$, $\alpha = 1, \ldots, n$, are independently and identically distributed normal p-vectors with mean $\mathbf{0}$ and covariance matrix $C\Sigma C'$, and hence the theorem. Q.E.D.

THEOREM 6.4.3 Let the $p \times p$ symmetric random matrix $S = (S_{ij})$ be distributed as $W_p(n, \Sigma)$. The characteristic function of S (i.e., the characteristic function of $S_{11}, S_{22}, \ldots, S_{pp}, 2S_{12}, 2S_{13}, \ldots, 2S_{p-1,p}$) is given by

$$E(\exp(i \operatorname{tr} \theta S)) = (\det(I - 2i\Sigma\theta))^{-n/2} \tag{6.50}$$

where $\theta = (\theta_{ij})$ is a real symmetric matrix of dimension $p \times p$.

Proof S is distributed as $\sum_{\alpha=1}^{n} \mathbf{Y}^{\alpha}(\mathbf{Y}^{\alpha})'$ where the $\mathbf{Y}^{\alpha}$, $\alpha = 1, \ldots, n$, have the same distribution as in Theorem 6.4.2. Hence

$$\begin{aligned} E(\exp(i \operatorname{tr} \theta S)) &= E\left(\exp\left(i \operatorname{tr} \theta \sum_{\alpha=1}^{n} \mathbf{Y}^{\alpha}(\mathbf{Y}^{\alpha})'\right)\right) \\ &= \prod_{\alpha=1}^{n} E(\exp(i \operatorname{tr} \theta \mathbf{Y}^{\alpha}(\mathbf{Y}^{\alpha})')) = (E \exp(i \operatorname{tr} \theta \mathbf{Y}(\mathbf{Y})'))^{n}, \end{aligned} \tag{6.51}$$

where $\mathbf{Y}$ has p-dimensional normal distribution with mean $\mathbf{0}$ and covariance matrix Σ. Since θ is real and Σ is positive definite there exists a nonsingular matrix C such that

$$C'\Sigma^{-1}C = I \qquad \text{and} \qquad C'\theta C = D,$$

where D is a diagonal matrix of diagonal elements d_{ii}. Let $\mathbf{Y} = C\mathbf{Z}$. Then $\mathbf{Z}$ has a p-dimensional normal distribution with mean $\mathbf{0}$ and covariance matrix I. Hence

$$\begin{aligned} E(\exp(i \operatorname{tr} \mathbf{Y}'\theta\mathbf{Y})) &= E(\exp(i \operatorname{tr} \mathbf{Z}'D\mathbf{Z})) \\ &= \prod_{j=1}^{p} E(\exp(id_{jj}Z_j^2)) = \prod_{j=1}^{p} (1 - 2id_{jj})^{-1/2} \\ &= (\det(I - 2iD))^{-1/2} = (\det(I - 2iC'\theta C))^{-1/2} \\ &= (\det(C'C))^{-1/2}(\det(\Sigma^{-1} - 2i\theta))^{-1/2} \\ &= (\det(I - 2i\theta\Sigma))^{-1/2}. \end{aligned}$$

Hence

$$E(\exp(i \operatorname{tr} \theta S)) = (\det(I - 2i\theta\Sigma))^{-n/2}. \qquad \text{Q.E.D.}$$

From this it follows that

$$E(S) = n\Sigma, \qquad \operatorname{cov}(S) = 2n\Sigma \otimes \Sigma. \tag{6.52}$$

THEOREM 6.4.4 If S_i, $i = 1, \ldots, k$, are independently distributed as $W_p(n_i, \Sigma)$, then $\sum_1^k S_i$ is distributed as $W_p(\sum_1^k n_i, \Sigma)$.

Proof Since S_i, $i = 1, \ldots, k$, are independently distributed as Wishart we can write

$$S_i = \sum_{\alpha = n_1 + \cdots + n_{i-1} + 1}^{n_1 + \cdots + n_i} \mathbf{Y}^\alpha (\mathbf{Y}^\alpha)', \qquad i = 1, \ldots, k,$$

where $\mathbf{Y}^\alpha = (Y_{\alpha 1}, \ldots, Y_{\alpha p})'$, $\alpha = 1, \ldots, \sum_1^k n_i$, are independently distributed p-dimensional normal random vectors with mean $\mathbf{0}$ and covariance matrix Σ. Hence

$$\sum_1^k S_i = \sum_{\alpha=1}^{n_1 + \cdots + n_k} \mathbf{Y}^\alpha (\mathbf{Y}^\alpha)'$$

is distributed as $W_p(\sum_1^k n_i, \Sigma)$. Q.E.D.

6.5 THE NONCENTRAL WISHART DISTRIBUTION

Let $\mathbf{X}^\alpha = (X_{\alpha 1}, \ldots, X_{\alpha p})'$, $\alpha = 1, \ldots, N$, be independently distributed normal p-vectors with mean $\boldsymbol{\mu}^\alpha = (\mu_{\alpha 1}, \ldots, \mu_{\alpha p})'$ and the same covariance matrix Σ. Let

$$X = (\mathbf{X}^1, \ldots, \mathbf{X}^N), \qquad D = XX', \qquad M = (\boldsymbol{\mu}^1, \ldots, \boldsymbol{\mu}^N).$$

The probability density function of X is given by

$$f_X(x) = (2\pi)^{-Np/2} (\det \Sigma)^{-N/2} \exp\{-\tfrac{1}{2} \operatorname{tr} \Sigma^{-1} (x - M)(x - M)'\}. \tag{6.53}$$

The distribution of D is called the noncentral Wishart distribution. Its probability density function in its most general form was first derived by James, 1954, 1955, 1964; see also Constantine, 1963; Anderson, 1945, 1946, and it involves the characteristic roots of $\Sigma^{-1} M M'$. The noncentral Wishart distribution is said to belong to the linear case if the rank of M is 1, and to the planar case if the rank of M is 2. In particular, if $\Sigma = I$, the probability density function of D can be written as (with respect to the Lebesgue measure)

$$\begin{aligned} f_D(d) = {} & 2^{-Np/2} \pi^{-p(p-1)/4} \left[\prod_{i=1}^p \Gamma\left(\frac{N - i + 1}{2} \right) \right]^{-1} \\ & \times \exp\{-\tfrac{1}{2}(\operatorname{tr} MM' + \operatorname{tr} d)\} (\det d)^{(N-p-1)/2} \\ & \times \int_{O(N)} \exp\{-\operatorname{tr} M' x \theta\}\, d\theta \end{aligned} \tag{6.54a}$$

where $O(N)$ is the group of $N \times N$ orthogonal matrices θ and $d\theta$ is the Lebesgue measure in the space of $O(N)$. In particular, if $\Sigma = I$ and

$$M = \begin{pmatrix} \mu_1 & \cdots & \mu_N \\ 0 & \cdots & 0 \\ 0 & \cdots & 0 \end{pmatrix},$$

then the distribution of $D = (D_{ij})$ is given by $[d = (d_{ij})]$

$$f_D(d) = \exp\{-\tfrac{1}{2}\lambda^2\} \sum_{\alpha=0}^{\infty} \frac{(\lambda^2/2)^\alpha}{\alpha!} \frac{\Gamma(N/2)}{\Gamma(N/2 + \alpha)} \left(\frac{d_{11}}{2}\right)^\alpha, \tag{6.54b}$$

where $\lambda^2 = \sum_1^N \mu_i^2$. This is called the canonical form of the noncentral Wishart distribution in the linear case.

6.6 GENERALIZED VARIANCE

For the p-variate normal distribution with mean $\boldsymbol{\mu}$ and covariance matrix Σ, det Σ is called the generalized variance of the distribution (see Wilks, 1932). Its estimate, based on sample observations $\mathbf{x}^\alpha = (x_{\alpha 1}, \ldots, x_{\alpha p})'$, $\alpha = 1, \ldots, N$,

$$\det\left(\frac{1}{N-1} \sum_{\alpha=1}^{N} (\mathbf{x}^\alpha - \bar{\mathbf{x}})(\mathbf{x}^\alpha - \bar{\mathbf{x}})'\right) = \frac{1}{(N-1)^p} \det(s)$$

is called the sample generalized variance or the generalized variance of the sample observations $\mathbf{x}^\alpha$, $\alpha = 1, \ldots, N$. The sample generalized variance occurs in many test criteria of statistical hypotheses concerning the means and covariance matrices of multivariate normal distributions. We will now consider the distribution of det S where S is distributed as $W_p(n, \Sigma)$.

THEOREM 6.6.1 Let S be distributed as $W_p(n, \Sigma)$. Then det S is distributed as $(\det \Sigma) \prod_{i=1}^p \chi^2_{n+1-i}$, where χ^2_{n+1-i}, $i = 1, \ldots, p$, are independent central chi-square random variables.

Note $W_1(n, 1)$ is a central chi-square random variable with n degrees of freedom.

Proof Since Σ is positive definite there exists a nonsingular matrix C such that $C\Sigma C' = I$. Let $S^* = CSC'$. Then S^* is distributed as Wishart $W_p(n, I)$. Now det $S^* = (\det \Sigma^{-1})(\det S)$. Hence det S is distributed as $(\det \Sigma)(\det S^*)$. Write $S^* = (S^*_{ij})$ as

$$S^* = \begin{pmatrix} S^*_{(11)} & S^*_{(12)} \\ S^*_{(21)} & S^*_{pp} \end{pmatrix}$$

where S^*_{pp} is 1×1. Then $\det S^* = S^*_{pp} \det(S^*_{(11)} - S^*_{(12)}S^{*-1}_{pp}S_{(21)})$. By Theorem 6.4.1 S^*_{pp} is distributed as $W_1(n, 1)$ independently of $(S^*_{(11)} - S^*_{(21)}S^{*-1}_{pp}S^*_{(12)})$ and $S^*_{(11)} - S^*_{(21)}S^{*-1}_{pp}S^*_{(12)}$ is distributed as $W_{p-1}(n-1, I_{p-1})$, where I_{p-1} is the identity matrix of dimension $(p-1) \times (p-1)$. Thus $\det(W_p(n, I_p))$ is distributed as the product of χ_n^2 and $\det(W_{p-1}(n-1, I_{p-1}))$, where χ_n^2 and $W_{p-1}(n-1, I_{p-1})$ are independent. Repeating this argument $p-1$ times we conclude that $\det S^*$ is distributed as $\prod_{i=1}^p \chi^2_{n+1-i}$ where χ^2_{n+1-i}, $i = 1, \ldots, p$, are independent chi-square random variables. Q.E.D.

6.7 DISTRIBUTION OF THE BARTLETT DECOMPOSITION (RECTANGULAR COORDINATES)

(*Rectangular Coordinates*) Let S be distributed as $W_p(n, \Sigma)$ and $n \geq p$. As we have observed earlier, S is positive definite with probability 1. Let $B = (B_{ij})$, $B_{ij} = 0$, $i < j$, be the unique lower triangular matrix with positive diagonal elements such that

$$S = BB' \tag{6.55}$$

(see Theorem 1.6.5). By Theorem 2.4.6 the Jacobian of the transformation $S \to B$ is given by $[s = (s_{ij}),\ b = (b_{ij})]$

$$\det\left(\frac{\partial s}{\partial b}\right) = 2^p \prod_{i=1}^{p} (b_{ii})^{p+1-i}. \tag{6.56}$$

From (6.32), (6.55), and (6.56) the probability density function of B with respect to the Lebesgue measure db is given by

$$\begin{aligned} f_B(b) &= K2^p(\det \Sigma)^{-n/2}(\det b)^{n-p-1} \exp\{-\tfrac{1}{2}\operatorname{tr} \Sigma^{-1}bb'\} \prod_{i=1}^{p} (b_{ii})^{p+1-i} \\ &= K2^p(\det \Sigma)^{-n/2} \prod_{i=1}^{p} (b_{ii})^{n-i} \exp\{-\tfrac{1}{2}\operatorname{tr} \Sigma^{-1}bb'\} \end{aligned} \tag{6.57}$$

where K is given by (6.37).

Let $T = (T_{ij})$ be a nonsingular lower triangular matrix (not necessarily with positive diagonal elements). Then we can write $T = B\theta$ where θ is a diagonal matrix with diagonal entries ± 1. Since the Jacobian of the transformation $B \to T = B\theta$ is unity, from (6.57) the probability density function of T is given by (with respect to the Lebesgue measure dt)

$$\begin{aligned} f_T(t) &= K2^p(\det \Sigma)^{-n/2}(\det(TT'))^{(n-p-1)/2} \\ &\quad \times \exp\{-\tfrac{1}{2}\operatorname{tr} \Sigma^{-1}TT'\} \prod_{i=1}^{p} |t_{ii}|^{p+1-i}, \end{aligned} \tag{6.58}$$

$t = (t_{ij})$, where K is given by (6.37). If $\Sigma = I$, (6.58) reduces to

$$f_T(t) = K2^p \exp\left\{-\frac{1}{2}\sum_{i=1}^{p}\sum_{j=1}^{i} t_{ij}^2\right\} \prod_{i=1}^{p} (t_{ii}^2)^{(n-i)/2}. \tag{6.59}$$

From (6.59) it is obvious that in this particular case the T_{ij} are independently distributed and T_{ii}^2 is distributed as central chi-square with $n - i + 1$ degrees of freedom ($i = 1, \ldots, p$), and T_{ij} ($i \neq j$) is normally distributed with mean 0 and variance 1.

6.8 DISTRIBUTION OF HOTELLING'S T^2 AND A RELATED DISTRIBUTION

Let $\mathbf{X}^\alpha = (X_{\alpha 1}, \ldots, X_{\alpha p})'$, $\alpha = 1, \ldots, N$, be independently distributed p-variate normal random variables with the same mean $\boldsymbol{\mu}$ and the same positive definite covariance matrix Σ. Let

$$\bar{\mathbf{X}} = \frac{1}{N}\sum_{1}^{N} \mathbf{X}^\alpha, \qquad S = \sum_{\alpha=1}^{N} (\mathbf{X}^\alpha - \bar{\mathbf{X}})(\mathbf{X}^\alpha - \bar{\mathbf{X}})'.$$

We have observed that $\sqrt{N}\,\bar{\mathbf{X}}$ has a p-variate normal distribution with mean $\sqrt{N}\,\boldsymbol{\mu}$ and covariance matrix Σ and that it is independent of S, which is distributed as $\sum_{\alpha=1}^{N-1} \mathbf{Y}^\alpha(\mathbf{Y}^\alpha)'$, where $\mathbf{Y}^\alpha = (Y_{\alpha 1}, \ldots, Y_{\alpha p})'$, $\alpha = 1, \ldots, N-1$, are independently and identically distributed normal p-vectors with mean $\mathbf{0}$ and covariance matrix Σ. We will prove the following theorem (due to Bowker).

THEOREM 6.8.1 $N\bar{\mathbf{X}}'S^{-1}\bar{\mathbf{X}}$ is distributed as

$$\frac{\chi_p^2(N\boldsymbol{\mu}'\Sigma^{-1}\boldsymbol{\mu})}{\chi_{N-p}^2} \tag{6.60}$$

where $\chi_p^2(N\boldsymbol{\mu}'\Sigma^{-1}\boldsymbol{\mu})$ and χ_{N-p}^2 are independent.

Proof Since Σ is positive definite there exists a nonsingular matrix C such that $C\Sigma C' = I$. Define $\mathbf{Z} = \sqrt{N}\,C\bar{\mathbf{X}}$, $A = CSC'$, and $\mathbf{v} = \sqrt{N}\,C\boldsymbol{\mu}$. Then $\mathbf{Z}$ is normally distributed with mean $\mathbf{v}$ and covariance matrix I, and A is distributed as $\sum_{\alpha=1}^{N-1} \mathbf{Z}^\alpha(\mathbf{Z}^\alpha)'$ where $\mathbf{Z}^\alpha = (Z_{\alpha 1}, \ldots, Z_{\alpha p})'$, $\alpha = 1, \ldots, N-1$, are independently and identically distributed normal p-vectors with mean $\mathbf{0}$ and covariance matrix I. Furthermore, A and $\mathbf{Z}$ are independent.

Consider a random orthogonal matrix Q of dimension $p \times p$ whose first row is $\mathbf{Z}'(\mathbf{Z}'\mathbf{Z})^{-1/2}$ and whose remaining $p - 1$ rows are defined arbitrarily. Let

$$\mathbf{U} = (U_1, \ldots, U_p)' = Q\mathbf{Z}, \qquad B = (B_{ij}) = QAQ'.$$

Obviously,

$$U_1 = (\mathbf{Z}'\mathbf{Z})^{1/2}, \qquad U_i = 0, \qquad i = 2, \ldots, p,$$

and

$$N\bar{\mathbf{X}}'S^{-1}\bar{\mathbf{X}} = \mathbf{Z}'A^{-1}\mathbf{Z} = \mathbf{U}'B^{-1}\mathbf{U} = U_1{}^2/(B_{11} - B_{(12)}B_{(22)}^{-1}B_{(21)}) \tag{6.61}$$

where

$$B = \begin{pmatrix} B_{11} & B_{(12)} \\ B_{(21)} & B_{(22)} \end{pmatrix}.$$

Since the conditional distribution of B given Q is Wishart with $N - 1$ degrees of freedom and parameter I, by Theorem 6.4.1, the conditional distribution of $B_{11} - B_{(12)}B_{(22)}^{-1}B_{(21)}$ given Q is central chi-square with $N - p$ degrees of freedom. As this conditional distribution does not depend on Q, the unconditional distribution of $B_{11} - B_{(12)}B_{(22)}^{-1}B_{(21)}$ is also central chi-square with $N - p$ degrees of freedom. By the results presented in Section 6.1, $\mathbf{Z}'\mathbf{Z}$ is distributed as a noncentral chi-square with p degrees of freedom and the noncentrality parameter $\mathbf{v}'\mathbf{v} = N\boldsymbol{\mu}'\Sigma^{-1}\boldsymbol{\mu}$. The independence of $\mathbf{Z}'\mathbf{Z}$ and $B_{11} - B_{(12)}B_{(22)}^{-1}B_{(21)}$ is obvious. Q.E.D.

We now need the following lemma to demonstrate the remaining results in this section.

LEMMA 6.8.1 For any p-vector $\mathbf{Y} = (Y_1, \ldots, Y_p)'$ and any $p \times p$ positive definite matrix A

$$\mathbf{Y}'(A + \mathbf{Y}\mathbf{Y}')^{-1}\mathbf{Y} = \frac{\mathbf{Y}'A^{-1}\mathbf{Y}}{1 + \mathbf{Y}'A^{-1}\mathbf{Y}}. \tag{6.62}$$

Proof Let

$$(A + \mathbf{Y}\mathbf{Y}')^{-1} = A^{-1} + C.$$

Then

$$I = (A^{-1} + C)(A + \mathbf{Y}\mathbf{Y}') = I + A^{-1}\mathbf{Y}\mathbf{Y}' + CA + C\mathbf{Y}\mathbf{Y}'.$$

Since $(A + \mathbf{Y}\mathbf{Y}')$ is positive definite,

$$C = -A^{-1}\mathbf{Y}\mathbf{Y}'(A + \mathbf{Y}\mathbf{Y}')^{-1}.$$

Now

$$\mathbf{Y}'(A + \mathbf{Y}\mathbf{Y}')^{-1}\mathbf{Y} = \mathbf{Y}'A^{-1}\mathbf{Y} - (\mathbf{Y}'A^{-1}\mathbf{Y})(\mathbf{Y}'(A + \mathbf{Y}\mathbf{Y}')^{-1}\mathbf{Y}),$$

or

$$\mathbf{Y}'(A + \mathbf{Y}\mathbf{Y}')^{-1}\mathbf{Y} = \frac{\mathbf{Y}'A^{-1}\mathbf{Y}}{1 + \mathbf{Y}'A^{-1}\mathbf{Y}}. \qquad \text{Q.E.D.}$$

Notations For any p-vector $\mathbf{Y} = (Y_1, \ldots, Y_p)'$ and any $p \times p$ matrix $A = (a_{ij})$ we shall write for $i = 1, \ldots, k$ and $k \leq p$

$$\mathbf{Y} = (\mathbf{Y}_{(1)}, \ldots, \mathbf{Y}_{(k)})', \qquad A = \begin{pmatrix} A_{(11)} & \cdots & A_{(1k)} \\ \vdots & & \vdots \\ A_{(k1)} & \cdots & A_{(kk)} \end{pmatrix},$$

$$\mathbf{Y}_{[i]} = (Y_{(1)}, \ldots, Y_{(i)})',$$

$$A_{[ij]} = (A_{(i1)}, \ldots, A_{(ij)}),$$

$$A_{[ji]} = (A_{(1i)}, \ldots, A_{(ji)}), \qquad A_{[ii]} = \begin{pmatrix} A_{(11)} & \cdots & A_{(1i)} \\ \vdots & & \vdots \\ A_{(i1)} & \cdots & A_{(ii)} \end{pmatrix},$$

where $\mathbf{Y}_{(i)}$ are subvectors of $\mathbf{Y}$ of dimension $p_i \times 1$, and $A_{(ii)}$ are submatrices of A of dimension $p_i \times p_i$, where the p_i are arbitrary integers including zero such that $\sum_1^k p_i = p$. Let us now define $R_1, \ldots, R_k$ by

$$\sum_{j=1}^{i} R_j = N\bar{\mathbf{X}}'_{[i]}(S_{[ii]} + N\bar{\mathbf{X}}_{[i]}\bar{\mathbf{X}}'_{[i]})^{-1}\bar{\mathbf{X}}_{[i]}$$

$$= \frac{N\bar{\mathbf{X}}'_{[i]}S_{[ii]}^{-1}\bar{\mathbf{X}}_{[i]}}{1 + N\bar{\mathbf{X}}'_{[i]}S_{[ii]}^{-1}\bar{\mathbf{X}}_{[i]}}, \qquad i = 1, \ldots, k. \tag{6.63}$$

Since S is positive definite with probability 1 (we shall assume $N > p$), $S_{[ii]}$, $i = 1, \ldots, k$, are positive definite and hence $R_i \geq 0$ for $i = 1, \ldots, k$ with probability 1. We are interested here in showing that the joint probability density function of $R_1, \ldots, R_k$ is given by

$$f_{R_1, \ldots, R_k}(r_1, \ldots, r_k) = \Gamma\left(\frac{N}{2}\right)\left[\Gamma\left(\tfrac{1}{2}\left(N - \sum_1^k p_i\right)\right)\prod_{i=1}^{k}(\Gamma(\tfrac{1}{2}p_i))\right]^{-1}$$

$$\times \prod_{i=1}^{k}(r_i)^{p_i/2-1}\left(1 - \sum_{i=1}^{k} r_i\right)^{(N-\Sigma_1^k p_i)/2-1}$$

$$\times \exp\left\{-\frac{1}{2}\sum_1^k \delta_i^2 + \frac{1}{2}\sum_{j=1}^{k} r_j \sum_{i>j}^{k} \delta_i^2\right\}$$

$$\times \prod_{i=1}^{k} \phi(\tfrac{1}{2}(N - \sigma_{i-1}), \tfrac{1}{2}p_i; \tfrac{1}{2}r_i\delta_i^2) \tag{6.64}$$

where

$$\sigma_i = \sum_{j=1}^{i} p_j \qquad \text{with} \quad \sigma_0 = 0$$

$$\sum_{j=1}^{i} \delta_j^2 = N\boldsymbol{\mu}'_{[i]}\Sigma_{[ii]}^{-1}\boldsymbol{\mu}_{[i]}, \qquad i = 1, \ldots, k$$

and $\phi(a, b; x)$ is the confluent hypergeometric function given by

$$\phi(a, b; x) = 1 + \frac{a}{b}x + \frac{a(a+1)}{b(b+1)}\frac{x^2}{2!} + \frac{a(a+1)(a+2)}{b(b+1)(b+2)}\frac{x^3}{3!} + \cdots. \tag{6.65}$$

For $k = 1$,

$$R_1 = \frac{N\bar{\mathbf{X}}'S^{-1}\bar{\mathbf{X}}}{1 + N\bar{\mathbf{X}}'S^{-1}\bar{\mathbf{X}}}, \qquad \delta_1^2 = N\boldsymbol{\mu}'\Sigma^{-1}\boldsymbol{\mu}.$$

From (6.60) its probability density function is given by

$$f_{R_1}(r_1) = \frac{\Gamma(\frac{1}{2}N)}{\Gamma(\frac{1}{2}(N-p))\Gamma(\frac{1}{2}p)} r_1^{p/2-1}(1 - r_1)^{(N-p)/2-1} \times \exp\{-\tfrac{1}{2}\delta_1^2\}\phi(\tfrac{1}{2}N, \tfrac{1}{2}p; \tfrac{1}{2}r_1\delta_1^2) \tag{6.66}$$

which agrees with (6.64). To prove (6.64) in general we first consider the case $k = 2$ and then use this result for the case $k = 3$. The desired result for the general case will then follow from these cases. The statistics $R_1, \ldots, R_k$ play an important role in tests of hypotheses concerning means of multivariate normal distributions with unknown covariance matrices (see Chapter VII) and tests of hypotheses concerning discriminant coefficients of discriminant analysis (see Chapter IX).

Let us now prove (6.64) for $k = 2$, $p_1 + p_2 = p$. Let

$$S = \begin{pmatrix} S_{(11)} & S_{(12)} \\ S_{(21)} & S_{(22)} \end{pmatrix}$$

$$W = S_{(22)} - S_{(21)}S_{(11)}^{-1}S_{(12)}, \qquad U = S_{(21)}S_{(11)}^{-1}, \qquad V = S_{(11)}. \tag{6.67}$$

Identifying $S_{(22)}$ with $S_{(11)}$, $S_{(21)}$ with $S_{(12)}$, and vice versa in Theorem 6.4.1 we obtain: W is distributed as $W_{p_2}(N - 1 - p_1, \Sigma_{(22)} - \Sigma_{(21)}\Sigma_{(11)}^{-1}\Sigma_{(12)})$, the conditional distribution of U, given that $S_{(11)} = s_{(11)}$ is normal with mean $\Sigma_{(21)}\Sigma_{(11)}^{-1}$ and covariance matrix $(\Sigma_{(22)} - \Sigma_{(21)}\Sigma_{(11)}^{-1}\Sigma_{(12)}) \otimes s_{(11)}^{-1}$; V is distributed as $W_{p_1}(N - 1, \Sigma_{(11)})$; and W is independent of (U, V).

Hence the conditional distribution of

$$\sqrt{N}\, S_{(21)}S_{(11)}^{-1}\bar{\mathbf{X}}_{(1)}$$

given that $\bar{\mathbf{X}}_{(1)} = \bar{\mathbf{x}}_{(1)}$, $S_{(11)} = s_{(11)}$ is a p_2-variate normal with mean $\sqrt{N}\,\Sigma_{(21)}\Sigma_{(11)}^{-1}\bar{\mathbf{x}}_{(1)}$ and covariance matrix

$$(N\bar{\mathbf{x}}_{(1)}'s_{(11)}^{-1}\bar{\mathbf{x}}_{(1)})(\Sigma_{(22)} - \Sigma_{(21)}\Sigma_{(11)}^{-1}\Sigma_{(12)}).$$

Now let

$$W_1 = N\bar{\mathbf{X}}'_{(1)}S^{-1}_{(11)}\bar{\mathbf{X}}_{(1)} = R_1(1 - R_1)^{-1}, \tag{6.68}$$

$$\begin{aligned} W_2 &= \frac{N(\bar{\mathbf{X}}_{(2)} - S_{(21)}S^{-1}_{(11)}\bar{\mathbf{X}}_{(1)})'(S_{(22)} - S_{(21)}S^{-1}_{(11)}S_{(12)})^{-1}(\bar{\mathbf{X}}_{(2)} - S_{(21)}S^{-1}_{(11)}\bar{\mathbf{X}}_{(1)})}{1 + W_1} \\ &= \{(R_1 + R_2)(1 - R_1 - R_2)^{-1} - R_1(1 - R_1)^{-1}\}(1 + R_1(1 - R_1)^{-1})^{-1} \\ &= R_2(1 - R_1 - R_2)^{-1}. \end{aligned}$$

Then

$$\begin{aligned} N\bar{\mathbf{X}}'S^{-1}\bar{\mathbf{X}} &= N\bar{\mathbf{X}}'_{(1)}S^{-1}_{(11)}\bar{\mathbf{X}}_{(1)} \\ &\quad + N(\bar{\mathbf{X}}_{(2)} - S_{(21)}S^{-1}_{(11)}\bar{\mathbf{X}}_{(1)})'(S_{(22)} - S_{(21)}S^{-1}_{(11)}S_{(12)})^{-1} \\ &\quad \times (\bar{\mathbf{X}}_{(2)} - S_{(21)}S^{-1}_{(11)}\bar{\mathbf{X}}_{(1)}) \\ &= W_1 + W_2(1 + W_1). \end{aligned}$$

Similarly, from $N\boldsymbol{\mu}'\Sigma^{-1}\boldsymbol{\mu} = \delta_1{}^2 + \delta_2{}^2$, $\delta_1{}^2 = N\boldsymbol{\mu}'_{(1)}\Sigma^{-1}_{(11)}\boldsymbol{\mu}_{(1)}$, we get

$$\begin{aligned} \delta_2{}^2 &= N(\boldsymbol{\mu}_{(2)} - \Sigma_{(21)}\Sigma^{-1}_{(11)}\boldsymbol{\mu}_{(1)})'(\Sigma_{(22)} - \Sigma_{(21)}\Sigma^{-1}_{(11)}\Sigma_{(12)})^{-1} \\ &\quad \times (\boldsymbol{\mu}_{(2)} - \Sigma_{(21)}\Sigma^{-1}_{(11)}\boldsymbol{\mu}_{(1)}). \end{aligned} \tag{6.69}$$

Since $\sqrt{N}\,\bar{\mathbf{X}}$ is independent of S and is distributed as a p-variate normal with mean $\sqrt{N}\,\boldsymbol{\mu}$ and covariance matrix Σ, the conditional distribution of $N\bar{\mathbf{X}}_{(2)}$ given that $S_{(11)} = s_{(11)}$ and $\bar{\mathbf{X}}_{(1)} = \bar{\mathbf{x}}_{(1)}$ is a p_2-variate normal with mean $\sqrt{N}(\boldsymbol{\mu}_{(2)} + \Sigma_{(21)}\Sigma^{-1}_{(11)}(\bar{\mathbf{x}}_{(1)} - \boldsymbol{\mu}_{(1)}))$ and covariance matrix $\Sigma_{(22)} - \Sigma_{(21)}\Sigma^{-1}_{(11)}\Sigma_{(12)}$. Furthermore this conditional distribution is independent of the conditional distribution of $\sqrt{N}\,S_{(21)}S^{-1}_{(11)}\bar{\mathbf{X}}_{(1)}$ given that $S_{(11)} = s_{(11)}$ and $\bar{\mathbf{X}}_{(1)} = \bar{\mathbf{x}}_{(1)}$. Hence the conditional distribution of

$$\sqrt{N}(\bar{\mathbf{X}}_{(2)} - S_{(21)}S^{-1}_{(11)}\bar{\mathbf{X}}_{(1)})(1 + W_1)^{-1/2}$$

given that $S_{(11)} = s_{(11)}$, $\bar{\mathbf{X}}_{(1)} = \bar{\mathbf{x}}_{(1)}$ is a p_2-variate normal with mean $\sqrt{N}(\boldsymbol{\mu}_{(2)} - \Sigma_{(21)}\Sigma^{-1}_{(11)}\boldsymbol{\mu}_{(1)})(1 + w_1)^{-1/2}$ (w_1 is the value of W_1 corresponding to $S_{(11)} = s_{(11)}$, $\bar{\mathbf{X}}_{(1)} = \bar{\mathbf{x}}_{(1)}$) and covariance matrix $\Sigma_{(22)} - \Sigma_{(21)}\Sigma^{-1}_{(11)}\Sigma_{(12)}$.

Since $S_{(22)} - S_{(21)}S^{-1}_{(11)}S_{(12)}$ is distributed independently of $(S_{(21)}, S_{(11)})$ and $\bar{\mathbf{X}}$ as $W_{p_2}(N - 1 - p_1, \Sigma_{(22)} - \Sigma_{(21)}\Sigma^{-1}_{(11)}\Sigma_{(12)})$, by Theorem 6.8.1, the conditional distribution of W_2 given that $S_{(11)} = s_{(11)}$, $\bar{\mathbf{X}}_{(1)} = \bar{\mathbf{x}}_{(1)}$, is

$$\chi^2_{p_2}(\delta_2{}^2(1 + w_1)^{-1})/\chi^2_{N-p_1-p_2} \tag{6.70}$$

where $\chi^2_{p_2}(\delta_2{}^2(1 + w_1)^{-1})$ and $\chi^2_{N-p_1-p_2}$ are independent. Furthermore by the same theorem W_1 is distributed as

$$\chi^2_{p_1}(\delta_1{}^2)/\chi^2_{N-p_1}, \tag{6.71}$$

where $\chi^2_{p_1}(\delta_1{}^2)$ and $\chi^2_{N-p_1}$ are independent. Thus the joint probability density function of (W_1, W_2) is given by

$$\begin{aligned} f_{W_1,W_2}(w_1, w_2) &= \exp\{-\tfrac{1}{2}\delta_2^{\,2}(1 + w_1)^{-1}\} \\ &\times \sum_{\beta=1}^{\infty} \frac{(\tfrac{1}{2}\delta_2^{\,2}(1 + w_1)^{-1})^{\beta}(w_2)^{p_2/2+\beta-1}\Gamma(\tfrac{1}{2}(N - p_1) + \beta)}{\beta!(1 + w_2)^{(N-p_1)/2+\beta}\Gamma(\tfrac{1}{2}p_2 + \beta)\Gamma(\tfrac{1}{2}(N - p))} \\ &\times \exp(-\tfrac{1}{2}\delta_1^{\,2}) \sum_{j=0}^{\infty} \frac{(\tfrac{1}{2}\delta_1^{\,2})^j w_1^{p_1/2+j-1}\Gamma(\tfrac{1}{2}N + j)}{j!(1 + w_1)^{N/2+j}\Gamma(\tfrac{1}{2}p_1 + j)\Gamma(\tfrac{1}{2}(N - p_1))}. \end{aligned} \tag{6.72}$$

Now transforming $(W_1, W_2) \to (R_1, R_2)$ as given by (6.68) the joint probability density function of R_1, R_2 is

$$\begin{aligned} f_{R_1,R_2}(r_1, r_2) &= \Gamma(\tfrac{1}{2}N)[\Gamma(\tfrac{1}{2}(N - p))\Gamma(\tfrac{1}{2}p_1)\Gamma(\tfrac{1}{2}p_2)]^{-1} \\ &\times (r_1)^{p_1/2-1}(r_2)^{p_2/2-1}(1 - r_1 - r_2)^{(N-p)/2-1} \\ &\times \exp\{-\tfrac{1}{2}(\delta_1^{\,2} + \delta_2^{\,2}) + \tfrac{1}{2}\delta_2^{\,2}r_1\} \prod_{i=1}^{2} \phi(N - \sigma_{i-1}, \tfrac{1}{2}p_i; \tfrac{1}{2}r_i\delta_i^{\,2}), \end{aligned} \tag{6.73}$$

which agrees with (6.64) for $k = 2$.

Let us now consider the case $k = 3$. Let

$$W_3 = (N\bar{\mathbf{X}}'S^{-1}\bar{\mathbf{X}} - N\bar{\mathbf{X}}'_{[2]}S_{[22]}^{-1}\bar{\mathbf{X}}_{[2]})/(1 + N\bar{\mathbf{X}}'_{[2]}S_{[22]}^{-1}\bar{X}_{[2]}). \tag{6.74}$$

Now $S_{(33)} - S_{[32]}S_{[22]}^{-1}S_{[23]}$ is distributed as

$$W_{p_3}(N - 1 - p_1 - p_2, (\Sigma_{(33)} - \Sigma_{[32]}\Sigma_{[22]}^{-1}\Sigma_{[23]}))$$

and is independent of $S_{[22]}$ and $S_{[32]}$. Also, the conditional distribution of $\sqrt{N}\,\bar{\mathbf{X}}_{(3)}$ given that $S_{[22]} = s_{[22]}$, $\bar{\mathbf{X}}_{[2]} = \bar{\mathbf{x}}_{[2]}$ is normal with mean $\sqrt{N}(\boldsymbol{\mu}_{(3)} - \Sigma_{[32]}\Sigma_{[22]}^{-1}(\bar{\mathbf{x}}_{[2]} - \boldsymbol{\mu}_{[2]}))$ and covariance matrix $\Sigma_{(33)} - \Sigma_{[32]}\Sigma_{[22]}^{-1}\Sigma_{[23]}$ and is independent of the conditional distribution of $\sqrt{N}\ S_{[32]}S_{[22]}^{-1}\bar{\mathbf{X}}_{[2]}$ given that $S_{[22]} = s_{[22]}$ and $\bar{\mathbf{X}}_{[2]} = \bar{\mathbf{x}}_{[2]}$, which is normal with mean $\sqrt{N}\,\Sigma_{[32]}\Sigma_{[22]}^{-1}\bar{\mathbf{x}}_{[2]}$ and covariance matrix $(N\bar{\mathbf{x}}'_{[2]}s_{[22]}^{-1}\bar{\mathbf{x}}_{[2]})(\Sigma_{(33)} - \Sigma_{[32]}\Sigma_{[22]}^{-1}\Sigma_{[23]})$. Hence as before the conditional distribution of W_3 given that $S_{[22]} = s_{[22]}$ and $\bar{\mathbf{X}}_{[2]} = \bar{\mathbf{x}}_{[2]}$ or, equivalently, given that $W_1 = w_1$, $W_2 = w_2$, is given by

$$f_{W_3|W_1,W_2}(w_3|w_1, w_2) = \chi_{p_3}^2(\delta_3^{\,2}(1 + w_1)(w_1 + w_2 + w_1w_2)^{-1})/\chi_{N-p}^2 \tag{6.75}$$

where $\chi_{p_3}^2(\cdot)$ and χ_{N-p}^2 are independent.

Thus the joint probability density function of W_1, W_2, W_3 is

$$\begin{aligned} f_{W_1,W_2,W_3}(w_1, w_2, w_3) &= \frac{\chi_{p_3}^2(\delta_3^{\,2}(1 + w_1)(w_1 + w_2 + w_1w_2)^{-1})}{\chi_{N-p}^2} \\ &\times \frac{\chi_{p_2}^2(\delta_2^{\,2}(1 + w_1)^{-1})}{\chi_{N-p_1-p_2}^2} \times \frac{\chi_{p_1}^2(\delta_1^{\,2})}{\chi_{N-p_1}^2}. \end{aligned} \tag{6.76}$$

Now replacing the W_i by R_i we get

$$W_1 = R_1(1 - R_1)^{-1}, \qquad W_2 = R_2(1 - R_1 - R_2)$$
$$W_3 = ((R_1 + R_2 + R_3)(1 - R_1 - R_2 - R_3)^{-1} - (R_1 + R_2)(1 - R_1 - R_2)^{-1})(1 - R_1 - R_2) = R_3(1 - R_1 - R_2 - R_3)^{-1}.$$

From (6.76) the joint probability density function of R_1, R_2, R_3 is given by

$$\begin{aligned} f_{R_1,R_2,R_3}(r_1, r_2, r_3) &= \Gamma(\tfrac{1}{2}N)[\Gamma(\tfrac{1}{2}(N - p)) \prod_{i=1}^{3} \Gamma(\tfrac{1}{2}p_i)]^{-1} \\ &\times \prod_{i=1}^{3} (r_i)^{p_i/2 - 1}\left(1 - \sum_1^3 r_i\right)^{(N-p)/2 - 1} \\ &\times \exp\left\{-\frac{1}{2}\sum_1^3 \delta_j^2 + \frac{1}{2}\sum_{j=1}^{3} r_j \sum_{i>j}^{3} \delta_i^2\right\} \\ &\times \prod_{i=1}^{3} \phi(\tfrac{1}{2}(N - \sigma_{i-1}), \tfrac{1}{2}p_i; \tfrac{1}{2}r_i\delta_i^2) \end{aligned} \tag{6.77}$$

which agrees with (6.64) for $k = 3$. Proceeding exactly in this fashion we get (6.64) for general k.

6.9 DISTRIBUTION OF MULTIPLE AND PARTIAL CORRELATION COEFFICIENTS

Let S be distributed as $W_p(N - 1, \Sigma)$ and let

$$S = (S_{ij}) = \begin{pmatrix} S_{11} & S_{(12)} \\ S_{(21)} & S_{(22)} \end{pmatrix}, \qquad \Sigma = (\Sigma_{ij}) = \begin{pmatrix} \Sigma_{11} & \Sigma_{(12)} \\ \Sigma_{(21)} & \Sigma_{(22)} \end{pmatrix}.$$

We shall first find the distribution of

$$R_2 = \frac{S_{(12)}S_{(22)}^{-1}S_{(21)}}{S_{11}}. \tag{6.78}$$

From this

$$\begin{aligned} \frac{R^2}{1 - R^2} &= \frac{S_{(12)}S_{(22)}^{-1}S_{(21)}}{S_{11} - S_{(12)}S_{(22)}^{-1}S_{(21)}} \\ &= \left(\frac{S_{(12)}S_{(22)}^{-1}S_{(21)}}{\Sigma_{11} - \Sigma_{(12)}\Sigma_{(22)}^{-1}\Sigma_{(21)}}\right)\Big/\left(\frac{S_{11} - S_{(12)}S_{(22)}^{-1}S_{(21)}}{\Sigma_{11} - \Sigma_{(12)}\Sigma_{(22)}^{-1}\Sigma_{(21)}}\right) = \frac{X}{Y}, \end{aligned} \tag{6.79}$$

where

$$X = \frac{S_{(12)}S_{(22)}^{-1}S_{(21)}}{\Sigma_{11} - \Sigma_{(12)}\Sigma_{(22)}^{-1}\Sigma_{(21)}}, \qquad Y = \frac{S_{(11)} - S_{(12)}S_{(22)}^{-1}S_{(21)}}{\Sigma_{11} - \Sigma_{(12)}\Sigma_{(22)}^{-1}\Sigma_{(21)}}.$$

From Theorem 6.4.1, Y is distributed as central chi-square with $N - p$ degrees of freedom and is independent of $(S_{(12)}, S_{(22)})$ and the conditional distribution of $S_{(12)}S_{(22)}^{-1/2}$ given that $S_{(22)} = s_{(22)}$ is a $(p-1)$-variate normal distribution with mean $\Sigma_{(12)}\Sigma_{(22)}^{-1}s_{(22)}^{1/2}$ and covariance matrix $(\Sigma_{11} - \Sigma_{(12)}\Sigma_{(22)}^{-1}\Sigma_{(21)})I$. Hence the conditional distribution of Y given that $S_{(22)} = s_{(22)}$ is noncentral chi-square

$$\chi_{p-1}^2\left(\frac{\Sigma_{(12)}\Sigma_{(22)}^{-1}s_{(22)}\Sigma_{(22)}^{-1}\Sigma_{(21)}}{\Sigma_{11} - \Sigma_{(12)}\Sigma_{(22)}^{-1}\Sigma_{(21)}}\right). \tag{6.80}$$

Since $S_{(22)}$ is distributed as $W_{p-1}(N-1, \Sigma_{(22)})$ (see Theorem 6.4.1) by Exercise 4,

$$\frac{\Sigma_{(12)}\Sigma_{(22)}^{-1}S_{(22)}\Sigma_{(22)}^{-1}\Sigma_{(21)}}{\Sigma_{11} - \Sigma_{(12)}\Sigma_{(22)}^{-1}\Sigma_{(21)}} \tag{6.81}$$

is distributed as χ_{N-1}^2. Since

$$\frac{\Sigma_{(12)}\Sigma_{(22)}^{-1}\Sigma_{(21)}}{\Sigma_{(11)} - \Sigma_{(12)}\Sigma_{(22)}^{-1}\Sigma_{(21)}} = \frac{\rho^2}{1-\rho^2}, \tag{6.82}$$

where

$$\rho^2 = \frac{\Sigma_{(12)}\Sigma_{(22)}^{-1}\Sigma_{(21)}}{\Sigma_{11}}, \tag{6.83}$$

$R^2/(1-R^2)$ is distributed as the ratio (of independent random variables) X/Y, where Y is distributed as χ_{N-p}^2 and X is distributed as $\chi_{p-1}^2((\rho^2/(1-\rho^2))\chi_{N-1}^2)$ with random noncentrality parameter $(\rho/(1-\rho^2))\chi_{N-1}^2$. Since, from (6.5), a noncentral chi-square $Z = \chi_N{}^2(\lambda)$ is distributed as χ_{N+2K}^2 where K is a Poisson random variable with parameter $\lambda/2$, i.e., its probability density function is given by

$$f_Z(z) = \sum_{k=0}^{\infty} f_{\chi_{N+2k}^2}(z)p_K(k),$$

where $p_K(k)$ is the Poisson probability mass function with parameter λ, it follows that the conditional distribution of X given that $\chi_{N-1}^2 = t$. χ_{p-1+2K}^2, where the conditional distribution of K given that $\chi_{N-1}^2 = t$ is Poisson with parameter $\frac{1}{2}t(\rho^2/(1-\rho^2))$.

Let $\lambda/2 = \frac{1}{2}(\rho^2/(1-\rho^2))$. The unconditional probability mass function of K is given by

$$p_K(k) = \int_0^\infty \exp\{-\tfrac{1}{2}\lambda t\}\frac{(\lambda t/2)^k}{k!}\frac{t^{((N-1)/2)-1}\exp\{-\frac{1}{2}t\}\,dt}{2^{(N-1)/2}\Gamma(\frac{1}{2}(N-1))}$$

$$= \frac{\Gamma(\frac{1}{2}(N-1)+k)}{k!\Gamma(\frac{1}{2}(N-k))}(\rho^2)^k(1-\rho^2)^{(N-1)/2}, \qquad k = 0, 1, 2, \ldots. \tag{6.84}$$

This implies that the unconditional distribution of K is negative binomial. Hence the probability density function of X is given by

$$f_X(x) = \sum_{k=0}^{\infty} f_{\chi^2_{p-1+2k}}(x)p_K(k), \tag{6.85}$$

where $f_{\chi^2_{p-1+2k}}(x)$ is the probability density function of χ^2_{p-1+2k} and $p_K(k)$ is given by (6.84). Thus we get the following theorem.

THEOREM 6.9.1 The probability density function of $R^2/(1-R^2)$ is given by the probability density function of the ratio X/Y of two independently distributed random variables X, Y, where X is distributed as a chi-square random variable χ^2_{p-1+2K} with K a negative binomial random variable with probability mass function given in (6.84) and Y is distributed as χ^2_{N-p}.

It is now left to the reader as an exercise to verify that the probability density function of R^2 is given by

$$f_{R^2}(r^2) = \begin{cases} \dfrac{(1-\rho^2)^{(N-1)/2}(1-r^2)^{(N-p-2)/2}}{\Gamma(\frac{1}{2}(N-1))\Gamma(\frac{1}{2}(N-p))} \\ \times \displaystyle\sum_{j=0}^{\infty} \frac{(\rho^2)^j(r^2)^{(p-1)/2+j-1}\Gamma^2(\frac{1}{2}(N-1)+j)}{j!\Gamma(\frac{1}{2}(p-1)+j)}, \\ \qquad\qquad \text{if } r^2 \geq 0, \\ 0 \qquad \text{otherwise.} \end{cases} \tag{6.86}$$

This derivation is due to Fisher (1928). For $p = 2$, the special case studied by Fisher in 1915, the probability density function of R is given by

$$f_R(r) = \frac{2^{N-3}(1-\rho^2)^{(N-1)/2}(1-r^2)^{(N-4)/2}}{(N-3)!\pi}\sum_{j=0}^{\infty}\frac{(2\rho r)^j}{j!}\Gamma^2(\tfrac{1}{2}(N-1)+j)$$

$$= \frac{(1-\rho^2)^{(N-1)/2}(1-r^2)^{(N-4)/2}}{\pi(N-3)!}\left[\frac{d^{n-1}}{dx^{n-1}}\left\{\frac{\cos^{-1}(-x)}{(1-x^2)^{1/2}}\right\}\bigg|_{x=r\rho}\right], \tag{6.87}$$

which follows from (6.86) with $p = 2$ and the fact that

$$\Gamma(n)\Gamma(n+\tfrac{1}{2}) = \sqrt{\pi}\,\Gamma(2n)/2^{2n-1}. \tag{6.88}$$

It is well known that as $N \to \infty$, the distribution of $(\sqrt{N}(R - \rho))/(1 - \rho^2)$ tends to normal distribution with mean 0 and variance 1.

Let $\mathbf{X} = (X_1, \ldots, X_p)'$ be normally distributed with mean $\boldsymbol{\mu}$ and covariance matrix $\Sigma = (\sigma_{ij})$ and let $\rho_{ij} = \sigma_{ij}/(\sigma_{ii}\sigma_{jj})^{1/2}$. It is now obvious that the distribution of the sample correlation coefficient R_{ij} between the ith and jth components of $\mathbf{X}$, based on a random sample of size N, is obtained from (6.87) by replacing ρ by ρ_{ij}.

Let $\mathbf{X}^\alpha = (X_{\alpha 1}, \ldots, X_{\alpha p})'$, $\alpha = 1, \ldots, N$, be a random sample of size N from the distribution of $\mathbf{X}$ and let

$$S = \sum_{\alpha=1}^{N} (\mathbf{X}^\alpha - \bar{\mathbf{X}})(\mathbf{X}^\alpha - \bar{\mathbf{X}})'$$

be partitioned as

$$S = \begin{pmatrix} S_{(11)} & S_{(12)} \\ S_{(21)} & S_{(22)} \end{pmatrix},$$

where $S_{(11)}$ is a $q \times q$ submatrix of S. We observed in Chapter V that the sample partial correlation coefficients $r_{ij.1,\ldots,q}$ can be computed from $s_{(22)} - s_{(21)}s_{(11)}^{-1}s_{(12)}$ in the same way that the sample simple correlation coefficients r_{ij} are computed from s. Furthermore, we observed that to obtain the distribution of R_{ij} (random variable corresponding to r_{ij}) we needed the fact that S is distributed as $W_p(N - 1, \Sigma)$.

Since, from Theorem 6.4.1, $S_{(22)} - S_{(21)}S_{(11)}^{-1}S_{(12)}$ is distributed as $W_{p-q}(N - q - 1, \Sigma_{(22)} - \Sigma_{(21)}\Sigma_{(11)}^{-1}\Sigma_{(12)})$ and is independent of $(S_{(11)}, S_{(12)})$, it follows that the distribution $R_{ij.1,\ldots,q}$ based on N observations is the same as that of the simple correlation coefficient R_{ij} based on $N - q$ observations with a corresponding population parameter $\rho_{ij.1,\ldots,q}$.

For more relevent results in connection with multivariate distributions we refer to Anderson (1958), Bartlett (1933), Giri (1965, 1971, 1972, 1973), Giri, Kiefer and Stein (1963), Karlin and Traux (1960), Kabe (1964, 1965), Khatri (1959), Khirsagar (1972), Mahalanobis, Bose and Roy (1937), Olkin and Rubin (1964), Roy and Ganadisekan (1959), Stein (1969), Wijsman (1957), Wishart (1948), and Wishart and Bartlett (1932, 1933).

EXERCISES

1. Show that if a quadratic form is distributed as a noncentral chi-square, the noncentrality parameter is the value of the quadratic form when the variables are replaced by their expected values.

2. Show that the sufficient condition of Theorem 6.1.3 is also necessary for the independence of the quadratic form $\mathbf{X}'A\mathbf{X}$ and the linear form $B\mathbf{X}$.

Let $\mathbf{X} = (X_1, \ldots, X_p)'$ be normally distributed with mean $\boldsymbol{\mu}$ and covariance matrix Σ. Show that two quadratic forms $\mathbf{X}'A\mathbf{X}$ and $\mathbf{X}'B\mathbf{X}$ are independent if $A\Sigma B = 0$.

4. Let S be distributed as $W_p(n, \Sigma)$. Show that for any nonnull p-vector $\boldsymbol{l} = (l_1, \ldots, l_p)'$

(a) $\boldsymbol{l}'S\boldsymbol{l}/\boldsymbol{l}'\Sigma\boldsymbol{l}$ is distributed as $\chi_n{}^2$,

(b) $\boldsymbol{l}'\Sigma^{-1}\boldsymbol{l}/\boldsymbol{l}'S^{-1}\boldsymbol{l}$ is distributed as χ^2_{n-p+1}.

5. Let $S = (S_{ij})$ be distributed as $W_p(n, \Sigma)$, $n \geq p$, $\Sigma = (\sigma_{ij})$. Show that

$$E(S_{ij}) = n\sigma_{ij}, \qquad \mathrm{var}(S_{ij}) = n(\sigma_{ij}^2 + \sigma_{ii}\sigma_{jj}),$$
$$\mathrm{cov}(S_{ij}, S_{kl}) = n(\sigma_{ik}\sigma_{jl} + \sigma_{il}\sigma_{jk}).$$

6. Let $S_0, S_1, \ldots, S_k$ be independently distributed Wishart random variables $W_p(n_i, I)$, $i = 1, \ldots, k$, and let $V_j = S_0^{-1/2}S_jS_0^{-1/2}$, $j = 1, \ldots, k$. Show that the joint probability density function of $V_1, \ldots, V_k$ is given by

$$C \prod_{i=1}^{k} (\det V_i)^{(n_i-p-1)/2} \left(\det\left(I + \sum_1^k V_i\right)\right)^{-(\Sigma_1^k n_i)/2}$$

where C is the normalizing constant.

7. Let $S_0, S_1, \ldots, S_k$ be independently distributed as $W_p(n_i, \Sigma)$, $i = 0, 1, \ldots, k$.

(a) Let, for $j = 1, \ldots, k$,

$$W_j = \left(\sum_0^k S_j\right)^{-1/2} S_j \left(\sum_0^k S_j\right)^{-1/2}, \qquad V_j = S_0^{-1/2}S_jS_0^{-1/2}$$

$$Z_j = \left(I + \sum_1^k V_j\right)^{-1/2} V_j \left(I + \sum_1^k V_i\right)^{-1/2}.$$

Show that the joint probability density function of $W_1, \ldots, W_k$ is given by (with respect to the Lebesgue measure $\prod_{j=1}^k dw_j$)

$$f_{W_1,\ldots,W_k}(w_1, \ldots, w_k) = C \prod_{j=1}^{k} (\det w_j)^{(n_i-p-1)/2} \left(\det\left(I - \sum_{j=1}^{k} w_j\right)\right)^{(n_0-p-1)/2}$$

where C is the normalizing constant. Also verify that the joint probability density function of $Z_1, \ldots, Z_k$ is the same as that of $W_1, \ldots, W_k$.

(b) Let T_j be a lower triangular nonsingular matrix such that

$$S_1 + \cdots + S_j = T_jT_j', \qquad j = 1, \ldots, k-1,$$

and let

$$W_j = T_j^{-1}S_{j+1}T_j'^{-1}, \qquad j = 1, \ldots, k-1.$$

Show that $W_1, \ldots, W_{k-1}$ are independently distributed.

(c) Let

$$Y_j = (S_1 + \cdots + S_{j+1})^{-1/2} S_{j+1} (S_1 + \cdots + S_{j+1})^{-1/2}, \qquad j = 1, \ldots, k-1.$$

Show that $Y_1, \ldots, Y_{k-1}$ are stochastically independent.

8. Let S be distributed as $W_p(n, \Sigma)$, $n \geq p$, let $T = (T_{ij})$ be a lower triangular matrix such that $S = TT'$, and let

$$X_{ii}^2 = T_{ii}^2 + \sum_{j=1}^{i-1} T_{ij}^2, \qquad i = 1, \ldots, p,$$

$$X_{ii-1} = (T_{ii}^2 - T_{i1}^2 - \cdots - T_{ii-2}^2)^{1/2} \left(\frac{T_{ii-1}/(n-i+1)^{1/2}}{1 + T_{ii-1}^2/(n-i+1)} \right),$$

$$X_{ii-2} = (T_{ii}^2 - T_{i1}^2 - \cdots - T_{ii-3}^2)^{1/2} \left(\frac{T_{ii-2}/(n-i+2)^{1/2}}{1 + T_{ii-2}^2/(n-i+2)} \right),$$

$$\vdots$$

$$X_{i1} = T_{ii} \frac{T_{i1}/(n-1)^{1/2}}{1 + T_{i1}^2/(n-1)}.$$

Obtain the joint probability density function of X_{ii}^2, $i = 1, \ldots, p$, and all t_{ij}, $i \neq j$, $i < j$. Show that the X_{ii}^2 are distributed as central chi-squares whereas the t_{ij} have Student's t-distributions.

9. Let Y be a $k \times n$ matrix and let D be a $(p-k) \times n$ matrix, $n > p$. Show that

$$\int_{YY' = G,\, YD' = V} dY = 2^{-k} \left(\prod_{i=1}^{k} C(n - p + i) \right) (\det(DD'))^{-p/2} \times (\det(G - V(DD')^{-1} V'))^{(n-p-1)/2},$$

where $C(n)$ is the surface area of a unit n-dimensional sphere.

10. Let S be distributed as $W_p(n, \Sigma)$, $n \geq p$, and let S and Σ be similarly partitioned into

$$S = \begin{pmatrix} S_{(11)} & S_{(12)} \\ S_{(21)} & S_{(22)} \end{pmatrix}, \qquad \Sigma = \begin{pmatrix} \Sigma_{(11)} & \Sigma_{(12)} \\ \Sigma_{(21)} & \Sigma_{(22)} \end{pmatrix}.$$

Show that if $\Sigma_{(12)} = 0$, then

$$\frac{\det S}{(\det S_{(11)})(\det S_{(22)})}$$

is distributed as a product of independent beta random variables.

11. Let S_i, $i = 1, 2, \ldots, k$, be independently distributed Wishart random variables $W_p(n_i, \Sigma)$, $n_i \geq p$. Show that

(a) the characteristic roots of $\det(S_1 - \lambda(S_1 + S_2)) = 0$ are independent of $S_1 + S_2$;

(b) $(\det S_1)/\det(S_1 + S_2)$ is distributed independently of $S_1 + S_2$;

(c) $(\det S_1)/\det(S_1 + S_2), \det(S_1 + S_2)/\det(S_1 + S_2 + S_3), \ldots$ are all independently distributed.

12. Let $\bar{\mathbf{X}}$, S be based on a random sample of size N from a p-variate normal population with mean $\boldsymbol{\mu}$ and covariance matrix Σ, $N > p$, and let $\mathbf{X}$ be an additional random observation from this population. Find the distribution of

(a) $\mathbf{X} - \bar{\mathbf{X}}$,

(b) $(N/(N+1))(\mathbf{X} - \bar{\mathbf{X}})'S^{-1}(\mathbf{X} - \bar{\mathbf{X}})$.

13. Show that given the joint probability density function of $R_1, \ldots, R_k$ as in (6.64), the marginal probability density function of $R_1, \ldots, R_j$, $i \leq j \leq k$, can be obtained from (6.64) by replacing k by j. Also show that for $k = 2$, $\delta_1^2 = 0, \delta_2^2 = 0, (1 - R_1 - R_2)/(1 - R_1)$ is distributed as the beta random variable with parameter $(\frac{1}{2}(N - p_1 - p_2), \frac{1}{2}p_2)$.

14. (*Square root of Wishart*) Let S be distributed as $W_p(n, \Sigma)$, $n \geq p$, and let $S = CC'$ where C is a nonsingular matrix of dimension $p \times p$. Show that the probability density function of C with respect to the Lebesgue measure dc in the space of all nonsingular matrices c of dimension $p \times p$ is given by

$$K(\det \Sigma)^{-n/2} \exp\{-\tfrac{1}{2} \operatorname{tr} \Sigma^{-1}cc'\}(\det(cc'))^{(n-p)/2}.$$

[*Hint*: Write $C = T\theta$ where T is the unique lower triangular matrix with positive diagonal elements such that $S = TT'$ and θ is a random orthogonal matrix distributed independently of T. The Jacobian of the transformation $C \to (T, \theta)$ is $\prod_{i=1}^{p}(t_{ii})^{p-i}h(\theta)$ where $h(\theta)$ is a function of θ only (see Roy, 1959).]

15. (a) Let G be the set of all $p \times r$ $(r \geq p)$ real matrices g and let $\boldsymbol{\alpha} = (R, 0, \ldots, 0)'$ be a real p-vector, $\boldsymbol{\beta} = (\delta, 0, \ldots, 0)'$ a real r-vector. Show that for $k > 0$ (dg stands for the Lebesgue measure on G)

$$\int_G (\det(gg'))^k \exp\{-\tfrac{1}{2}\operatorname{tr}(gg' - 2g\boldsymbol{\beta}\boldsymbol{\alpha}')\}\, dg$$

$$= \exp\{-\tfrac{1}{2}R^2\delta^2\}(2\pi)^{pr/2}E(\chi_r^{\,2}(R^2\delta^2))^k \prod_{i=1}^{p-1} E(\chi_{r-i}^2)^k.$$

(b) Let $x = (x_1, \ldots, x_p)'$, $y = (y_1, \ldots, y_r)'$, $\delta = (\mathbf{y}'\mathbf{y})^{1/2}$, $R = (\mathbf{x}'\mathbf{x})^{1/2}$. Show that for $k > 0$

$$\int_G (\det(gg'))^k \exp\{-\tfrac{1}{2}\operatorname{tr}(gg' - 2g\mathbf{y}\mathbf{x}')\}\, dg$$

$$= \exp\{-\tfrac{1}{2}R^2\delta^2\}(2\pi)^{pr/2}E(\chi_r^{\,2}(R^2\delta^2))^k \prod_{i=1}^{p-1} E(\chi_{r-i}^2)^k.$$

16. Consider Exercise 12. Let Σ be a positive definite matrix of dimension $p \times p$. Show that for $k > 0$,

$$\int_G (\det(gg'))^k \exp\{-\tfrac{1}{2} \operatorname{tr} \Sigma(gg' - 2g\mathbf{y}\mathbf{x}')\}\, dg$$
$$= (\det \Sigma)^{-(2k-p)/2} \exp\{-\tfrac{1}{2}(\mathbf{x}'\Sigma\mathbf{x})(\mathbf{y}'\mathbf{y})\}$$
$$\times \int_G (\det(gg'))^k \exp\{-\tfrac{1}{2} \operatorname{tr}(g - \mathbf{z}\mathbf{y}')(g - \mathbf{z}\mathbf{y}')'\}\, dg$$

where $\mathbf{z} = C\mathbf{x}$ and C is a nonsingular matrix such that $\Sigma = CC'$.

Let B be the unique lower triangular matrix with positive diagonal elements such that $S = BB'$, where S is distributed independently of $\sqrt{N}\,\bar{\mathbf{X}}$ (normal with mean $\sqrt{N}\,\boldsymbol{\mu}$ and covariance Σ), as $W_p(N-1, \Sigma)$, and let $V = B^{-1}\bar{\mathbf{X}}$. Show that the probability density function of $\mathbf{V}$ is given by

$$f_{\mathbf{V}}(\mathbf{v}) = 2^p C \int_{G_T} \exp\{-\tfrac{1}{2}(gg' + N(g\mathbf{v} - \rho)(g\mathbf{v} - \rho)')\} \prod_{i=1}^{p} |g_{ii}|^{N-i} \prod_{i \geq j} dg_{ij},$$

where

$$C = N^{p/2}\left[2^{Np/2}\pi^{p(p+1)/4} \prod_{i=1}^{p} \Gamma((N-i)/2)\right]^{-1}$$

and $g = (g_{ij}) \in G_T$ where G_T is the group of $p \times p$ nonsingular lower triangular matrices with positive diagonal elements. Use the distribution of $\mathbf{V}$ to find the probability density function of $R_1, \ldots, R_p$ as defined in (6.63) with $k = p$.

18. Let $\boldsymbol{\xi}$ be a p-variate complex normal random variable with mean $\boldsymbol{\alpha}$ and complex positive definite Hermitian covariance matrix Σ. Show that

$$2\boldsymbol{\xi}^*\Sigma^{-1}\boldsymbol{\xi}$$

is distributed as $\chi^2_{2p}(2\boldsymbol{\alpha}^*\Sigma^{-1}\boldsymbol{\alpha})$.

19. Let $\boldsymbol{\xi}^\alpha$, $\alpha = 1, \ldots, N$ $(N > p)$, be a random sample of size N from a p-variate complex normal distribution with mean $\boldsymbol{\alpha}$ and complex positive definite Hermitian matrix Σ, and let

$$S = \sum_{\alpha=1}^{N} (\boldsymbol{\xi}^\alpha - \bar{\boldsymbol{\xi}})(\boldsymbol{\xi}^\alpha - \bar{\boldsymbol{\xi}})^*, \qquad \bar{\boldsymbol{\xi}} = \frac{1}{N}\sum_{1}^{N} \boldsymbol{\xi}^\alpha.$$

(a) Show that the probability density function of S is given by

$$f_S(s) = K(\det \Sigma)^{-(N-1)}(\det s)^{N-p-1} \exp\{-\operatorname{tr} \Sigma^{-1}s\}$$

where $K^{-1} = \pi^{p(p-1)/2} \prod_{i=1}^{p} \Gamma(N-i)$.

(b) Show that $N\bar{\boldsymbol{\xi}}^* S^{-1}\bar{\boldsymbol{\xi}}$ is distributed as

$$\chi^2_{2p}(2N\boldsymbol{\alpha}^*\Sigma^{-1}\boldsymbol{\alpha})/\chi^2_{2(N-p)}.$$

(c) Let $S = (S_{ij})$, $\Sigma = (\Sigma_{ij})$ be similarly partitioned into submatrices

$$S = \begin{pmatrix} S_{11} & S_{(12)} \\ S_{(21)} & S_{(22)} \end{pmatrix}, \qquad \Sigma = \begin{pmatrix} \Sigma_{11} & \Sigma_{(12)} \\ \Sigma_{(21)} & \Sigma_{(22)} \end{pmatrix}.$$

Show that the probability density function of

$$R_c^{\,2} = \frac{S_{(12)} S_{(22)}^{-1} S_{(21)}}{S_{11}}$$

is given by

$$f_{R_c{}^2}(r_c^{\,2}) = \frac{\Gamma(N-1)}{\Gamma(p-1)\Gamma(N-p)} (1-\rho_c^{\,2})^{N-1} (r_c^{\,2})^{p-2} (1-r_c^{\,2})^{N-p-1}$$
$$\times F(N-1, N-1; p-1; r_c^{\,2}\rho_c^{\,2}),$$

where

$$\rho_c^{\,2} = \frac{\Sigma_{(12)} \Sigma_{(22)}^{-1} \Sigma_{(21)}}{\Sigma_{11}}$$

$$F(a, b; c; x) = 1 + \frac{ab}{c} x + \frac{a(a+1)b(b+1)}{c(c+1)} \frac{x^2}{2!} + \cdots.$$

(d) Let $T = (T_{ij})$ be a complex upper triangular matrix with positive real diagonal elements T_{ii} such that $T^*T = S$. Show that the probability density function of T is given by

$$K(\det \Sigma)^{N-1} \prod_{j=1}^{p} (T_{jj})^{2n-(2j-1)} \exp\{-\operatorname{tr} \Sigma^{-1} T^* T\}.$$

(e) Define $R_1, \ldots, R_k$ in terms of S (complex) and $\boldsymbol{\xi}, \delta_1, \ldots, \delta_k$ in terms of $\boldsymbol{\alpha}$, and Σ in the same way as in (6.63) and (6.64) for the real case. Show that the joint probability density function of $R_1, \ldots, R_k$ is given by $(k \le p)$

$$f_{R_1, \ldots, R_k}(r_1, \ldots, r_k) = \Gamma(N) \left[\Gamma(N-p) \prod_{i=1}^{k} \Gamma(p_i) \right]^{-1} \left(1 - \sum_1^k r_i \right)^{N-p-1} \prod_{i=1}^{k} r_i^{p_i - 1}$$
$$\times \exp\left\{ -\sum_1^k \delta_j + \sum_1^k r_j \sum_{i>j} \delta_i \right\} \prod_{i=1}^{k} \phi(N - \sigma_{i-1}, p_i; r_i \delta_i).$$

REFERENCES

Anderson, T. W. (1945). The Noncentral Wishart Distribution and its Application to Problems of Multivariate Analysis. Ph.D. Thesis, Princeton Univ., Princeton, New Jersey.

Anderson, T. W. (1946). The noncentral Wishart distribution and certain problems of multivariate analysis, *Ann. Math. Statist.* **17**, 409–431.

Anderson, T. W. (1958). "Introduction to Multivariate Statistical Analysis." Wiley, New York.

Bartlett, M. S. (1933). On the theory of statistical regression, *Proc. R. Soc. Edinburgh* **33**, 260–283.

Cochran, W. G. (1934). The distribution of quadratic form in a normal system with applications to the analysis of covariance, *Proc. Cambridge Phil. Soc.* **30**, 178–191.

Constantine, A. G. (1963). Some noncentral distribution problems in multivariate analysis, *Ann. Math. Statist.* **34**, 1270–1285.

Eaton, M. L. (1972). Multivariate Statistical Analysis, Inst. of Math. Statist., Univ. of Copenhagen, Denmark.

Fisher, R. A. (1915). Frequency distribution of the values of the correlation coefficient in samples from an indefinitely large population, *Biometrika* **10**, 507–521.

Fisher, R. A. (1928). The general sampling distribution of the multiple correlation coefficient, *Proc. R. Soc. A* **121**, 654–673.

Giri, N. (1965). On the complex analogues of T^2- and R^2-tests, *Ann. Math. Statist.* **36**, 644–670.

Giri, N. (1971). On the distribution of a multivariate statistic, *Sankhya* **A33**, 207–210.

Giri, N. (1972). On testing problems concerning the mean of a multivariate complex Gaussian distribution, *Ann. Inst. Statist. Math.* **24**, 245–250.

Giri, N. (1973). An integral—its evaluation and applications, *Sankhya* **A35**, 334–340.

Giri, N. (1974). "Introduction to Probability and Statistics," Part I, Probability. Dekker, New York.

Giri, N., Kiefer, J., and Stein, C. (1963). Minimax character of Hotelling's T^2-test in the simplest case, *Ann. Math. Statist.* **34**, 1524–1535.

Graybill, F. A. (1961). "Introduction to Linear Statistical Model." McGraw-Hill, New York.

Hogg, R. V., and Craig, A. T. (1958). On the decomposition of certain χ^2 variables, *Ann. Math. Statist.* **29**, 608–610.

Ingham, A. E. (1933). An integral that occurs in statistics, *Proc. Cambridge Phil. Soc.* **29**, 271–276.

James, A. T. (1954). Normal multivariate analysis and the orthogonal group, *Ann. Math. Statist.* **25**, 40–75.

James, A. T. (1955). The noncentral Wishart distribution, *Proc. R. Soc. London A* **229**, 364–366.

James, A. T. (1964). The distribution of matrix variates and latent roots derived from normal samples, *Ann. Math. Statist.* **35**, 475–501.

Karlin, S., and Traux, D. (1960). Slippage problems, *Ann. Math. Statist.* **31**, 296–324.

Kabe, D. G. (1964). Decomposition of Wishart distribution, *Biometrika* **51**, 267.

Kabe, D. G. (1965). On the noncentral distribution of Rao's *U*-statistic, *Ann. Math. Statist.* **17**, 15.

Khatri, C. G. (1959). On the conditions of the forms of the type $\mathbf{X}'A\mathbf{X}$ to be distributed independently or to obey Wishart distribution, *Calcutta Statist. Assoc. Bull.* **8**, 162–168.

Khatri, C. G. (1963). Wishart distribution, *J. Indian Statist. Assoc.* **1**, 30.

Khirsagar, A. M. (1959). Bartlett decomposition and Wishart distribution, *Ann. Math. Statist.* **30**, 239.

Khirsagar, A. M. (1972). "Multivariate Analysis." Dekker, New York.

Mahalanobis, P. C., Bose, R. C., and Roy, S. N. (1937). Normalization of variates and the use of rectangular coordinates in the theory of sampling distributions, *Sankhya* **3**, 1–40.

Nachbin, L. (1965). "The Haar Integral." Van Nostrand–Reinhold, Princeton, New Jersey.

Ogawa, J. (1949). On the independence of linear and quadratic forms of a random sample from a normal population, *Ann. Inst. Math. Statist.* **1**, 83–108.

Olkin, I., and Roy, S. N. (1954). On multivariate distribution theory, *Ann. Math. Statist.* **25**, 325–339.

Olkin, I., and Rubin, H. (1964). Multivariate beta distributions and independence properties of Wishart distribution, *Ann. Math. Statist.* **35**, 261–269.

Rao, C. R. (1965). "Linear Statistical Inference and its Applications." Wiley, New York.

Rasch, G. (1948). A functional equation for Wishart distribution, *Ann. Math. Statist.* **19**, 262–266.

Roy, S. N. (1957). "Some Aspects of Multivariate Analysis." Wiley, New York.

Roy, S. N., and Ganadesikan, R. (1959). Some contributions to ANOVA in one or more dimensions II, *Ann. Math. Statist.* **30**, 318–340.

Stein, C. (1969). Multivariate analysis I (Notes recorded by M. L. Eaton), Dept. of Statist., Stanford Univ., California.

Sverdrup, E. (1947). Derivation of the Wishart distribution of the second order sample moments by straightforward integration of a multiple integral, *Skand. Akturarietidskr.* **30**, 151–166.

Wijsman, R. A. (1957). Random orthogonal transformations and their use in some classical distribution problems in multivariate analysis, *Ann. Math. Statist.* **28**, 415–423.

Wilks, S. S. (1932). Certain generalizations in the analysis of variance, *Biometrika* **24**, 471–494.

Wishart, J. (1928). The generalized product moment distribution from a normal multivariate distribution, *Biometrika* **20A**, 32–52.

Wishart, J. (1948). Proof of the distribution law of the second order moment statistics, *Biometrika* **35**, 55–57.

Wishart, J., and Bartlett, M. S. (1932). The distribution of second order moment statistics in a normal system, *Proc. Cambridge Phil. Soc.* **28**, 455–459.

Wishart, J., and Bartlett, M. S. (1933). The generalized product moment distribution in a normal system, *Proc. Cambridge Phil. Soc.* **29**, 260–270.

CHAPTER VII

Tests of Hypotheses of Mean Vectors

7.0 INTRODUCTION

This chapter deals with testing problems concerning mean vectors of multivariate normal distributions. Using the same developments of the appropriate test criteria we will also construct the confidence region for a mean vector. It will not be difficult for the reader to construct the confidence regions for the other cases discussed in this chapter. The population covariance matrix Σ is rarely known in most practical problems and tests of hypotheses concerning the mean vectors must be based on an appropriate estimate of Σ. However, in cases of long experience with the same experimental variables, we can sometimes assume Σ to be known. In deriving suitable test criteria for different testing problems we shall use mainly the well-known likelihood ratio principle and the approach of invariance as outlined in Chapter III. The heuristic approach of Roy's union–intersection principle of test construction also leads to suitable test criteria. We shall include it as an exercise. For further material on this the reader is referred to Anderson (1958), Giri (1965), Giri and Behara (1971), Kshirsagar (1972), and Roy (1953, 1957). Nandi (1965) has shown that the test statistic obtained from Roy's union–intersection principle is consistent if the component tests (univariate) are so, unbiased under certain conditions, and admissible if again the component tests are admissible.

7.1 TESTS AND CONFIDENCE REGION FOR MEAN VECTORS WITH KNOWN COVARIANCE MATRICES

Let $\mathbf{X}^\alpha = (X_{\alpha 1}, \ldots, X_{\alpha p})'$, $\alpha = 1, \ldots, N$, be a random sample of size N from a p-variate normal population with mean $\boldsymbol{\mu}$ and positive definite covariance matrix Σ. We will consider the problem of testing the hypothesis $H_0 : \boldsymbol{\mu} = \boldsymbol{\mu}_0$ (specified) and a related problem of finding the confidence region for $\boldsymbol{\mu}$ under the assumption that Σ is known. In the univariate case ($p = 1$) we use the fact that the difference between the sample mean and the population mean is normally distributed with mean 0 and known variance and use the existing table of standard normal distributions to determine the significance points or the confidence interval. In the multivariate case such a difference has a p-variate normal distribution with mean $\mathbf{0}$ and known covariance matrix, and hence we can set up the confidence interval or prescribe the test for each component as in the univariate case.

Such a solution has several drawbacks. First, the choice of confidence limits is somewhat arbitrary. Second, for testing purposes it may lead to a test whose performance may be poor against some alternatives. Finally, and probably most important for $p > 2$ detailed tables for multivariate normal distributions are not available. The procedure suggested below can be computed easily and can be given a general intuitive and theoretical justification. Let $\bar{\mathbf{X}} = (1/N) \sum_{\alpha=1}^{N} \mathbf{X}^\alpha$. By Theorem 6.2.2, under H_0, $N(\bar{\mathbf{X}} - \boldsymbol{\mu}_0)'\Sigma^{-1}(\bar{\mathbf{X}} - \boldsymbol{\mu}_0)$ has central chi-square distribution with p degrees of freedom and hence the test which rejects $H_0 : \boldsymbol{\mu} = \boldsymbol{\mu}_0$ whenever

$$N(\bar{\mathbf{x}} - \boldsymbol{\mu}_0)'\Sigma^{-1}(\bar{\mathbf{x}} - \boldsymbol{\mu}_0) \geq \chi^2_{p,\alpha}, \tag{7.1}$$

where $\chi^2_{p,\alpha}$ is a constant such that $P(\chi_p{}^2 \geq \chi^2_{p,\alpha}) = \alpha$, has the power function which increases monotonically with the noncentrality parameter $N(\boldsymbol{\mu} - \boldsymbol{\mu}_0)'\Sigma^{-1}(\boldsymbol{\mu} - \boldsymbol{\mu}_0)$. Thus the power function of the test given in (7.1) has the minimum value α (level of significance) when $\boldsymbol{\mu} = \boldsymbol{\mu}_0$ and its power is greater than α when $\boldsymbol{\mu} \neq \boldsymbol{\mu}_0$. For a given sample mean $\bar{\mathbf{x}}$, consider the inequality

$$N(\bar{\mathbf{x}} - \boldsymbol{\mu})'\Sigma^{-1}(\bar{\mathbf{x}} - \boldsymbol{\mu}) \leq \chi^2_{p,\alpha}. \tag{7.2}$$

The probability is $1 - \alpha$ that the mean of a sample of size N from a p-variate normal distribution with mean $\boldsymbol{\mu}$ and known positive definite covariance matrix Σ satisfies (7.2). Thus the set of values of $\boldsymbol{\mu}$ satisfying (7.2) gives the confidence region for $\boldsymbol{\mu}$ with confidence coefficient $1 - \alpha$, and represents the interior and the surface of an ellipsoid with center $\bar{\mathbf{x}}$, with shape depending on Σ and size depending on Σ and $\chi^2_{p,\alpha}$.

For the case of two p-dimensional normal populations with mean vectors $\boldsymbol{\mu}$, $\mathbf{v}$ but with the same known positive definite covariance matrix Σ we now consider the problem of testing the hypothesis $H_0:\boldsymbol{\mu} - \mathbf{v} = \mathbf{0}$ and the problem of setting a confidence region for $\boldsymbol{\mu} - \mathbf{v}$ with confidence coefficient $1 - \alpha$. Let $\mathbf{X}^\alpha = (X_{\alpha 1}, \ldots, X_{\alpha p})'$, $\alpha = 1, \ldots, N_1$, be a random sample of size N_1 from the normal distribution with mean $\boldsymbol{\mu}$ and covariance matrix Σ, and let $\mathbf{Y}^\alpha = (Y_{\alpha 1}, \ldots, Y_{\alpha p})'$, $\alpha = 1, \ldots, N_2$, be a random sample of size N_2 (independent of $\mathbf{X}^\alpha$, $\alpha = 1, \ldots, N_1$) from the other normal distribution with mean $\mathbf{v}$ and the same covariance matrix Σ. If

$$\bar{\mathbf{X}} = \frac{1}{N_1} \sum_{\alpha=1}^{N_1} \mathbf{X}^\alpha, \qquad \bar{\mathbf{Y}} = \frac{1}{N_2} \sum_{\alpha=1}^{N_2} \mathbf{Y}^\alpha,$$

then by Theorem 6.2.2, under H_0,

$$\frac{N_1 N_2}{N_1 + N_2} (\bar{\mathbf{X}} - \bar{\mathbf{Y}})' \Sigma^{-1} (\bar{\mathbf{X}} - \bar{\mathbf{Y}})$$

is distributed as chi-square with p degrees of freedom. Given sample observations $\mathbf{x}^\alpha$, $\alpha = 1, \ldots, N_1$, and $\mathbf{y}^\alpha$, $\alpha = 1, \ldots, N_2$, the test which rejects H_0 whenever

$$\frac{N_1 N_2}{N_1 + N_2} (\bar{\mathbf{x}} - \bar{\mathbf{y}})' \Sigma^{-1} (\bar{\mathbf{x}} - \bar{\mathbf{y}}) \geq \chi^2_{p,\alpha}, \tag{7.3}$$

has a power function which increases monotonically with the noncentrality parameter

$$\frac{N_1 N_2}{N_1 + N_2} (\boldsymbol{\mu} - \mathbf{v})' \Sigma^{-1} (\boldsymbol{\mu} - \mathbf{v}); \tag{7.4}$$

its power is greater than α (the level of significance) whenever $\boldsymbol{\mu} \neq \mathbf{v}$, and the power function attains its minimum value α whenever $\boldsymbol{\mu} = \mathbf{v}$. Given $\mathbf{x}^\alpha$, $\alpha = 1, \ldots, N_1$, and $\mathbf{y}^\alpha$, $\alpha = 1, \ldots, N_2$, the confidence region for $\boldsymbol{\mu} - \mathbf{v}$ with confidence coefficient $1 - \alpha$ is given by the set of values of $\boldsymbol{\mu} - \mathbf{v}$ satisfying

$$\frac{N_1 N_2}{N_1 + N_2} (\bar{\mathbf{x}} - \bar{\mathbf{y}} - (\boldsymbol{\mu} - \mathbf{v}))' \Sigma^{-1} (\bar{\mathbf{x}} - \mathbf{y} - (\boldsymbol{\mu} - \mathbf{v})) \leq \chi^2_{p,\alpha}, \tag{7.5}$$

which is an ellipsoid with center $\bar{\mathbf{x}} - \bar{\mathbf{y}}$ and whose shape depends on Σ. In this context it is worth noting that the quantity

$$(\boldsymbol{\mu} - \mathbf{v})' \Sigma^{-1} (\boldsymbol{\mu} - \mathbf{v}) \tag{7.6}$$

is called the *Mahalanobis distance* between two p-variate normal populations with the same positive definite covariance matrix Σ but with different mean vectors.

Consider now k p-variate normal populations with the same known covariance matrix Σ but with different mean vectors $\boldsymbol{\mu}_i$, $i = 1, \ldots, k$. Let $\bar{\mathbf{X}}_i$ be the mean of a random sample of size N_i from the ith population and let $\bar{\mathbf{x}}_i$ be its sample value. An appropriate test for the hypothesis

$$H_0: \sum_{i=1}^{k} \beta_i \boldsymbol{\mu}_i = \boldsymbol{\mu}_0, \tag{7.7}$$

where the β_i are known constants and $\boldsymbol{\mu}_0$ is a known p-vector, rejects H_0 whenever

$$C\left(\sum_1^k \beta_i \bar{\mathbf{x}}_i - \boldsymbol{\mu}_0\right)' \Sigma^{-1} \left(\sum_1^k \beta_i \bar{\mathbf{x}}_i - \boldsymbol{\mu}_0\right) \geq \chi^2_{p,\alpha}, \tag{7.8}$$

where the constant C is given by

$$C^{-1} = \sum_{i=1}^{k} \frac{\beta_i^2}{N_i}. \tag{7.9}$$

Obviously

$$C\left(\sum_{i=1}^k \beta_i \bar{\mathbf{X}}_i - \boldsymbol{\mu}_0\right)' \Sigma^{-1} \left(\sum_{i=1}^k \beta_i \bar{\mathbf{X}}_i - \boldsymbol{\mu}_0\right)$$

is distributed as noncentral chi-square with p degrees of freedom and with noncentrality parameter $C(\boldsymbol{\mu} - \boldsymbol{\mu}_0)'\Sigma^{-1}(\boldsymbol{\mu} - \boldsymbol{\mu}_0)$ where $\sum_1^k \beta_i \boldsymbol{\mu}_i = \boldsymbol{\mu}$. Given $\bar{\mathbf{x}}_i$, $i = 1, \ldots, k$, the $(1 - \alpha)100\%$ confidence region for $\boldsymbol{\mu}$ is given by the ellipsoid

$$C\left(\sum_{i=1}^k \beta_i \bar{\mathbf{x}}_i - \boldsymbol{\mu}\right)' \Sigma^{-1} \left(\sum_{i=1}^k \beta_i \bar{\mathbf{x}}_i - \boldsymbol{\mu}\right) \leq \chi^2_{p,\alpha} \tag{7.10}$$

with center $\sum_{i=1}^k \beta_i \bar{\mathbf{x}}_i$.

7.2 TESTS AND CONFIDENCE REGION FOR MEAN VECTORS WITH UNKNOWN COVARIANCE MATRICES

In most practical problems concerning mean vectors the covariance matrices are rarely known and statistical testing of hypotheses about mean vectors has to be carried out assuming that the covariance matrices are unknown. We shall first consider testing problems concerning the mean $\boldsymbol{\mu}$ of a p-variate normal population with unknown covariance matrix Σ. Testing problems concerning mean vectors of more than one multivariate normal population with unknown covariance matrices will be treated as applications of these problems.

7.2.1 Hotelling's T^2-Test

Let $\mathbf{x}^\alpha = (x_{\alpha 1}, \ldots, x_{\alpha p})'$, $\alpha = 1, \ldots, N$, be a sample of size $N (N > p)$ from a p-variate normal distribution with unknown mean $\boldsymbol{\mu}$ and unknown positive definite covariance matrix Σ. On the basis of these observations we are interested in testing the hypothesis $H_0 : \boldsymbol{\mu} = \boldsymbol{\mu}_0$ against the alternatives $H_1 : \boldsymbol{\mu} \neq \boldsymbol{\mu}_0$ where Σ is unknown and $\boldsymbol{\mu}_0$ is specified. In the univariate case ($p = 1$) this is a basic problem in statistics with applications in every branch of applied science, and the well-known Student t-test is its optimum solution. For the general multivariate case we shall show that a multivariate analog of Student's t is an optimum solution. This problem is commonly known as Hotelling's problem since Hotelling (1931) first proposed the extension of Student's t-statistic for the two-sample multivariate problem and derived its distribution under the null hypothesis. We shall now derive the likelihood ratio test of this problem. The likelihood of the observations $\mathbf{x}^\alpha$, $\alpha = 1, \ldots, N$, is given by

$$L(\mathbf{x}^1, \ldots, \mathbf{x}^N | \boldsymbol{\mu}, \Sigma)$$
$$= (2\pi)^{-Np/2} (\det \Sigma)^{-N/2} \exp\left\{ -\tfrac{1}{2} \operatorname{tr} \Sigma^{-1} \left(\sum_{\alpha=1}^{N} (\mathbf{x}^\alpha - \boldsymbol{\mu})(\mathbf{x}^\alpha - \boldsymbol{\mu})' \right) \right\}. \quad (7.11)$$

Given $\mathbf{x}^\alpha$, $\alpha = 1, \ldots, N$, the likelihood L is a function of $\boldsymbol{\mu}$, Σ, for simplicity written as $L(\boldsymbol{\mu}, \Sigma)$. Let Ω be the parametric space of $(\boldsymbol{\mu}, \Sigma)$ and let ω be the subspace of Ω when $H_0 : \boldsymbol{\mu} = \boldsymbol{\mu}_0$ is true. Under ω the likelihood function reduces to

$$(2\pi)^{-Np/2} (\det \Sigma)^{-N/2} \exp\left\{ -\tfrac{1}{2} \operatorname{tr} \Sigma^{-1} \sum_{\alpha=1}^{N} (\mathbf{x}^\alpha - \boldsymbol{\mu}_0)(\mathbf{x}^\alpha - \boldsymbol{\mu}_0)' \right\}. \quad (7.12)$$

By Lemma 5.1.1, we obtain from (7.12)

$$\max_{\omega} L(\boldsymbol{\mu}, \Sigma) = (2\pi)^{-Np/2} \left[\det\left(\frac{1}{N} \sum_{\alpha=1}^{N} (\mathbf{x}^\alpha - \boldsymbol{\mu}_0)(\mathbf{x}^\alpha - \boldsymbol{\mu}_0)' \right) \right]^{-N/2}$$
$$\times \exp\{ -\tfrac{1}{2} Np \}. \quad (7.13)$$

We observed in Chapter V that under Ω, $L(\boldsymbol{\mu}, \Sigma)$ is maximum when

$$\boldsymbol{\mu} = \frac{1}{N} \sum_{\alpha=1}^{N} \mathbf{x}^\alpha = \bar{\mathbf{x}}, \qquad \Sigma = \frac{1}{N} \sum_{\alpha=1}^{N} (\mathbf{x}^\alpha - \bar{\mathbf{x}})(\mathbf{x}^\alpha - \bar{\mathbf{x}})' = \frac{s}{N}.$$

Hence

$$\max_{\Omega} L(\boldsymbol{\mu}, \Sigma) = (2\pi)^{-Np/2}\left[\det\left(\frac{1}{N}\sum_{\alpha=1}^{N}(\mathbf{x}^{\alpha}-\overline{\mathbf{x}})(\mathbf{x}^{\alpha}-\overline{\mathbf{x}})'\right)\right]^{-N/2} \times \exp\{-\tfrac{1}{2}Np\}. \tag{7.14}$$

From (7.13) and (7.14) the likelihood ratio test criterion for testing $H_0:\boldsymbol{\mu} = \boldsymbol{\mu}_0$ is given by

$$\begin{aligned}\lambda &= \frac{\max_{\omega} L(\boldsymbol{\mu}, \Sigma)}{\max_{\Omega} L(\boldsymbol{\mu}, \Sigma)} = \left[\frac{\det s}{\det(\sum_{\alpha=1}^{N}(\mathbf{x}^{\alpha}-\boldsymbol{\mu}_0)(\mathbf{x}^{\alpha}-\boldsymbol{\mu}_0)')}\right]^{N/2} \\ &= \left[\frac{\det s}{\det(s + N(\overline{\mathbf{x}}-\boldsymbol{\mu}_0)(\overline{\mathbf{x}}-\boldsymbol{\mu}_0)')}\right]^{N/2} \\ &= (1 + N(\overline{\mathbf{x}}-\boldsymbol{\mu}_0)'s^{-1}(\overline{\mathbf{x}}-\boldsymbol{\mu}_0))^{-N/2}.\end{aligned} \tag{7.15}$$

The right-hand side of (7.15) follows from Exercise 1.12. Since λ is a monotonically decreasing function of $N(\overline{\mathbf{x}}-\boldsymbol{\mu}_0)'s^{-1}(\overline{\mathbf{x}}-\boldsymbol{\mu}_0)$, the likelihood ratio test of $H_0:\boldsymbol{\mu} = \boldsymbol{\mu}_0$ when Σ is unknown rejects H_0 whenever

$$N(N-1)(\overline{\mathbf{x}}-\boldsymbol{\mu}_0)'s^{-1}(\overline{\mathbf{x}}-\boldsymbol{\mu}_0) \geq c, \tag{7.16}$$

where c is a constant depending on the level of significance α of the test.

Note In connection with tests we shall use c as the generic notation for the significance point of the test.

From (6.60) the distribution of

$$T^2 = N(N-1)(\overline{\mathbf{x}}-\boldsymbol{\mu}_0)'s^{-1}(\overline{\mathbf{x}}-\boldsymbol{\mu}_0)$$

is given by

$$f_{T^2}(t^2|\delta^2) = \frac{\exp\{-\frac{1}{2}\delta^2\}}{(N-1)\Gamma(\frac{1}{2}(N-p))} \times \sum_{j=0}^{\infty}\frac{(\frac{1}{2}\delta^2)^j(t^2/(N-1))^{p/2+j-1}\Gamma(\frac{1}{2}N+j)}{j!\Gamma(\frac{1}{2}p+j)(1+t^2/(N-1))^{N/2+j}}, \quad t^2 \geq 0 \tag{7.17}$$

where $\delta^2 = N(\boldsymbol{\mu}-\boldsymbol{\mu}_0)'\Sigma^{-1}(\boldsymbol{\mu}-\boldsymbol{\mu}_0)$. This is often called the distribution of T^2 with $N-1$ degrees of freedom. Under H_0, $\boldsymbol{\mu} = \boldsymbol{\mu}_0$, $\delta^2 = 0$, and $(T^2/(N-1))((N-p)/p)$ is distributed as central F with parameter $(p, N-p)$. Thus for any given level of significance α, $0 < \alpha < 1$, the constant c of (7.16) is given by

$$c = \frac{(N-1)p}{N-p}F_{p,N-p,\alpha}, \tag{7.18}$$

where $F_{p,N-p,\alpha}$ is the $(1-\alpha)100\%$ point of the F-distribution with degrees of freedom $(p, N-p)$. Tang (1938) has tabulated the type II error (1-power) of this test for various values of δ^2, p, $N-p$ and for $\alpha = 0.05$ and 0.01. Lehmer (1944) has computed values of δ^2 for given values of α and type II error. This table is useful for finding the value of δ^2 (or, equivalently, the value of N for given $\boldsymbol{\mu}$ and Σ) needed to make the probability of accepting H_0 very small whenever H_0 is false. Hsu (1938) and Bose and Roy (1938) have also derived the distribution of T^2 by different methods.

Another equivalent test procedure for testing H_0 rejects H_0 whenever

$$r_1 = \frac{N\overline{\mathbf{x}}'s^{-1}\overline{\mathbf{x}}}{1 + N\overline{\mathbf{x}}'s^{-1}\overline{\mathbf{x}}} \leq c. \tag{7.19}$$

From (6.66) the probability density function of R_1 (random variable corresponding to r_1) is

$$f_{R_1}(r_1) = \frac{\Gamma(\frac{1}{2}N)}{\Gamma(\frac{1}{2}p)\Gamma(\frac{1}{2}(N-p))} r_1^{p/2-1}(1-r_1)^{(N-p)/2-1}$$
$$\times \exp\{-\tfrac{1}{2}\delta^2\}\phi(\tfrac{1}{2}(N-p), \tfrac{1}{2}p; \tfrac{1}{2}r_1\delta), \qquad 0 < r_1 < 1. \tag{7.20}$$

Thus under H_0, R_1 has a central beta distribution with parameter $(\frac{1}{2}p, \frac{1}{2}(N-p))$. The significance points for the test based on R_1 are given by Tang (1938). From (7.17) and (7.20) it is obvious that the power of Hotelling's T^2-test or its equivalent depends only on the quantity δ^2 and increases monotonically with δ^2.

7.2.2 Optimum Invariant Properties of the T^2-Test

To examine various optimum properties of the T^2-test, we need to verify that the statistic T^2 is the maximal invariant in the sample space under the group of transformations acting on the sample space which leaves the present testing problem invariant. In effect we will prove a more general result since it will be useful for other testing problems concerning mean vectors considered in this chapter. It is also convenient to take $\boldsymbol{\mu}_0 = \mathbf{0}$, which we can assume without any loss of generality.

Let

$$\overline{\mathbf{X}} = \frac{1}{N}\sum_1^N \mathbf{X}^\alpha, \qquad S = \sum_{\alpha=1}^N (\mathbf{X}^\alpha - \overline{\mathbf{X}})(\mathbf{X}^\alpha - \overline{\mathbf{X}})'$$

be partitioned as

$$\overline{\mathbf{X}} = (\overline{\mathbf{X}}_{(1)}, \ldots, \overline{\mathbf{X}}_{(k)})', \qquad S = \begin{pmatrix} S_{(11)} & \cdots & S_{(1k)} \\ \vdots & & \vdots \\ S_{(k1)} & \cdots & S_{(kk)} \end{pmatrix}, \tag{7.21}$$

where the $\bar{\mathbf{X}}_{(i)}$ are subvectors of $\bar{\mathbf{X}}$ of dimension $p_i \times 1$ and the $S_{(ij)}$ are submatrices of S of dimension $p_i \times p_j$ such that $\sum_1^k p_i = p$. Let

$$\bar{\mathbf{X}}_{[i]} = (\bar{\mathbf{X}}_{(1)}, \ldots, \bar{\mathbf{X}}_{(i)})', \qquad S_{[ii]} = \begin{pmatrix} S_{(11)} & \cdots & S_{(1i)} \\ \vdots & & \vdots \\ S_{(i1)} & \cdots & S_{(ii)} \end{pmatrix}. \tag{7.22}$$

We shall denote the space of values of $\bar{\mathbf{X}}$ by $\mathscr{X}_1$ and the space of values of S by $\mathscr{X}_2$ and write $\mathscr{X} = \mathscr{X}_1 \times \mathscr{X}_2$, the product space of $\mathscr{X}_1$, $\mathscr{X}_2$. Let G_{BT} be the multiplicative group of nonsingular lower triangular matrices g

$$g = \begin{pmatrix} g_{(11)} & 0 & 0 & \cdots & 0 \\ g_{(21)} & g_{(22)} & 0 & \cdots & 0 \\ \vdots & \vdots & & & \vdots \\ g_{(k1)} & g_{(k2)} & & \cdots & g_{(kk)} \end{pmatrix} \tag{7.23}$$

of dimension $p \times p$ where the $g_{(ij)}$ are submatrices of g of dimension $p_i \times p_j$, $i, j = 1, \ldots, k$; and let G_{BT} operate on $\mathscr{X}$ as

$$(\bar{\mathbf{X}}, S) \to (g\bar{\mathbf{X}}, gSg'), \qquad g \in G_{\mathrm{BT}}.$$

Define $R_1^*, \ldots, R_k^*$ by

$$\sum_{j=1}^{i} R_j^* = N\bar{\mathbf{X}}_{[i]}' S_{[ii]}^{-1} \bar{\mathbf{X}}_{[i]}, \qquad i = 1, \ldots, k. \tag{7.24}$$

Since $N > p$ by assumption, S is positive definite with probability 1 and hence $R_i^* > 0$ for all i with probability 1. It may be observed that if $p_1 = p$, $p_i = 0$, $i = 2, \ldots, k$, then $R_1^* = T^2/(N-1)$.

LEMMA 7.2.1 The statistic $(R_1^*, \ldots, R_k^*)$ is a maximal invariant under G_{BT}, operating as

$$(\bar{\mathbf{X}}, S) \to (g\bar{\mathbf{X}}, gSg') \qquad \text{for} \quad g \in G_{\mathrm{BT}}.$$

Proof We shall prove the lemma for the case $k = 2$, the general case following obviously from this. First let us observe the following.

(a) If $(\bar{\mathbf{X}}, S) \to (g\bar{\mathbf{X}}, gSg')$, $g \in G_{\mathrm{BT}}$, then

$$(\bar{\mathbf{X}}_{(1)}, S_{(11)}) \to (g_{(11)}\bar{\mathbf{X}}_{(1)}, g_{(11)}S_{(11)}g'_{(11)}).$$

Thus $(R_1^*, R_1^* + R_2^*)$ is invariant under G_{BT}.

(b) Since

$$\begin{aligned} N\bar{\mathbf{X}}'S^{-1}\bar{\mathbf{X}} &= N\bar{\mathbf{X}}_{(1)}'S_{(11)}^{-1}\bar{\mathbf{X}}_{(1)} \\ &\quad + N(\bar{\mathbf{X}}_{(2)} - S_{(21)}S_{(11)}^{-1}\bar{\mathbf{X}}_{(1)})'(S_{(22)} - S_{(21)}S_{(11)}^{-1}S_{(12)})^{-1} \\ &\quad \times (\bar{\mathbf{X}}_{(2)} - S_{(21)}S_{(11)}^{-1}\bar{\mathbf{X}}_{(1)}), \\ R_2^* &= N(\bar{\mathbf{X}}_{(2)} - S_{(21)}S_{(11)}^{-1}\bar{\mathbf{X}}_{(1)})'(S_{(22)} - S_{(21)}S_{(11)}^{-1}S_{(12)})^{-1} \\ &\quad \times (\bar{\mathbf{X}}_{(2)} - S_{(21)}S_{(11)}^{-1}\bar{\mathbf{X}}_{(1)}). \end{aligned} \tag{7.25}$$

(c) For any two p-vectors $\mathbf{X}$, $\mathbf{Y} \in E^p$, $\mathbf{X}'\mathbf{X} = \mathbf{Y}'\mathbf{Y}$ if and only if there exists an orthogonal matrix O of dimension $p \times p$ such that $\mathbf{X} = O\mathbf{Y}$.

Let $\bar{\mathbf{X}}$, $\bar{\mathbf{Y}} \in \mathscr{X}_1$ and S, $T \in \mathscr{X}_2$ be similarly partitioned and let

$$N\bar{\mathbf{X}}'_{(1)}S^{-1}_{(11)}\bar{\mathbf{X}}_{(1)} = N\bar{\mathbf{Y}}'_{(1)}T^{-1}_{(11)}\bar{\mathbf{Y}}_{(1)} \tag{7.26}$$

$$N\bar{\mathbf{X}}'S^{-1}\bar{\mathbf{X}} = N\bar{\mathbf{Y}}'T^{-1}\bar{\mathbf{Y}}. \tag{7.27}$$

To show that (R_1^*, R_2^*) is a maximal invariant under G_{BT} we must show that there exists a $g_1 \in G_{\mathrm{BT}}$ such that

$$\bar{\mathbf{X}} = g_1\bar{\mathbf{Y}}, \qquad S = g_1 T g_1'.$$

Choose

$$g = \begin{pmatrix} g_{(11)} & 0 \\ g_{(21)} & g_{(22)} \end{pmatrix}$$

with $g_{(11)} = S^{-1/2}_{(11)}$, $g_{(22)} = (S_{(22)} - S_{(21)}S^{-1}_{(11)}S_{(12)})^{-1/2}$, and

$$g_{(21)} = -g_{(22)}S_{(21)}S^{-1}_{(11)}.$$

Then

$$gSg' = I. \tag{7.28}$$

Similarly, choose $h \in G_{\mathrm{BT}}$ such that

$$hTh' = I. \tag{7.29}$$

Since (7.26) implies

$$(g_{(11)}\bar{\mathbf{X}}_{(1)})'(g_{(11)}\bar{\mathbf{X}}_{(1)}) = (h_{(11)}\bar{\mathbf{Y}}_{(1)})'(h_{(11)}\bar{\mathbf{Y}}_{(1)}) \tag{7.30}$$

from (c) we conclude that there exists an orthogonal matrix θ_1 of dimension $p_1 \times p_1$ such that

$$g_{(11)}\bar{\mathbf{X}}_{(1)} = \theta_1 h_{(11)}\bar{\mathbf{Y}}_{(1)}. \tag{7.31}$$

From (7.27), (7.28), and (7.29) we get

$$\begin{aligned}(g\bar{\mathbf{X}})'(g\bar{\mathbf{X}}) &= \|g_{(11)}\bar{\mathbf{X}}_{(1)}\|^2 + \|g_{(21)}\bar{\mathbf{X}}_{(1)} + g_{(22)}\bar{\mathbf{X}}_{(2)}\|^2 \\ &= (h\bar{\mathbf{Y}})'(h\bar{\mathbf{Y}}) = \|h_{(11)}\bar{\mathbf{Y}}_{(1)}\|^2 + \|h_{(21)}\bar{\mathbf{Y}}_{(1)} + h_{(22)}\bar{\mathbf{Y}}_{(2)}\|^2\end{aligned}$$

where $\| \ \|$ denotes the norm, and hence from (7.30) we obtain

$$\|g_{(21)}\bar{\mathbf{X}}_{(1)} + g_{(22)}\bar{\mathbf{X}}_{(2)}\|^2 = \|h_{(21)}\bar{\mathbf{Y}}_{(1)} + h_{(22)}\bar{\mathbf{Y}}_{(2)}\|^2. \tag{7.32}$$

From this we conclude that there exists an orthogonal matrix θ_2 of dimension $p_2 \times p_2$ such that

$$g_{(21)}\bar{\mathbf{X}}_{(1)} + g_{(22)}\bar{\mathbf{X}}_{(2)} = \theta_2(h_{(21)}\bar{\mathbf{Y}}_{(1)} + h_{(22)}\bar{\mathbf{Y}}_{(2)}). \tag{7.33}$$

Letting

$$\theta = \begin{pmatrix} \theta_1 & 0 \\ 0 & \theta_2 \end{pmatrix},$$

we get from (7.31) and (7.33)

$$\bar{\mathbf{X}} = g^{-1}\theta h\bar{\mathbf{Y}} = g_1\bar{\mathbf{Y}},$$

where $g_1 = g^{-1}\theta h \in G_{\text{BT}}$, and from

$$gSg' = I = hTh' = \theta hTh'\theta'$$

we get $S = g_1 T g_1'$. Hence $(R_1{}^*, R_1{}^* + R_2{}^*)$ or, equivalently, $(R_1{}^*, R_2{}^*)$ is a maximal invariant under G_{BT} on $\mathscr{X}$.

The proof for the general case is established by showing that $(R_1{}^*, R_1{}^* + R_2{}^*, \ldots, R_1{}^* + \cdots + R_k{}^*)$ is a maximal invariant under G_{BT}. The orthogonal matrix θ needed is a diagonal matrix in the block form. Q.E.D.

It may be remarked that the statistic $(R_1, \ldots, R_k)$ defined in Chapter VI is a one-to-one transformation of $(R_1{}^*, \ldots, R_k{}^*)$, and hence $(R_1, \ldots, R_k)$ is also a maximal invariant under G_{BT}.

The induced transformation G_{BT}^T on the parametric space Ω corresponding to G_{BT} on $\mathscr{X}$ is identically equal to G_{BT} and is defined by

$$(\boldsymbol{\mu}, \Sigma) \to (g\boldsymbol{\mu}, g\Sigma g'), \qquad g \in G_{\text{BT}} = G_{\text{BT}}^T.$$

Thus a corresponding maximal invariant in Ω under G_{BT} is $(\delta_1{}^2, \ldots, \delta_k{}^2)$, where

$$\sum_{j=1}^{i} \delta_j{}^2 = N\boldsymbol{\mu}_{[i]}'\Sigma_{[ii]}^{-1}\boldsymbol{\mu}_{[i]}, \qquad i = 1, \ldots, k. \tag{7.34}$$

The problem of testing the hypothesis $H_0 : \boldsymbol{\mu} = \mathbf{0}$ against the alternatives $H_1 : \boldsymbol{\mu} \neq \mathbf{0}$ on the basis of observations $\mathbf{x}^\alpha$, $\alpha = 1, \ldots, N$ $(N > p)$, remains invariant under the group G of linear transformations g (set of all $p \times p$ nonsingular matrices) which transform each $\mathbf{x}^\alpha$ to $g\mathbf{x}^\alpha$. These transformations induce on the space of the sufficient statistic $(\bar{\mathbf{X}}, S)$ the transformations

$$(\bar{\mathbf{x}}, s) \to (g\bar{\mathbf{x}}, gsg').$$

Obviously $G = G_{\text{BT}}$ if $k = 1$ and $p_1 = p$. A maximal invariant in the space of $(\bar{\mathbf{X}}, S)$ is $(N - 1)R_1{}^* = T^2 = N(N - 1)\bar{\mathbf{X}}'S^{-1}\bar{\mathbf{X}}$. The corresponding maximal invariant in the parametric space Ω under G is $\delta_1{}^2 = N\boldsymbol{\mu}'\Sigma^{-1}\boldsymbol{\mu} = \delta^2$ (say). Its probability density function is given in (7.17).

The following two theorems give the optimum character of the T^2-test among the class of all invariant level α tests for $H_0 : \boldsymbol{\mu} = \mathbf{0}$. To state them we need the following definition of a statistical test.

DEFINITION 7.2.1 *Statistical test* A statistical test ϕ is a function of the random sample $\mathbf{X}^\alpha$, $\alpha = 1, \ldots, N$, which takes values between 0 and 1 inclusive such that $E(\phi(\mathbf{X}^1, \ldots, \mathbf{X}^N)) = \alpha$, the level of the test when H_0 is true.

In this terminology $\phi(\mathbf{x}^1, \ldots, \mathbf{x}^N)$ is the probability of rejecting H_0 when $\mathbf{x}^1, \ldots, \mathbf{x}^N$ are observed.

THEOREM 7.2.1 Let $\mathbf{x}^\alpha, \alpha = 1, \ldots, N$, be a sequence of N observations from the p-variate normal distribution with mean $\boldsymbol{\mu}$ and unknown positive definite covariance matrix Σ. Among all (statistical) tests $\phi(\mathbf{X}^1, \ldots, \mathbf{X}^N)$ of level α for testing $H_0:\boldsymbol{\mu} = \mathbf{0}$ against the alternatives $H_1:\boldsymbol{\mu} \neq \mathbf{0}$ which are invariant with respect to the group of transformations G transforming $\mathbf{x}^\alpha \to g\mathbf{x}^\alpha$, $\alpha = 1, \ldots, N$, $g \in G$, Hotelling's T^2-test or its equivalent (7.19) is uniformly most powerful.

Proof Let $\phi(\mathbf{X}^1, \ldots, \mathbf{X}^N)$ be a statistical test which is invariant with respect to G. Since $(\bar{\mathbf{X}}, S)$ is sufficient for $(\boldsymbol{\mu}, \Sigma)$, $E(\phi(\mathbf{X}^1, \ldots, \mathbf{X}^N | \bar{\mathbf{X}} = \bar{\mathbf{x}}, S = s)$ is independent of $(\boldsymbol{\mu}, \Sigma)$ and depends only on $(\bar{\mathbf{x}}, s)$. Since ϕ is invariant, i.e., $\phi(\mathbf{X}^1, \ldots, \mathbf{X}^N) = \phi(g\mathbf{X}^1, \ldots, g\mathbf{X}^N)$, $g \in G$, $E(\phi | \bar{X} = \bar{x}, S = s)$ is invariant under G. Since $E(E(\phi | \bar{\mathbf{X}}, S)) = E(\phi)$, $E(\phi | \bar{\mathbf{X}}, S)$ and ϕ have the same power function. Thus each test in the larger class of level α tests which are functions of $\mathbf{X}^\alpha$, $\alpha = 1, \ldots, N$, can be replaced by one in the smaller class of tests which are functions of $(\bar{\mathbf{X}}, S)$ having identical power functions. By Lemma 7.2.1 and Theorem 3.2.2 the invariant test $E(\phi | \bar{\mathbf{X}}, S)$ depends on $(\bar{\mathbf{X}}, S)$ only through the maximal invariant T^2. Since the distribution of T^2 depends only on $\delta^2 = N\boldsymbol{\mu}'\Sigma^{-1}\boldsymbol{\mu}$, the most powerful level α invariant test of $H_0:\delta^2 = 0$ against the simple alternatives $\delta^2 = \delta_0{}^2$, where $\delta_0{}^2$ is specified, rejects H_0 (by the Neyman–Pearson fundamental lemma) whenever

$$\frac{f_{T^2}(t^2 | \delta_0{}^2)}{f_{T^2}(t^2 | 0)} = \frac{\Gamma(\frac{1}{2}p) \exp(-\frac{1}{2}\delta_0{}^2)}{\Gamma(\frac{1}{2}N)} \sum_{j=0}^{\infty} \frac{(\frac{1}{2}\delta_0{}^2)^j \Gamma(\frac{1}{2}N + j)}{j!\Gamma(\frac{1}{2}p + j)}$$

$$\times \left(\frac{t^2/(N-1)}{1 + t^2/(N-1)} \right)^j \geq c, \tag{7.35}$$

where the constant c is chosen such that the test has level α. Since the left-hand side of this inequality is a monotonically increasing function of $t^2/(N - 1 + t^2)$ and hence of t^2, the most powerful level α test of H_0 against the simple alternative $\delta^2 = \delta_0{}^2$ ($\delta_0{}^2 \neq 0$) rejects H_0 whenever $t^2 \geq c$, where the constant c depends on the level α of the test. Obviously this conclusion holds good for any nonzero value of δ^2 instead of $\delta_0{}^2$. Hence Hotelling's T^2-test which rejects H_0 whenever $t^2 \geq c$ is uniformly most powerful invariant for testing $H_0:\boldsymbol{\mu} = \mathbf{0}$ against the alternatives $\boldsymbol{\mu} \neq \mathbf{0}$. Q.E.D.

The power function of any invariant test depends only on the maximal invariant in the parametric space. However, in general, the class of tests whose power function depends on the maximal invariant δ^2 contains the class of invariant tests as a subclass. The following theorem proves a stronger optimum property of the T^2-test than the one proved in Theorem 7.2.1. Theorem 7.2.2 is due to Semika (1941), although the proof presented here differs from the original proof.

THEOREM 7.2.2 On the basis of the observations $\mathbf{x}^\alpha$, $\alpha = 1, \ldots, N$, from the p-variate normal distribution with mean $\boldsymbol{\mu}$ and positive definite covariance matrix Σ, among all tests of $H_0 : \boldsymbol{\mu} = \mathbf{0}$ against the alternatives $H_1 : \boldsymbol{\mu} \neq \mathbf{0}$ with power functions depending only on $\delta^2 = N\boldsymbol{\mu}'\Sigma^{-1}\boldsymbol{\mu}$, the T^2-test is uniformly most powerful.

Proof In Theorem 7.2.1 we observed that each test in the larger class of tests that are functions of $\mathbf{X}^\alpha$, $\alpha = 1, \ldots, N$, can be replaced by one in the smaller class of tests that are functions of the sufficient statistic $(\bar{\mathbf{X}}, S)$, having the identical power function. Let $\phi(\bar{\mathbf{X}}, S)$ be a test with power function depending on δ^2. Since δ^2 is a maximal invariant in the parametric space of $(\boldsymbol{\mu}, \Sigma)$ under the transformation $(\boldsymbol{\mu}, \Sigma) \to (g\boldsymbol{\mu}, g\Sigma g')$, $g \in G$, we get

$$E_{\boldsymbol{\mu},\Sigma}\phi(\bar{\mathbf{X}}, S) = E_{g^{-1}\boldsymbol{\mu},g^{-1}\Sigma g^{-1\prime}}\phi(\bar{\mathbf{X}}, S) = E_{\boldsymbol{\mu},\Sigma}\phi(g\bar{\mathbf{X}}, gSg'). \tag{7.36}$$

Since the distribution of $(\bar{\mathbf{X}}, S)$ is boundedly complete (see Chapter V) and

$$E_{\boldsymbol{\mu},\Sigma}(\phi(\bar{\mathbf{X}}, S) - \phi(g\bar{\mathbf{X}}, gSg')) = 0 \tag{7.37}$$

identically in $\boldsymbol{\mu}$, Σ we conclude that

$$\phi(\bar{\mathbf{X}}, S) - \phi(g\bar{\mathbf{X}}, gSg') = 0$$

almost everywhere (may depend on particular g) in the space of $(\bar{\mathbf{X}}, S)$. In other words, ϕ is almost invariant with respect to G (see Definition 3.2.6). As explained in Chapter III if the group G is such that there exists a right invariant measure on G, then almost invariance implies invariance. Such a right invariant measure on G is given in Example 3.2.6. Hence if the power of the test $\phi(\bar{\mathbf{X}}, S)$ depends only on $\delta^2 = N\boldsymbol{\mu}'\Sigma^{-1}\boldsymbol{\mu}$, then $\phi(\bar{\mathbf{X}}, S)$ is almost invariant under G, which for our problem implies that $\phi(\bar{\mathbf{X}}, S)$ is invariant with respect to G transforming $(\bar{\mathbf{X}}, S) \to (g\bar{\mathbf{X}}, gSg')$, $g \in G$. Since by Theorem 7.2.1 the T^2-test is uniformly most powerful among the class of tests which are invariant with respect G, we conclude the proof of the theorem. Q.E.D.

7.2.3 Admissibility and Minimax Property of T^2

We shall now consider the optimum properties of the T^2-test among the class of all level α tests. In almost all standard hypothesis testing problems in multivariate analysis—in particular, in normal ones—no meaningful non-asymptotic (in the sample size N) optimum properties are known either for the classical tests or for any other tests. The property of being best invariant under a group of transformations that leave the problem invariant, which is often possessed by some of these tests, is often unsatisfactory because the Hunt–Stein theorem (see Chapter III) is not valid. In particular, for the case of the T^2-test the property of being uniformly most powerful invariant under the full linear group G causes the same difficulty since G does not satisfy the conditions of the Hunt–Stein theorem. The following demonstration is due to Stein as reported by Lehmann (1959, p. 338).

Let $\mathbf{X} = (X_1, \ldots, X_p)'$, $Y = (Y_1, \ldots, Y_p)'$ be independently distributed normal p-vectors with the same mean $\mathbf{0}$ and with positive definite covariance matrices Σ, $\delta\Sigma$, respectively, where δ is an unknown scalar constant. The problem of testing the hypothesis $H_0 : \delta = 1$ against the alternatives $H_1 : \delta > 1$ remains invariant under the full linear group G transforming $\mathbf{X} \to g\mathbf{X}$, $\mathbf{Y} \to g\mathbf{Y}$, $g \in G$. Since the full linear group G is transitive (see Chapter II) over the space of values of $(\mathbf{X}, \mathbf{Y})$ with probability 1, the uniformly most powerful level α invariant test under G is the trivial test $\phi(\mathbf{x}, \mathbf{y}) = \alpha$ which rejects H_0 with constant probability α for all values $(\mathbf{x}, \mathbf{y})$ of $(\mathbf{X}, \mathbf{Y})$. Thus the maximum power that can be achieved over the alternatives H_1 by any invariant test under G is also α. On the other hand, consider the test which rejects H_0 whenever $x_1{}^2/y_1{}^2 \geq c$ for any observed $\mathbf{x}$, $\mathbf{y}$ (c depending on α). This test has strictly increasing power function $\beta(\delta)$ whose minimum over the set $\delta \geq \delta_1 > 1$ is $\beta(\delta_1) > \beta(1) = \alpha$.

The admissibility of various classical tests in the univariate and multivariate situations is established by using (1) the Bayes procedure, (2) exponential structure of the parametric space, (3) invariance, and (4) local properties. For a comprehensive presentation of this the reader is referred to Keifer and Schwartz (1965). The admissibility of the T^2-test was first proved by Stein (1956) using the exponential structure of the parametric space and by showing that no other test of the same level is superior to T^2 when $\delta^2 = N\boldsymbol{\mu}'\Sigma^{-1}\boldsymbol{\mu}$ is the large (very far from H_0), and later by Kiefer and Schwartz (1965) using the Bayes procedure. It is the latter method of proof that we reproduce here. A special feature of this proof is that it yields additional information on the behavior of the T^2-test closer to H_0. The technique is to select suitable priors (probability measures or positive constant multiples thereof) Π_1 and Π_0 (say) for the parameters $(\boldsymbol{\mu}, \Sigma)$ under H_1 and for Σ under H_0 so that the T^2-test can be identified as the unique Bayes test which, by standard theory, is then admissible. The T^2-test can be written as

$$\mathbf{X}'(YY' + \mathbf{X}\mathbf{X}')^{-1}\mathbf{X} \geq c$$

where $\mathbf{X} = \sqrt{N}\,\mathbf{X}$, $S = \sum_{\alpha=1}^{N-1} \mathbf{Y}^\alpha \mathbf{Y}^{\alpha\prime}$, $Y = (\mathbf{Y}^1, \ldots, \mathbf{Y}^{N-1})$, $\mathbf{Y}^\alpha$s are independently and identically distributed normal p-vectors with mean $\mathbf{0}$ and covariance matrix Σ. It may now be recalled that if $\theta = (\boldsymbol{\mu}, \Sigma)$ and the Lebesgue density function of $V = (\mathbf{X}, Y)$ on a Euclidean set is denoted by $f_V(v|\theta)$, then every Bayes rejection region for the 0–1 loss function is of the form

$$\left\{v:\left[\int f_V(v|\theta)\Pi_1\,(d\theta)\bigg/\int f_V(v|\theta)\Pi_0\,(d\theta)\right] \geq c\right\} \tag{7.38}$$

for some c $(0 \leq c \leq \infty)$. Since in our case the subset of this set corresponding to equal to c has probability 0 for all θ in the parametric space, our Bayes procedure will be essentially unique and hence admissible.

Let both Π_1 and Π_0 assign all their measure to the θ for which $\Sigma^{-1} = I + \boldsymbol{\eta}\boldsymbol{\eta}'$ for some random p-vector $\boldsymbol{\eta}$ under both H_0 and H_1, and for which $\boldsymbol{\mu} = \mathbf{0}$ under H_0 and $\boldsymbol{\mu} = \Sigma\boldsymbol{\eta}$ with probability 1 under H_1. Regarding the distribution of $\boldsymbol{\eta}$ on the p-dimensional Euclidean space E^p we assume that

$$\begin{aligned}
&\text{under } H_1: && \frac{d\Pi_1(\boldsymbol{\eta})}{d\boldsymbol{\eta}} \propto [\det(I + \boldsymbol{\eta}\boldsymbol{\eta}')]^{-N/2} \exp\{\tfrac{1}{2}\boldsymbol{\eta}'(I + \boldsymbol{\eta}\boldsymbol{\eta}')^{-1}\boldsymbol{\eta}\},\\
&\text{under } H_0: && \frac{d\Pi_0(\boldsymbol{\eta})}{d\boldsymbol{\eta}} \propto [\det(I + \boldsymbol{\eta}\boldsymbol{\eta}')]^{-N/2}.
\end{aligned} \tag{7.39}$$

That these priors represent bona fide probability measures follows from the fact that $\boldsymbol{\eta}'(I + \boldsymbol{\eta}\boldsymbol{\eta}')^{-1}\boldsymbol{\eta}$ is bounded by unity and $\det(I + \boldsymbol{\eta}\boldsymbol{\eta}') = 1 + \boldsymbol{\eta}'\boldsymbol{\eta}$ so that

$$\int_{E^p} (1 + \boldsymbol{\eta}'\boldsymbol{\eta})^{-N/2}\,d\boldsymbol{\eta} < \infty \tag{7.40}$$

if and only if $N > p$ (which is our assumption). Since in our case

$$f_V(v|\theta) = f_{\mathbf{X}}(\mathbf{x}|\boldsymbol{\mu}, \Sigma) f_Y(y|\Sigma) \tag{7.41}$$

where

$$\begin{aligned}
f_{\mathbf{X}}(\mathbf{x}|\boldsymbol{\mu}, |\Sigma) &= (2\pi)^{-p/2}(\det \Sigma)^{-1/2} \exp\{-\tfrac{1}{2}\operatorname{tr}\Sigma^{-1}(\mathbf{x} - \boldsymbol{\mu})(\mathbf{x} - \boldsymbol{\mu})'\},\\
f_Y(y|\Sigma) &= (2\pi)^{-(N-1)p/2}(\det \Sigma)^{-(N-1)/2} \exp\{-\tfrac{1}{2}\operatorname{tr}\Sigma^{-1}yy'\},
\end{aligned}$$

it follows from (7.39) that

$$\begin{aligned}
&\frac{\int f_V(v|\theta)\Pi_1\,(d\theta)}{\int f_V(v|\theta)\Pi_0\,(d\theta)}\\
&= \frac{\int (\det(I + \boldsymbol{\eta}\boldsymbol{\eta}'))^{N/2} \times \exp\{-\tfrac{1}{2}\operatorname{tr}(I + \boldsymbol{\eta}\boldsymbol{\eta}')(\mathbf{x}\mathbf{x}' + yy') + \boldsymbol{\eta}\mathbf{x}' - \tfrac{1}{2}(I + \boldsymbol{\eta}\boldsymbol{\eta}')^{-1}\boldsymbol{\eta}\boldsymbol{\eta}'\}\Pi_1\,(d\eta)}{\int (\det(I + \boldsymbol{\eta}\boldsymbol{\eta}'))^{N/2} \exp\{-\tfrac{1}{2}\operatorname{tr}(I + \boldsymbol{\eta}\boldsymbol{\eta}')(\mathbf{x}\mathbf{x}' + yy')\}\Pi_0\,(d\eta)}\\
&= \exp\{\tfrac{1}{2}\operatorname{tr}(\mathbf{x}\mathbf{x}' + yy')^{-1}\mathbf{x}\mathbf{x}'\}\\
&\quad \times \frac{\int \exp\{-\tfrac{1}{2}\operatorname{tr}(\mathbf{x}\mathbf{x}' + yy')^{-1}(\boldsymbol{\eta} - (\mathbf{x}\mathbf{x}' + yy')^{-1}\mathbf{x})(\boldsymbol{\eta} - (\mathbf{x}\mathbf{x}' + yy')^{-1}\mathbf{x})'\}\,d\boldsymbol{\eta}}{\int \exp\{-\tfrac{1}{2}\operatorname{tr}(\mathbf{x}\mathbf{x}' + yy')\boldsymbol{\eta}\boldsymbol{\eta}\}\,d\boldsymbol{\eta}}\\
&= \exp\{-\tfrac{1}{2}\operatorname{tr}(\mathbf{x}\mathbf{x}' + yy')^{-1}\mathbf{x}\mathbf{x}'\}.
\end{aligned} \tag{7.42}$$

But $\mathbf{X}'(\mathbf{X}\mathbf{X}' + YY')^{-1}\mathbf{X} = c$ has probability 0 for all θ. Hence we conclude the following.

THEOREM 7.2.3 For each $c \geq 0$ the rejection region $\mathbf{X}'(\mathbf{X}\mathbf{X}' + YY')^{-1}\mathbf{X} \geq c$ or, equivalently, the T^2-test is admissible for testing $H_0:\boldsymbol{\mu} = \mathbf{0}$ against $H_1:\boldsymbol{\mu} \neq \mathbf{0}$.

We shall now examine the minimax property of the T^2-test for testing $H_0:\boldsymbol{\mu} = \mathbf{0}$ against the alternatives $N\boldsymbol{\mu}'\Sigma^{-1}\boldsymbol{\mu} > 0$. As shown earlier the full linear group G does not satisfy the conditions of the Hunt–Stein theorem. But the subgroup G_{T} (G_{BT} with $k = p$), the multiplicative group of $p \times p$ nonsingular lower triangular matrices which leaves the present problem invariant operating as

$$(\bar{\mathbf{X}}, S; \boldsymbol{\mu}, \Sigma) \rightarrow (g\bar{\mathbf{X}}, gSg'; g\boldsymbol{\mu}, g\Sigma g'), \qquad g \in G_{\mathrm{T}},$$

satisfies the conditions of the Hunt–Stein theorem (see Kiefer, 1957; or Lehmann, 1959, p. 345). We observed in Chapter III that on G_{T} there exists a right invariant measure. Thus there is a test of level α which is almost invariant under G_{T}, and hence in the present problem there is such a test which is invariant under G_{T} and which maximizes among all level α tests the minimum power over H_1. Whereas T^2 was a maximal invariant under G with a single distribution under each of H_0 and H_1 for each δ^2, the maximal invariant under G_{T} is the p-dimensional statistic $(R_1^*, \ldots, R_p^*)$ as defined in Section 7.2.1 with $k = p$, $p_1 = \cdots = p_k = 1$, or its equivalent statistic $(R_1, \ldots, R_p)$ as defined in Chapter VI with $k = p$. The distribution of $\mathbf{R} = (R_1, \ldots, R_p)$ has been worked out in Chapter VI. As we have observed there, under H_0 ($\delta_1^2 = \cdots = \delta_p^2 = 0$), $\mathbf{R}$ has a single distribution, but under H_1 with δ^2 fixed, it depends continuously on a $(p-1)$-dimensional vector $\boldsymbol{\Delta} = \{(\delta_1^2, \ldots, \delta_p^2): \delta_i^2 \geq 0, \sum_1^p \delta_i^2 = \delta^2\}$ for each fixed δ^2. Thus for $N > p > 1$ there is no uniformly most powerful invariant test under G_{T} for testing H_0 against $H_1: N\boldsymbol{\mu}'\Sigma^{-1}\boldsymbol{\mu} > 0$. Let $f_{\mathbf{R}}(\mathbf{r}|\Delta)$, $f_{\mathbf{R}}(\mathbf{r}|\mathbf{0})$ denote the probability density function of $\mathbf{R}$ under H_1 (for fixed δ^2) and H_0, respectively. Because of the compactness of the reduced parametric spaces $\{\mathbf{0}\}$ under H_0 and

$$\Gamma = \{(\delta_1^2, \ldots, \delta_p^2): \delta_i^2 \geq 0, \textstyle\sum_1^p \delta_i^2 = \delta^2\}$$

under H_1 and the continuity of $f_{\mathbf{R}}(\mathbf{r}|\boldsymbol{\Delta})$ in $\boldsymbol{\Delta}$, it follows that (see Wald, 1950) every minimax test for the reduced problem in terms of $\mathbf{R}$ is Bayes. In particular, Hotelling's test which rejects H_0 whenever $\sum_1^p r_i \geq c$, which has constant power on each contour $N\boldsymbol{\mu}'\Sigma^{-1}\boldsymbol{\mu} = \delta^2$ (fixed) and which is also G_{T} invariant, maximizes the minimum power over H_1 for each fixed δ^2 if and only if there is a probability density measure λ on Γ such that for some constant K

$$\int_\Gamma \frac{f_{\mathbf{R}}(\mathbf{r}|\Delta)}{f_{\mathbf{R}}(\mathbf{r}|\mathbf{0})}\lambda\,(d\Delta)\begin{Bmatrix} > \\ = \\ < \end{Bmatrix} K \tag{7.43}$$

according as

$$\sum_1^p r_i \begin{Bmatrix} > \\ = \\ < \end{Bmatrix} c$$

except possibly for a set of measure 0. Obviously c depends on the level of significance α and the measure λ, and the constant K may depend only on c and the specific value of δ^2.

From (6.64) with $k = p$, we get

$$\frac{f_{\mathbf{R}}(\mathbf{r}|\Delta)}{f_{\mathbf{R}}(\mathbf{r}|\mathbf{0})} = \exp\left\{-\tfrac{1}{2}\delta^2 + \sum_{j=1}^p r_j \sum_{i>j} \delta_i^2/2\right\} \prod_{i=1}^p \phi(\tfrac{1}{2}(N-i+1), \tfrac{1}{2}; \tfrac{1}{2}r_i\delta_i^2).$$

An examination of the integrand in this expression allows us to replace (7.43) by its equivalent

$$\int_\Gamma \frac{f_{\mathbf{R}}(\mathbf{r}|\boldsymbol{\Delta})}{f_{\mathbf{R}}(\mathbf{r}|\mathbf{0})}\lambda\,(d\boldsymbol{\Delta}) = K \qquad \text{if} \quad \sum_{i=1}^p r_i = c. \tag{7.44}$$

Clearly (7.43) implies (7.44). On the other hand, if there are a λ and a K for which (7.44) is satisfied and if $\mathbf{r}^* = (r_1^*, \ldots, r_p^*)'$ is such that $\sum_{i=1}^p r_i^* = c' > c$, writing $f(\mathbf{r}) = f_{\mathbf{R}}(\mathbf{r}|\boldsymbol{\Delta})/f_{\mathbf{R}}(\mathbf{r}|\mathbf{0})$ and $\mathbf{r}^{**} = c\mathbf{r}^*/c'$, we see at once that $f(\mathbf{r}^*) = f(c'\mathbf{r}^{**}/c) > f(\mathbf{r}^{**}) = K$, because of the form of f and the fact that $c'/c > 1$ and $\sum_{i=1}^p r_i^{**} = c$. This and a similar argument for the case $c' < c$ show that (7.44) implies (7.43). [Of course we do not assert that the left-hand side of (7.44) still depends only on $\sum_{i=1}^p r_i$ if $\sum_{i=1}^p r_i \neq c$.]

The computations in the next section are somewhat simplified by the fact that for fixed c and δ^2 we can at this point compute the unique value of K for which (7.44) can possibly be satisfied. Let $\hat{\mathbf{R}} = (R_1, \ldots, R_{p-1})'$ and write $f_{\hat{\mathbf{R}}}(\hat{\mathbf{r}}|\boldsymbol{\Delta}, u)$ for the version of the conditional Lebesgue density of $\hat{\mathbf{R}}$ given that $\sum_{i=1}^p R_i = u$ which is continuous in $\hat{\mathbf{r}}$ and u for $r_i > 0$, $\sum_{i=1}^{p-1} r_i < u < 1$, and is zero elsewhere. Write $f_U(u|\delta^2)$ for the probability density function of $U = \sum_{i=1}^p R_i$ which depends on $\boldsymbol{\Delta}$ only through δ^2, and is continuous for $0 < u < 1$ and vanishes elsewhere. Then (7.44) can be written as

$$\int f_{\hat{\mathbf{R}}}(\hat{\mathbf{r}}|\boldsymbol{\Delta}, c)\lambda\,(d\boldsymbol{\Delta}) = \left[K\frac{f_U(c|0)}{f_U(c|\delta^2)}\right] f_{\hat{\mathbf{R}}}(\hat{\mathbf{r}}|\mathbf{0}, c) \tag{7.45}$$

for $r_i > 0$, $\sum_{i=1}^{p-1} r_i < c$. The integral of (7.45), being a probability mixture of probability densities, is itself a probability density in $\hat{\mathbf{r}}$, as is $f_{\hat{\mathbf{R}}}(\hat{\mathbf{r}}|\mathbf{0}, c)$. Hence the expression in brackets equals 1. It is well known that, for $0 < c < 1$ (see Theorem 6.8.1),

$$f_U(c|\delta^2) = \frac{\Gamma(\frac{1}{2}N)\exp\{-\frac{1}{2}\delta^2\}}{\Gamma(\frac{1}{2}p)\Gamma(\frac{1}{2}(N-p))} c^{(p-2)/2}(1-c)^{(N-p-2)/2}\phi(\tfrac{1}{2}N, \tfrac{1}{2}p; \tfrac{1}{2}c\delta^2).$$

Hence (7.44) becomes

$$\int_\Gamma \exp\left\{\sum_{j=1}^{p} r_j \sum_{i>j} \tfrac{1}{2}\sigma_i^2\right\} \prod_{i=1}^{p} \phi(\tfrac{1}{2}(N-i+1), \tfrac{1}{2}; \tfrac{1}{2}r_i\delta_i^2)\lambda\,(d\boldsymbol{\Delta})$$

$$= \phi(\tfrac{1}{2}N, \tfrac{1}{2}p; \tfrac{1}{2}c\delta^2) \qquad \text{if } \sum_{i=1}^{p} r_i = c. \tag{7.46}$$

For $p = 2$, $N = 3$, writing

$$\gamma = c\delta^2, \qquad \beta_i = \delta_i^2/\delta^2, \qquad t_i = \gamma r_i/c,$$

$$\Gamma_1 = \left\{(\beta_1, \ldots, \beta_p): \beta_i \geq 0, \sum_{i=1}^{p} \beta_i = 1\right\}$$

λ^* for the measure associated with λ on Γ $[\lambda^*(A) = \lambda(\delta^2 A)]$ and noting that $\phi(\frac{3}{2}, \frac{1}{2}; \frac{1}{2}x) = (1+x)\exp\{\frac{1}{2}x\}$, we obtain from (7.46)

$$\int_0^1 [1 + (\gamma - t_2)(1-\beta_2)]\phi(1, \tfrac{1}{2}; \tfrac{1}{2}\beta_2 t_2)\, d\lambda^*(\beta_2)$$

$$= \exp\{\tfrac{1}{2}(t_2 - \gamma)\}\phi(\tfrac{3}{2}, \tfrac{1}{2}; \tfrac{1}{2}\gamma). \tag{7.47}$$

Writing

$$B = \exp\{-\tfrac{1}{2}\gamma\}\phi(\tfrac{3}{2}, 1; \tfrac{1}{2}\gamma), \qquad \mu_i = \int_0^1 \beta^i\, d\lambda^*(\beta),$$

$0 \leq i < \infty$, for the ith moment of λ^* we obtain from (7.47)

$$1 + \lambda - \lambda\mu_1 = B,$$

$$-(2r-1)\mu_{r-1} + (2r+\gamma)\mu_r - \gamma\mu_{r+1} = B\left[\frac{\Gamma(r+\frac{1}{2})}{r!\Gamma(\frac{1}{2})}\right], \quad r \geq 1. \tag{7.48}$$

Giri, *et al.* (1963) after lengthy calculations showed that there exists an absolutely continuous probability measure λ^* whose derivative $m_\gamma(x)$ is given by

$$m_\gamma(x) = \frac{\exp\{-\frac{1}{2}\gamma x\}}{2\pi x^{1/2}(1-x)^{1/2}}\left\{\int_0^\infty \exp\{-\tfrac{1}{2}\gamma u\}\left[\frac{B}{1+u} - \frac{u^{1/2}}{(1+u)^{3/2}}\right] du \right.$$

$$\left. + B\int_0^x \frac{\exp(\frac{1}{2}\gamma u)}{1-u}\, du\right\}, \tag{7.49}$$

proving that for $p = 2$, $N = 3$, the T^2-test is minimax for testing H_0 against H_1. Later Salaevski (1968), using this reduction of the problem, after voluminous computations was able to show that there exists a probability measure λ for general p and N, establishing that the T^2-test is minimax in general.

Giri and Kiefer (1962) developed the theory of local (near the null hypothesis) and asymptotic (far in distance from the null hypothesis) minimax tests for general multivariate problems. This theory serves two purposes. First, there is the obvious point of demonstrating such properties for their own sake, though well-known and valid doubts have been raised as to the extent of meaningfulness of such properties. Second, local and asymptotic minimax properties can give an indication of what to look for in the way of genuine minimax or admissibility properties of certain test procedures, even though the latter do not follow from these properties. We present in the following section the theory of local and asymptotic minimax tests as developed by Giri and Kiefer (1962) and use them to show that the T^2-test possesses both of these properties for every α, N, p. This lends to the conjecture that the T^2-test is minimax for all N, p. For relevant further results in connection with the minimax property of the T^2-test the reader is also referred to Linnik *et al.* (1966) and Linnik (1966). For a more complete presentation of minimax tests in the multivariate setup the reader is referred to Giri (1975).

7.2.4 Locally and Asymptotically Minimax Tests

Locally Minimax Tests

Let $\mathscr{X}$ be a space with an associated σ-field which, along with the other obvious measurability considerations, we will not mention in what follows. For each point (δ^2, η) in the parametric space Ω ($\delta^2 \geq 0$ and η may be a vector or matrix) suppose that $f(\cdot\,; \delta^2, \eta)$ is a probability density function on $\mathscr{X}$ with respect to a σ-finite measure μ. The range of η may depend on δ^2. For fixed α, $0 < \alpha < 1$, we shall be interested in testing, at level α, the null hypothesis $H_0\!:\delta^2 = 0$ against the alternative $H_1\!:\delta^2 = \lambda$, where λ is a specified positive value. This is a local theory in the sense that $f(x; \lambda, \eta)$ is close to $f(x; 0, \eta)$ when λ is small. Throughout this presentation, such expressions as $o(1)$, $o(h(\lambda))$, ..., are to be interpreted as $\lambda \to 0$.

For each α, $0 < \alpha < 1$, we shall consider rejection regions of the form $R = \{x\!: U(x) > C_\alpha\}$, where U is bounded and positive and has a continuous distribution function for each (δ^2, η), equicontinuous in (δ^2, η) for $\delta^2 <$ some δ^2 and where

$$P_{0,\eta}\{R\} = \alpha, \qquad P_{\lambda,\eta}\{R\} = \alpha + h(\lambda) + q(\lambda, \eta), \tag{7.50}$$

where $q(\lambda, \eta) = o(h(\lambda))$ uniformly in η, with $h(\lambda) > 0$ for $\lambda > 0$ and $h(\lambda) = o(1)$.

We shall also be concerned with probability measures $\xi_{0,\lambda}$ and $\xi_{1,\lambda}$ on the sets $\delta^2 = 0$ and $\delta^2 = \lambda$, respectively, for which

$$\frac{\int f(x; \lambda, \eta)\xi_{1,\lambda}(d\eta)}{\int f(x; \lambda, \eta)\xi_{0,\lambda}(d\eta)} = 1 + h(\lambda)[g(\lambda) + r(\lambda)U(x)] + B(x, \lambda), \tag{7.51}$$

where $0 < C_1 < r(\lambda) < C_2 < \infty$ for λ sufficiently small, and where $g(\lambda) = O(1)$ and $B(x, \lambda) = o(h(\lambda))$ uniformly in x.

THEOREM 7.2.4 *Locally minimax* If R satisfies (7.50) and if for sufficiently small λ there exist $\xi_{0,\lambda}$ and $\xi_{1,\lambda}$ satisfying (7.51), then R is locally minimax of level α for testing $H_0 : \delta^2 = 0$ against $H_1 : \delta^2 = \lambda$ as $\lambda \to 0$; that is,

$$\lim_{\lambda \to 0} \frac{\inf_\eta P_{\lambda,\eta}\{R\} - \alpha}{\sup_{\phi_\lambda \in Q_\alpha} \inf_\eta P_{\lambda,\eta}\{\phi_\lambda \text{ rejects } H_0\} - \alpha} = 1, \tag{7.52}$$

where Q_α is the class of tests of level α.

Proof Write

$$\tau_\lambda = 1/\{2 + h(\lambda)[g(\lambda) + C_\alpha r(\lambda)]\}, \tag{7.53}$$

so that

$$(1 - \tau_\lambda)/\tau_\lambda = 1 + h(\lambda)[g(\lambda) + C_\alpha r(\lambda)]. \tag{7.54}$$

A Bayes rejection region relative to the a priori distribution $\xi_\lambda = (1 - \tau_\lambda)\xi_{0,\lambda} + \tau_\lambda \xi_{1,\lambda}$ (for 0–1 losses) is, by (7.51) and (7.54),

$$B_\lambda = \{x : U(x) + \frac{B(x, \lambda)}{r(\lambda)h(\lambda)} > C_\alpha\}. \tag{7.55}$$

Write

$$P^*_{0,\lambda}\{A\} = \int P_{0,\lambda}\{A\}\xi_{0,\lambda}(d\eta), \qquad P^*_{1,\lambda}\{A\} = \int P_{\lambda,\eta}\{A\}\xi_{1,\lambda}(d\eta).$$

Let $V_\lambda = R_\lambda - B_\lambda$ and $W_\lambda = B_\lambda - R$. Using the fact that $\sup_x|B(x, \lambda)/h(\lambda)| = o(1)$ and our continuity assumption on the distribution function of U, we have

$$P^*_{0,\lambda}\{V_\lambda + W_\lambda\} = o(1). \tag{7.56}$$

Also, for $U_\lambda = V_\lambda$ or W_λ,

$$P^*_{1,\lambda}\{U_\lambda\} = P^*_{0,\lambda}\{U_\lambda\}[1 + O(h(\lambda))]. \tag{7.57}$$

Write $r^*_{1,\lambda}(A) = (1 - \tau_\lambda)P^*_{0,\lambda}\{A\} + \tau_\lambda(1 - P^*_{1,\lambda}\{A\})$. From (7.53) and (7.57), the integrated Bayes risk relative to ξ_λ is then

$$\begin{aligned} r_\lambda^*(B_\lambda) &= r_\lambda^*(R) + (1 - \tau_\lambda)(P^*_{0,\lambda}\{W_\lambda\} - P^*_{0,\lambda}\{V_\lambda\}) \\ &\quad + \tau_\lambda(P^*_{1,\lambda}\{V_\lambda\} - P^*_{1,\lambda}\{W_\lambda\}) \\ &= r_\lambda^*(R) + (1 - 2\tau_\lambda)(P^*_{0,\lambda}\{W_\lambda\} - P^*_{0,\lambda}\{V_\lambda\}) \\ &\quad + P^*_{0,\lambda}\{V_\lambda + W_\lambda\}O(h(\lambda)) \\ &= r_\lambda^*(R) + o(h(\lambda)). \end{aligned} \tag{7.58}$$

If (7.52) is false, we could, by (7.53), find a family of tests $\{\phi_\lambda\}$ of level α such that ϕ_λ has power function $\alpha + g(\lambda, \eta)$ on the set $\delta^2 = \lambda$, with

$$\limsup_{\lambda\to 0} \left(\frac{[\inf_\eta g(\lambda, \eta) - h(\lambda)]}{h(\lambda)}\right) > 0.$$

The integrated risk r_λ' of ϕ_λ with respect to ξ_λ would then satisfy

$$\limsup_{\lambda\to 0} \left(\frac{r_\lambda^*(R) - r_\lambda'}{h(\lambda)}\right) > 0,$$

thus contradicting (7.58). Q.E.D.

Asymptotically Minimax Tests

Here we treat the case $\lambda \to \infty$, and expressions such as $o(1)$, $o(H(\lambda))$ are to be interpreted in this light. Suppose that in place of (7.50) R satisfies

$$P_{0,\eta}\{R\} = \alpha, \qquad P_{\lambda,\eta}\{R\} = 1 - \exp\{-H(\lambda)(1 + o(1))\}, \tag{7.59}$$

where $H(\lambda) \to \infty$ with λ and the $o(1)$ term is uniform in η. Suppose, replacing (7.51), that

$$\frac{\int f(x; \lambda, \eta)\xi_{1,\lambda}(d\eta)}{\int f(x; 0, \eta)\xi_{0,\lambda}(d\eta)} = \exp\{H(\lambda)[G(\lambda) + R(\lambda)U(x)] + B(x, \lambda)\}, \tag{7.60}$$

where $\sup_x|B(x, \lambda)| = o(H(\lambda))$ and $0 < C_1 < R(\lambda) < C_2 < \infty$. Our only other regularity assumption is that C_α is a point of increase from the left of the distribution function of U, when $\delta^2 = 0$, uniformly in η; that is,

$$\inf_\eta P_{0,\eta}\{U \geq C_\alpha - \varepsilon\} > \alpha \tag{7.61}$$

for every $\varepsilon > 0$.

THEOREM 7.2.5 If R satisfies (7.59) and (7.61) and if for sufficiently large λ there exist $\xi_{0,\lambda}$ and $\xi_{1,\lambda}$ satisfying (7.60), then R is asymptotically logarithmically minimax of level α for testing $H_0:\delta^2 = 0$ against $H_1:\delta^2 = \lambda$

so that $\lambda \to \infty$; that is,

$$\lim_{\lambda\to\infty} \frac{\inf_\eta\{-\log[1 - P_{\lambda,\eta}\{R\}]\}}{\sup_{\phi_\lambda \in Q_\alpha} \inf_\eta\{-\log[1 - P_{\lambda,\eta}\{\phi_\lambda \text{ rejects } H_0\}]\}} = 1. \tag{7.62}$$

Proof Suppose, contrary to (7.62), that there is an $\varepsilon > 0$ and an unbounded sequence Γ of values λ with corresponding tests ϕ_λ in Q_α for which

$$P_{\lambda,\eta}\{R\} > 1 - \exp\{-H(\lambda)(1 + 5\varepsilon)\} \qquad \text{for all} \quad \eta. \tag{7.63}$$

There are two cases: (7.64) and (7.67). If $\lambda \in \Gamma$ and

$$-1 - G(\lambda) \leq R(\lambda)C_\alpha + 2\varepsilon, \tag{7.64}$$

consider the a priori distribution given by $\xi_{i,\lambda}$ and by τ_λ satisfying

$$\tau_\lambda/(1 - \tau_\lambda) = \exp\{H(\lambda)(1 + 4\varepsilon)\}. \tag{7.65}$$

The integrated risk of any Bayes procedure B_λ must satisfy

$$\begin{aligned} r_\lambda^*(B_\lambda) \leq r_\lambda^*(\phi_\lambda) &\leq (1 - \tau_\lambda)\alpha + \tau_\lambda \exp\{-H(\lambda)(1 + 5\varepsilon)\} \\ &= (1 - \tau_\lambda)[\alpha + \exp(-\varepsilon H(\lambda))], \end{aligned} \tag{7.66}$$

by (7.63) and (7.65). But from (7.60) a Bayes critical region is

$$B_\lambda = \left\{x: \frac{U(x) + B(x, \lambda)}{R(\lambda)H(\lambda)} \geq \frac{-(1 + 4\varepsilon) - G(\lambda)}{R(\lambda)}\right\}.$$

Hence if λ is so large that $\sup_x|B(x, \lambda)/H(\lambda)R(\lambda)| < \varepsilon/C_2$, we get from (7.64)

$$B_\lambda \supset \{x: U(x) > C_\alpha - \varepsilon/C_2\} = B_\lambda' \qquad \text{(say).}$$

The assumption (7.61) implies that

$$P_{0,\eta}\{B_\lambda'\} > \alpha + \varepsilon'$$

with $\varepsilon' > 0$, contradicting (7.66) for large λ. On the other hand, if $\lambda \in \Gamma$ and

$$-1 - G(\lambda) > R(\lambda)C_\alpha + 2\varepsilon, \tag{7.67}$$

let

$$\tau_\lambda/(1 - \tau_\lambda) = \exp\{H(\lambda)(1 + \varepsilon)\}. \tag{7.68}$$

Then by (7.60)

$$B_\lambda = \left\{x: \frac{U(x) + B(x, \lambda)}{R(\lambda)H(\lambda)} \geq \frac{-(1 + \varepsilon) - G(\lambda)}{R(\lambda)}\right\}.$$

Hence if $\sup_x|B(x, \lambda)/H(\lambda)R(\lambda)| < \varepsilon/2C_2$, we conclude from (7.67) that $B_\lambda \subset R$, so that, by (7.59) and (7.68),

$$r^*(B_\lambda) > \tau_\lambda \exp\{-H(\lambda)[1 + o(1)]\} = (1 - \tau_\lambda)\exp\{H(\lambda)(\varepsilon - o(1))\}. \tag{7.69}$$

But

$$r^*(B_\lambda) \le r^*(\phi_\lambda) \le (1 - \tau_\lambda)\alpha + \tau_\lambda \exp\{-H(\lambda)(1 + 5\varepsilon)\}$$
$$= (1 - \tau_\lambda)[\alpha + \exp\{-4\varepsilon H(\lambda)\}],$$

which contradicts (7.69) for sufficiently large λ. Q.E.D.

THEOREM 7.2.6 For every p, N, and α, Hotelling's T^2-test is locally minimax for testing $H_0:\delta^2 = 0$ against $H_1:\delta^2 = \lambda$ as $\lambda \to 0$.

Proof In our search for a locally minimax test as $\lambda \to 0$ we look for a level α test which is almost invariant under G_T and which minimizes among all level α tests the minimum power under H_1 (as discussed in the case of the genuine minimax property of the T^2-test). So we restrict our attention to the space of $\mathbf{R} = (R_1, \ldots, R_p)'$, the maximal invariant under G_T in the space of $(\bar{\mathbf{X}}, S)$. We now verify the assumption of Theorem 7.2.4 with $x = \mathbf{r}, \eta_i = \delta_i^2/\delta^2$, $\eta = \boldsymbol{\eta} = (\eta_1, \ldots, \eta_p)'$, and $U(x) = \sum_{i=1}^p r_i$. We can take $h(\lambda) = b\lambda$ with b a positive constant. Of course, $P_{\lambda,\boldsymbol{\eta}}\{R\}$ does not depend on $\boldsymbol{\eta}$. From (6.66)

$$\frac{f(\mathbf{r};\lambda,\boldsymbol{\eta})}{f(\mathbf{r};0,\mathbf{0})} = 1 + \frac{\lambda}{2}\left\{-1 + \sum_{j=1}^{p} r_j\left[\sum_{i>j}\eta_i + (N-j+1)\eta_j\right]\right\}$$
$$+ B(\mathbf{r},\boldsymbol{\eta},\lambda), \tag{7.70}$$

where $B(\mathbf{r}, \boldsymbol{\eta}, \lambda) = o(\lambda)$ uniformly in $\mathbf{r}$ and $\boldsymbol{\eta}$. Here the set $\{\delta^2 = 0\}$ is a single point. Also the set $\{\delta^2 = \lambda\}$ is a convex finite-dimensional Euclidian set where in each component η_i is $O(h(\lambda))$. If there exists any $\xi_{1,\lambda}$ satisfying (7.51), the degenerate $\xi'_{1,\lambda}$ which assigns measure 1 to the mean of $\xi_{1,\lambda}$ also satisfies (7.51), and (7.51) is satisfied by letting $\xi_{0,\lambda}$ give measure 1 to the single point $\boldsymbol{\eta} = \mathbf{0}$, whereas $\xi_{1,\lambda}$ gives measure 1 to the single point η^* (say) whose jth coordinate is $(N-j)^{-1}(N-j+1)^{-1}p^{-1}N(N-p)$, so that $\sum_{i>j}\eta_i^* + (N-j+1)\eta_j^* = N/p$ for all j. Applying Theorem 7.2.4 we get the result. Q.E.D.

THEOREM 7.2.7 For every α, p, N, Hotelling's T^2-test is asymptotically (logarithmically) minimax for testing $H_0:\delta^2 = 0$ against the alternative $H_1:\delta^2 = \lambda$ as $\lambda \to \infty$.

Proof From (6.64) [since $\phi(a, b; x) = \exp(x(1 + o(1)))$ as $x \to \infty$] we get

$$\frac{f(\mathbf{r};\lambda,\boldsymbol{\eta})}{f(\mathbf{r};0,\boldsymbol{\eta})} = \exp\left\{\frac{\lambda}{2}\left[-1 + \sum_{j=1}^{p} r_j \sum_{i\ge j}\eta_i\right](1 + B(\mathbf{r},\boldsymbol{\eta},\lambda))\right\} \tag{7.71}$$

with $\sup_{\mathbf{r},\boldsymbol{\eta}}|B(\mathbf{r},\boldsymbol{\eta},\lambda)| = o(1)$ as $\lambda \to \infty$. From this and the smoothness of $f(\mathbf{r}; 0, \boldsymbol{\eta})$ we see (e.g., putting $\eta_p = 1$, the density of U being independent of $\boldsymbol{\eta}$) that

$$P_{\lambda,\boldsymbol{\eta}}\{U < C_\alpha\} = \exp\{\tfrac{1}{2}\lambda(C_\alpha - 1)[1 + o(1)]\} \tag{7.72}$$

as $\lambda \to \infty$. Thus (7.59) is satisfied with $H(\lambda) = \frac{1}{2}(1 - C_\alpha)$. Next, letting $\xi_{1,\lambda}$ assign measure 1 to the point $\eta_1 = \cdots = \eta_{p-1} = 0, \eta_p = 1$, and $\xi_{0,\lambda}$ assign measure 1 to $(0, \mathbf{0})$, we obtain (7.60). Finally (7.61) is trivial. Applying Theorem 7.2.5 we get the result. Q.E.D.

Suppose, for a parameter set $\Omega' = \{(\theta, \eta):\theta \in \Theta, \eta \in H\}$ with associated distributions, with Θ a Euclidean set, that every test ϕ has a power function $\beta_\phi(\theta, \eta)$ which, for each η is twice continuously differentiable in the components of θ at $\theta = 0$, an interior point of Θ. Let Q_α be the class of locally strictly unbiased level α tests of $H_0:\theta = 0$ against $H_1:\theta \neq 0$; our assumption on β_ϕ implies that all tests in Q_α are similar and that $\partial\beta_\phi/\partial\theta_i|_{\theta=0} = 0$ for ϕ in Q_α. Let $\Delta_\phi(\eta)$ be the determinant of the matrix $B_\phi(\eta)$ of second derivatives of $\beta_\phi(\theta, \eta)$ with respect to the components of θ at $\theta = 0$. We assume the parametrization to be such that $\Delta_{\phi'}(\eta) > 0$ for all η for at least one ϕ' in Q_α.

A test ϕ^* is said to be of type E if $\phi^* \in Q_\alpha$ and $\Delta_{\phi^*}(\eta) = \max_{\phi \in Q_\alpha} \Delta_\phi(\eta)$ for all η. If H is a single point, ϕ^* is said to be of type D.

Write

$$\bar{\Delta}(\eta) = \max_{\phi \in Q_\alpha} \Delta_\phi(\eta).$$

A test ϕ^* will be said to be of type D_A if $\phi \in Q_\alpha$ and

$$\max_\eta[\bar{\Delta}(\eta) - \Delta_{\phi^*}(\eta)] = \min_{\phi \in Q_\alpha} \max_\eta [\bar{\Delta}(\eta) - \Delta_\phi(\eta)]$$

and of type D_M if

$$\max_\eta[\bar{\Delta}(\eta)/\Delta_{\phi^*}(\eta)] = \min_{\phi \in Q_\alpha} \max_\eta [\bar{\Delta}(\eta)/\Delta_\phi(\eta)].$$

The notion of type D and E regions is due to Isaacson (1951). The D_A and D_M criteria resemble stringency and regret criteria employed elsewhere in statistics.

The reader is referred to Giri and Kiefer (1964) for the proof that the T^2-test is not of type D among all G_T invariant tests and hence is not of type D_A or D_M or E among all tests.

7.2.5 Applications of the T^2-Test

Confidence Region of Mean Vector

Let $\mathbf{x}^\alpha = (x_{\alpha 1}, \ldots, x_{\alpha p})'$, $\alpha = 1, \ldots, N$, be a sample of size N from a p-variate normal distribution with unknown mean $\boldsymbol{\mu}$ and unknown positive definite covariance matrix Σ. Let

$$\bar{\mathbf{x}} = \frac{1}{N}\sum_1^N \mathbf{x}^\alpha, \qquad s = \sum_1^N (\mathbf{x}^\alpha - \bar{\mathbf{x}})(\mathbf{x}^\alpha - \bar{\mathbf{x}})'.$$

For the corresponding random sample $\mathbf{X}^\alpha$, $\alpha = 1, \ldots, N$, $N(N-1)(\bar{\mathbf{X}} - \boldsymbol{\mu})'S^{-1}(\bar{\mathbf{X}} - \boldsymbol{\mu})$ is distributed as Hotelling's T^2 with $N-1$ degrees of freedom. Let $T_0{}^2(\alpha)$, for $0 < \alpha < 1$, be such that $P(T^2 \geq T_0{}^2(\alpha)) = \alpha$. Then the probability of drawing a sample $\mathbf{x}^\alpha$, $\alpha = 1, \ldots, N$, of size N with mean $\bar{\mathbf{x}}$ and sample covariance s such that

$$N(N-1)s^{-1}(\bar{\mathbf{x}} - \boldsymbol{\mu})'s^{-1}(\bar{\mathbf{x}} - \boldsymbol{\mu}) \leq T_0{}^2(\alpha)$$

is $1 - \alpha$. Hence given $\mathbf{x}^\alpha$, $\alpha = 1, \ldots, N$, the $100(1-\alpha)\%$ confidence region of $\boldsymbol{\mu}$ consists of all p-vectors $\mathbf{m}$ satisfying

$$N(N-1)(\bar{\mathbf{x}} - \mathbf{m})'s^{-1}(\bar{\mathbf{x}} - \mathbf{m}) \leq T_0{}^2(\alpha). \tag{7.73}$$

The boundary of this region is an ellipsoid whose center is at the point $\bar{\mathbf{x}}$ and whose size and shape depend on s and α.

Test for the Equality of Two Mean Vectors

Let $\mathbf{X}^\alpha = (X_{\alpha 1}, \ldots, X_{\alpha p})'$, $\alpha = 1, \ldots, N_1$, be a random sample of size N_1 from a p-variate normal population with mean vector $\boldsymbol{\mu}$ and positive definite covariance matrix Σ, and let $\mathbf{Y}^\alpha = (Y_{\alpha 1}, \ldots, Y_{\alpha p})'$, $\alpha = 1, \ldots, N_2$, be a random sample of size N_2 from another independent normal population with mean $\mathbf{v}$ and positive definite covariance matrix Σ. Let

$$\bar{\mathbf{X}} = \frac{1}{N_1}\sum_1^{N_1} \mathbf{X}^\alpha, \qquad \bar{\mathbf{Y}} = \frac{1}{N_2}\sum_1^{N_2} \mathbf{Y}^\alpha,$$

$$S = \sum_1^{N_1} (\mathbf{X}^\alpha - \bar{\mathbf{X}})(\mathbf{X}^\alpha - \bar{\mathbf{X}})' + \sum_1^{N_2} (\mathbf{Y}^\alpha - \bar{\mathbf{Y}})(\mathbf{Y}^\alpha - \bar{\mathbf{Y}})'.$$

It can be verified that $(\bar{\mathbf{X}}, \bar{\mathbf{Y}}, S)$ is a complete sufficient statistic for $(\boldsymbol{\mu}, \mathbf{v}, \Sigma)$, $(N_1N_2/(N_1 + N_2))^{1/2}(\bar{\mathbf{X}} - \bar{\mathbf{Y}})$ has p-variate normal distribution with mean $(N_1N_2/(N_1 + N_2))^{1/2}(\boldsymbol{\mu} - \mathbf{v})$ and positive definite covariance matrix Σ, and S is distributed as Wishart $W_p(N_1 + N_2 - 2, \Sigma)$ independently of $(\bar{\mathbf{X}}, \bar{\mathbf{Y}})$. The problem of testing the hypothesis $H_0 : \boldsymbol{\mu} - \mathbf{v} = \mathbf{0}$ against the alternatives $H_1 : \boldsymbol{\mu} - \mathbf{v} \neq 0$ remains invariant under the group of affine transformations $\mathbf{X}^\alpha \to g\mathbf{X}^\alpha + \mathbf{b}$, $\alpha = 1, \ldots, N_1$, $\mathbf{Y}^\alpha \to g\mathbf{Y}^\alpha + \mathbf{b}$, $\alpha = 1, \ldots, N_2$, where $g \in G$, $\mathbf{b} \in E^p$ (Euclidean p-space). The maximal invariant under the group of affine transformations in the space of $(\bar{\mathbf{X}}, \bar{\mathbf{Y}}, S)$ is given by

$$T^2 = (N_1 + N_2 - 2)(N_1N_2/(N_1 + N_2))(\bar{\mathbf{X}} - \bar{\mathbf{Y}})'S^{-1}(\bar{\mathbf{X}} - \bar{\mathbf{Y}})$$

and T^2 is distributed as Hotelling's T^2 with $N_1 + N_2 - 2$ degrees of freedom and the noncentrality parameter

$$\delta^2 = (N_1N_2/(N_1 + N_2))(\boldsymbol{\mu} - \mathbf{v})'\Sigma^{-1}(\boldsymbol{\mu} - \mathbf{v}).$$

An optimum test for this problem is the Hotelling's T^2-test which rejects H_0 for large values of T^2. This test possesses all the properties of the T^2-test discussed above.

EXAMPLE 7.2.1 Consider Example 5.3.1 and assume that the two p-variate normal populations have the same positive definite covariance matrix Σ (unknown). Let the mean of population I (1971) be $\boldsymbol{\mu}$ and that of population II (1972) be $\mathbf{v}$. We are interested here in testing the hypothesis $H_0 : \boldsymbol{\mu} - \mathbf{v} = \mathbf{0}$. Here $N_1 = N_2 = 27$:

$$\bar{\mathbf{x}} = (84.89, 186.30, 9.74, 13.46, 304.37, 13.63)'$$

$$\bar{\mathbf{y}} = (77.14, 167.18, 10.45, 13.10, 361.55, 14.76)'$$

$$\frac{s}{52} = \begin{array}{c} \\ 1 \\ 2 \\ 3 \\ 4 \\ 5 \\ 6 \end{array}\!\!\begin{pmatrix} 1 & 2 & 3 & 4 & 5 & 6 \\ 1143.07 & & & & & \\ 57.40 & 3.84 & & & & \\ 70.16 & 4.25 & 25.54 & & & \\ 79.48 & 0.66 & 23.62 & 326.56 & & \\ 15.28 & 0.77 & 1.18 & 2.40 & 0.30 & \\ 21.60 & 1.04 & 2.56 & 4.14 & 0.39 & 0.83 \end{pmatrix}.$$

The value of

$$t^2 = (N_1 N_2/(N_1 + N_2))(N_1 + N_2 - 2)(\bar{\mathbf{x}} - \bar{\mathbf{y}})' s^{-1}(\bar{\mathbf{x}} - \bar{\mathbf{y}}) = 217.55.$$

The 1% significance value of T^2 is 21.21. Thus we reject the hypothesis that the means of the two populations are equal.

Problem of Symmetry and Tests of Significance of Contrasts

Let $\mathbf{x}^\alpha = (x_{\alpha 1}, \ldots, x_{\alpha p})'$, $\alpha = 1, \ldots, N$, be a sample of size N from a p-variate normal population with mean $\boldsymbol{\mu} = (\mu_1, \ldots, \mu_p)'$ and covariance matrix Σ. We are interested in testing the hypothesis

$$H_0 : \mu_1 = \cdots \mu_p = \gamma \qquad \text{(unknown)}.$$

Let $\mathbf{E} = (1, \ldots, 1)'$ be a p-vector with components all equal to unity. A matrix C of dimension $(p - 1) \times p$ is called a contrast matrix if $C\mathbf{E} = \mathbf{0}$.

EXAMPLE 7.2.2 The $(p - 1) \times p$ matrix C_1

$$C_1 = \begin{pmatrix} 1 & -1 & 0 & \cdots & 0 & 0 \\ 0 & 1 & -1 & \cdots & 0 & 0 \\ \vdots & \vdots & \vdots & & \vdots & \vdots \\ 0 & 0 & 0 & \cdots & 1 & -1 \end{pmatrix}$$

is a contrast matrix of rank $p-1$. The $(p-1) \times p$ matrix C_2

$$C_2 = \begin{pmatrix} \frac{1}{(1.2)^{1/2}} & \frac{-1}{(1.2)^{1/2}} & 0 & \cdots & 0 \\ \frac{1}{(2.3)^{1/2}} & \frac{1}{(2.3)^{1/2}} & \frac{-2}{(2.3)^{1/2}} & \cdots & 0 \\ \vdots & \vdots & \vdots & & \vdots \\ \frac{1}{((p-1)p)^{1/2}} & \frac{1}{((p-1)p)^{1/2}} & \frac{1}{((p-1)p)^{1/2}} & \cdots & \frac{-(p-1)}{((p-1)p)^{1/2}} \end{pmatrix}$$

is an orthogonal contrast matrix of rank $(p-1)$ and is known as a Helmert matrix.

Obviously from the relation $C\mathbf{E} = \mathbf{0}$ we conclude that all rows of C are orthogonal to $\mathbf{E}$ and the sum of the elements of any row of C is zero. Furthermore any two contrast matrices C_1, C_2 are related by

$$C_1 = DC_2, \tag{7.74}$$

where D is a nonsingular matrix of dimension $(p-1) \times (p-1)$. Under H_0, $\boldsymbol{\mu} = \gamma\mathbf{E}$ and hence $E(C\mathbf{X}^\alpha) = \mathbf{0}$ for any contrast matrix C. Conversely, if $E(C\mathbf{X}^\alpha) = 0$ for some contrast matrix C (for each α), we have $C\boldsymbol{\mu} = 0$. But on account of (7.46) $C = DC_1$, where C_1 is defined in Example 7.2.2, and hence $\mathbf{0} = DC_1\boldsymbol{\mu}$, which implies $C_1\boldsymbol{\mu} = \mathbf{0}$, and thus $\mu_1 = \cdots = \mu_p$.

Furthermore, for any contrast matrix C of dimension $(p-1) \times p$ (of rank $p-1$), the matrix $\binom{E}{C}$ is a nonsingular matrix and hence $C\mathbf{X}^\alpha$, $\alpha = 1, \ldots, N$, are independently and identically distributed $(p-1)$-dimensional normal vectors with mean $C\boldsymbol{\mu}$ and positive definite covariance matrix $C\Sigma C'$. Hence the appropriate test for $H_0 : C\boldsymbol{\mu} = 0$ rejects H_0 if

$$t^2 = N(N-1)(C\bar{\mathbf{x}})'(CsC')^{-1}(C\bar{\mathbf{x}}) \geq k,$$

where CSC' is distributed independently of $C\bar{\mathbf{X}}$ as Wishart $W_{p-1}(N-1, C\Sigma C')$ and the constant k is chosen such that the test has level α. Obviously the statistic T^2 (in this case) is distributed as Hotelling's T^2 based on a random sample $C\mathbf{X}^\alpha$, $\alpha = 1, \ldots, N$, of size N. It may be noted that T^2 does not depend on the particular choice of the contrast matrix C. As for any other contrast matrix C_1 we can write $C_1 = DC$ where D is nonsingular and

$$T^2 = (N-1)N(C_1\bar{\mathbf{X}})'(C_1SC_1')^{-1}(C_1\bar{\mathbf{X}}) = (N-1)N(C\bar{\mathbf{X}})'(CSC')^{-1}(C\bar{\mathbf{X}}).$$

The noncentrality parameter of this distribution is

$$N(C\boldsymbol{\mu})'(C\Sigma C')^{-1}(C\boldsymbol{\mu}).$$

EXAMPLE 7.2.3 An interesting application of this was given by Rao (1965) in the case of a four-dimensional normal vector $\mathbf{X} = (X_1, X_2, X_3, X_4)'$, where $X_1, \ldots, X_4$ represent the thickness of cork borings on trees in the four directions north, south, east, and west, respectively. The hypothesis in this case is that of equal bark deposit in every direction. The contrast matrix C in this case is

$$C = \begin{pmatrix} 1 & 1 & -1 & -1 \\ 1 & -1 & 0 & 0 \\ 0 & 0 & 1 & -1 \end{pmatrix}.$$

For numerical data and the results the reader is referred to Rao (1965).

EXAMPLE 7.2.4 *Randomized block design with correlated observations* Consider a randomized block design with N blocks and p treatments. Let y_{ij} denote the yield of the ith treatment in the jth block and let Y_{ij} be the corresponding random variables. Assume that the Y_{ij} are normally distributed with

$$E(Y_{ij}) = \mu + \mu_i + \beta_j,$$

$$\operatorname{cov}(Y_{ij}, Y_{i'j'}) = \begin{cases} \sigma_{ii'} & \text{if } j = j', \\ 0 & \text{otherwise,} \end{cases}$$

$$\operatorname{var}(Y_{ij}) = \sigma_{ii},$$

$i = 1, \ldots, p$; $j = 1, \ldots, N$, where μ_i is the ith treatment effect, and β_j is the jth block effect. Such a case arises when, for example the β_j are random variables (random effect model). Write $Y = (\mathbf{Y}^1, \ldots, \mathbf{Y}^N)$, $\mathbf{Y}^\alpha = (Y_{\alpha 1}, \ldots, Y_{\alpha p})'$, $\alpha = 1, \ldots, N$. Y is a $p \times N$ random matrix of elements Y_{ij} and Σ is a $p \times p$ matrix of elements $\sigma_{ii'}$. Then $\operatorname{cov}(Y) = \Sigma \otimes I$ where I is the identity matrix of dimension $N \times N$. The usual hypothesis in this case is $H_0 : \mu_1 = \cdots = \mu_p$. With the contrast matrix C_1 in Example 7.2.2, under H_0,

$$E(C_1 Y) = \begin{pmatrix} \mu_1 - \mu_2 \\ \mu_2 - \mu_3 \\ \vdots \\ \mu_{p-1} - \mu_p \end{pmatrix} \mathbf{E} = 0,$$

where $\mathbf{E}$ is an N vector with all components equal to unity and $\operatorname{cov}(C_1 Y) = (C_1 \Sigma C_1') \otimes I$. Under the assumption of normality the column vectors of $C_1 Y$ are independently distributed $(p - 1)$-variate normal vectors with mean $\mathbf{0}$ under H_0 and with covariance matrix $C_1 \Sigma C_1'$. The appropriate test statistic for testing H_0 rejects H_0 when

$$t^2 = (N - 1)N(C_1 \bar{\mathbf{y}})'(C_1 s C_1')^{-1}(C_1 \bar{\mathbf{y}}) \geq c,$$

where c is a constant depending on the level α of the test and $\bar{\mathbf{y}} = (1/N)\sum_1^N \mathbf{y}^\alpha$, $s = \sum_1^N(\mathbf{y}^\alpha - \bar{\mathbf{y}})(\mathbf{y}^\alpha - \bar{\mathbf{y}})'$. It is easy to see that $C_1 S_1 C_1'$ $(N > p)$ is distributed independently of $\bar{\mathbf{Y}}$ as $W_{p-1}(N-1, C_1\Sigma C_1')$. Thus T^2 is distributed as Hotelling's T^2 with the noncentrality parameter

$$\delta^2 = N(\mu_1 - \mu_2, \mu_2 - \mu_3, \ldots, \mu_{p-1} - \mu_p)'(C_1\Sigma C_1')^{-1} \times (\mu_1 - \mu_2, \mu_2 - \mu_3, \ldots, \mu_{p-1} - \mu_p).$$

Paired T^2-Test and the Multivariate Analog of the Behren–Fisher Problem

Let $\mathbf{X}^\alpha = (X_{\alpha 1}, \ldots, X_{\alpha p})'$, $\alpha = 1, \ldots, N_1$, be a random sample of size N_1 from a p-variate normal population with mean $\boldsymbol{\mu}$ and positive definite covariance matrix Σ_1, and let $\mathbf{Y}^\alpha = (Y_{\alpha 1}, \ldots, Y_{\alpha p})'$, $\alpha = 1, \ldots, N_2$, be a random sample of size N_2 from another independent p-variate normal population with mean $\mathbf{v}$ and positive definite covariance matrix Σ_2. We are interested here in testing the hypothesis $H_0 : \boldsymbol{\mu} = \mathbf{v}$. It is well known that even for $p = 1$ the likelihood ratio test is very complicated and is not suitable for practical use. If $\Sigma_1 = \Sigma_2$, we have shown that the T^2-test is the appropriate solution. However, if $\Sigma_1 \neq \Sigma_2$ but $N_1 = N_2 = N$, a suitable solution is reached by using the following paired device. Define $\mathbf{Z}^\alpha = \mathbf{X}^\alpha - \mathbf{Y}^\alpha$, $\alpha = 1, \ldots, N$. Obviously $\mathbf{Z}^\alpha$, $\alpha = 1, \ldots, N$, constitute a random sample of size N from a p-variate normal distribution with mean $\boldsymbol{\theta} = \boldsymbol{\mu} - \mathbf{v}$ and positive definite covariance matrix $\Sigma_1 + \Sigma_2 = \Sigma$ (say). The testing problem reduces to that of testing $H_0 : \boldsymbol{\theta} = \mathbf{0}$ when Σ is unknown. Define

$$\bar{\mathbf{Z}} = \frac{1}{N}\sum_1^N \mathbf{Z}^\alpha, \qquad S = \sum_1^N (\mathbf{Z}^\alpha - \bar{\mathbf{Z}})(\mathbf{Z}^\alpha - \bar{\mathbf{Z}})'.$$

On the basis of sample observations $\mathbf{Z}^\alpha$, $\alpha = 1, \ldots, N$, the likelihood ratio test of H_0 rejects H_0 whenever

$$t^2 = (N-1)N\bar{\mathbf{z}}' s^{-1}\bar{\mathbf{z}} \geq c,$$

where the constant c depends on the level α of the test, and it possesses all the optimum properties of Hotelling's T^2-test (obviously in the class of tests based only on the differences $\mathbf{Z}^\alpha$, $\alpha = 1, \ldots, N$).

When $\Sigma_1 \neq \Sigma_2$, the multivariate analog of Scheffé's solution (Scheffé, 1943) gives an appropriate solution. This extension is due to Bennet (1951). Assume without any loss of generality that $N_1 < N_2$. Define

$$\mathbf{Z}^\alpha = \mathbf{X}^\alpha - \left(\frac{N_1}{N_2}\right)^{1/2} \mathbf{Y}^\alpha + \frac{1}{(N_1 N_2)^{1/2}} \sum_1^{N_1} \mathbf{Y}^\alpha - \frac{1}{N_2}\sum_1^{N_2} \mathbf{Y}^\alpha, \qquad \alpha = 1, \ldots, N_1.$$

It is easy to verify that $\mathbf{Z}^\alpha$, $\alpha = 1, \ldots, N_1$, are independently distributed normal p-vectors with the same mean $\boldsymbol{\mu} - \mathbf{v}$ and the same covariance matrix

$\Sigma_1 + (N_1/N_2)\Sigma_2$. Let

$$\bar{\mathbf{Z}} = \frac{1}{N_1}\sum_1^{N_1} \mathbf{Z}^\alpha, \qquad S = \sum_1^{N_1} (\mathbf{Z}^\alpha - \bar{\mathbf{Z}})(\mathbf{Z}^\alpha - \bar{\mathbf{Z}})'.$$

Obviously $\bar{\mathbf{Z}}$ and S are independent, $\bar{\mathbf{Z}}$ has a p-variate normal distribution with mean $\boldsymbol{\mu} - \boldsymbol{\nu}$ and with positive definite covariance matrix $(\Sigma_1 + (N_1/N_2)\Sigma_2)$, and S is distributed as $W_p(N_1 - 1, \Sigma_1 + (N_1/N_2)\Sigma_2)$. An appropriate solution for testing $H_0:\boldsymbol{\mu} - \boldsymbol{\nu} = \mathbf{0}$ is given by

$$t^2 = (N_1 - 1)N_1\bar{\mathbf{z}}'s^{-1}\bar{\mathbf{z}} \geq c,$$

where c depends on the level α of the test and T^2 has Hotelling's T^2-distribution with $N_1 - 1$ degrees of freedom and the noncentrality parameter $N_1(\boldsymbol{\mu} - \boldsymbol{\nu})'(\Sigma_1 + (N_1/N_2)\Sigma_2)^{-1}(\boldsymbol{\mu} - \boldsymbol{\nu})$.

7.3 TESTS OF HYPOTHESES CONCERNING SUBVECTORS OF $\boldsymbol{\mu}$

Let $\mathbf{X}^\alpha = (X_{\alpha 1}, \ldots, X_{\alpha p})'$, $\alpha = 1, \ldots, N$, be a random sample of size N from a p-variate normal distribution with mean $\boldsymbol{\mu}$ and positive definite covariance matrix Σ. We shall use the notations of Section 7.2.2 for the presentation of this section. We shall consider the following two testing problems concerning subvectors of $\boldsymbol{\mu}$. The two-sample analogs of these problems are obvious and their appropriate solutions can be easily obtained from the one-sample results presented here.

(a) In the notation of Section 7.2.2, let $k = 2$, $p_1 + p_2 = p$. We are interested here in testing the hypothesis $H_0:\boldsymbol{\mu}_{(1)} = \mathbf{0}$ when Σ is unknown. Let Ω be the parametric space of $(\boldsymbol{\mu}, \Sigma)$ and $\omega = \{(\mathbf{0}, \boldsymbol{\mu}_{(2)}), \Sigma\}$ be the subspace of Ω when H_0 is true. The likelihood of the observations $\mathbf{x}^\alpha$, $\alpha = 1, \ldots, N$, on $\mathbf{X}^\alpha$, $\alpha = 1, \ldots, N$, is

$$L(\boldsymbol{\mu}, \Sigma) = (2\pi)^{-Np/2}(\det \Sigma)^{-N/2} \exp\left\{-\tfrac{1}{2} \operatorname{tr} \Sigma^{-1} \sum_{\alpha=1}^{N} (\mathbf{x}^\alpha - \boldsymbol{\mu})(\mathbf{x}^\alpha - \boldsymbol{\mu})'\right\}.$$

Obviously

$$\max_\Omega L(\boldsymbol{\mu}, \Sigma) = (2\pi)^{-Np/2}[\det(s/N)]^{-N/2} \exp\{-\tfrac{1}{2}Np\}. \tag{7.75}$$

It is also easy to verify that

$$\max_\omega L(\boldsymbol{\mu}, \Sigma) = (2\pi)^{-Np/2}\left[\det\left(\frac{s_{(11)} + N\bar{\mathbf{x}}_{(1)}\bar{\mathbf{x}}_{(1)}'}{N}\right)\right]^{-N/2}$$
$$\times \left[\det\left(\frac{s_{(22)} - s_{(21)}s_{(11)}^{-1}s_{(12)}}{N}\right)\right]^{-N/2} \exp\{-\tfrac{1}{2}Np\}. \tag{7.76}$$

The likelihood ratio criterion for testing H_0 is given by

$$\lambda = \frac{\max_\omega L(\boldsymbol{\mu}, \Sigma)}{\max_\Omega L(\boldsymbol{\mu}, \Sigma)} = \left[\frac{\det s_{(11)}}{\det(s_{(11)} + N\bar{\mathbf{x}}_{(1)}\bar{\mathbf{x}}'_{(1)})}\right]^{-N/2} = (1 + r_1{}^*)^{-N/2}. \quad (7.77)$$

Thus the likelihood ratio test of H_0 rejects H_0 whenever

$$(N-1)r_1{}^* > c,$$

where the constant c depends on the level of significance α of the test. In terms of the statistic R_1, this is also equivalent to rejecting H_0 whenever $r_1 \geq c$. From Chapter VI the probability density function of R_1 is given by

$$f_{R_1}(r_1|\delta_1{}^2) = \frac{\Gamma(\frac{1}{2}N)}{\Gamma(\frac{1}{2}p_1)\Gamma(\frac{1}{2}(N-p_1))} r_1^{p_1/2-1}(1-r_1)^{(N-p_1)/2-1}$$
$$\times \exp\{-\tfrac{1}{2}\delta_1{}^2\}\phi(\tfrac{1}{2}N, \tfrac{1}{2}p_1; \tfrac{1}{2}r_1\delta_1{}^2)$$

provided $r_1 \geq 0$ and is zero elsewhere, where $\delta_1{}^2 = N\boldsymbol{\mu}'_{(1)}\Sigma^{-1}_{(11)}\boldsymbol{\mu}_{(1)}$ and $R_1{}^*$ is a Hotelling's T^2-statistic based on the random sample $\mathbf{X}^\alpha_{(1)} = (X_{\alpha 1}, \ldots, X_{\alpha p_1})'$, $\alpha = 1, \ldots, N$, from a p_1-variate normal distribution with mean $\boldsymbol{\mu}_{(1)}$ and positive definite covariance matrix $\Sigma_{(11)}$.

Let T_1 be the translation group such that $\mathbf{t}_1 \in T_1$ translates the last p_2 components of each $\mathbf{X}^\alpha$ and let G_{BT} be as defined in Section 7.2.2 with $k = 2$. This problem remains invariant under the affine group (G_{BT}, T_1) transforming

$$\mathbf{X}^\alpha = g\mathbf{X}^\alpha + \mathbf{t}_1, \qquad \alpha = 1, \ldots, N_1, \quad g \in G_{\mathrm{BT}}, \quad \mathbf{t}_1 \in T_1.$$

Note that $\mathbf{t}_1$ can be regarded as a p-vector with its first p_1 components equal to zero. A maximal invariant in the space of $(\bar{\mathbf{X}}, S)$ is R_1 and the corresponding maximal invariant in the parametric space Ω is $\delta_1{}^2$. From the computations in connection with the T^2-test it is now obvious that this test possesses the same optimum properties as those of Hotelling's T^2-test (Theorems 7.2.1, 7.2.2, 7.2.3, and the minimax property).

(b) In the notation of Section 7.2.2 let $k = 3$, $p_1 + p_2 + p_3 = p$. We are interested here in testing the hypothesis $H_0: \mu_{[2]} = \mathbf{0}$ when $\boldsymbol{\mu}, \Sigma$ are unknown and the parametric space $\Omega = \{(\mathbf{0}, \boldsymbol{\mu}_{(2)}, \boldsymbol{\mu}_{(3)}), \Sigma\}$. It may be verified that

$$\max_\Omega L(\boldsymbol{\mu}, \Sigma) = (2\pi)^{-Np/2}[\det(s/N)]^{-N/2}(1 + N\bar{\mathbf{x}}'_{(1)}s^{-1}_{(11)}\bar{\mathbf{x}}_{(1)})^{-N/2}$$
$$\times \exp\{-\tfrac{1}{2}Np\}, \quad (7.78)$$

$$\max_\omega L(\boldsymbol{\mu}, \Sigma) = (2\pi)^{-Np/2}[\det(s/N)]^{-N/2}(1 + N\bar{\mathbf{x}}'_{[2]}s^{-1}_{[22]}\bar{\mathbf{x}}_{[2]})^{-N/2}$$
$$\times \exp\{-\tfrac{1}{2}Np\}, \quad (7.79)$$

where ω is the subspace of Ω when H_0 is true. Hence the likelihood ratio criterion λ is

$$\lambda = \frac{\max_\omega L(\boldsymbol{\mu}, \Sigma)}{\max_\Omega L(\boldsymbol{\mu}, \Sigma)} = \left(\frac{1 + N\overline{\mathbf{x}}'_{(1)} s_{(11)}^{-1} \overline{\mathbf{x}}_{(1)}}{1 + N\overline{\mathbf{x}}'_{[2]} s_{[22]}^{-1} \overline{\mathbf{x}}_{[2]}}\right)^{-N/2}$$

$$= \left(\frac{1 + r_1^* + r_2^*}{1 + r_1^*}\right)^{N/2} = \left(\frac{1 - r_1}{1 - r_1 - r_2}\right)^{N/2} \tag{7.80}$$

Hence the likelihood ratio test of $H_0 : \boldsymbol{\mu}_{[2]} = \mathbf{0}$ rejects H_0 whenever $(1 - r_1 - r_2)/(1 - r_1) \geq c$, where c is a constant depending on the level of significance α. From Chapter VI the joint probability density function of (R_1, R_2) is given by

$$f_{R_1,R_2}(r_1, r_2 | \delta_1^2, \delta_2^2) = \frac{\Gamma(\frac{1}{2}N)}{\Gamma(\frac{1}{2}p_1)\Gamma(\frac{1}{2}(N - p_1))} r_1^{p_1/2 - 1} r_2^{p_2/2 - 1}$$
$$\times (1 - r_1 - r_2)^{(N - p_1 - p_2)/2 - 1}$$
$$\times \exp\{-\tfrac{1}{2}(\delta_1^2 + \delta_2^2) + \tfrac{1}{2} r_1 \delta_2^2\}$$
$$\times \phi(\tfrac{1}{2}N, \tfrac{1}{2}p_1; \tfrac{1}{2}r_1\delta_1^2)\phi(\tfrac{1}{2}(N - p_1), \tfrac{1}{2}p_2; \tfrac{1}{2}r_2\delta_2^2) \tag{7.81}$$

provided $r_1 \geq 0$, $r_2 \geq 0$ and

$$\delta_1^2 = N\boldsymbol{\mu}'_{(1)}\Sigma_{(11)}^{-1}\boldsymbol{\mu}_{(1)}, \qquad \delta_1^2 + \delta_2^2 = N\boldsymbol{\mu}'_{[2]}\Sigma_{[22]}^{-1}\boldsymbol{\mu}_{[2]}.$$

Under H_0, $\delta_1^2 = \delta_2^2 = 0$. From (7.81) it follows that under H_0

$$Z = (1 - R_1 - R_2)/(1 - R_1)$$

is distributed as a central beta random variable with parameter

$$(\tfrac{1}{2}(N - p_1 - p_2), \tfrac{1}{2}p_2).$$

Let T_2 be the transformation group which translates the last p_3 components of each $\mathbf{X}^\alpha$, and let G_{BT} be as defined in Section 7.2.2 with $k = 3$, $p_1 + p_2 + p_3 = p$. This problem remains invariant under the group (G_{BT}, T_2) of affine transformations, transforming

$$\mathbf{X}^\alpha \to g\mathbf{X}^\alpha + \mathbf{t}, \qquad \alpha = 1, \ldots, N,$$

$g \in G_{BT}$ (with $k = 3$), $\mathbf{t} \in T_2$ ($\mathbf{t}$ can be considered as a p-vector with the first $p_1 + p_2$ components equal to zero). A maximal invariant in the space of $(\overline{\mathbf{X}}, S)$ [the induced transformation on $(\overline{\mathbf{X}}, S)$ is $(\overline{\mathbf{X}}, S) \to (g\overline{\mathbf{X}} + \mathbf{t}, gSg')$] is (R_1, R_2) [also its equivalent statistic (R_1^*, R_2^*)]. A corresponding maximal invariant in Ω is (δ_1^2, δ_2^2). Under H_0, $\delta_1^2 = \delta_2^2 = 0$ and under the alternatives H_1, $\delta_2^2 > 0$, $\delta_1^2 = 0$.

From (7.81) it follows that the likelihood ratio test is not uniformly most powerful (optimum) invariant for this problem and that there is no uniformly most powerful invariant test for the problem. However, for fixed p, the likelihood ratio is nearly optimum as N becomes large (Wald, 1943). Thus, if p

is not large, it seems likely that the sample size occurring in practice will usually be large enough for this result to be relevant. However, if the dimension p is large, it may be that the sample size N must be extremely large for this result to apply. Giri (1961) has shown that the difference of the powers of the likelihood ratio test and the best invariant test is $o(N^{-1})$ when p_1, p are both equal to $O(N)$ and $\delta_1{}^2 = O(\sqrt{N})$. For the minimax property Giri (1968) has shown that no invariant test under (G_{BT}, T_2) is minimax for testing H_0 against $H_1 : \delta_1{}^2 = \lambda$ for every choice of λ.

EXERCISES

1. Prove (7.48).

2. Prove (7.50) and (7.51).

3. Test the hypothesis H_0 given in (7.7) when Σ is unknown.

4. Let T^2 be distributed as Hotelling's T^2 with $N - 1$ degrees of freedom. Show that $((N - p)/p)(T^2/(N - 1))$ is distributed as a noncentral F with $(p, N - p)$ degrees of freedom.

5. Let $\sqrt{N}\,\bar{\mathbf{X}}$ be distributed as a p-dimensional normal random variable with mean $\sqrt{N}\,\boldsymbol{\mu}$ and positive definite covariance matrix Σ and let S be distributed, independently of $\bar{\mathbf{X}}$, as $W_p(N - 1, \Sigma)$. Show that the distribution of $T^2 = N(N - 1)\bar{\mathbf{X}}'S^{-1}\bar{\mathbf{X}}$ remains unchanged if $\boldsymbol{\mu}$ is replaced by $(\delta, 0, \ldots, 0)'$ and Σ by I where $\delta^2 = \boldsymbol{\mu}'\Sigma^{-1}\boldsymbol{\mu}$.

6. (*Test of symmetry of biological organs*) Let $\mathbf{X}^\alpha$, $\alpha = 1, \ldots, N$, be a random sample of size N from a p-variate normal population with mean $\boldsymbol{\mu}$ and positive definite covariance matrix Σ. Assume that p is an even integer, $p = 2k$. Let $\boldsymbol{\mu} = (\boldsymbol{\mu}_{(1)}, \boldsymbol{\mu}_{(2)})$, $\boldsymbol{\mu}_{(1)} = (\mu_1, \ldots, \mu_k)'$. On the basis of the observations $\mathbf{x}^\alpha$ on $\mathbf{X}^\alpha$, $\alpha = 1, \ldots, N$, find the appropriate T^2-test of $H_0 : \boldsymbol{\mu}_{(1)} = \boldsymbol{\mu}_{(2)}$. *Note*: In many anthropological problems $x_1, \ldots, x_k$ represent measurements on characters on the left side and $x_{k+1}, \ldots, x_p$ represent measurements on the same characters on the right side.

7. (*Profile analysis*) Suppose a battery of p psychological tests is administered to a group and $\mu_1, \ldots, \mu_p$ are their expected scores. The profile of the group is defined as the graph obtained by joining the points (i, μ_i), $i = 1, \ldots, p$, successively. For two different groups with expected scores $(\mu_1, \ldots, \mu_p)$ and $(v_1, \ldots, v_p)$, respectively, for the same battery of tests we obtain two different profiles, one obtained from the points (i, μ_i) and the other obtained from the points (i, v_i), $i = 1, \ldots, p$. Two profiles are said to be similar if line segments joining the points (i, μ_i), $(i + 1, \mu_{i+1})$ are parallel to the corresponding line segments joining the points (i, v_i), $(i + 1, v_{i+1})$. For two groups of sizes N_1, N_2, respectively, let $\mathbf{x}^\alpha = (x_{\alpha 1}, \ldots, x_{\alpha p})'$, $\alpha = 1, \ldots, N_1$, be the scores of N_1 individuals from the first group and let

$\mathbf{y}^\alpha = (y_{\alpha 1}, \ldots, y_{\alpha p})', \alpha = 1, \ldots, N_2$, be the scores of N_2 individuals from the second group. Assume that they are samples from two independent p-variate normal populations with different mean vectors $\boldsymbol{\mu} = (\mu_1, \ldots, \mu_p)'$, $\mathbf{v} = (v_1, \ldots, v_p)'$ and the same covariance matrix Σ. On the basis of these observations test the hypothesis

$$H_0 : \mu_i - \mu_{i+1} = v_i - v_{i+1}, \qquad i = 1, \ldots, p.$$

Hint: Let C_1 be the contrast matrix as defined in Example 7.2.2. The hypothesis H_0 is equivalent to testing the hypothesis that $E(C_1 \mathbf{X}^\alpha) = E(C_1 \mathbf{Y}^\beta)$, $\alpha = 1, \ldots, N_1$, $\beta = 1, \ldots, N_2$.

8. (*Union–intersection principle*) Let $\mathbf{X}^\alpha = (X_{\alpha 1}, \ldots, X_{\alpha p})'$, $\alpha = 1, \ldots, N$, be a random sample of size N from a p-variate normal distribution with mean $\boldsymbol{\mu}$ and positive definite covariance matrix Σ. The hypothesis $H_0 : \boldsymbol{\mu} = \mathbf{0}$ is true if and only if $H_{\boldsymbol{l}} : \boldsymbol{l}'\boldsymbol{\mu} = 0$ for any nonnull vector $\boldsymbol{l} \in E^p$ is true. Thus H_0 will be rejected if at least one of the hypotheses $H_{\boldsymbol{l}}$, $\boldsymbol{l} \in L = E^p - \{\mathbf{0}\}$, is rejected and hence $H_0 = \bigcap_{\boldsymbol{l} \in L} H_{\boldsymbol{l}}$. Let $\omega_{\boldsymbol{l}}$ denote the rejection region of the hypothesis $H_{\boldsymbol{l}}$. Obviously the rejection region of H_0 is $\omega = \bigcup_{\boldsymbol{l} \in L} \omega_{\boldsymbol{l}}$. The sizes of $\omega_{\boldsymbol{l}}$ should be such that ω is a size α rejection region of H_0. This is known as the union–intersection principle of Roy.

It is evident that $H_{\boldsymbol{l}} : \boldsymbol{l}'\boldsymbol{\mu} = 0$ is the hypothesis about the scalar mean of the random variables $\boldsymbol{l}'\mathbf{X}^\alpha$, $\alpha = 1, \ldots, N$, with variances $\boldsymbol{l}'\Sigma\boldsymbol{l}$, and that the optimum test for this univariate problem is Student's t-test. Show that the union–intersection principle for testing H_0 leads to the T^2-test.

9. Let $\mathbf{X}_i^\alpha = (X_{i\alpha 1}, \ldots, X_{i\alpha p})'$, $\alpha = 1, \ldots, N_i$, $i = 1, \ldots, k$, be a random sample of size N_i from k independent p-variate normal populations with mean vectors $\boldsymbol{\mu}^i = (\mu_{i1}, \ldots, \mu_{ip})'$ and positive definite covariance matrix Σ_i. Let $N_1 = \min_i N_i$. Define for known scalar constants $\beta_1, \ldots, \beta_k$

$$\mathbf{Y}^\alpha = \beta_1 \mathbf{X}_1^\alpha + \sum_{i=2}^{k} \beta_i \left(\frac{N_1}{N_i}\right)^{1/2}$$
$$\times \left(\mathbf{X}_i^\alpha - \frac{1}{N_1}\sum_{\beta=1}^{N_1} \mathbf{X}_i^\beta + \frac{1}{(N_1 N_i)^{1/2}} \sum_{\gamma=1}^{N_i} \mathbf{X}_i^\gamma\right), \qquad \alpha = 1, \ldots, N_1$$

$$\overline{\mathbf{Y}} = \frac{1}{N_1}\sum_1^{N_1} \mathbf{Y}^\alpha, \qquad S = \sum_{\alpha=1}^{N_1} (\mathbf{Y}^\alpha - \overline{\mathbf{Y}})(\mathbf{Y}^\alpha - \overline{\mathbf{Y}})'.$$

Consider the problem of testing $H_0 : \sum_1^k \beta_i \boldsymbol{\mu}^i = \boldsymbol{\mu}_0$ (specified). Show that

$$T^2 = N_1(N_1 - 1)(\overline{\mathbf{Y}} - \boldsymbol{\mu}_0)' S^{-1} (\overline{\mathbf{Y}} - \boldsymbol{\mu}_0)$$

is distributed as Hotelling's T^2 with $N_1 - 1$ degrees of freedom under H_0.

10. [*Giri, 1965*] Let $\mathbf{Z} = (Z_1, \ldots, Z_p)'$ be a complex p-dimensional Gaussian random variable with mean $\boldsymbol{\alpha} = (\alpha_1, \ldots, \alpha_p)'$ and positive definite

Hermitian complex covariance matrix $\Sigma = E(\mathbf{Z} - \boldsymbol{\alpha})(\mathbf{Z} - \boldsymbol{\alpha})^*$, and let $\mathbf{Z}^\alpha = (Z_{\alpha 1}, \ldots, Z_{\alpha p})'$, $\alpha = 1, \ldots, N$, be a sample of size N from the distribution of $\mathbf{Z}$. On the basis of these observations find the likelihood ratio test of the following testing problems.

(a) To test the hypothesis $H_0 : \boldsymbol{\alpha} = \mathbf{0}$, when Σ is unknown.

(b) To test the hypothesis $H_0 : \alpha_1 = \cdots = \alpha_{p_1} = 0$, $p_1 < p$, when Σ is unknown.

(c) To test the hypothesis $H_0 : \alpha_1 = \cdots = \alpha_{p_1 + p_2} = 0$, $p_1 + p_2 < p$, when it is given that $\alpha_1 = \cdots = \alpha_{p_1} = 0$, when Σ is unknown.

11. Let $\mathbf{X}^\alpha = (X_{\alpha 1}, \ldots, X_{\alpha p})'$, $\alpha = 1, \ldots, N_1$, be a random sample of size N_1 from a p-dimensional normal distribution with mean $\boldsymbol{\mu} = (\mu_1, \ldots, \mu_p)'$ and positive definite covariance matrix Σ_1 (unknown), and let $\mathbf{Y}^\alpha = (Y_{\alpha 1}, \ldots, Y_{\alpha p})'$, $\alpha = 1, \ldots, N_2$, be a random sample of size N_2 from another independent p-dimensional normal distribution with mean $\mathbf{v} = (v_1, \ldots, v_p)'$ and positive definite covariance matrix Σ_2 (unknown). Find the appropriate test of $H_0 : \mu_i - v_i = 0$, $i = 1, \ldots, p_1 < p$.

REFERENCES

Anderson, T. W. (1958). "An Introduction to Multivariate Statistical Analysis. Wiley, New York.

Bennet, B. M. (1951). Note on a solution of the generalized Behrens–Fisher problem, *Ann. Inst. Statist. Math.* **2**, 87–90.

Bose, R. C., and Roy, S. N. (1938). The distribution of Studentized D^2-statistic, *Sankhya* **4**, 19.

Giri, N., Kiefer, J., and Stein, C. (1963). Minimax character of Hotelling's T^2-test in the simplest case, *Ann. Math. Statist.* **34**, 1524–1535.

Giri, N., and Kiefer, J. (1962). Minimax property of Hotelling's and certain other multivariate tests (abstract), *Ann. Math. Statist.* **33**, 1490–1491.

Giri, N., and Kiefer, J. (1964). Local and asymptotic minimax properties of multivariate tests, *Ann. Math. Statist.* **35**, 21–35.

Giri, N. (1961). On the Likelihood Ratio Tests of Some Multivariate Problems, Ph.D. Thesis, Stanford Univ.

Giri, N. (1965). On the complex analogues of T^2- and R^2-tests, *Ann. Math. Statist.* **36**, 664–670.

Giri, N. (1968). Locally and asymptotically minimax tests of a multivariate problem, *Ann. Math. Statist.* **39**, 171–178.

Giri, N. (1972). On testing problem concerning mean of multivariate complex Gaussian distribution, *Ann. Inst. Statist. Math.* **24**, 245–250.

Giri, N., and Behara, M. (1971). Locally and asymptotically minimax tests of some multivariate decision problems, *Arch. Math.* **4**, 436–441.

Giri, N. (1975). Invariance and statistical minimax tests, "Selecta Statistica Canadiana," Vol. 3. Hindusthan Publ. Corp., India.

Hotelling, H. (1931). The generalization of Student's ratio, *Ann. Math. Statist.* **2**, 360–378.

Hsu, P. L. (1938). Notes on Hotelling's generalized *T*, *Ann. Math. Statist.* **16**, 231–243.

Isaacson, S. L. (1951). On the theory of unbiased tests of simple statistical hypotheses specifying the values of two or more parameters, *Ann. Math. Statist.* **22**, 217–234.

Kiefer, J. (1957). Invariance, minimax sequential estimation and continuous time processes, *Ann. Math. Statist.* **28**, 573–601.

Kiefer, J., and Schwartz, R. (1965). Admissible Bayes character of T^2- and R^2- and other fully invariant tests for classical multivariate normal problems, *Ann. Math. Statist.* **36**, 747–770.

Kshirsagar, A. M. (1972). "Multivariate Analysis." Dekker, New York.

Lehmann, E. (1959). "Testing Statistical Hypothesis." Wiley, New York.

Lehmer, E. (1944). Inverse tables of probabilities of errors of the second kind, *Ann. Math. Statist.* **15**, 388–398.

Linnik, Ju. Vo. (1966). *Teor. Verojatn. Ee. primen.* 561, MR 34.

Linnik, Ju. Vo., Pliss, V. A., and Salaevskie, O. V. (1966). *Dokl. Akad. Nauk SSSR* 168.

Nandi, H. K. (1965). On some properties of Roy's union-intersection tests, *Calcutta Statist. Assoc. Bull.* **4**, 9–13.

Rao, C. R. (1965). "Linear Statistical Inference and its Applications." Wiley, New York.

Roy, S. N. (1953). On a heuristic method of test construction and its use in multivariate analysis, *Ann. Math. Statist.* **24**, 220–238.

Roy, S. N. (1957). "Some Aspects of Multivariate Analysis." Wiley, New York.

Salaevskii, O. V. (1968). Minimax character of Hotelling's T^2-test, *Sov. Math. Dokl.* **9**, 733–735.

Scheffé, H. (1943). On solutions of Behrens–Fisher problem based on the t-distribution, *Ann. Math. Statist.* **14**, 35–44.

Semika, J. B. (1941). An optimum property of two statistical tests, *Biometrika* **32**, 70–80.

Stein, C. (1956). The admissibility of Hotelling's T^2-test, *Ann. Math. Statist.* **27**, 616–623.

Tang, P. C. (1938). The power functions of the analysis of variance tests with tables and illustration of their uses, *Statist. Res. mem.* **2**, 126–157.

Wald, A. (1943). Tests of statistical hypotheses concerning several parameters when the number of observations is large, *Trans. Amer. Math. Soc.* **54**, 426–482.

Wald, A. (1950). "Statistical Decision Functions." Wiley, New York.

CHAPTER VIII

Tests Concerning Covariance Matrices and Mean Vectors

8.0 INTRODUCTION

In this chapter we develop techniques for testing hypotheses concerning covariance matrices and testing covariance matrices and mean vectors of several p-variate normal populations. The tests discussed are invariant tests, and most of the problems and tests considered are generalizations of univariate ones. In Section 8.1 we discuss the problem of testing the hypothesis that the covariance matrix of a p-variate normal population is a given matrix. In Section 8.2 we consider the sphericity test, that is, where the covariance matrix is equal to a given matrix except for an unknown proportionality factor, which has only a trivial corresponding univariate hypothesis. In Section 8.3 we divide the set of p-variates having a joint multivariate normal distribution into k subsets and study the problem of mutual independence of these subsets. We consider, in detail, the special case of two subsets where the first subset has only one component and where the R^2-test is the appropriate test statistic. Section 8.4 deals with the multivariate general linear hypothesis. In Section 8.5 we study problems of testing hypotheses of equality of covariance matrices and equality of both covariance matrices and mean vectors. The asymptotic distribution of the likelihood ratio test statistic under the null hypothesis is given for each problem.

8.1 HYPOTHESIS THAT A COVARIANCE MATRIX IS EQUAL TO A GIVEN MATRIX

Let $\mathbf{X}^\alpha = (X_{\alpha 1}, \ldots, X_{\alpha p})'$, $\alpha = 1, \ldots, N$, be a random sample of size N $(N > p)$ from a p-variate normal distribution with unknown mean $\boldsymbol{\mu}$ and unknown positive definite covariance matrix Σ. As usual we assume throughout that $N > p$, so that the sample covariance matrix S is positive definite with probability 1. We are interested in testing the null hypothesis $H_0 : \Sigma = \Sigma_0$ against the alternatives $H_1 : \Sigma \neq \Sigma_0$ where Σ_0 is a fixed positive definite matrix. Since Σ_0 is positive definite there exists a nonsingular matrix $g \in G_l(p)$, the full linear group, such that $g\Sigma_0 g' = I$. In particular, we can take $g = \Sigma_0^{1/2}$ where $\Sigma_0^{1/2}$ is a symmetric matrix such that $\Sigma_0 = \Sigma_0^{1/2}\Sigma_0^{1/2}$.

Let $\mathbf{Y}^\alpha = g\mathbf{X}^\alpha$, $\alpha = 1, \ldots, N$, $\mathbf{v} = g\boldsymbol{\mu}$, and $\Sigma^* = g\Sigma g'$. Then Y^α, $\alpha = 1, \ldots, N$, constitute a random sample of size N from a p-variate normal distribution with unknown mean $\mathbf{v}$ and unknown positive definite covariance matrix Σ^*. The problem is transformed to testing the null hypothesis $H_0 : \Sigma^* = I$ against alternatives that $\Sigma^* \neq I$ on the basis of sample observations $\mathbf{y}^\alpha$ on $\mathbf{Y}^\alpha$, $\alpha = 1, \ldots, N$. The parametric space $\Omega = \{(\mathbf{v}, \Sigma^*)\}$ is the space of $\mathbf{v}$ and Σ^*, and under H_0 it reduces to the subspace $\omega = \{(\mathbf{v}, I)\}$. Let

$$\bar{\mathbf{x}} = \frac{1}{N}\sum_{\alpha=1}^{N} \mathbf{x}^\alpha, \qquad \bar{\mathbf{y}} = \frac{1}{N}\sum_{\alpha=1}^{N} \mathbf{y}^\alpha, \qquad s = \sum_{\alpha=1}^{N} (\mathbf{x}^\alpha - \bar{\mathbf{x}})(\mathbf{x}^\alpha - \bar{\mathbf{x}})',$$

$$b = \sum_{\alpha=1}^{N} (\mathbf{y}^\alpha - \bar{\mathbf{y}})(\mathbf{y}^\alpha - \bar{\mathbf{y}})'.$$

Obviously $\bar{\mathbf{y}} = g\bar{\mathbf{x}}$, $b = gsg'$. The likelihood of the observations $\mathbf{y}_\alpha$, $\alpha = 1, \ldots, N$, is

$$L(\mathbf{v}, \Sigma^*) = (2\pi)^{-Np/2}(\det \Sigma^*)^{-N/2} \times \exp\left(-\tfrac{1}{2} \operatorname{tr} \Sigma^{*-1}\left(\sum_{\alpha=1}^{N} (\mathbf{y}^\alpha - \mathbf{v})(\mathbf{y}^\alpha - \mathbf{v})'\right)\right\}. \tag{8.1}$$

By Lemma 5.1.1,

$$\max_{\Omega} L(\mathbf{v}, \Sigma^*) = (2\pi/N)^{-Np/2}(\det(b))^{-N/2} \exp\{-\tfrac{1}{2}Np\}. \tag{8.2}$$

Under H_0, $L(\mathbf{v}, \Sigma^*)$ reduces to

$$L(\mathbf{v}, I) = (2\pi)^{-Np/2} \exp\left\{-\tfrac{1}{2} \operatorname{tr} \sum_{\alpha=1}^{N} (\mathbf{y}^\alpha - \mathbf{v})(\mathbf{y}^\alpha - \mathbf{v})'\right\}, \tag{8.3}$$

so

$$\max_{\omega} L(\mathbf{v}, \Sigma^*) = (2\pi)^{-Np/2} \exp\{-\tfrac{1}{2} \operatorname{tr} b\}. \tag{8.4}$$

Hence the likelihood ratio criterion for testing $H_0 : \Sigma^* = I$ ($\Sigma = \Sigma_0$) is given by

$$\begin{aligned}\lambda &= \frac{\max_{\omega} L(\mathbf{v}, \Sigma^*)}{\max_{\Omega} L(\mathbf{v}, \Sigma^*)} = \left(\frac{e}{N}\right)^{-Np/2} (\det(b))^{N/2} \exp\{-\tfrac{1}{2} \operatorname{tr} b\} \\ &= \left(\frac{e}{N}\right)^{-Np/2} (\det(\Sigma_0^{-1} s))^{N/2} \exp\{-\tfrac{1}{2} \operatorname{tr} \Sigma_0^{-1} s\} \end{aligned} \tag{8.5}$$

as $g'g = \Sigma_0^{-1}$. Thus we get the following theorem.

THEOREM 8.1.1 The likelihood ratio test of $H_0 : \Sigma = \Sigma_0$ rejects H_0 whenever

$$\lambda = (e/N)^{-Np/2} (\det \Sigma_0^{-1} s)^{N/2} \exp\{-\tfrac{1}{2} \operatorname{tr} \Sigma_0^{-1} s\} \leq C,$$

where the constant C is chosen in such a way that the test has size α.

To evaluate the constant C, we need the distribution of λ under H_0. Let B be the random matrix corresponding to b; that is, $B = \sum_{\alpha=1}^{N} (\mathbf{Y}^\alpha - \bar{\mathbf{Y}})(\mathbf{Y}^\alpha - \bar{\mathbf{Y}})'$. Then B has a Wishart distribution with parameter I and $N - 1$ degrees of freedom when H_0 is true. The characteristic function $\phi(t)$ of $-2 \log \lambda$ under H_0 is given by (see Anderson, 1958)

$$\phi(\lambda) = E(\exp\{-it2 \log \lambda\}) = E(\lambda^{-2it}) = \prod_{j=1}^{p} \phi_j(t) \tag{8.6}$$

where $\phi_j(t)$, $j = 1, \ldots, p$, is given by

$$\phi_j(t) = \frac{(2e/N)^{-iNt}(1 - 2it)^{-(N-1-2iNt)/2}\Gamma(\tfrac{1}{2}(N - j) - iNt)}{\Gamma(\tfrac{1}{2}(N - j))}. \tag{8.7}$$

But as $N \to \infty$, using Stirling's approximation for the gamma function, we obtain

$$\begin{aligned}\phi_j(t) &\sim 2^{-iNt} e^{-iNt} (1 - 2it)^{(2iNt - N + 1)/2} \\ &\quad \times \frac{\exp\{-[\tfrac{1}{2}(N - j) - iNt]\}[\tfrac{1}{2}(N - j - 2) - iNt]^{(N-j-1)/2 - iNt}}{\exp\{-[\tfrac{1}{2}(N - j)]\}[\tfrac{1}{2}(N - j - 2)]^{(N-j-1)/2}} \\ &= (1 - 2it)^{-j/2} \left(1 - \frac{it(j+2)}{\tfrac{1}{2}(N - j - 2)(1 - 2it)}\right)^{(N-j)/2 - 1/2} \\ &\quad \times \left(1 - \frac{it(j+2)}{itN(1 - 2it)}\right)^{-iNt} \\ &\to (1 - 2it)^{-j/2}. \end{aligned} \tag{8.8}$$

Thus as $N \to \infty$

$$\phi(t) \to \prod_{j=1}^{p} (1 - 2it)^{-j/2}. \tag{8.9}$$

Since $(1 - 2it)^{-j/2}$ is the characteristic function of a chi-square random variable χ_j^2 with j degrees of freedom, as $N \to \infty$, $-2 \log \lambda$ is distributed as $\sum_{j=1}^{p} \chi_j^2$, where the χ_j^2 are independent whenever H_0 is true. Thus $-2 \log \lambda$ is distributed as $\chi^2_{p(p+1)/2}$ when H_0 is true and $N \to \infty$.

The problem of testing $H_0:\Sigma^* = I$ against the alternatives $H_1:\Sigma^* \neq I$ remains invariant under the affine group $G = (O(p), E^p)$ where $O(p)$ is the multiplicative group of $p \times p$ orthogonal matrices, and E^p is the translation group, operating as

$$\mathbf{Y}^\alpha \to g\mathbf{Y}^\alpha + \mathbf{a}, \qquad g \in O(p), \quad \mathbf{a} \in E^p, \quad \alpha = 1, \ldots, N.$$

This induces in the space of the sufficient statistic $(\bar{\mathbf{Y}}, B)$ the transformations $(\bar{\mathbf{Y}}, B) \to (g\bar{\mathbf{Y}} + \mathbf{a}, gBg')$.

LEMMA 8.1.1 A set of maximal invariants in the space of $(\bar{\mathbf{Y}}, B)$ under the affine group G comprises the characteristic roots of B, that is, the roots of

$$\det(B - \lambda I) = 0. \tag{8.10}$$

Proof Since $\det(g(B - \lambda I)g') = \det(B - \lambda I)$, the roots of the equation $\det(B - \lambda I) = 0$ are invariant under G. To see that they are maximal invariant suppose that $\det(B - \lambda I) = 0$ and $\det(B^* - \lambda I) = 0$ have the same roots, where B, B^* are two symmetric positive definite matrices; we want to show that there exists a $g \in O(p)$ such that $B^* = gBg'$. Since B, B^* are symmetric positive definite matrices there exist orthogonal matrices g_1, $g_2 \in O(p)$ such that

$$g_1 B g_1' = \Delta, \qquad g_2 B^* g_2' = \Delta,$$

where Δ is a diagonal matrix whose diagonal elements are the roots of (8.10). Since $g_1 B g_1' = g_2 B^* g_2'$ we get

$$B^* = g_2' g_1 B g_1' g_2 = gBg'$$

where $g = g_2' g_1 \in O(p)$. Q.E.D.

We shall denote the characteristic roots of B by $R_1, \ldots, R_p$. Similarly the corresponding maximal invariants in the parametric space of $(\mathbf{v}, \Sigma^*)$ under G are $\theta_1, \ldots, \theta_p$, the roots of $\det(\Sigma^* - \lambda I) = 0$. Under H_0 all $\theta_i = 1$, and under H_1 at least one $\theta_i \neq 1$.

The likelihood ratio test criterion λ in terms of the maximal invariants $(R_1, \ldots, R_p)$ can be written as

$$\lambda = (e/N)^{-Np/2} \prod_{i=1}^{p} (r_i)^{N/2} \exp\left\{-\frac{1}{2}\sum_{i=1}^{p} r_i\right\}. \tag{8.11}$$

The modified likelihood ratio test for testing $H_0:\Sigma = \Sigma_0$ rejects H_0 when

$$(e/N)^{-Np/2}(\det \Sigma_0^{-1}s)^{(N-1)/2} \exp\{-\tfrac{1}{2}\operatorname{tr}\Sigma_0^{-1}s\} \le C',$$

where the constant C' depends on the size α of the test. Note that the modified likelihood ratio test is obtained from the corresponding likelihood ratio test by replacing the sample size N by $N - 1$. Since e/N is constant, we do not change the constant term in the modified likelihood ratio test for the sake of convenience only.

It is well known that (see, e.g., Lehmann, 1959, p. 165) for $p = 1$ the rejection region of the likelihood ratio test is not unbiased. The same result also holds in this case (Das Gupta, 1969). However, the modified likelihood ratio test is unbiased. The following theorem is due to Sugiura and Nagao (1968).

THEOREM 8.1.2 For testing $H_0:\Sigma = \Sigma_0$ against the alternatives $\Sigma \neq \Sigma_0$ for unknown $\boldsymbol{\mu}$, the modified likelihood ratio test is unbiased.

Proof Let $g \in O(p)$ be such that $g\Sigma_0^{-1/2}\Sigma\Sigma_0^{-1/2}g'$ is a diagonal matrix Γ where $\Sigma_0^{-1/2}$ is the inverse matrix of the symmetric matrix $\Sigma_0^{1/2}$ such that $\Sigma_0^{1/2}\Sigma_0^{1/2} = \Sigma_0$. As indicated earlier we can assume, without any loss of generality, that $\Sigma_0 = I$ and $\Sigma = \Gamma$, the diagonal matrix whose diagonal elements are the characteristic roots of $\Sigma_0^{-1/2}\Sigma\Sigma_0^{-1/2}$. Hence S has a Wishart distribution with parameter Γ when H_1 is true. Let ω be the acceptance region of the modified likelihood ratio test, that is,

$$\begin{aligned}\omega = \{s|s &\text{ is positive definite and } (e/N)^{-Np/2}(\det \Sigma_0^{-1}s)^{-(N-1)/2} \\ &\times \exp\{-\tfrac{1}{2}\operatorname{tr}\Sigma_0^{-1}s\} > C'\}.\end{aligned}$$

Then the probability of accepting H_0 when H_1 is true is given by [see (6.32)]

$$\begin{aligned}P\{\omega|H_1\} &= \int_{\omega} C_{n,p}(\det s)^{(N-p-2)/2}(\det\Gamma)^{-(N-1)/2}\exp\{-\tfrac{1}{2}\operatorname{tr}\Gamma^{-1}s\}\,ds \\ &= \int_{\omega^*} C_{n,p}(\det u)^{(N-p-2)/2}\exp\{-\tfrac{1}{2}\operatorname{tr}u\}\,du, \end{aligned} \tag{8.12a}$$

where $u = \Gamma^{-1/2}s\Gamma^{-1/2}$ and ω^* is the set of all positive definite matrices u such that $\Gamma^{1/2}u\Gamma^{1/2}$ belongs to ω. Note that the Jacobian is $\det(\partial u/\partial s) = (\det\Gamma)^{-(p+1)/2}$. Since $\omega^* = \omega$ when H_0 is true and in the region ω

$$(\det u)^{(N-p-2)/2}\exp\{-\tfrac{1}{2}\operatorname{tr}u\} \ge C'(e/N)^{Np/2}(\det u)^{-(p+1)/2}, \tag{8.12b}$$

we get

$$\int_{\omega - \omega\cap\omega^*} (\det u)^{(N-p-2)/2} \exp\{-\tfrac{1}{2} \operatorname{tr} u\}\, du \quad \text{exists,} \tag{8.13}$$

$$\int_{\omega^* - \omega\cap\omega^*} (\det u)^{(N-p-2)/2} \exp\{-\tfrac{1}{2} \operatorname{tr} u\}\, du \geq C'\left(\frac{e}{N}\right)^{Np/2} \int_{\omega - \omega\cap\omega^*} (\det u)^{-(p+1)/2}\, du, \tag{8.14}$$

and

$$-\int_{\omega^* - \omega\cap\omega^*} (\det u)^{(N-p-2)/2} \exp\{-\tfrac{1}{2} \operatorname{tr} u\}\, du \geq -C'\left(\frac{e}{N}\right)^{Np/2} \int_{\omega^* - \omega\cap\omega^*} (\det u)^{-(p+1)/2}\, du. \tag{8.15}$$

Combining (8.14) and (8.15) with the fact that

$$\int_{\omega\cap\omega^*} (\det u)^{-(p+1)/2}\, du < \infty,$$

we get

$$\begin{aligned} P\{\omega|H_0\} &- P\{\omega|H_1\} \\ &= C_{n,p}\left\{\int_{\omega - \omega\cap\omega^*} - \int_{\omega^* - \omega\cap\omega^*}\right\}(\det u)^{(N-p-2)/2} \exp\{-\tfrac{1}{2} \operatorname{tr} u\}\, du \\ &\geq C_{n,p} C'\left(\frac{e}{N}\right)^{Np/2} \left\{\int_{\omega - \omega\cap\omega^*} - \int_{\omega^* - \omega\cap\omega^*}\right\}(\det u)^{-(p+1)/2}\, du \\ &= C_{n,p} C'\left(\frac{e}{N}\right)^{Np/2} \left\{\int_{\omega} - \int_{\omega^*}\right\}(\det u)^{-(p+1)/2}\, du = 0. \end{aligned} \tag{8.16}$$

The last inequality follows from the fact (see Example 3.2.8) that $(\det u)^{-(p+1)/2}$ is the invariant measure in the space of the u under the full linear group $G_l(p)$ transforming $u \to gug'$, $g \in G_l(p)$; that is,

$$\int_{\omega^*} (\det u)^{-(p+1)/2}\, du = \int_{\omega} (\det u)^{-(p+1)/2}\, du,$$

and hence the result. Q.E.D.

The acceptance region of the likelihood ratio test does not possess this property, and within the acceptance region we have

$$(\det u)^{(N-p-2)/2} \exp\{-\tfrac{1}{2} \operatorname{tr} u\} \geq C(e/N)^{Np/2}(\det u)^{-(p+2)/2}$$

instead of (8.12b).

Anderson and Das Gupta (1964a) showed that (this will follow trivially from Theorem 8.5.1) any invariant test for this problem (obviously it depends only on $r_1, \ldots, r_p$) with the acceptance region such that if $(r_1, \ldots, r_p)$ is in the region, so also is $(\bar{r}_1, \ldots, \bar{r}_p)$ with $\bar{r}_i \le r_i$, $i = 1, \ldots, p$, has a power function that is an increasing function of each θ_i where $\theta_1, \ldots, \theta_p$ are the characteristic roots of Σ^*.

Das Gupta (1969) obtained the following results.

THEOREM 8.1.3 The likelihood ratio test for $H_0:\Sigma = \Sigma_0$ (i) is biased against $H_1:\Sigma \ne \Sigma_0$, and (ii) has a power function $\beta(\theta)$ that increases as the absolute deviation $|\theta_i - 1|$ increases for each i.

Proof As in Theorem 8.1.2 we take $\Sigma_0 = I$ and $\Sigma = \Gamma$, the diagonal matrix with diagonal elements $(\theta_1, \ldots, \theta_p)$. S has a Wishart distribution with parameter Γ and $N - 1$ degrees of freedom. Let $S = (S_{ij})$. Then

$$(\det s)^{N/2} \exp\{-\tfrac{1}{2} \operatorname{tr} s\} = \left[\frac{\det s}{\prod_{i=1}^{p} s_{ii}}\right]^{N/2} \left[\prod_{i=1}^{p} s_{ii}^{N/2} \exp\{-\tfrac{1}{2} s_{ii}\}\right]. \tag{8.17}$$

From (6.32), since Γ is a diagonal matrix, S_{ii}/θ_i, $i = 1, \ldots, p$, are independently distributed χ^2_{N-1} random variables and for any $k\,(>0)$ $[\det S/\prod_{i=1}^{p} S_{ii}]^k$ and the S_{ii} (or any function thereof) are mutually independent. Furthermore, the distribution of $[\det S/\prod_{i=1}^{k} S_{ii}]^k$ is independent of $\theta_1, \ldots, \theta_p$. From Exercise 5 it follows that there exists a constant $\theta_p{}^*$ such that $1 < \theta_p{}^* < N/(N - 1)$ and

$$P\{S_{pp}^{N/2} \exp(-\tfrac{1}{2} S_{pp}) \ge C|\theta_p = 1\} < P\{S_{pp}^{N/2} \exp(-\tfrac{1}{2} S_{pp}) \ge C|\theta_p = \theta_p{}^*\} \tag{8.18}$$

irrespective of the value of C chosen. Hence if we evaluate the probability with respect to S_{pp}, keeping $S_{11}, \ldots, S_{p-1,p-1}$ and $[\det S/\prod_{i=1}^{p} S_{ii}]^{N/2}$ fixed, we obtain

$$\begin{aligned} &P\{(\det S)^{N/2} \exp\{-\tfrac{1}{2} \operatorname{tr} S\} \ge C|H_1\} \\ &\quad > P\{(\det S)^{N/2} \exp\{-\tfrac{1}{2} \operatorname{tr} S\} \ge C|H_0\}. \end{aligned} \tag{8.19}$$

Thus the acceptance region ω of the likelihood ratio test satisfies $P(\omega|H_1) - P(\omega|H_0) > 0$. Hence the likelihood ratio test is biased.

From Exercise 5 it follows that if $2r = m$, then $\beta(\theta)$ increases as $|\theta - 1|$ increases. Hence from the fact noted in connection with the proof of (i) we get the proof of (ii). Q.E.D.

Das Gupta and Giri (1973) proved the following theorem. Consider the class of rejection regions $C(r)$ for $r \ge 0$, given by

$$C(r) = \{s|s \text{ is positive definite and } (\det \Sigma_0^{-1} s)^{r/2} \exp\{-\tfrac{1}{2} \operatorname{tr} \Sigma_0^{-1} s\} \le k\}. \tag{8.20a}$$

THEOREM 8.1.4 For testing $H_0:\Sigma = \Sigma_0$ against the alternatives $H_1:\Sigma \ne \Sigma_0$: (a) $P\{C(r)|H_1\}$ increases monotonically as each θ_i (characteristic root of $\Sigma_0^{-1/2}\Sigma\Sigma_0^{-1/2}$) deviates from $r/(N - 1)$ either in the positive or in the negative direction; (b) $C(r)$ for which $(1 - \det(\Sigma_0^{-1/2}\Sigma\Sigma_0^{-1/2}))(r - n) \geq 0$ is unbiased for H_0 against H_1.

The proof follows from Theorems 8.1.2 and 8.1.3 and the fact that $C(r)$ (with $\Sigma_0 = I$) can also be written as

$$(\det s^*)^{(N-1)/2} \exp\{-\tfrac{1}{2} \operatorname{tr} s^*\} \leq k^*,$$

where $s^* = ((N-1)/r)s$.

Using the techniques of Kiefer and Schwartz (1965), Das Gupta and Giri (1969) have observed that the following rejection regions are unique (almost everywhere), Bayes, and hence admissible for this problem whenever $N - 1 > p$:

(i) $(\det \Sigma_0^{-1}s)^{r/2} \exp\{-\frac{1}{2} \operatorname{tr} \Sigma_0^{-1}s\} \leq k,\ 1 < r < \infty$,
(ii) $(\det \Sigma_0^{-1}s)^{r/2} \exp\{-\frac{1}{2} \operatorname{tr} \Sigma_0^{-1}s\} \geq k,\ -\infty < r < 0$.

For this problem Kiefer and Schwartz (1965) have shown that the test which rejects H_0 whenever

$$(\det \Sigma_0^{-1}s) \leq C, \tag{8.20b}$$

where the constant C is chosen such that the test has the level of significance α, is admissible Bayes against the alternatives that $\Sigma_0 - \Sigma$ is negative definite. The value of C can be determined from Theorem 6.6.1. Note that $\det(\Sigma_0^{-1}S) = \det(\Sigma_0^{-1/2}S\Sigma_0^{-1/2})$. They have also shown that for testing $H_0{:}\Sigma = \Sigma_0$, the test which rejects H_0 whenever

$$\operatorname{tr} \Sigma_0^{-1}s \geq C_1 \quad \text{or} \quad \leq C_2, \tag{8.21}$$

where C_1, C_2 are constants depending on the level of significance α of the test, is admissible against the alternatives $H_1{:}\Sigma \neq \Sigma_0$. This is in the form that is familiar to us when $p = 1$. It is easy to see that $\operatorname{tr}(\Sigma_0^{-1}S)$ has a χ^2 distribution with $(N-1)p$ degrees of freedom when H_0 is true.

8.2 THE SPHERICITY TEST

Let $\mathbf{X}^\alpha = (X_{\alpha 1}, \ldots, X_{\alpha p})'$, $\alpha = 1, \ldots, N$, be a random sample of size N $(N > p)$ from a p-variate normal population with unknown mean $\boldsymbol{\mu}$ and unknown positive definite covariance matrix Σ. We are interested here in testing the null hypothesis $H_0{:}\Sigma = \sigma^2\Sigma_0$ against the alternatives $H_1{:}\Sigma \neq \sigma^2\Sigma_0$ where Σ_0 is a fixed positive definite matrix and σ^2, $\boldsymbol{\mu}$ are unknown. Since Σ_0 is positive definite, there exists a $g \in G_l(p)$ such that $g\Sigma_0 g' = I$. Let $\mathbf{Y}^\alpha = g\mathbf{X}^\alpha$, $\alpha = 1, \ldots, N$; $\mathbf{v} = g\boldsymbol{\mu}$, $\Sigma^* = g\Sigma g'$. Then $\mathbf{Y}^\alpha$, $\alpha = 1, \ldots, N$, constitute a random sample of size N from a p-variate normal population with mean $\mathbf{v}$ and positive definite covariance matrix Σ^*. The problem is reduced to testing $H_0:\Sigma^* = \sigma^2 I$ against $H_1:\Sigma^* \neq \sigma^2 I$ when σ^2, $\boldsymbol{\mu}$ are unknown. Since under $H_0:\Sigma^* = \sigma^2 I$, the ellipsoid $(\mathbf{y} - \mathbf{v})'\Sigma^{*^{-1}}(\mathbf{y} - \mathbf{v}) = \text{const}$ reduces to the sphere $(\mathbf{y} - \mathbf{v})'(\mathbf{y} - \mathbf{v})/\sigma^2 = \text{const}$, the hypothesis is called the hypothesis of sphericity. Let $\bar{\mathbf{X}}$, S, $\bar{\mathbf{Y}}$, B be as defined as in Section 8.1. The likelihood of the observations $\mathbf{y}^\alpha$ on $\mathbf{Y}^\alpha$, $\alpha = 1, \ldots, N$, is given by

$$L(\mathbf{v}, \Sigma^*) = (2\pi)^{-Np/2}(\det \Sigma^*)^{-N/2} \times \exp\left\{-\tfrac{1}{2} \operatorname{tr} \Sigma^{*^{-1}}\left(\sum_{\alpha=1}^{N} (\mathbf{y}^\alpha - \mathbf{v})(\mathbf{y}^\alpha - \mathbf{v})'\right)\right\}. \tag{8.22}$$

The parametric space $\Omega = \{(\mathbf{v}, \Sigma^*)\}$ is the space of $\mathbf{v}$ and Σ^*. Under H_0 it reduces to $\omega = \{(\mathbf{v}, \sigma^2 I)\}$. From (8.22)

$$L(\mathbf{v}, \sigma^2 I) = (2\pi)^{-Np/2}(\sigma^2)^{-Np/2} \exp\left\{-\frac{1}{2\sigma^2} \operatorname{tr}\left(\sum_{\alpha=1}^{N} (\mathbf{y}^\alpha - \mathbf{v})(\mathbf{y}^\alpha - \mathbf{v})'\right)\right\}. \tag{8.23}$$

Hence

$$\max_{\omega} L(\mathbf{v}, \sigma^2 I) = (2\pi)^{-Np/2}\left(\frac{\operatorname{tr} b}{Np}\right)^{-Np/2} \exp\{-\tfrac{1}{2}Np\}, \tag{8.24}$$

since the maximum likelihood estimate of σ^2 is $\operatorname{tr}(b)/Np$. Thus we get

$$\lambda = \frac{\max_\omega L(\mathbf{v}, \sigma^2 I)}{\max_\Omega L(\mathbf{v}, \Sigma^*)} = \left[\frac{\det b}{(\operatorname{tr} b)/p}\right]^{N/2} = \left[\frac{\det \Sigma_0^{-1} s}{(\operatorname{tr} \Sigma_0^{-1} s)/p}\right]^{N/2}. \tag{8.25}$$

THEOREM 8.2.1 For testing $H_0:\Sigma = \sigma^2\Sigma_0$ where σ^2, $\boldsymbol{\mu}$ are unknown and Σ_0 is a fixed positive definite matrix, the likelihood ratio test of H_0 rejects H_0 whenever

$$\lambda = \frac{(\det \Sigma_0^{-1} s)^{N/2}}{((\operatorname{tr} \Sigma_0^{-1} s)/p)^{Np/2}} \leq C, \tag{8.26}$$

where the constant C is chosen such that the test has the required size α.

The corresponding modified likelihood ratio test of this problem is obtained from the likelihood ratio test by replacing N by $N - 1$.

To find the constant C we need the probability density function of λ when H_0 is true. Mauchly (1940) first derived the test criterion and obtained various moments of this criterion under the null hypothesis. Writing $W = \lambda^{2/N}$, Mauchly showed that

$$E(W^k) = p^{kp} \frac{\Gamma(\frac{1}{2}p(N - 1))}{\Gamma(\frac{1}{2}p(N - 1) + pk)} \times \prod_{j=1}^{p} \frac{\Gamma(\frac{1}{2}(N - j) + k)}{\Gamma(\frac{1}{2}(N - j))}, \qquad k = 0, 1, \ldots. \tag{8.27}$$

For $p = 2$ (8.27) reduces to

$$E(W^k) = \frac{N-2}{N-2+2k} = (N-2)\int_0^1 (z)^{N-3+2k}\,dz. \tag{8.28}$$

Thus under H_0, W is distributed as Z^2 where the probability density function of Z is

$$f_Z(z) = \begin{cases} (N-2)z^{N-3} & 0 \le z \le 1 \\ 0 & \text{otherwise.} \end{cases} \tag{8.29}$$

Khatri and Srivastava (1971) obtained the exact distribution of W in terms of zonal polynomials and Meijer's G-function. Consul (1968) obtained the null distribution of W.

From Anderson (1958, Section 8.6) we obtain

$$\begin{aligned} P\{-(N-1)\rho \log W \le z\} = P\{\chi_f{}^2 \le z\} + \omega_2[P\{\chi_{f+4}^2 \le z\} \\ - P\{\chi_f{}^2 \le z\}] + O(1/N^3), \end{aligned} \tag{8.30}$$

where

$$1 - \rho = \frac{2p^2 + p + 2}{6p(N-1)}, \qquad \omega_2 = \frac{(p+2)(p-1)(p-2)(2p^3 + 6p^2 + 3p + 2)}{288p^2(N-1)^2p^2},$$

$$f = \tfrac{1}{2}(p)(p+1) + 1.$$

Thus for large N

$$P\{-(N-1)\rho \log W \le z\} = P\{\chi_f{}^2 \le z\}.$$

The problem of testing $H_0:\Sigma^* = \sigma^2 I$ against the alternatives $H_1:\Sigma^* \ne \sigma^2 I$ remains invariant under the group G of affine transformations $(g, \mathbf{a})$, $g \in O(p)$, $\mathbf{a} \in E^p$, transforming each $\mathbf{Y}^\alpha \to g\mathbf{Y}^\alpha + \mathbf{a}$, $\alpha = 1, \ldots, N$. A set of maximal invariants under G in the space of $(\bar{\mathbf{Y}}, B)$ is $(R_1, \ldots, R_p)$, the characteristic roots of B, and the corresponding maximal invariant in the parametric space Ω under the induced group G is $(\theta_1, \ldots, \theta_p)$, the characteristic roots of Σ^*. Under H_0 all $\theta_i = \sigma^2$, and under H_1 at least one $\theta_i \ne \sigma^2$. We shall now prove that the modified likelihood ratio test for this problem is unbiased. This was first proved by Glesser (1966) and then by Sugiura and Nagao (1968), whose proof we present here.

THEOREM 8.2.2 For testing $H_0:\Sigma = \sigma^2\Sigma_0$ against the alternatives $H_1:\Sigma \ne \sigma^2\Sigma_0$, where σ^2 is an unknown positive quantity, $\boldsymbol{\mu}$ is unknown, and Σ_0 is a fixed positive definite matrix, the modified likelihood ratio test with the acceptance region

$$\omega = \left\{ s : s \text{ is positive definite and } \frac{(\det \Sigma_0^{-1} s)^{(N-1)/2}}{((\operatorname{tr} \Sigma_0^{-1} s)/p)^{(N-1)p/2}} \geq C' \right\} \qquad (8.31)$$

is unbiased.

Proof As in Theorem 8.1.2, considering $g\Sigma_0^{-1/2} s \Sigma_0^{-1/2} g'$ instead of s where $g \in O(p)$ such that $g\Sigma_0^{-1/2}\Sigma\Sigma_0^{-1/2}g' = \Gamma$, we can without any loss of generality assume that $\Sigma_0 = I$ and $\Sigma = \Gamma$, the diagonal matrix whose diagonal elements are the p characteristic roots $\theta_1, \ldots, \theta_p$ of $(\Sigma_0^{-1/2}\Sigma\Sigma_0^{-1/2})$. Thus S has a Wishart distribution with parameter Γ and $N - 1$ degrees of freedom. Hence

$$\begin{aligned} P\{\omega | H_1\} &= C_{n,p} \int_\omega (\det s)^{(N-p-2)/2} (\det \Gamma)^{-(N-1)/2} \exp\{-\tfrac{1}{2} \operatorname{tr} \Gamma^{-1} s\}\, ds \\ &= C_{n,p} \int_{\omega^*} (\det u)^{(N-p-2)/2} \exp\{-\tfrac{1}{2} \operatorname{tr} u\}\, du, \end{aligned} \qquad (8.32)$$

where u and ω^* are defined as in Theorem 8.1.2. Transform u to $v_{11}v$ where the symmetric matrix v is given by

$$v = \begin{pmatrix} 1 & v_{12} & \cdots & v_{1p} \\ v_{21} & v_{22} & \cdots & v_{2p} \\ \vdots & \vdots & & \vdots \\ v_{p1} & v_{p2} & \cdots & v_{pp} \end{pmatrix}.$$

The Jacobian of this transformation is $v_{11}^{p(p+1)/2-1}$. Since the region remains invariant under the transformation $u \to cu$, where c is a positive real number, we get

$$\begin{aligned} P\{\omega | H_1\} &= C_{n,p} \int_{\omega^*} (v_{11})^{(N-1)p/2-1} (\det v)^{(N-p-2)/2} \\ &\quad \times \exp\{-\tfrac{1}{2} \operatorname{tr}(v_{11} v)\}\, dv_{11}\, dv \\ &= C_{n,p} 2^{(N-1)p/2} \Gamma(\tfrac{1}{2}(N-1)p) \int_{\omega^{**}} (\det v)^{(N-p-2)/2} \\ &\quad \times (\operatorname{tr}(v))^{-(N-1)p/2}\, dv \end{aligned} \qquad (8.33)$$

where ω^{**} is the set of positive definite matrices v such that $\Gamma^{1/2} v \Gamma^{1/2}$ belongs to ω. Now proceeding as in Theorem 8.1.2

$$\begin{aligned} P\{\omega | H_0\} - P\{\omega | H_1\} &\geq 2^{p(N-1)/2} \Gamma(\tfrac{1}{2} p(N-1)) C_{n,p} C' \\ &\quad \times \left\{ \int_\omega - \int_{\omega^{**}} \right\} (\det v)^{-(p+1)/2}\, dv. \end{aligned} \qquad (8.34)$$

Transform $v \to x = \theta_1^{-1}\Gamma^{1/2}v\Gamma^{1/2}$ in the second integral of (8.34). Since the Jacobian of this transformation is $(\det \Gamma)^{(p+1)/2}\theta_1^{-p(p+1)/2}$, we get

$$\int_{\omega^{**}} (\det v)^{-(p+1)/2}\, dv = \int_{\omega} (\det x)^{-(p+1)/2}\, dx,$$

and hence the result. Q.E.D.

Kiefer and Schwartz (1965) showed that the likelihood ratio test for this problem is admissible Bayes whenever $N - 1 > p$. Das Gupta (1969) showed that the likelihood ratio test for this problem is also unbiased for testing H_0 against H_1. The proof proceeds in the same way as that of Theorem 8.1.3.

8.3 TESTS OF INDEPENDENCE AND THE R^2-TEST

Let $\mathbf{X} = (X_1, \ldots, X_p)'$ be a normally distributed p-vector with unknown mean $\boldsymbol{\mu}$ and positive definite covariance matrix Σ. Let $\mathbf{X}^\alpha = (X_{\alpha 1}, \ldots, X_{\alpha p})'$, $\alpha = 1, \ldots, N$, be a random sample of size N $(N > p)$ from this population. Let

$$\bar{\mathbf{X}} = \frac{1}{N}\sum_{\alpha=1}^{N} \mathbf{X}^\alpha, \qquad S = \sum_{\alpha=1}^{N} (\mathbf{X}^\alpha - \bar{\mathbf{X}})(\mathbf{X}^\alpha - \bar{\mathbf{X}})'.$$

We shall use the notation of Section 7.2.2. Partition $\bar{\mathbf{X}}, \boldsymbol{\mu}, S, \Sigma$ as

$$\begin{aligned} \boldsymbol{\mu} &= (\boldsymbol{\mu}_{(1)}, \ldots, \boldsymbol{\mu}_{(k)})', \\ \bar{\mathbf{X}} &= (\bar{\mathbf{X}}_{(1)}, \ldots, \bar{\mathbf{X}}_{(k)})', \\ \mathbf{X} &= (\mathbf{X}_{(1)}, \ldots, \mathbf{X}_{(k)})', \end{aligned} \quad \Sigma = \begin{pmatrix} \Sigma_{(11)} & \cdots & \Sigma_{(1k)} \\ \vdots & & \vdots \\ \Sigma_{(k1)} & \cdots & \Sigma_{(kk)} \end{pmatrix}, \quad S = \begin{pmatrix} S_{(11)} & \cdots & S_{(1k)} \\ \vdots & & \vdots \\ S_{(k1)} & \cdots & S_{(kk)} \end{pmatrix}.$$

We are interested in testing the null hypothesis that the subvectors $\mathbf{X}_{(1)}, \ldots, \mathbf{X}_{(k)}$ are mutually independent. The null hypothesis can be stated, equivalently, as

$$H_0 : \Sigma_{(ij)} = 0 \qquad \text{for all} \quad i \neq j. \tag{8.35}$$

Note that both the problems considered earlier in this chapter can be transformed into the problem of independence of components of $\mathbf{X}$.

Let Ω be the parametric space of $(\boldsymbol{\mu}, \Sigma)$. Under H_0, Ω is reduced to $\omega = \{(\boldsymbol{\mu}, \Sigma_D)\}$ where Σ_D is a diagonal matrix in the block form with unknown diagonal elements $\Sigma_{(ii)}$, $i = 1, \ldots, k$. The likelihood of the sample observations $\mathbf{x}^\alpha$ on $\mathbf{X}^\alpha$, $\alpha = 1, \ldots, N$, is given by

$$L(\boldsymbol{\mu}, \Sigma) = (2\pi)^{-Np/2}(\det \Sigma)^{-N/2} \times \exp\left\{-\tfrac{1}{2} \operatorname{tr} \Sigma^{-1}\left(\sum_{\alpha=1}^{N} (\mathbf{x}^\alpha - \boldsymbol{\mu})(\mathbf{x}^\alpha - \boldsymbol{\mu})'\right)\right\}. \tag{8.36}$$

Hence

$$\max_{\Omega} L(\boldsymbol{\mu}, \Sigma) = (2\pi)^{-Np/2}[\det(s/N)]^{-N/2} \exp\{-\tfrac{1}{2}Np\}. \tag{8.37}$$

Under H_0,

$$L(\boldsymbol{\mu}, \Sigma_D) = \prod_{i=1}^{k} (2\pi)^{-Np_i/2}(\det \Sigma_{(ii)})^{-N/2}$$

$$\times \exp\left\{-\tfrac{1}{2} \operatorname{tr} \Sigma_{(ii)}^{-1}\left(\sum_{\alpha=1}^{N} (\mathbf{x}_{\alpha(i)} - \boldsymbol{\mu}_{(i)})(\mathbf{x}_{\alpha(i)} - \boldsymbol{\mu}_{(i)})'\right)\right\}, \quad (8.38a)$$

where $\mathbf{x}_\alpha = (\mathbf{x}_{\alpha(1)}, \ldots, \mathbf{x}_{\alpha(k)})'$, and $x_{\alpha(i)}$ is $p_i \times 1$. Now

$$\max_\omega L(\boldsymbol{\mu}, \Sigma_D) = \prod_{i=1}^{k} \max_{\Sigma_{(ii)}, \boldsymbol{\mu}_{(i)}} \Bigg[(2\pi)^{-Np_i/2}(\det \Sigma_{(ii)})^{-N/2}$$

$$\times \exp\left\{-\tfrac{1}{2} \operatorname{tr} \Sigma_{(ii)}^{-1} : \sum_{\alpha=1}^{N} (\mathbf{x}_{\alpha(i)} - \boldsymbol{\mu}_{(i)})(\mathbf{x}_{\alpha(i)} - \boldsymbol{\mu}_{(i)})'\right\}\Bigg]$$

$$= \prod_{i=1}^{k} \{(2\pi)^{-Np_i/2}[\det(s_{(ii)}/N)]^{-N/2} \exp\{-\tfrac{1}{2}Np_i)\}\}. \quad (8.38b)$$

From (8.37) and (8.38b), the likelihood ratio criterion λ for testing H_0 is given by

$$\lambda = \frac{\max_\omega L(\boldsymbol{\mu}, \Sigma_D)}{\max_\Omega L(\boldsymbol{\mu}, \Sigma)} = \left[\frac{\det s}{\prod_{i=1}^{k} \det s_{(ii)}}\right]^{N/2} = v^{N/2}, \quad (8.39)$$

where $v = (\det s)/(\prod_{i=1}^{k} \det s_{(ii)})$. Hence we have the following theorem.

THEOREM 8.3.1 For testing $H_0:\Sigma = \Sigma_D$, the likelihood ratio test rejects H_0 whenever $\lambda \le C'$ or, equivalently, $v \le C$, where C' or C is chosen such that the test has level of significance α.

Let $s = (s_{ij})$. Writing $r_{ij} = s_{ij}/(s_{ii}s_{jj})^{1/2}$, the matrix r of sample correlation coefficients r_{ij} is

$$r = \begin{pmatrix} 1 & r_{12} & \cdots & r_{1p} \\ r_{21} & 1 & \cdots & r_{2p} \\ \vdots & \vdots & & \vdots \\ r_{p1} & r_{p2} & \cdots & 1 \end{pmatrix}. \quad (8.40)$$

Obviously $\det s = (\prod_{i=1}^{p} s_{ii}) \det r$. Let us now partition r into submatrices $r_{(ij)}$ similar to s as

$$r = \begin{pmatrix} r_{(11)} & \cdots & r_{(1k)} \\ r_{(21)} & \cdots & r_{(2k)} \\ \vdots & & \vdots \\ r_{(k1)} & \cdots & r_{(kk)} \end{pmatrix}. \quad (8.41)$$

Then

$$\det s_{(ii)} = \det(r_{(ii)}) \prod_{j=p_1+\cdots+p_{i-1}+1}^{p_1+\cdots+p_i} s_{jj}.$$

Thus

$$v = \frac{\det(r)}{\prod_{i=1}^{k} \det(r_{(ii)})} \tag{8.42}$$

gives a representation of v in terms of sample correlation coefficients.

Let G_{BD} be the group of $p \times p$ nonsingular block diagonal matrices g of the form

$$g = \begin{pmatrix} g_{(11)} & 0 & \cdots & 0 \\ 0 & g_{(22)} & \cdots & 0 \\ \vdots & \vdots & & \vdots \\ 0 & 0 & \cdots & g_{(kk)} \end{pmatrix},$$

where $g_{(ii)}$ is a $p_i \times p_i$ submatrix of g and $\sum_1^k p_i = p$. The problem of testing $H_0:\Sigma = \Sigma_{\mathrm{D}}$ against the alternatives $H_1:\Sigma \neq \Sigma_{\mathrm{D}}$ remains invariant under the group G of affine transformations $(g, \mathbf{a})$, $g \in G_{\mathrm{BD}}$ and $\mathbf{a} \in E^p$, transforming each $\mathbf{X}$ to $g\mathbf{X}^\alpha + \mathbf{a}$. The corresponding induced group of transformations in the space of $(\bar{\mathbf{X}}, S)$ is given by $(\bar{\mathbf{X}}, S) \to (g\bar{\mathbf{X}} + \mathbf{a}, gSg')$. Obviously this implies that

$$\bar{\mathbf{X}}_{(i)} \to g_{(ii)}\bar{\mathbf{X}}_{(i)} + \mathbf{a}_{(i)}, \qquad S_{(ii)} \to g_{(ii)}S_{(ii)}g'_{(ii)},$$

and hence

$$\frac{\det s}{\prod_{i=1}^{k} \det s_{(ii)}} = \frac{\det(gsg')}{\prod_{i=1}^{k} \det(g_{(ii)}s_{(ii)}g'_{(ii)})}.$$

To determine the likelihood ratio or the test based on v we need the distribution of V under H_0. Under H_0, S has a Wishart distribution with parameter Σ_D and $N - 1$ degrees of freedom; the $\mathbf{X}_{(i)}$ are mutually independent; the marginal distribution of $S_{(ii)}$ is Wishart with parameter $\Sigma_{(ii)}$ and $N - 1$ degrees of freedom; and $S_{(ii)}$ is distributed independently of $S_{(jj)}$ $(i \neq j)$. Using these facts it can be shown that (for details see Anderson, 1958), under H_0,

$$E(V^h) = \frac{\prod_{i=1}^{p} \Gamma(\tfrac{1}{2}(N-i)+h) \prod_{i=1}^{k}\{\prod_{j=1}^{p_i} \Gamma(\tfrac{1}{2}(N-j))\}}{\prod_{i=1}^{p} \Gamma(\tfrac{1}{2}(N-i)) \prod_{i=1}^{k}\{\prod_{j=1}^{p_i} \Gamma(\tfrac{1}{2}(N-j)+h)\}}, \quad h = 0, 1, \ldots. \tag{8.43}$$

Since $0 \leq V \leq 1$, these moments determine the distribution of V uniquely. Since these moments are independent of Σ_D when H_0 is true, from (8.43) it follows (see Anderson, 1958, Section 9.4) that when H_0 is true V is distributed as $\prod_{i=2}^{k}\{\prod_{j=1}^{p_i} X_{ij}\}$ where the X_{ij} are independently distributed central beta random variables with parameters

$$(\tfrac{1}{2}(N - \delta_{i-1} - j), \tfrac{1}{2}\delta_{i-1}) \qquad \text{with} \quad \delta_j = \sum_{i=1}^{j} p_i, \quad \delta_0 = 0.$$

If all the p_i are even, $p_i = 2r_i$ (say), then under H_0, V is distributed as $\prod_{i=2}^{k}\{\prod_{j=1}^{r_i} Y_{ij}^2\}$ where the Y_{ij} are independently distributed central beta random variables with parameters $((N - \delta_{i-1} - 2j), \delta_{i-1})$.

Wald and Brookner (1941) have given a method for deriving the distribution when the p_i are odd. For further results on the distribution we refer the reader to Anderson (1958, Section 9.4).

Let

$$f = \tfrac{1}{2}\left[p(p+1) - \sum_{i=1}^{k} p_i(p_i+1)\right],$$

$$\rho = 1 - \frac{2(p^3 - \sum_{i=1}^{k} p_i^3) + 9(p^2 - \sum_{i=1}^{k} p_i^2)}{6N(p^2 - \sum_{i=1}^{k} p_i^2)},$$

$$\alpha = \rho N, \qquad \gamma_2 = \frac{p^4 - \sum_{i=1}^{k} p_i^4}{48} - \frac{5(p^2 - \sum_{i=1}^{k} p_i^2)}{96} - \frac{(p^3 - \sum_{i=1}^{k} p_i^3)^2}{72(p^2 - \sum_{i=1}^{k} p_i^2)}.$$

Using Box (1949), we obtain

$$P\{-a \log V \le z\} = P\{\chi_f^2 \le z\} + \frac{\gamma_2}{a^2}[P\{\chi_{f+4}^2 \le z\} - P\{\chi_f^2 \le z\}] + O(a^{-3}).$$

Thus for large N

$$P\{-a \log V \le z\} \simeq P\{\chi_f^2 \le z\}. \tag{8.44}$$

8.3.1 The R^2-Test

If $k = 2$, $p_1 = 1$, $p_2 = p - 1$, then the likelihood ratio test criterion λ is given by

$$\lambda = \left(\frac{\det s}{s_{11}\det(s_{(22)})}\right)^{N/2} = \left(\frac{s_{11} - s_{(12)}s_{(22)}^{-1}s_{(21)}}{s_{11}}\right)^{N/2}$$
$$= (1 - r^2)^{N/2} \tag{8.45}$$

where $r^2 = s_{(12)}s_{(22)}^{-1}s_{(21)}/s_{11}$ is the square of the sample multiple correlation coefficient between X_1 and $(X_2, \ldots, X_p)$. The distribution of $R^2 = S_{(12)}S_{(22)}^{-1}S_{(21)}/S_{11}$ is given in (6.86), and depends on $\rho^2 = \Sigma_{(12)}\Sigma_{(22)}^{-1}\Sigma_{(21)}/\Sigma_{11}$, the square of the population multiple correlation coefficient between X_1 and $(X_2, \ldots, X_p)$. Since $\Sigma_{(22)}$ is positive definite, $\rho^2 = 0$ if and only if $\Sigma_{(12)} = \mathbf{0}$. From (6.86) under H_0, $((N - p)/(p - 1))(R^2/(1 - R^2))$ is distributed as a central $F_{p-1,N-p}$ with $(p - 1, N - p)$ degrees of freedom.

THEOREM 8.3.2 The likelihood ratio test of $H_0:\rho^2 = 0$ rejects H_0 whenever

$$\frac{N-p}{p-1}\frac{r^2}{1-r^2} \geq F_{p-1,N-p,\alpha}$$

where $F_{p-1,N-p,\alpha}$ is the upper significance point corresponding to the level of significance α.

Observe that this is also equivalent to rejecting H_0 whenever $r^2 \geq C$, where the constant C depends on the level of significance α of the test.

EXAMPLE 8.3.1 Consider the data given in Example 5.3.1. Let ρ^2 be the square of the population multiple correlation coefficient between X_6 and $(X_1, \ldots, X_5)$. The square of the sample multiple correlation coefficient r^2 based on 27 observations for each year's data is given by

$$r^2 = 0.85358 \qquad \text{for 1971 observations,}$$
$$r^2 = 0.80141 \qquad \text{for 1972 observations.}$$

We wish to test the hypothesis at $\alpha = 0.01$ that the wheat yield is independent of the variables plant height at harvesting (X_1), number of effective tillers (X_2), length of ear (X_3), number of fertile spikelets per 10 ears (X_4), and number of grains per 10 ears (X_5). We compare the value of $(21/5)(r^2/(1-r^2))$ with $F_{5,21,0.01} = 9.53$ for each year's data. Obviously for each year's data $(21/5)(r^2/(1-r^2)) > 9.53$, which implies that the result is highly significant. Thus the wheat yield is highly dependent on $(X_1, \ldots, X_5)$.

As stated earlier the problem of testing $H_0:\Sigma_{(12)} = \mathbf{0}$ against $H_1:\Sigma_{(12)} \neq \mathbf{0}$ remains invariant under the group G of affine transformations $(g, \mathbf{a})$, $g \in G_{BD}$, with $k = 2$, $p_1 = 1$, $p_2 = p-1$, $\mathbf{a} \in E^p$, transforming $(\bar{\mathbf{X}}, S; \boldsymbol{\mu}, \Sigma) \to (g\bar{\mathbf{X}} + \mathbf{a}, gSg'; g\boldsymbol{\mu} + \mathbf{a}, g\Sigma g')$. A maximal invariant in the space of $(\bar{\mathbf{X}}, S)$ under G is

$$R^2 = \frac{S_{(12)}S_{(22)}^{-1}S_{(21)}}{S_{11}},$$

and the corresponding maximal invariant in the parametric space Ω is

$$\rho^2 = \frac{\Sigma_{(12)}\Sigma_{(22)}^{-1}\Sigma_{(21)}}{\Sigma_{11}}.$$

Under H_0, $\rho^2 = 0$ and under H_1, $\rho^2 > 0$. The probability density function of R^2 is given in (6.86).

THEOREM 8.3.3 On the basis of observations $\mathbf{x}^i = (x_{i1}, \ldots, x_{ip})'$, $i = 1, \ldots, N$ $(N > p)$, from a p-variate normal distribution with unknown mean $\boldsymbol{\mu}$ and unknown positive definite covariance matrix Σ, among all tests $\phi(\mathbf{X}^1, \ldots, \mathbf{X}^N)$ of $H_0:\Sigma_{(12)} = \mathbf{0}$ against the alternatives $H_1:\Sigma_{(12)} \neq \mathbf{0}$ which are invariant under the group of affine transformations G, the test which rejects H_0 whenever the square of the sample multiple correlation coefficient $r^2 > C$, where the constant C depends on the level of significance α of the test (or equivalently the likelihood ratio test), is uniformly most powerful.

Proof Let $\phi(\mathbf{X}^1, \ldots, \mathbf{X}^N)$ be an invariant test with respect to the group of affine transformations G. Since $(\overline{\mathbf{X}}, S)$ is sufficient for $(\boldsymbol{\mu}, \Sigma)$, $E(\phi(\mathbf{X}^1, \ldots, \mathbf{X}^N)|\overline{\mathbf{X}} = \overline{\mathbf{x}}, S = s)$ is independent of $(\boldsymbol{\mu}, \Sigma)$ and depends only on $(\overline{\mathbf{x}}, s)$. As ϕ is invariant under G, $E(\phi|\overline{\mathbf{X}} = \overline{\mathbf{x}}, S = s)$ is invariant under G, and ϕ, $E(\phi|\overline{\mathbf{X}}, S)$ have the same power function. Thus each test in the larger class of level α tests which are functions of $\mathbf{X}^i$, $i = 1, \ldots, N$, can be replaced by one in the smaller class of tests which are functions of $(\overline{\mathbf{X}}, S)$ having identical power functions. Since R^2 is a maximal invariant in the space of $(\overline{\mathbf{X}}, S)$ under G, the invariant test $E(\phi|\overline{\mathbf{X}}, S)$ depends on $(\overline{\mathbf{X}}, S)$ only through R^2, whose distribution depends on $(\boldsymbol{\mu}, \Sigma)$ only through ρ^2. The most powerful level α invariant test of $H_0:\rho^2 = 0$ against the simple alternative $\rho^2 = {\rho_0}^2$, where ${\rho_0}^2$ is a fixed positive number, rejects H_0 whenever [from (6.86)]

$$(1 - {\rho_0}^2)^{(N-1)/2} \sum_{j=0}^{\infty} \frac{({\rho_0}^2)^j (r^2)^{j-1} \Gamma^2(\frac{1}{2}(N-1)+j)\Gamma(\frac{1}{2}(p-1))}{j!\,\Gamma^2(\frac{1}{2}(N-1))\Gamma(\frac{1}{2}(p-1)+j)} \geq C', \tag{8.46}$$

where the constant C' is so chosen that the test has level α. From (8.46) it is now obvious that the R^2-test which rejects H_0 whenever $r^2 \geq C$ is uniformly most powerful among all invariant level α tests for testing $H_0:\rho^2 = 0$ against the alternatives $H_1:\rho^2 > 0$. Q.E.D.

Simaika (1941) proved the following stronger optimum property of the R^2-test than the one presented in Theorem 8.3.3.

THEOREM 8.3.4 On the basis of observations $\mathbf{x}^i$, $i = 1, \ldots, N$, from the p-variate normal distribution with unknown mean $\boldsymbol{\mu}$ and unknown positive definite covariance matrix Σ, among all tests (level α) of $H_0:\rho^2 = 0$ against $H_1:\rho^2 > 0$ with power functions depending only on ρ^2, the R^2-test is uniformly most powerful.

This theorem can be proved from Theorem 8.3.3 in the same way as Theorem 7.2.2 is proved from Theorem 7.2.1. It may be added that the proof suggested here differs from Simaika's original proof.

8.3.2 Admissibility of the Test of Independence and the R^2-Test

The development in this section follows the approach of Section 7.2.2. To prove the admissibility of the R^2-test we first prove the admissibility of the likelihood ratio test of independence using the approach of Kiefer and Schwartz (1965) and then give the modifications needed to prove the admissibility of the R^2-test.

Let $V = (Y, \mathbf{X})$, where $\mathbf{X} = \sqrt{N}\,\bar{\mathbf{X}}$, $Y = (\mathbf{Y}^1, \ldots, \mathbf{Y}^{N-1})$ are such that $S = YY' = \sum_{i=1}^{N-1} \mathbf{Y}^i\mathbf{Y}^{i'}$, and $\mathbf{Y}^1, \ldots, \mathbf{Y}^{N-1}$ are independently and identically distributed normal p-vectors with mean $\mathbf{0}$ and covariance matrix Σ, and $\mathbf{X}$ is distributed, independently of $\mathbf{Y}^1, \ldots, \mathbf{Y}^{N-1}$, as p-variate normal with mean $\mathbf{v} = \sqrt{N}\,\boldsymbol{\mu}$ and covariance matrix Σ. It may be recalled that if $\theta = (\boldsymbol{\mu}, \Sigma)$ and the Lebesgue density function of V on a Euclidean set is denoted by $f_V(v|\theta)$, then every Bayes rejection region for the 0–1 loss function is of the form

$$\left\{ v: \frac{\int f_V(v|\theta)\pi_0\,(d\theta)}{\int f_V(v|\theta)\pi_1\,(d\theta)} \leq C \right\} \tag{8.47}$$

for some constant C $(0 \leq C \leq \infty)$ where π_1 and π_0 are the probability measures (or positive constant multiples thereof) for the parameter θ under H_1 and H_0, respectively. Since in our case the subset of this set corresponding to equality sign C has probability 0 for all θ in the parametric space, our Bayes procedures will be essentially unique and hence admissible.

Write $Y' = (Y'_{(1)}, \ldots, Y'_{(k)})$, where the $Y'_{(i)}$ are submatrices of dimension $(N-1) \times p_i$. Then $S_{(ii)} = Y_{(i)}Y'_{(i)}$ and the likelihood ratio test of independence rejects H_0 whenever

$$\det(yy') \Big/ \prod_{i=1}^{k} \det(y_{(i)}y'_{(i)}) \leq C. \tag{8.48}$$

Let π_1 assign all its measure to values of θ for which $\Sigma^{-1} = I + \boldsymbol{\eta}\boldsymbol{\eta}'$ for some random p-vector $\boldsymbol{\eta}$ and $\mathbf{v} = \Sigma\boldsymbol{\eta}Z$ for some random variable Z. Let the conditional (a priori) distribution of $\mathbf{v}$ given Σ under H_1 be such that with a priori probability 1, $\Sigma^{-1}\mathbf{v} = \boldsymbol{\eta}Z$ where Z is normally distributed with mean 0 and variance $(1 - \boldsymbol{\eta}'(I + \boldsymbol{\eta}\boldsymbol{\eta}')^{-1}\boldsymbol{\eta})^{-1}$, and let the marginal distribution $\pi_1{}^*$ of Σ under H_1 be given by

$$\frac{d\pi^*(\boldsymbol{\eta})}{d\boldsymbol{\eta}} = \left[\det(I + \boldsymbol{\eta}\boldsymbol{\eta}')\right]^{-(N-1)/2}, \tag{8.49}$$

which is integrable on E^p (Euclidean p-space) provided $N - 1 > p$. Let π_0 assign all its measure to values of θ for which $\Sigma = \Sigma_{\mathrm{D}}$ with

$$\Sigma_{\mathrm{D}}^{-1} = \begin{pmatrix} I_{(1)} + \boldsymbol{\eta}_{(1)}\boldsymbol{\eta}'_{(1)} & 0 & 0 & 0 \\ 0 & I_{(2)} + \boldsymbol{\eta}_{(2)}\boldsymbol{\eta}'_{(2)} & 0 & 0 \\ \vdots & \vdots & \vdots & \vdots \\ 0 & 0 & 0 & I_{(k)} + \boldsymbol{\eta}_{(k)}\boldsymbol{\eta}'_{(k)} \end{pmatrix}$$

for some random vector $\boldsymbol{\eta} = (\boldsymbol{\eta}_{(1)}, \ldots, \boldsymbol{\eta}_{(k)})'$ where the $\boldsymbol{\eta}_{(i)}$ are subvectors of dimension $p_i \times 1$ with $\sum_{i=1}^{k} p_i = p$ and $\mathbf{v} = \Sigma_{\mathrm{D}}^{-1}\boldsymbol{\eta}Z$ for some random variable Z. Let the conditional a priori distribution of $\mathbf{v}$ under H_0 given Σ_{D} be such that with a priori probability 1, $\Sigma_{\mathrm{D}}^{-1}\mathbf{v} = \boldsymbol{\eta}Z$ where Z is normally distributed with mean 0 and variance $(1 - \sum_{i=1}^{k}[\boldsymbol{\eta}'_{(i)}(I_{(i)} + \boldsymbol{\eta}_{(i)}\boldsymbol{\eta}'_{(i)})^{-1}\boldsymbol{\eta}_{(i)}])^{-1}$, and let the marginal (a priori) distribution of Σ under H_0 be given by

$$\frac{d\pi_0{}^*(\boldsymbol{\eta})}{d\boldsymbol{\eta}} = \prod_{i=1}^{k} [\det(I_{(i)} + \boldsymbol{\eta}_{(i)}\boldsymbol{\eta}'_{(i)})]^{-(N-1)/2}, \tag{8.50}$$

which is integrable on E^p provided $N - 1 > p$. The fact that these priors represent bona fide probability measures follows from Exercise 8.4. Since in our case

$$\begin{aligned} f_V(v|\theta) &= f_{\mathbf{X}}(\mathbf{x}|\mathbf{v}, \Sigma) f_Y(y|\Sigma) \\ &= (2\pi)^{-Np/2}(\det \Sigma)^{-N/2} \exp\{-\tfrac{1}{2} \operatorname{tr} \Sigma^{-1}(yy' + (\mathbf{x} - \mathbf{v})(\mathbf{x} - \mathbf{v})')\}, \end{aligned} \tag{8.51}$$

$$\begin{aligned} &\int f_V(v|\theta)\pi_1 (d\theta) \\ &\quad = \int [(2\pi)^{-(Np+1)/2}(\det(I + \boldsymbol{\eta}\boldsymbol{\eta}'))^{N/2} \\ &\qquad \times \exp\{-\tfrac{1}{2} \operatorname{tr}[(I + \boldsymbol{\eta}\boldsymbol{\eta}')(yy' + \mathbf{x}\mathbf{x}') + \boldsymbol{\eta}\mathbf{x}'z - \tfrac{1}{2}(I + \boldsymbol{\eta}\boldsymbol{\eta}')^{-1}\boldsymbol{\eta}\boldsymbol{\eta}'z^2]\} \\ &\qquad \times (\det(I + \boldsymbol{\eta}\boldsymbol{\eta}'))^{-(N-1)/2}(1 - \boldsymbol{\eta}'(I + \boldsymbol{\eta}\boldsymbol{\eta}')^{-1}\boldsymbol{\eta})^{1/2} \\ &\qquad \times \exp\{-\tfrac{1}{2}z^2(1 - \boldsymbol{\eta}'(I + \boldsymbol{\eta}\boldsymbol{\eta}')^{-1}\boldsymbol{\eta})\}\, d\boldsymbol{\eta}\, dz \\ &\quad = A \exp\{-\tfrac{1}{2}\mathbf{x}\mathbf{x}'\} \int \exp\{-\tfrac{1}{2} \operatorname{tr}(I + \boldsymbol{\eta}\boldsymbol{\eta}')(yy')\}\, d\boldsymbol{\eta}, \end{aligned}$$

where A is a constant independent of $\boldsymbol{\eta}$. Similarly,

$$\begin{aligned} \int f_V(v|\theta)\pi_0 (d\theta) &= A \exp\{-\tfrac{1}{2}\mathbf{x}\mathbf{x}'\} \\ &\quad \times \prod_{i=1}^{k} \int \exp\{-\tfrac{1}{2} \operatorname{tr}(I_{(i)} + \boldsymbol{\eta}_{(i)}\boldsymbol{\eta}'_{(i)})y_{(i)}y'_{(i)}\}\, d\boldsymbol{\eta}_{(i)} \\ &= A \exp\{-\tfrac{1}{2} \operatorname{tr}(yy' + \mathbf{x}\mathbf{x}')\} \\ &\quad \times \prod_{i=1}^{k} \int \exp\{-\tfrac{1}{2} \operatorname{tr}(\boldsymbol{\eta}_{(i)}\boldsymbol{\eta}'_{(i)}y_{(i)}y'_{(i)})\}\, d\boldsymbol{\eta}_{(i)}. \end{aligned} \tag{8.52}$$

From (8.51) and (8.52), using the results of Exercise 8.4 we obtain

$$\frac{\int f_V(v|\theta)\pi_0 (d\theta)}{\int f_V(v|\theta)\pi_1 (d\theta)} = \left[\frac{\det(yy')}{\prod_{i=1}^{k} \det(y_{(i)}y'_{(i)})}\right]^{1/2} = \left[\frac{\det s}{\prod_{i=1}^{k} \det(s_{(ii)})}\right]^{1/2}. \tag{8.53}$$

Hence we get the following theorem.

THEOREM 8.3.5 For testing $H_0 : \Sigma = \Sigma_D$ against the alternatives $H_1 : \Sigma \neq \Sigma_D$ when $\boldsymbol{\mu}$ is unknown the likelihood ratio test that rejects H_0 whenever $[\det(s)/\prod_{i=1}^{k} \det(s_{(ii)})]^{N/2} \leq C$, where the constant C depends on the level of significance α of the test, is admissible Bayes whenever $N - 1 > p$.

This approach does not handle the case of the minimum sample size $(N - 1 = p)$. In the special case $k = 2$, $p_1 = 1$, $p_2 = p - 1$, a slightly different trick, used by Lehmann and Stein (1948), will work even when

$N - 1 = p$. Let π_1 assign all its measure under H_1 to values of θ for which

$$\Sigma^{-1} = I + \begin{pmatrix} 1 & \boldsymbol{\eta}' \\ \boldsymbol{\eta} & \boldsymbol{\eta}\boldsymbol{\eta}' \end{pmatrix},$$

where $\boldsymbol{\eta}$ is a $(p-1) \times 1$ random vector and the marginal (a priori) distribution of Σ under H_1 is

$$\frac{d\pi_1{}^*(\boldsymbol{\eta})}{d\boldsymbol{\eta}} = \left[\det\left(I + \begin{pmatrix} 1 & \boldsymbol{\eta}' \\ \boldsymbol{\eta} & \boldsymbol{\eta}\boldsymbol{\eta}' \end{pmatrix}\right)\right]^{-p/2}, \tag{8.54}$$

which is integrable on E^{p-1}, and let the conditional distribution of $\mathbf{v}$ given Σ under H_1 remain the same as the general case above.

Let π_0 assign all its measure to Σ^{-1}, which is of the form

$$\Sigma^{-1} = I + \begin{pmatrix} 1-b & 0 \\ 0 & \boldsymbol{\eta}\boldsymbol{\eta}' \end{pmatrix},$$

where $\boldsymbol{\eta}$ is a $(p-1) \times 1$ random vector, $0 \le b \le 1$, and the marginal (a priori) distribution of Σ under H_0 is

$$\frac{d\pi_0{}^*(\boldsymbol{\eta})}{d\boldsymbol{\eta}} = \left[\det\left(I + \begin{pmatrix} 1-b & 0 \\ 0 & \boldsymbol{\eta}\boldsymbol{\eta}' \end{pmatrix}\right)\right]^{-p/2}, \tag{8.55}$$

which is integrable on E^{p-1}, and let the conditional distribution of $\mathbf{v}$ given Σ under H_0 remain the same as the general case above.

Consider the particular Bayes test which rejects H_0 whenever

$$\frac{\int f_V(v|\theta)\pi_0\,(d\theta)}{\int f_V(v|\theta)\pi_1\,(d\theta)} \le 1.$$

Carrying out the integration as in the general case with the modified marginal distribution of Σ under H_0, H_1, we obtain the rejection region

$$\exp\{\tfrac{1}{2}by_{(1)}y'_{(1)}\}/\exp\{\tfrac{1}{2}y_{(1)}y'_{(2)}(y_{(2)}y'_{(2)})^{-1}y_{(2)}y'_{(1)}\} \le 1.$$

Taking logarithms of both sides we finally get the rejection region

$$\frac{s_{(12)}s_{(22)}^{-1}s_{(21)}}{s_{11}} \ge b,$$

which in the special case is equivalent to (8.53). Thus we have the following theorem.

THEOREM 8.3.6 For testing $H_0:\rho^2 = 0$ against the alternatives $H_1:\rho^2 > 0$, the R^2-test (based on the square of the sample multiple correlation coefficient R^2) which rejects H_0 whenever $r^2 \ge C$, the constant C depending on the level α of the test, is admissible Bayes.

8.3.3 Minimax Character of the R^2-Test

The solution presented here is due to Giri and Kiefer (1964b) and parallels that of Giri *et al.* (1963), as discussed in Section 7.2.3 for the corresponding T^2-results, the steps are the same, the detailed calculations in this case being slightly more complicated. The reader is referred back to Section 7.2.3 for the discussion of the Hunt–Stein theorem, its validity under the group of real lower triangular nonsingular matrices, and its failure under the full linear group.

We have already proved that among all tests based on the sufficient statistic $(\bar{\mathbf{X}}, S)$, the R^2-test is best invariant for testing $H_0 : \rho^2 = 0$ against the simple alternative $\rho^2 = \rho_0{}^2$ (>0) under the group of affine transformations G. For $p > 2$, this does not imply our minimax result because of the failure of the Hunt–Stein theorem.

We consider, without any loss of generality, test functions which depend on the statistic $(\bar{\mathbf{X}}, S)$. It can be verified that the group H of translations $(\bar{\mathbf{X}}, S; \boldsymbol{\mu}, \Sigma) \to (\bar{\mathbf{X}} + \mathbf{a}, S; \boldsymbol{\mu} + \mathbf{a}; \Sigma)$ leaves the testing problem in question invariant, that H is a normal subgroup in the group G^* generated by H and the group G_{T}, the multiplicative group of $p \times p$ nonsingular lower triangular matrices whose first column contains only zeros except for the first element, and that G_{T} and H (and hence G^*) satisfy the Hunt–Stein conditions. Furthermore it is obvious that the action of the transformations in H is to reduce the problem to that where $\boldsymbol{\mu} = \mathbf{0}$ (known) and $S = \sum_{\alpha=1}^{N} \mathbf{X}^\alpha \mathbf{X}^{\alpha'}$ is sufficient for Σ, where N has been reduced by unity from what it was originally. Using the standard method of reduction in steps, we can therefore treat the latter formulation, considering $\mathbf{X}^1, \ldots, \mathbf{X}^N$ to have $\mathbf{0}$ mean. We assume also that $N \geq p \geq 2$ (note that N is really $N - 1$ when the mean vector is not $\mathbf{0}$). Furthermore with this formulation, we need only consider test functions which depend on the sufficient statistic $S = \sum_{\alpha=1}^{N} \mathbf{X}^\alpha \mathbf{X}^{\alpha'}$, the Lebesgue density of which is given in (6.32).

We now consider the group G_{T} (of nonsingular matrices). A typical element $g \in G_{\mathrm{T}}$ can be written as

$$g = \begin{pmatrix} g_{11} & 0 \\ 0 & g_{(22)} \end{pmatrix}$$

where $g_{(22)}$ is $(p-1) \times (p-1)$ lower triangular. It is easily seen that the group G_{T} operating as $(S, \Sigma) \to (gSg', g\Sigma g')$ leaves this reduced problem invariant.

We now compute a maximal invariant in the space of S under G_{T} in the usual fashion. If a test function ϕ (of S) is invariant under G_{T}, then $\phi(S) = \phi(gSg')$ for all $g \in G_{\mathrm{T}}$ and for all S. Since S is symmetric, writing

$$S = \begin{pmatrix} S_{11} & S_{(12)} \\ S_{(21)} & S_{(22)} \end{pmatrix},$$

we get

$$\phi(S_{11}, S_{(12)}, S_{(22)}) = \phi(g_{11}S_{11}g_{11}, g_{11}S_{(12)}g_{(22)}, g_{(22)}S_{(22)}g'_{(22)}).$$

Since S is symmetric and positive definite with probability 1 for all Σ, there is an F in G_T with positive diagonal elements such that

$$FF' = \begin{pmatrix} S_{11} & 0 \\ 0 & S_{(22)} \end{pmatrix}.$$

Let $g = LF^{-1}$ where L is any diagonal matrix with values ± 1 in any order on the main diagonal. Then ϕ is a function only of $L_{(22)}F_{(22)}^{-1}S_{(21)}L_{11}/F_{11}$, and hence because of the freedom of choice of L, of $|F_{(22)}^{-1}S_{(21)}/F_{11}|$, or equivalently, of the $(p-1)$-vector whose ith component Z_i $(2 \leq i \leq p)$ is the sum of squares of the first i components of $|F_{(22)}^{-1}S_{(21)}/F_{11}|$ (whose components are indexed $2, 3, \ldots, p$). Write $\mathbf{b}_{[i]}$ for the $(i-1)$-vector consisting of the first $i-1$ components of the $(p-1)$-vector $\mathbf{b}$ and $C_{[i]}$ for the upper left-hand $(i-1) \times (i-1)$ submatrix of a $(p-1) \times (p-1)$ matrix C. Then Z_i can be written as

$$Z_i = \frac{S_{(12)[i]}(F_{(22)[i]}^{-1})'(F_{(22)[i]}^{-1})S'_{(12)[i]}}{S_{11}} = \frac{S_{(12)[i]}S_{(22)[i]}S'_{(12)[i]}}{S_{11}}. \tag{8.56}$$

The vector $\mathbf{Z} = (Z_2, \ldots, Z_p)'$ is thus a maximal invariant under G_T if it is invariant under G_T, and it is easily seen to be the latter. Z_i is essentially the squared sample multiple correlation coefficient computed from the first i coordinates of $\mathbf{X}^j, j = 1, \ldots, N$. Let us define a $(p-1)$-vector $\mathbf{R} = (R_2, \ldots, R_p)'$ by

$$\sum_{j=1}^{i} R_j = Z_i, \qquad 2 \leq i \leq p. \tag{8.57}$$

Obviously $R_i = Z_i - Z_{i-1}$, where we define $Z_1 = 0$. It now follows trivially that $\mathbf{R}$ is a maximal invariant under G_T and $R_i \geq 0$ for each i, $\sum_{i=2}^{p} R_i \leq 1$, and of course

$$\sum_{i=2}^{p} R_i = \frac{S_{(12)}S_{(22)}^{-1}S_{(21)}}{S_{11}} = R^2. \tag{8.58}$$

We shall find it more convenient to work with the equivalent statistic $\mathbf{R}$ instead of with $\mathbf{Z}$. A corresponding maximal invariant $\mathbf{\Delta} = (\delta_2^2, \ldots, \delta_p^2)'$ in the parametric space of Σ under G_T, when H_1 is true, is given by

$$\sum_{j=2}^{i} \delta_j^2 = \frac{\Sigma_{(12)[i]}\Sigma_{(22)[i]}^{-1}\Sigma'_{(12)[i]}}{\Sigma_{11}}, \qquad 2 \leq i \leq p. \tag{8.59}$$

It is clear that $\delta_j^2 \geq 0$ and $\sum_{j=2}^{p} \delta_j^2 = \rho^2$. The corresponding maximal invariant under H_0 takes on the single value $\mathbf{0}$. Thus the Lebesgue density function $f_{\mathbf{R}}(\mathbf{r}|\mathbf{\Delta})$ depends only on $\mathbf{\Delta}$ under H_1 and is a fixed $f_{\mathbf{R}}(\mathbf{r}|\mathbf{0})$ under H_0.

We can assume $\Sigma_{11} = 1$, $\Sigma_{(22)} = I$ [the $(p-1) \times (p-1)$ identity matrix], and $\Sigma_{(21)} = (\delta_2, \ldots, \delta_p)' = \boldsymbol{\delta}^*$ in (6.32), since $f_{\mathbf{R}}(\mathbf{r}|\boldsymbol{\Delta})$ depends only on $\boldsymbol{\Delta}$. With this choice of Σ (Σ^*, say) we can write (6.32) as [also denote it by $f(s_{11}, s_{(12)}, s_{(22)}|\Sigma)$]

$$\begin{aligned} W_p(N, \Sigma^*) = {} & K(1 - \rho^2)^{-N/2} \\ & \times \exp\{-\tfrac{1}{2}\operatorname{tr}[(1 - \rho^2)^{-1} s_{11} - 2(1 - \rho^2)^{-1} \boldsymbol{\delta}^{*\prime} s_{(21)} \\ & + (I - \boldsymbol{\delta}^* \boldsymbol{\delta}^{*\prime})^{-1} s_{(22)}]\}(\det s)^{(N-p-2)/2}. \end{aligned} \tag{8.60}$$

Let B be the unique lower triangular matrix belonging to G_{T} with positive diagonal elements B_{ii} ($1 \leq i \leq p$) such that $S_{(22)} = B_{(22)} B'_{(22)}$, $S_{11} = B_{11}^2$, and let $\mathbf{V} = B_{(22)}^{-1} S_{(21)}$. One can easily compute the Jacobians

$$\frac{\partial S_{(22)}}{\partial B_{(22)}} = 2^{p-1} \prod_{i=2}^{p} (B_{ii})^{p+1-i}, \qquad \frac{\partial S_{(21)}}{\partial \mathbf{V}} = \prod_{i=2}^{p} B_{ii}, \qquad \frac{\partial S_{11}}{\partial B_{11}} = 2B_{11}, \tag{8.61}$$

so the joint probability density of B_{11}, $\mathbf{V}$, and $B_{(22)}$ is

$$h(b_{11}, \mathbf{v}, b_{(22)}|\Sigma^*) = 2^p f(b_{11}^2, \mathbf{v}' b'_{(22)}, b_{(22)} b'_{(22)}|\Sigma^*, b_{11}) \prod_{i=2}^{p} b_{ii}^{p+2-i}. \tag{8.62}$$

Putting $\mathbf{W} = (W_2, \ldots, W_p)'$ with $W_i = |V_i|$ ($2 \leq i \leq p$), and noting that the $(p-1)$-vector $\mathbf{W}$ can arise from any of the 2^{p-1} vectors $\mathbf{V} = M_{(22)} \mathbf{V}$ where $M_{(22)}$ is a $(p-1) \times (p-1)$ diagonal matrix with diagonal entries ± 1, we write $g = bM$, where with $M_{11} = \pm 1$,

$$M = \begin{pmatrix} M_{11} & 0 \\ 0 & M_{(22)} \end{pmatrix},$$

g ranging over all matrices in G_{T}. We obtain for the density of W, writing g_{ij} ($i \geq j \geq 2$) for the elements of $g_{(22)}$,

$$\begin{aligned} f_{\mathbf{W}}(\boldsymbol{\omega}|\Sigma^*) &= 2^p \int f(g_{11}^2, \boldsymbol{\omega}' g_{(22)}, g_{(22)} g'_{(22)}) \prod_{i=2}^{p} |g_{ii}|^{p+2-i} \\ &\quad \times |g_{11}| \prod_{i \geq j \geq 2} dg_{ij}\, dg_{11} \\ &= (1 - \rho^2)^{-N/2} 2^p K \int \exp\{-\tfrac{1}{2}(1 - \rho^2)^{-1} \\ &\quad \times \operatorname{tr}(g_{11}^2 - \boldsymbol{\delta}^* \boldsymbol{\omega}' g'_{(22)} - \boldsymbol{\delta}^{*\prime} g_{(22)} \boldsymbol{\omega} \\ &\quad + (1 - \rho^2)(I - \boldsymbol{\delta}^* \boldsymbol{\delta}^{*\prime})^{-1} g_{(22)} g'_{(22)})\} \\ &\quad \times \prod_{i=2}^{p} |g_{ii}|^{N+1-i} |g_{11}|^{N-p} (1 - \boldsymbol{\omega}'\boldsymbol{\omega}/g_{11}^2)^{(N-p-1)/2} \\ &\quad \times \prod_{i \geq j \geq 2} dg_{ij}\, dg_{11}. \end{aligned} \tag{8.63}$$

Writing $\mathbf{W} = g_{11}\mathbf{U}$ and $R_j = U_j^2$ $(2 \le j \le p)$ we obtain from (8.63) that the probability density function of $\mathbf{R} = (R_2, \ldots, R_p)'$ is

$$f_{\mathbf{R}}(\mathbf{r}|\Delta) = \frac{(1-\rho^2)^{-N/2}2K}{\prod_{i=2}^{p} r_i^{1/2}}$$
$$\times \int \exp\{-\tfrac{1}{2}(1-\rho^2)^{-1}\operatorname{tr}(g_{11}^2 - 2g_{11}\boldsymbol{\delta}^* g_{(22)}\mathbf{r}^*$$
$$+ (1-\rho^2)(I - \boldsymbol{\delta}^*\boldsymbol{\delta}^{*\prime})^{-1} g_{(22)}g'_{(22)})\}$$
$$\times \left(1 - \sum_{j=2}^{p} r_j\right)^{(N-p-2)/2} |g_{11}|^{N-1} \prod_{i=2}^{p} |g_{ii}|^{N+1-i}$$
$$\times \prod_{i \ge j \ge 2} dg_{ij}\, dg_{11}, \tag{8.64}$$

where $\mathbf{r}^* = (r_2^{1/2}, \ldots, r_p^{1/2})'$. Let $C = (1-\rho^2)^{-1}(I - \boldsymbol{\delta}^*\boldsymbol{\delta}^{*\prime})$. Since C is positive definite, there exists a lower triangular $(p-1)\times(p-1)$ matrix T with positive diagonal elements T_{ii} $(2 \le i \le p)$ such that $TCT' = I$. Writing $h = Tg_{(22)}$, we obtain

$$\frac{\partial h}{\partial g_{(22)}} = \prod_{i=2}^{p} T_{ii}^{i-1}.$$

Let us define for $2 \le i \le p$,

$$\gamma_i = 1 - \sum_{j=2}^{i} \delta_j^2, \qquad \gamma_1 = 1 \qquad (\gamma_p = 1 - \rho^2),$$
$$\alpha_i = (\delta_i^2\gamma_p/\gamma_{i-1}\gamma_i)^{1/2}, \qquad \boldsymbol{\alpha} = (\alpha_2, \ldots, \alpha_p)'. \tag{8.65}$$

A simple calculation yields $(T_{[i]}\boldsymbol{\delta}_{[1]}^*)'(T_{[i]}\boldsymbol{\delta}_{[i]}^*) = \gamma_p(1-\gamma_i)/\gamma_i$, so that $\boldsymbol{\alpha} = T\boldsymbol{\delta}^*$. Since $C\boldsymbol{\delta}^* = \boldsymbol{\delta}^*$, by direct computation, we obtain

$$\boldsymbol{\alpha} = TC\boldsymbol{\delta}^* = (T^{-1})'\boldsymbol{\delta}^*.$$

From this and the fact that $\det C = (1-\rho^2)^{2-p}$, we obtain

$$f_{\mathbf{R}}(\mathbf{r}|\Delta) = 2K(1-\rho^2)^{-N(p-1)/2} \prod_{i=2}^{p} r_i^{-1/2}\left(1 - \sum_{j=2}^{p} r_j\right)^{(N-p-1)/2}$$
$$\times \int \exp\{-\tfrac{1}{2}(1-\rho^2)^{-1}g_{11}^2\}|g_{11}|^{N-1}$$
$$\times \left\{\int \exp\left\{-\tfrac{1}{2}(1-\rho^2)^{-1} \sum_{i \ge j \ge 2} [h_{ij}^2 - 2\alpha_i r_j^{1/2} h_{ij} g_{11}]\right\}\right.$$
$$\left.\times \prod_{i=2}^{p} |h_{ii}|^{N+1-i} \prod_{i \ge j \ge 2} dh_{ij}\right\} dg_{11}, \tag{8.66}$$

the integration being from $-\infty$ to ∞ in each variable. For $i > j$ the integration with respect to h_{ij} yields a factor

$$(2\pi)^{1/2}(1 - \rho^2)^{1/2} \exp\{\alpha_i{}^2 r_j g_{11}^2/2(1 - \rho^2)\}. \tag{8.67}$$

For $i = j$ we obtain a factor

$$\begin{aligned}(2\pi)^{1/2}&(1 - \rho^2)^{(N+2-i)/2} \exp[\alpha_i{}^2 r_i g_{11}^2/2(1 - \rho^2)] \\ &\times E(\chi_1{}^2(\alpha_i{}^2 r_i g_{11}^2/(1 - \rho^2))^{(N+1-i)/2}) \\ &= [2(1 - \rho^2)]^{(N+2-i)/2}\Gamma(\tfrac{1}{2}(N - i + 2)) \\ &\quad\times \phi(\tfrac{1}{2}(N - i + 2), \tfrac{1}{2}; r_i\alpha_i{}^2 g_{11}^2/2(1 - \rho^2))\end{aligned} \tag{8.68}$$

where $\chi_1{}^2(\beta)$ is a noncentral chi-square with one degree of freedom and noncentrality parameter β, and ϕ is the confluent hypergeometric function. Integrating with respect to g_{11} we obtain, from (8.66)–(8.68), that the probability density function of $\mathbf{R}$ is (for $\mathbf{r} \in H = \{\mathbf{r}: r_i \geq 0,\ 2 \leq i \leq p,\ \sum_{i=2}^p r_i < 1\}$)

$$\begin{aligned}f_{\mathbf{R}}(\mathbf{r}|\boldsymbol{\Delta}) &= \frac{(1-\rho^2)^{N/2}(1-\sum_{i=2}^p r_i)^{(N-p-1)/2}}{(1+\sum_{i=2}^p r_i(1-\rho^2)/\gamma_i - 1)^{N/2}\Gamma(\frac{1}{2}(N-p+1))\pi^{(p-1)/2}} \\ &\times \frac{1}{\prod_{i=2}^p\{r_i^{1/2}\Gamma(\frac{1}{2}(N-i+2))\}} \sum_{\beta_2=0}^{\infty} \cdots \sum_{\beta_p=0}^{\infty} \frac{\Gamma(\sum_{j=2}^p \beta_j + \frac{1}{2}N)}{\Gamma(\frac{1}{2}N)} \\ &\times \prod_{i=2}^p \left\{\frac{\Gamma(\frac{1}{2}(N-i+2)+\beta_i)}{\Gamma(\frac{1}{2}(N-i+2))(2\beta_i)!}\right. \\ &\times \left.\left[\frac{4r_i\alpha_i{}^2}{1+\sum_{j=2}^p r_j((1-\delta^2)/\gamma_j - 1)}\right]^{\beta_i}\right\}.\end{aligned} \tag{8.69}$$

The continuity of $f_{\mathbf{R}}(\mathbf{r}|\boldsymbol{\Delta})$ in $\boldsymbol{\Delta}$ over its compact domain $\Gamma = \{(\delta_2{}^2, \ldots, \delta_p{}^2): \delta_i{}^2 \geq 0, \sum_{j=2}^p \delta_j{}^2 = \delta^2\}$ is evident. As in the case of the T^2-test, we conclude here also that the minimax character of the critical region $\sum_{j=2}^p R_j \geq C$ is equivalent to the existence of a probability measure λ satisfying

$$\int_\Gamma \frac{f_{\mathbf{R}}(\mathbf{r}|\boldsymbol{\Delta})}{f_{\mathbf{R}}(\mathbf{r}|\mathbf{0})}\,\lambda\,(d\boldsymbol{\Delta})\begin{Bmatrix}>\\=\\<\end{Bmatrix}K, \tag{8.70}$$

according to whether $\sum_{i=2}^p r_i$ is greater than, equal to, or less than C, except possibly for a set of measure 0. We can replace (8.70) by its equivalent

$$\int_\Gamma \frac{f_{\mathbf{R}}(\mathbf{r}|\boldsymbol{\Delta})}{f_{\mathbf{R}}(\mathbf{r}|\mathbf{0})}\,\lambda\,(d\boldsymbol{\Delta}) = K \quad \text{if} \quad \sum_{i=2}^p r_i = C. \tag{8.71}$$

Clearly (8.70) implies (8.71). On the other hand, if there are a λ and a constant K satisfying (8.71) and if $\bar{\mathbf{r}} = (\bar{r}_2, \ldots, \bar{r}_p)'$ is such that $\sum_{i=2}^p \bar{r}_i = C' > C$, writing

$$f(\mathbf{r}) = [f_{\mathbf{R}}(\mathbf{r}|\Delta)/f_{\mathbf{R}}(\mathbf{r}|\mathbf{0})] \qquad \text{and} \qquad \bar{\bar{\mathbf{r}}} = (C/C')\bar{\mathbf{r}},$$

we see at once that

$$f(\bar{\mathbf{r}}) = f(C'\bar{\bar{\mathbf{r}}}/C) > f(\bar{\bar{\mathbf{r}}}) = K,$$

because of the form of f and the fact that $C'/C > 1$ and $\sum_{i=2}^{p} \bar{\bar{r}}_i = C$ [note that $\gamma_i^{-1}(1 - \rho^2) - 1 = -\sum_{j>1} \delta_j^2/\gamma_i$ and that $\gamma_i > 0$]. This and a similar argument for the case $C' < C$ show that (8.70) implies (8.71).

Using the same argument as in the case of the T^2-test, we can similarly show that the value of K which satisfies (8.71) is given by

$$K = (1 - \delta^2)^{N/2} F(\tfrac{1}{2}N, \tfrac{1}{2}N; \tfrac{1}{2}(p - 1); C\delta^2), \tag{8.72}$$

where $F(a, b; c; x)$ is the ordinary $({}_2F_1)$ hypergeometric series, given by

$$F(a, b; c; x) = \sum_{\alpha=0}^{\infty} \frac{x^\alpha \Gamma(a + \alpha)\Gamma(b + \alpha)\Gamma(c)}{\alpha!\Gamma(a)\Gamma(b)\Gamma(c + \alpha)}. \tag{8.73}$$

Giri and Kiefer (1964b) considered the case $p = 3$, $N = 3$ (or $N = 4$ if $\boldsymbol{\mu}$ is unknown). Proceeding exactly the same way as in the T^2-test they showed that there exists a probability measure λ whose derivative is given by

$$\begin{aligned} m_z(x) = \frac{(1 - zx)^{1/2}}{2\pi x^{1/2}(1 - x)^{1/2}} \Bigg\{ & B_z \int_0^x \frac{du}{(1 - u)(1 - zu)^{3/2}} \\ & + \int_0^\infty \Bigg[\frac{B_z u^{1/2}}{(1 + u)(z + u)^{3/2}} + \frac{1}{[u(1 + u)(z + u)]^{1/2}} \\ & - 2\frac{u^{1/2}}{(1 + u)^{1/2}(z + u)^{3/2}} \Bigg] du \Bigg\} \end{aligned} \tag{8.74}$$

where $z = c\delta^2$, $B_z = (1 - z)^{5/2} F(\tfrac{3}{2}, \tfrac{3}{2}; 1; z)$. The reader is referred to the original references for details of the proof of (8.74) and the other results that follow in this section. Taking (8.74) for granted we have proved the following theorem.

THEOREM 8.3.7 For testing $H_0 : \rho^2 = 0$ against the alternatives $H_1 : \rho^2 > 0$, the R^2-test is minimax for the case $p = 3$, $N = 3$ (or $N = 4$ if $\boldsymbol{\mu}$ is unknown).

Let us examine the local minimax property of the R^2-test in the sense of Giri and Kiefer (1964a) as outlined in Chapter VII. We shall be interested in testing at level α the hypothesis $H_0 : \rho^2 = 0$ against the alternatives $H_1 : \rho^2 > \lambda$, as $\lambda \to 0$.

Let

$$\eta_i = \delta_i^2/\delta^2, \qquad \boldsymbol{\eta} = (\eta_2, \ldots, \eta_p)', \qquad \delta^2 > 0.$$

From (8.69), as $\lambda \to 0$,

$$\frac{f_{\mathbf{R}}(\mathbf{r}|\lambda, \boldsymbol{\eta})}{f_{\mathbf{R}}(\mathbf{r}|0, \boldsymbol{\eta})} = 1 + \tfrac{1}{2}N\lambda\left\{-1 + \sum_{j=2}^{p} r_j\left[\sum_{i>j} \eta_j + (N - j + 2)\eta_j\right]\right\}$$
$$+ B(\mathbf{r}, \boldsymbol{\eta}, \lambda), \tag{8.75}$$

where $B(\mathbf{r}, \boldsymbol{\eta}, \lambda) = o(\lambda)$ uniformly in $\boldsymbol{\eta}$ and $\mathbf{r}$. As in the case of the T^2-test (Chapter VII) we see that the assumptions of Theorem 7.2.4 are again satisfied with $U = \sum_{i=2}^{p} R_i = R^2$ with $h(\lambda) = b\lambda$, and $\xi_{1,\lambda}$ assigns measure 1 to the point $\boldsymbol{\eta}$ whose jth coordinate $(2 \le j \le p)$ is

$$(N - j + 1)^{-1}(N - j + 2)^{-1}(p - 1)^{-1}N(N - p + 1).$$

Hence we have the following theorem.

THEOREM 8.3.8 For every p, N, and α, the rejection region of the R^2-test is locally minimax for testing $H_0\colon\rho^2 = 0$ against $H_1\colon\rho^2 = \lambda$ as $\lambda \to 0$.

The asymptotic minimax property of the T^2-test (Chapter VII) is obviously related to the underlying exponential structure which yields it to the Stein (1956) admissibility result. It is interesting to note that the same departure from this structure (in behavior as $\rho^2 \to 1$) which prevents Stein's method from proving the admissibility of the R^2-test, also prevents us from applying the asymptotic (as $\rho^2 \to 1$) minimax theory in the R^2-test.

8.4 MULTIVARIATE GENERAL LINEAR HYPOTHESIS

In this section we generalize the univariate general linear hypothesis and analysis of variance with fixed effect model to vector variates. The algebra is essentially the same as that of the univariate case. Unlike the univariate general linear hypothesis, there is more latitude in the choice of the test criteria in the multivariate case, although the distributions of different test criteria are quite involved. The reader is referred to Giri (1975) for a treatment of the univariate general linear hypothesis, which is very appropriate for following the developments here, to Roy (1953, 1957) for the union–intersection approach for obtaining a suitable test criterion which is also appropriate for this problem, and to Constantine (1963) for some connected distribution results. We shall first state and solve the problem in the most general form and then give the formulation of the multivariate general linear hypothesis in terms of multiple regression. The latter formulation is useful for analyzing multivariate design models.

Let $\mathbf{X}^\alpha = (X_{\alpha 1}, \ldots, X_{\alpha p})'$, $\alpha = 1, \ldots, N$, be N independently distributed p-variate normal vectors with mean $E(\mathbf{X}^\alpha) = \boldsymbol{\mu}^\alpha = (\mu_{\alpha 1}, \ldots, \mu_{\alpha p})'$ and a common positive definite covariance matrix Σ. A multivariate linear hypothesis is defined in terms of two linear subspaces π_Ω, π_ω of dimensions $s\,(<N)$, $s - r\;(0 \le s - r < s)$, respectively. It is assumed throughout that all vectors $(\mu_{1i}, \ldots, \mu_{Ni})'$, $i = 1, \ldots, p$, lie in π_Ω, and it is desired to test the null hypothesis H_0 that they lie in π_ω. We shall also assume that $N - s \ge p$ so that we have enough degrees of freedom to estimate Σ.

EXAMPLE 8.4.1 Let $N = N_1 + N_2$ and let $\mathbf{X}^\alpha$, $\alpha = 1, \ldots, N_1$, be a random sample of size N_1 from a p-variate normal population with mean $\boldsymbol{\mu}^1 = (\mu_{11}, \ldots, \mu_{1p})'$ and covariance matrix Σ (unknown). Let $\mathbf{X}^\alpha$, $\alpha = N_1 + 1, \ldots, N$, be a random sample of size N_2 from another p-variate normal population with mean $\boldsymbol{\mu}^2 = (\mu_{21}, \ldots, \mu_{2p})'$ and the same covariance matrix Σ. We are interested in testing the null hypothesis $H_0 : \boldsymbol{\mu}^1 = \boldsymbol{\mu}^2$. Here $s = 2$ and $s - r = 1$.

Let

$$X = \begin{pmatrix} X_{11} & \cdots & X_{1p} \\ X_{21} & \cdots & X_{2p} \\ \vdots & & \vdots \\ X_{N1} & \cdots & X_{Np} \end{pmatrix} = \begin{pmatrix} \mathbf{X}^{1'} \\ \mathbf{X}^{2'} \\ \vdots \\ \mathbf{X}^{N'} \end{pmatrix}, \qquad \Sigma = (\sigma_{ij}). \tag{8.76}$$

This problem can be reduced to a canonical form by applying to each of the N vectors $(X_{1i}, \ldots, X_{Ni})'$, $i = 1, \ldots, p$, an orthogonal transformation which transforms X to $Y = OX$ where O is an $N \times N$ orthogonal matrix

$$O = \begin{pmatrix} O_{11} & \cdots & O_{1N} \\ O_{21} & \cdots & O_{2N} \\ \vdots & & \vdots \\ O_{N1} & \cdots & O_{NN} \end{pmatrix} \tag{8.77}$$

such that its first s row vectors $\mathbf{O}_1, \ldots, \mathbf{O}_s$ span π_Ω with $\mathbf{O}_{r+1}, \ldots, \mathbf{O}_s$ spanning π_ω. Write

$$Y = \begin{pmatrix} \mathbf{Y}^{1'} \\ \mathbf{Y}^{2'} \\ \vdots \\ \mathbf{Y}^{N'} \end{pmatrix}, \qquad \mathbf{Y}^\alpha = (Y_{\alpha 1}, \ldots, Y_{\alpha p})', \qquad \alpha = 1, \ldots, N. \tag{8.78}$$

Thus $E(\mathbf{Y}^\alpha) = \mathbf{0}$ for $\alpha = s + 1, \ldots, N$ if and only if all $(\mu_{1i}, \ldots, \mu_{Ni})' \in \pi_\Omega$, $i = 1, \ldots, p$; and $E(\mathbf{Y}^\alpha) = \mathbf{0}$, $\alpha = 1, \ldots, r, s + 1, \ldots, N$, if and only if all $(\mu_{1i}, \ldots, \mu_{N_i})' \in \pi_\omega$, $i = 1, \ldots, p$. Now the covariance of $Y_{\alpha i} = \sum_{\gamma=1}^N O_{\alpha\gamma} X_{\gamma i}$,

$Y_{\beta j} = \sum_{\delta=1}^{N} O_{\beta\delta} X_{\delta j}$ is

$$\operatorname{cov}(Y_{\alpha i}, Y_{\beta j}) = \sum_{\gamma=1}^{N} \sum_{\delta=1}^{N} O_{\alpha\gamma} O_{\beta\delta} \operatorname{cov}(X_{\gamma i} X_{\delta j}) = \sigma_{ij} \sum_{\gamma=1}^{N} O_{\alpha\gamma} O_{\beta\gamma}$$

$$= \begin{cases} \sigma_{ij} & \text{when} \quad \alpha = \beta \\ 0 & \text{when} \quad \alpha \neq \beta, \end{cases}$$

since $\operatorname{cov}(X_{\gamma i}, X_{\delta j}) = \sigma_{ij}$ when $\gamma = \delta$, $\operatorname{cov}(X_{\gamma i}, X_{\delta j}) = 0$ when $\gamma \neq \delta$. Thus the row vectors of Y are independent normal p-vectors with the same covariance matrix Σ and under π_Ω,

$$E(\mathbf{Y}^\alpha) = \begin{cases} \mathbf{v}^\alpha \quad \text{(say)}, & \alpha = 1, \ldots, s, \\ \mathbf{0}, & \alpha = s+1, \ldots, N, \end{cases}$$

and under π_ω,

$$E(\mathbf{Y}^\alpha) = \begin{cases} \mathbf{v}^\alpha, & \alpha = r+1, \ldots, s, \\ \mathbf{0}, & \alpha = 1, \ldots, r, s+1, \ldots, N. \end{cases}$$

Hence in the canonical form we have the following problem: $\mathbf{Y}^\alpha, \alpha = 1, \ldots, N$, are independently distributed normal p-vectors with the same positive definite covariance matrix Σ (unknown) and the means $E(\mathbf{Y}^\alpha) = \mathbf{0}$, $\alpha = s+1, \ldots, N$. It is desired to test the null hypothesis $H_0: E(\mathbf{Y}^\alpha) = \mathbf{0}$, $\alpha = 1, \ldots, r$.

The likelihood of the observations $\mathbf{y}^\alpha$ on $\mathbf{Y}^\alpha$, $\alpha = 1, \ldots, N$, is given by

$$L(\mathbf{v}^1, \ldots, \mathbf{v}^s, \Sigma) = (2\pi)^{-Np/2} (\det \Sigma^{-1})^{N/2} \times \exp\left\{-\tfrac{1}{2} \operatorname{tr} \Sigma^{-1}\left[\sum_{\alpha=1}^{s} (\mathbf{y}^\alpha - \mathbf{v}^\alpha)(\mathbf{y}^\alpha - \mathbf{v}^\alpha)' + \sum_{\alpha=s+1}^{N} \mathbf{y}^\alpha \mathbf{y}^{\alpha'}\right]\right\}. \tag{8.79}$$

Using Lemma 5.1.1 we obtain

$$\max_{\pi_\Omega} L(\mathbf{v}^1, \ldots, \mathbf{v}^s, \Sigma) = (2\pi/N)^{-Np/2} \left[\det\left(\textstyle\sum_{\alpha=s+1}^{N} \mathbf{y}^\alpha \mathbf{y}^{\alpha'}\right)\right]^{-N/2} \exp\{-\tfrac{1}{2}Np\}. \tag{8.80}$$

Under H_0, L is reduced to

$$L(\mathbf{v}^{r+1}, \ldots, \mathbf{v}^s, \Sigma) = (2\pi)^{-Np/2} (\det \Sigma^{-1})^{N/2} \times \exp\left\{-\tfrac{1}{2} \operatorname{tr} \Sigma^{-1}\left[\sum_{\alpha=1}^{r} \mathbf{y}^\alpha \mathbf{y}^{\alpha'} + \sum_{\alpha=r+1}^{s} (\mathbf{y}^\alpha - \mathbf{v}^\alpha)(\mathbf{y}^\alpha - \mathbf{v}^\alpha)' + \sum_{\alpha=s+1}^{N} \mathbf{y}^\alpha \mathbf{y}^{\alpha'}\right]\right\} \tag{8.81}$$

and

$$\max_{\pi_\omega} L(\mathbf{v}^{r+1}, \ldots, \mathbf{v}^s, \Sigma)$$
$$= \left(\frac{2\pi}{N}\right)^{-Np/2} \left[\det\left(\sum_{\alpha=1}^{r} \mathbf{y}^\alpha \mathbf{y}^{\alpha'} + \sum_{\alpha=s+1}^{N} \mathbf{y}^\alpha \mathbf{y}^{\alpha'}\right)\right]^{-N/2} \exp\{-\tfrac{1}{2}Np\}. \quad (8.82)$$

Hence the likelihood ratio test of H_0 rejects H_0 whenever

$$\lambda = \left[\frac{\det b}{\det(a+b)}\right]^{N/2} \leq c \quad (8.83)$$

or, equivalently,

$$u = \frac{\det b}{\det(a+b)} \leq c', \quad (8.84)$$

where c, c' are constants chosen in such a way that the corresponding test has size α and $a = \sum_{\alpha=1}^{r} \mathbf{y}^\alpha \mathbf{y}^{\alpha'}$, $b = \sum_{\alpha=s+1}^{N} \mathbf{y}^\alpha \mathbf{y}^{\alpha'}$. This result is due to Hsu (1941) Wilks (1932). From Sections 6.3 and 6.5 we conclude that the corresponding random variables $A = \sum_{\alpha=1}^{r} \mathbf{Y}^\alpha \mathbf{Y}^{\alpha'}$, $B = \sum_{\alpha=s+1}^{N} \mathbf{Y}^\alpha \mathbf{Y}^{\alpha'}$ are independently distributed Wishart matrices of dimension $p \times p$, and B has a central Wishart distribution with parameter Σ and $N - s$ degrees of freedom. Under H_0, A is distributed as central Wishart with parameter Σ and r degrees of freedom whereas under H_1 it is distributed as noncentral Wishart.

In application to specific problems it is not straightforward to carry out the reduction to the canonical form just given explicitly. The test statistic u can be expressed in terms of the original random variables X. Let $(\hat{\mu}_{1i}, \ldots, \hat{\mu}_{Ni})'$ and $(\hat{\hat{\mu}}_{1i}, \ldots, \hat{\hat{\mu}}_{Ni})'$ be the projections of the vector $(X_{1i}, \ldots, X_{Ni})'$ on π_Ω and π_ω, respectively. Then $\sum_{\alpha=1}^{N}(X_{\alpha i} - \hat{\mu}_{\alpha i})(X_{\alpha i} - \hat{\mu}_{\alpha i})$ is the inner product of two vectors, each of which is the difference of the given vector $(X_{1i}, \ldots, X_{Ni})'$ and its projection on π_Ω, and it remains unchanged under the orthogonal transformation of the coordinate system in which the variables are expressed. Now $O(X_{1i}, \ldots, X_{Ni})'$ can be interpreted as expressing $(X_{1i}, \ldots, X_{Ni})'$ in a new coordinate system with the first s coordinate axes lying in π_Ω. Hence the projection on π_Ω of the transformed vector $(Y_{1i}, \ldots, Y_{Ni})'$ is $(Y_{1i}, \ldots, Y_{si}, 0, \ldots, 0)'$ so that the difference between the vector and its projection is $(0, \ldots, 0, Y_{s+1,i}, \ldots, Y_{Ni})$. The (i, j)th element of $\sum_{\alpha=s+1}^{N} \mathbf{Y}^\alpha \mathbf{Y}^{\alpha'}$ is therefore given by

$$\sum_{\alpha=s+1}^{N} Y_{\alpha i} Y_{\alpha j} = \sum_{\alpha=1}^{N} (X_{\alpha i} - \hat{\mu}_{\alpha i})(X_{\alpha j} - \hat{\mu}_{\alpha j}). \quad (8.85)$$

Similarly, for the transformed vector $(Y_{1i}, \ldots, Y_{Ni})'$ the difference between its projections on π_Ω and π_ω is therefore $(Y_{1i}, \ldots, Y_{ri}, 0, \ldots, 0)'$. Thus

$\sum_{\alpha=1}^{r} Y_{\alpha i} Y_{\alpha j}$ is equal to the inner product (for the ith and the jth vectors) of the difference of these projections. Comparing this with the expression of the same inner product in the original coordinate system, we obtain

$$\sum_{\alpha=1}^{r} Y_{\alpha i} Y_{\alpha j} = \sum_{\alpha=1}^{N} (\hat{\mu}_{\alpha i} - \hat{\hat{\mu}}_{\alpha i})(\hat{\mu}_{\alpha j} - \hat{\hat{\mu}}_{\alpha j}). \tag{8.86}$$

In terms of the transformed variable Y the problem of testing H_0 against $H_1 : \Lambda \neq 0$, where $\Lambda = (\mathbf{v}^1, \ldots, \mathbf{v}^r)'$, remains invariant under the following three groups of transformations.

(1) The group of translations T which translates $\mathbf{Y}^\alpha \to \mathbf{Y}^\alpha + \mathbf{d}^\alpha$, $\alpha = r+1, \ldots, s$, and $\mathbf{d}^\alpha = (d_{\alpha 1}, \ldots, d_{\alpha p})' \in T$. The maximal invariant under T in the space of Y is $(\mathbf{Y}^1, \ldots, \mathbf{Y}^r, \mathbf{Y}^{s+1}, \ldots, \mathbf{Y}^N)$.

(2) Let Z be an $r \times p$ matrix such that $Z' = (\mathbf{Y}^1, \ldots, \mathbf{Y}^r)$, and let W be the $(N-s) \times p$ matrix such that $W' = (\mathbf{Y}^{s+1}, \ldots, \mathbf{Y}^N)$. The group of $r \times r$ orthogonal transformations $O(r)$ operating in the space of Z as $Z \to OZ$, $O \in O(r)$, and the group of $(N-s) \times (N-s)$ orthogonal transformations $O(N-s)$ operating in the space of W as $W \to OW, O \in O(N-s)$, affect neither the independence nor the covariance matrix of the row vectors of Z and W.

LEMMA 8.4.1 $Z'Z = \sum_{\alpha=1}^{r} \mathbf{Y}^\alpha \mathbf{Y}^{\alpha'}$ is a maximal invariant under $O(r)$ in the space of Z.

Proof Since $(OZ)'(OZ) = Z'Z$, the matrix $Z'Z$ will be a maximal invariant if we show that for any two elements Z^*, Z in the same space, $Z^{*\prime}Z^* = Z'Z$ implies the existence of an orthogonal matrix $O \in O(r)$ such that $Z^* = OZ$. Consider first the case $r = p$. Without any loss of generality we can assume that the p columns of Z are linearly independent (the exceptional set of Z's for which this does not hold has probability measure 0). Now $Z^{*\prime}Z^* = Z'Z$ implies that $O = Z^*Z^{-1}$ is an orthogonal matrix and that $Z^* = OZ$. Consider now the case $r > p$. Without any loss of generality we can assume that the columns of Z are linearly independent. Since for any two p-dimensional subspaces of the r-space there exists an orthogonal transformation transforming one to the other, we assume that after a suitable orthogonal transformation the p column vectors of Z and Z^* lie in the same subspace and the problem is reduced to the case $r = p$.

If $r < p$, the first r column vectors of Z can be assumed to be linearly independent. Write $Z = (Z_1, Z_2)$, where Z_1, Z_2 are submatrices of dimensions $r \times r$ and $r \times (p - r)$, respectively, and similarly for Z^*. Since $Z^{*\prime}Z^* = Z'Z$, we obtain

$$Z_1^{*\prime} Z_1^* = Z_1' Z_1, \qquad Z_1^{*\prime} Z_2^* = Z_1' Z_2, \qquad \text{and} \qquad Z_2^{*\prime} Z_2^* = Z_2' Z_2.$$

Now by the previous argument $Z_1^{*\prime}Z_1^* = Z_1'Z_1$ implies that there exists an orthogonal matrix $B = (Z_1^{*\prime})^{-1}Z_1'$ such that $Z_1^* = BZ_1$. Also $Z_1^{*\prime}Z_2^* = Z_1'Z_2$ implies that $Z_2^* = BZ_2$. Obviously $Z_2^{*\prime}Z_2^* = Z_2'Z_2$ with $Z_2^* = BZ_2$. Q.E.D.

Similarly a maximal invariant in the space of W under $O(N - s)$ is $W'W = \sum_{\alpha=s+1}^{N} \mathbf{Y}^\alpha \mathbf{Y}^{\alpha\prime}$.

Note that as $N - s \geq p$, by Lemma 5.1.2, $W'W$ is positive definite with probability 1.

(3) The problem remains invariant under the full linear group $G_l(p)$ (multiplicative group of $p \times p$ nonsingular matrices) of transformation g transforming Z to gZ, W to gW. The corresponding induced transformation in the space of (A, B) is given by $(A, B) \to (gAg', gBg')$. By Exercise 7 the roots of $\det(A - \lambda B) = 0$ (the characteristic roots of AB^{-1}) are maximal invariant in the space of (A, B) under $G_l(p)$. Let $R_1, \ldots, R_p$ denote the roots of $\det(A - \lambda B) = 0$. A corresponding maximal invariant in the parametric space is $(\theta_1, \ldots, \theta_p)$, the characteristic roots of $\Lambda\Lambda'\Sigma^{-1}$ where $\Lambda = E(Z')$. The test statistic U in (8.84) can be written as

$$\det(B(A + B)^{-1}) = \prod_{i=1}^{p} (1 + R_i)^{-1}. \tag{8.87}$$

Anderson (1958) called this statistic $U_{p,r,N-s}$. Some other invariant tests are also proposed for this problem. They are as follows. In all cases the constant c will depend on the level of significance α of the test.

1. Wilks' criterion (Wilks, 1932; Hsu, 1940):

$$\text{Reject } H_0 \text{ whenever } \det a(b + a)^{-1} \leq c. \tag{8.88}$$

2. Lawley's v (Lawley, 1938) and Hotelling's T_0^2 (Hotelling, 1951) criterion:

$$\text{Reject } H_0 \text{ whenever } v = \operatorname{tr} ab^{-1} = \frac{T_0^2}{N - p} \geq c. \tag{8.89}$$

3. The largest and the smallest root criteria of Roy (Roy, 1957):

$$\begin{aligned} &\text{Reject } H_0 \text{ whenever } \max_i r_i \geq c \\ &\text{Reject } H_0 \text{ whenever } \min_i r_i \geq c. \end{aligned} \tag{8.90}$$

4. Pillai's statistic (Pillai, 1955):

$$\text{Reject } H_0 \text{ whenever } \operatorname{tr} a(a + b)^{-1} \geq c. \tag{8.91}$$

These test statistics are functions of $R_1, \ldots, R_p$. Among these invariant tests, test 4 has received much less attention than the others. These tests 1–4, of course, reduce to Hotelling's T^2-test when $r = 1$, and if $r > 1$ and $\min(p, r) > 1$, there does not exist a uniformly most powerful invariant test. All these tests reduce to the univariate F-test when $p = 1$ and to the two-tailed t-test when $p = r = 1$. In theory, we would be able to derive the distribution of these statistics from the joint distribution of $R_1, \ldots, R_p$. Since for any $g \in G_l(p)$, $\det(gAg' - \lambda gBg') = \det(gg')\det(A - \lambda B)$, choosing g such that $g\Sigma g' = I$, we conclude that to find the joint distribution of $(R_1, \ldots, R_p)$ under H_0, we can without any loss of generality assume that $\Sigma = I$. In other words, the joint distribution is independent of Σ under H_0 and under $H_1: \Lambda \neq 0$ this distribution depends only on $\theta_1, \ldots, \theta_p$.

8.4.1 Distribution of $(R_1, \ldots, R_p)$ under H_0

From Section 6.3, B and A are independently distributed (Wishart matrices) as $W_p(\Sigma, N - s)$ and $W_p(\Sigma, r)$, respectively, provided $N - s \geq p$, $r \geq p$. Let $N - s = n_2$, $r = n_1$, let $R_1, \ldots, R_p$ be the characteristic roots of AB^{-1}, and let $R_1 > R_2 > \cdots > R_p > 0$ denote the ordered characteristic roots of AB^{-1} (the probability of two roots being equal is 0). Rather than finding the distribution of $(R_1, \ldots, R_p)$ directly, we will find it convenient to first find the joint distribution of $V_1, \ldots, V_p$ such that $V_i = R_i/(1 + R_i)$, $i = 1, \ldots, p$. Obviously $V_1, \ldots, V_p$ are the characteristic roots of $A(A + B)^{-1}$, that is, the roots of $\det(A - \lambda(A + B)) = 0$. Let V be a diagonal matrix with diagonal elements $V_1, \ldots, V_p$ and let $C = A + B$.

We can write

$$C = WW', \qquad A = WVW' \tag{8.92}$$

where $W = (W_{ij})$ is a nonsingular matrix of dimension $p \times p$. To determine W uniquely we require here that $W_{i1} \geq 0$, $i = 1, \ldots, p$ (the probability of $W_{i1} = 0$ is 0). Writing J for Jacobian, the Jacobian of the transformation $(A, B) \to (W, V)$ is equal to

$$\mathrm{J}[(A, B) \to (W, V)] = \mathrm{J}[(A, B) \to (A, C)] \times \mathrm{J}[(A, C) \to (W, V)]. \tag{8.93}$$

It is easily seen that $\mathrm{J}[(A, B) \to (A, C)] = 1$. By Exercise 8 [see also Olkin (1952)] the Jacobian of the transformation $(A, C) \to (W, V)$ is

$$2^p(\det W)^{p+2} \prod_{i<j} (V_i - V_j). \tag{8.94}$$

As indicated earlier we can take $\Sigma = I$, and hence the joint probability density function of A, B is (by Section 6.3)

$$C_{n_1,p} C_{n_2,p} (\det a)^{(n_1 - p - 1)/2} (\det b)^{(n_2 - p - 1)/2} \exp\{-\tfrac{1}{2}\operatorname{tr}(a + b)\}, \tag{8.95}$$

where $C_{n,p}$ is given by (6.32). From (8.93)–(8.95) the joint probability density function of (W, V) is

$$f_{W,V}(w, v) = C_{n_1,p}C_{n_2,p} \prod_{i=1}^{p} [v_i^{(n_1-p-1)/2}(1 - v_i)^{(n_2-p-1)/2}] \times \prod_{i<j} (v_i - v_j)(\det(ww'))^{(n_1+n_2-p)/2} \exp\{-\tfrac{1}{2} \operatorname{tr} ww'\}. \quad (8.96)$$

Now integrating out w in (8.96), we obtain the probability density function of V as

$$f_V(v) = KC_{n_1,p}C_{n_2,p} \prod_{i=1}^{p} v_i^{(n_1-p-1)/2}(1 - v_i)^{(n_2-p-1)/2} \prod_{i<j} (v_i - v_j), \quad (8.97)$$

where

$$K = (2\pi)^{p^2/2} \int \frac{1}{(2\pi)^{p^2/2}} 2^p[\det(ww')]^{(n_1+n_2-p)/2} \exp\{-\tfrac{1}{2}\operatorname{tr} ww'\}\, dw$$
$$= (2\pi)^{p^2/2} E[\det(WW')]^{(n_1+n_2-p)/2}$$

and $W = (W_{ij})$, the W_{ij} are independently distributed normal random variables with mean 0 and variance 1. Thus the $p \times p$ matrix $S = WW'$ is distributed as $W_p(I, p)$ and its probability density function [by (6.32)] is

$$f_S(s) = C_{p,p}(\det s)^{-1/2} \exp\{-\tfrac{1}{2} \operatorname{tr} s\}.$$

Hence

$$E(\det(WW'))^{(n_1+n_2-p)/2} = C_{p,p} \int (\det s)^{(n_1+n_2-p-1)/2} \exp\{-\tfrac{1}{2} \operatorname{tr} s\}\, ds = \frac{C_{p,p}}{C_{n_1+n_2,p}}.$$

Thus

$$K = \frac{(2\pi)^{p^2/2} C_{p,p}}{C_{n_1+n_2,p}}. \quad (8.98)$$

Since $dV_i = (1 + R_i^2)^{-1}\, dR_i$, from (8.97) the probability density of R, a diagonal matrix with diagonal elements $R_1, \ldots, R_p$, is

$$f_R(r) = C \prod_{i=1}^{p} r_i^{(n_1-p-1)/2}(1 + r_i)^{-(n_1+n_2)/2} \prod_{i<j} (r_i - r_j) \quad (8.99)$$

where

$$C = \frac{\pi^{p/2} \prod_{i=1}^{p} \Gamma(\tfrac{1}{2}(n_1 + n_2 - i + 1))}{\prod_{i=1}^{p} \Gamma(\tfrac{1}{2}(n_1 - i + 1))\Gamma(\tfrac{1}{2}(n_2 - i + 1))\Gamma(\tfrac{1}{2}(p + 1 - i))}.$$

Let us now consider the distribution of the characteristic roots of A where A is distributed as $W_p(I, n_1)$. Since B is distributed as $W_p(I, n_2)$, $B/n_2 \to I$ almost surely as $n_2 \to \infty$. Thus the roots of the equation $\det(A - \lambda(B/n_2)) = 0$ converge almost surely to the roots of $\det(A - \lambda I) = 0$. Let $\lambda_1 > \lambda_2 > \cdots > \lambda_p > 0$ be the ordered characteristic roots of A. To find the joint distribution of the λ_i, it is sufficient to find the limit as $n_2 \to \infty$ of the probability density function of the roots of $\det(A - \lambda(B/n_2)) = 0$. From (8.99), the probability density function of the roots $(\lambda_1, \ldots, \lambda_p)$ of $\det(A - \lambda(B/n_2)) = 0$ is given by

$$C(n_2)^{-n_1p/2} \prod_{i=1}^{p} \lambda_i^{(n_1-p-1)/2} \left(1 + \frac{\lambda_i}{n_2}\right)^{-(n_1+n_2)/2} \prod_{i<j} (\lambda_i - \lambda_j). \tag{8.100}$$

Since

$$\operatorname*{Lt}_{n_2 \to \infty} \prod_{i=1}^{p} \left(1 + \frac{\lambda_i}{n_2}\right)^{-(n_1+n_2)/2} = \exp\left\{-\frac{1}{2} \sum_{i=1}^{p} \lambda_i\right\},$$

$$\operatorname*{Lt}_{n_2 \to \infty} \frac{\Gamma(\frac{1}{2}(n_1 + n_2 - 1))}{(n_2)^{n_1/2}\Gamma(\frac{1}{2}(n_2 - j))} = 2^{-n_1/2},$$

we get

$$\begin{aligned}\operatorname*{Lt}_{n_2 \to \infty} C(n_2)^{-n_1p/2} &= \pi^{-p/2}\left[2^{n_1p/2} \prod_{i=1}^{p} \Gamma(\tfrac{1}{2}(n_1 - i + 1))\Gamma(\tfrac{1}{2}(p + 1 - i))\right]^{-1} \\ &= C' \qquad \text{(say)}.\end{aligned} \tag{8.101}$$

Thus the probability density function of the ordered characteristic roots $\lambda_1, \ldots, \lambda_p$ of A is (with λ a diagonal matrix with diagonal elements $\lambda_1, \ldots, \lambda_p$)

$$f_\lambda(\lambda) = C' \prod_{i=1}^{p} \lambda_i^{(n_1-p-1)/2} \exp\left\{-\frac{1}{2} \sum_{i=1}^{p} \lambda_i\right\} \prod_{i<j} (\lambda_i - \lambda_j). \tag{8.102}$$

8.4.2 Multivariate Regression Model

We now discuss a different formulation of the multivariate general linear hypothesis which is very appropriate for the analysis of design models. Let $\mathbf{X}^\alpha = (X_{\alpha 1}, \ldots, X_{\alpha p})'$, $\alpha = 1, \ldots, N$, be independently distributed normal p-vectors with means

$$E(\mathbf{X}^\alpha) = \beta z^\alpha, \qquad \alpha = 1, \ldots, N, \tag{8.103}$$

where $\mathbf{z}^\alpha = (z_{\alpha 1}, \ldots, z_{\alpha s})'$, $\alpha = 1, \ldots, N$, are known vectors and $\beta = (\beta_{ij})$ is a $p \times s$ matrix of unknown elements β_{ij}. As in the general formulation we shall assume that $N - s \geq p$, and that the rank of the $s \times N$ matrix $Z = (\mathbf{z}^1, \ldots, \mathbf{z}^N)$ is s. Let $\beta = (\beta_1, \beta_2)$, where β_1, β_2 are submatrices of dimensions $p \times r$ and $p \times (s - r)$, respectively. We are interested in testing the null hypothesis

$$H_0: \beta_1 = \beta_1{}^0 \qquad \text{(a fixed matrix)}$$

where β_2 and Σ are unknown. Here the dimension of π_Ω is s and that of π_ω is $s - r$. The likelihood of the sample observations $\mathbf{x}^\alpha$ on $\mathbf{X}^\alpha$, $\alpha = 1, \ldots, N$, is given by

$$L(\beta, \Sigma) = (2\pi)^{-Np/2} (\det \Sigma^{-1})^{N/2} \times \exp\left\{-\tfrac{1}{2} \operatorname{tr} \Sigma^{-1} \left(\sum_{\alpha=1}^{N} (\mathbf{x}^\alpha - \beta \mathbf{z}^\alpha)(\mathbf{x}^\alpha - \beta \mathbf{z}^\alpha)' \right)\right\}. \tag{8.104}$$

Let

$$A = ZZ' = \sum_{\alpha=1}^{N} \mathbf{z}^\alpha \mathbf{z}^{\alpha'}, \qquad C = xZ' = \sum_{\alpha=1}^{N} \mathbf{x}^\alpha \mathbf{z}^{\alpha'}, \qquad x = (\mathbf{x}^1, \ldots, \mathbf{x}^N).$$

Using Section 1.7, the maximum likelihood estimate $\hat{\beta}$ of β is given by

$$\hat{\beta} A = C. \tag{8.105}$$

Since the rank of Z is s, A is nonsingular and the unique maximum likelihood estimate of β is given by

$$\hat{\beta} = CA^{-1}. \tag{8.106}$$

Now using Lemma 5.1.1, the maximum likelihood estimate of Σ under π_Ω is

$$\hat{\Sigma} = \frac{1}{N} \sum_{\alpha=1}^{N} (\mathbf{x}^\alpha - \hat{\beta} \mathbf{z}^\alpha)(\mathbf{x}^\alpha - \hat{\beta} \mathbf{z}^\alpha)' = \frac{1}{N} \left(\sum_{\alpha=1}^{N} \mathbf{x}^\alpha \mathbf{x}^{\alpha'} - \hat{\beta} A \hat{\beta}' \right). \tag{8.107}$$

Thus

$$\max_{\beta, \Sigma} L(\beta, \Sigma) = (2\pi)^{-Np/2} \left\{ \det \left[\sum_{\alpha=1}^{N} \frac{(\mathbf{x}^\alpha - \hat{\beta} \mathbf{z}^\alpha)(\mathbf{x}^\alpha - \hat{\beta} \mathbf{z}^\alpha)'}{N} \right] \right\}^{-N/2} \times \exp\{-\tfrac{1}{2} Np\}. \tag{8.108}$$

To find the maximum of the likelihood function under H_0, let

$$\mathbf{z}^\alpha = \begin{pmatrix} z^\alpha_{(1)} \\ z^\alpha_{(2)} \end{pmatrix}, \qquad \mathbf{y}^\alpha = \mathbf{x}^\alpha - \beta_1{}^0 z^\alpha_{(1)}, \qquad \alpha = 1, \ldots, N,$$

where $z^\alpha_{(1)} = (z_{\alpha 1}, \ldots, z_{\alpha r})'$. Now $\mathbf{Y}^\alpha = \mathbf{X}^\alpha - \beta_1{}^0 z^\alpha_{(1)}$, $\alpha = 1, \ldots, N$, are independently normally distributed with mean $\beta_2 z^\alpha_{(2)}$ and the same covariance matrix Σ. Let $C = (C_1, C_2)$ with C_1 a $p \times r$ submatrix and

$$Z = \begin{pmatrix} Z_1 \\ Z_2 \end{pmatrix}, \qquad A = \begin{pmatrix} A_{(11)} & A_{(12)} \\ A_{(21)} & A_{(22)} \end{pmatrix}, \qquad \hat{\beta} = (\hat{\beta}_1, \hat{\beta}_2)$$

where Z_1 is $r \times N$, $A_{(11)}$ is $r \times r$, and $\hat{\beta}_1$ is $p \times r$.

Under H_0, the likelihood function can be written as

$$L(\beta_2, \Sigma) = (2\pi)^{-Np/2}(\det \Sigma^{-1})^{N/2} \times \exp\left\{-\tfrac{1}{2} \operatorname{tr} \Sigma^{-1}\left[\sum_{\alpha=1}^{N} (\mathbf{y}^\alpha - \beta_2 z_{(2)}^\alpha)(\mathbf{y}^\alpha - \beta_2 z_{(2)}^\alpha)'\right]\right\}. \quad (8.109)$$

Proceeding exactly in the same way as above we obtain the maximum likelihood estimates of β_2 and Σ under H_0 as

$$\hat{\hat{\beta}}_2 = \sum_{\alpha=1}^{N} \mathbf{y}^\alpha z_{(2)}^{\alpha'}\left(\sum_{\alpha=1}^{N} z_{(2)}^\alpha z_{(2)}^{\alpha'}\right)^{-1} = (C_2 - \beta_1{}^0 A_{(12)})A_{(22)}^{-1},$$

$$\hat{\hat{\Sigma}} = \frac{1}{N}\sum_{\alpha=1}^{N} (\mathbf{y}^\alpha - \hat{\hat{\beta}}_2 z_{(2)}^\alpha)(\mathbf{y}^\alpha - \hat{\hat{\beta}}_2 z_{(2)}^\alpha)'$$

$$= \frac{1}{N}\sum_{\alpha=1}^{N} (\mathbf{x}^\alpha - \beta_1{}^0 z_{(1)}^\alpha - \hat{\hat{\beta}}_2 z_{(2)}^\alpha)(\mathbf{x}^\alpha - \beta_1{}^0 z_{(1)}^\alpha - \hat{\hat{\beta}}_2 z_{(2)}^\alpha)'.$$

LEMMA 8.4.2

$$N\hat{\hat{\Sigma}} = N\hat{\Sigma} + (\hat{\beta}_1 - \beta_1{}^0)(A_{(11)} - A_{(12)}A_{(22)}^{-1}A_{(21)})(\hat{\beta}_1 - \beta_1{}^0)'.$$

Proof Since $C = \hat{\beta}A$,

$$C_2 = (\hat{\beta}_1, \hat{\beta}_2)\begin{pmatrix} A_{(12)} \\ A_{(22)} \end{pmatrix} = \hat{\beta}_1 A_{(12)} + \hat{\beta}_2 A_{(22)}.$$

Thus

$$\hat{\beta}_2 = C_2 A_{(22)}^{-1} - \hat{\beta}_1 A_{(12)} A_{(22)}^{-1}, \quad \hat{\hat{\beta}}_2 - \hat{\beta}_2 = (\hat{\beta}_1 - \beta_1{}^0)A_{(12)}A_{(22)}^{-1}.$$

Now under H_0

$$\begin{aligned} X - \beta Z &= X - \hat{\beta}Z + (\hat{\beta}_2 - \beta_2)Z_2 + (\hat{\beta}_1 - \beta_1{}^0)Z_1 \\ &= (X - \hat{\beta}Z) + (\hat{\hat{\beta}}_2 - \beta_2)Z_2 - (\hat{\hat{\beta}}_2 - \hat{\beta}_2)Z_2 + (\hat{\beta}_1 - \beta_1{}^0)Z_1 \\ &= (X - \hat{\beta}Z) + (\hat{\hat{\beta}}_2 - \beta_2)Z_2 + (\hat{\beta}_1 - \beta_1{}^0)(Z_1 - A_{(12)}A_{(22)}^{-1}Z_2). \end{aligned} \quad (8.110)$$

Now

$$(Z_1 - A_{(12)}A_{(22)}^{-1}Z_2)Z_2 = A_{(12)} - A_{(12)} = 0.$$

Since

$$(X - \hat{\beta}Z)Z' = XZ' - XZ'(ZZ')^{-1}ZZ' = XZ' - XZ' = 0,$$

which implies that

$$(X - \hat{\beta}Z)Z_i = 0, \qquad i = 1, 2,$$

we obtain

$$(X - \beta Z)(X - \beta Z)' = (X - \hat{\beta}Z)(X - \hat{\beta}Z)' + (\hat{\hat{\beta}}_2 - \beta_2)A_{(22)}(\hat{\hat{\beta}}_2 - \beta_2)' + (\hat{\beta}_1 - \beta_1{}^0)(A_{(11)} - A_{(12)}A_{(22)}^{-1}A_{(21)})(\hat{\beta}_1 - \beta_1{}^0)'. \quad (8.111)$$

Subtracting $(\hat{\beta}_2 - \beta_2)Z_2$ from both sides of (8.110), we obtain

$$(X - \beta_1{}^0 Z_1 - \hat{\hat{\beta}}_2 Z_2) = (X - \hat{\beta}Z) + (\hat{\beta}_1 - \beta_1{}^0)(Z_1 - A_{(12)}A_{(22)}^{-1}Z_2).$$

Thus

$$N\hat{\hat{\Sigma}} = (X - \hat{\beta}Z)(X - \hat{\beta}Z)' + (\hat{\beta}_1 - \beta_1{}^0)(A_{(11)} - A_{(12)}A_{(22)}^{-1}A_{(21)})(\hat{\beta}_1 - \beta_1{}^0)'. \qquad \text{Q.E.D.}$$

Using this lemma and (8.108)–(8.109), we conclude that the likelihood ratio test of $H_0: \beta_1 = \beta_1{}^0$ when β_2 and Σ are unknown rejects H_0 whenever

$$u = \frac{\det[\sum_{\alpha=1}^N (\mathbf{x}^\alpha - \hat{\beta}\mathbf{z}^\alpha)(\mathbf{x}^\alpha - \hat{\beta}z^\alpha)']}{\det[\sum_{\alpha=1}^N (\mathbf{x}^\alpha - \hat{\beta}\mathbf{z}^\alpha)(\mathbf{x}^\alpha - \hat{\beta}\mathbf{z}^\alpha)' + (\hat{\beta}_1 - \beta_1{}^0)(A_{(11)} - A_{(12)}A_{(22)}^{-1}A_{(21)})(\hat{\beta}_1 - \beta_1{}^0)']} \leq C, \quad (8.112)$$

where the constant C depends on the level of significance α of the test. We shall now show that the statistic U is distributed as the statistic U in (8.84). Wilks (1932) first derived the likelihood ratio test criterion for the special case of testing the equality of mean vectors of several populations. Wilks (1934) and Bartlett (1934) extended its use to regression coefficients.

In what follows we do not distinguish between an estimate and the corresponding estimator. For simplicity we shall use the same notation for both. For the maximum likelihood estimator $\hat{\beta}$

$$E(\hat{\beta}) = E\left(\sum_{\alpha=1}^N \mathbf{X}^\alpha \mathbf{z}^{\alpha\prime} A^{-1}\right) = \beta A A^{-1} = \beta, \quad (8.113)$$

and the covariance between the ith row vector $\hat{\boldsymbol{\beta}}_i$ and the jth row vector $\hat{\boldsymbol{\beta}}_j$ of $\hat{\beta}$ is given by

$$E(\hat{\boldsymbol{\beta}}_i - \boldsymbol{\beta}_i)(\hat{\boldsymbol{\beta}}_j - \boldsymbol{\beta}_j)' = A^{-1} E\left\{\sum_{\alpha=1}^N (X_{\alpha i} - E(X_{\alpha i}))\mathbf{z}^\alpha \left(\sum_{\gamma=1}^N (X_{\gamma i} - E(X_{\gamma i}))\right)\mathbf{z}^{\gamma\prime}\right\} A^{-1}$$

$$= A^{-1} \sum_{\alpha=1}^N \sigma_{ij} \mathbf{z}^\alpha z^{\alpha\prime} A^{-1} = \sigma_{ij} A^{-1}.$$

Obviously, thus, the row vectors $(\hat{\boldsymbol{\beta}}_1, \ldots, \hat{\boldsymbol{\beta}}_p)$ are normally distributed with mean $(\boldsymbol{\beta}_1, \ldots, \boldsymbol{\beta}_p)$ and covariance matrix

$$\begin{pmatrix} \sigma_{11}A^{-1} & \cdots & \sigma_{1p}A^{-1} \\ \sigma_{21}A^{-1} & \cdots & \sigma_{2p}A^{-1} \\ \vdots & & \vdots \\ \sigma_{p1}A^{-1} & \cdots & \sigma_{pp}A^{-1} \end{pmatrix} = \Sigma \otimes A^{-1}. \tag{8.114}$$

THEOREM 8.4.1 $N\hat{\Sigma} = \sum_{\alpha=1}^{N} \mathbf{X}^\alpha \mathbf{X}^{\alpha\prime} - \hat{\beta} A \hat{\beta}'$ is distributed independently of $\hat{\beta}$ as $W_p(\Sigma, N - s)$.

Proof Let F be an $s \times s$ nonsingular matrix such that $FAF' = I$. Let $E_2 = FZ$. Then $E_2 E_2' = FZZ'F' = I$. This implies that the s rows of E_2 are orthogonal and are of unit length. Thus it is possible to find an $(N - s) \times N$ matrix E_1 such that

$$E = \begin{pmatrix} E_1 \\ E_2 \end{pmatrix}$$

is an $N \times N$ orthogonal matrix. Let $Y = (\mathbf{Y}^1, \ldots, \mathbf{Y}^N) = XE'$. Then the columns of Y are independently distributed (normal vectors) with the same covariance matrix Σ and

$$E(Y) = \beta Z E' = \beta F^{-1} E_2 (E_1', E_2') = (O, \beta F^{-1}).$$

Since

$$XX' = \sum_{\alpha=1}^{N} \mathbf{X}^\alpha \mathbf{X}^{\alpha\prime} = YY' = \sum_{\alpha=1}^{N} \mathbf{Y}^\alpha \mathbf{Y}^{\alpha\prime}$$

$$\hat{\beta} A \hat{\beta}' = (XZ'A^{-1}) A (XZ'A^{-1})' = YEE_2'(F^{-1})' A^{-1} F^{-1} E_2 E' Y'$$

$$= Y \begin{pmatrix} 0 \\ I \end{pmatrix} (0, I) Y' = \sum_{\alpha=N-s+1}^{N} \mathbf{Y}^\alpha \mathbf{Y}^{\alpha\prime}, \tag{8.115}$$

we get

$$N\hat{\Sigma} = \sum_{\alpha=1}^{N-s} \mathbf{Y}^\alpha \mathbf{Y}^{\alpha\prime}, \tag{8.116}$$

where $\mathbf{Y}^\alpha$, $\alpha = 1, \ldots, N - s$, are independently distributed normal p-vectors with means $\mathbf{0}$ and the same covariance matrix Σ. From (8.115) and (8.116) $N\hat{\Sigma}$ is distributed as $W_p(N - s, \Sigma)$ independently of $\hat{\beta}$. Q.E.D.

THEOREM 8.4.2 Under $H_0, (\hat{\beta}_1 - \beta_1{}^0)(A_{(11)} - A_{(12)} A_{(22)}^{-1} A_{(21)})(\hat{\beta}_1 - \beta_1{}^0)'$ is dsitributed as $W_p(\Sigma, r)$ (independently of $N\hat{\Sigma}$).

Proof From (8.114) the covariance of the ith and the jth rows of the estimator $\hat{\beta}_1$ is $\sigma_{ij}(A_{(11)} - A_{(12)} A_{(22)}^{-1} A_{(21)})^{-1}$. Let E be an $r \times r$ nonsingular matrix such that

$$E(A_{(11)} - A_{(12)}A_{(22)}^{-1}A_{(21)})E' = I,$$

and let

$$\hat{\beta}_1 - \beta_1{}^0 = YE = (\mathbf{Y}^1, \ldots, \mathbf{Y}^r)E.$$

Then

$$(\hat{\beta}_1 - \beta_1{}^0)(A_{(11)} - A_{(12)}A_{(22)}^{-1}A_{(21)})(\hat{\beta}_1 - \beta_1{}^0)' = \sum_{\alpha=1}^{r} \mathbf{Y}^\alpha \mathbf{Y}^{\alpha'}.$$

Obviously under H_0 [$E(\cdot)$ denotes the expectation symbol]

$$E(Y) = E[(\hat{\beta}_1 - \beta_1{}^0)E^{-1}] = 0,$$

since $E(\hat{\beta}_1) = \beta_1{}^0$. Let the ith and the jth row of Y be $\mathbf{Y}_i$ and $\mathbf{Y}_j$, respectively, and let the ith and the jth row of $\hat{\beta}_1$ be $\hat{\boldsymbol{\beta}}_{i1}$ and $\hat{\boldsymbol{\beta}}_{j1}$, respectively. Then

$$\begin{aligned} E(\mathbf{Y}_i'\mathbf{Y}_j) &= E((E^{-1})'(\hat{\boldsymbol{\beta}}_{i1} - \boldsymbol{\beta}_{i1}^0)'(\hat{\boldsymbol{\beta}}_{j1} - \boldsymbol{\beta}_{j1}^0)E^{-1}) \\ &= \sigma_{ij}[E(A_{(11)} - A_{(12)}A_{(22)}^{-1}A_{(21)})E']^{-1} = \sigma_{ij}I. \end{aligned}$$

Thus $\sum_{\alpha=1}^{r} \mathbf{Y}^\alpha \mathbf{Y}^{\alpha'}$ is distributed as $W_p(\Sigma, r)$ when H_0 is true. Q.E.D.

Hence the statistics U as given in (8.84) and (8.112) have identical distributions.

8.4.3 The Distribution of U under H_0

Anderson (1958) called the statistic U, $U_{p,r,N-s}$. Computing various moments of U under H_0 (for details, see Anderson, 1958, Section 8.5), we can show that

$$E(U^k) = \prod_{i=1}^{p} E(X_i^k), \qquad k = 0, 1, \ldots, \tag{8.117}$$

where $X_1, \ldots, X_p$ are independently distributed central beta random variables with parameter $(\frac{1}{2}(N - s - i + 1), \frac{1}{2}r)$, $i = 1, \ldots, p$. Since U lies between 0 and 1, these moments determine the distribution of U (under H_0) uniquely. Thus, under H_0, U is distributed as

$$U = \prod_{i=1}^{p} X_i. \tag{8.118}$$

Furthermore, under H_0, $U_{p,r,N-s}$ and $U_{r,p,N-p-s+r}$ have the same distribution. From (8.117) it is easy to see that

$$\text{(i)} \quad \frac{1-U}{U}\left(\frac{N-s}{r}\right) \tag{8.119}$$

has central F-distribution with degrees of freedom $(r, N - s)$ when $p = 1$.

$$\text{(ii)} \quad \frac{1 - \sqrt{U}}{\sqrt{U}} \left(\frac{N - s - 1}{r} \right) \tag{8.120}$$

has central F-distribution with degrees of freedom $(2r, 2(N - s - 1))$ when $p = 2$.

Box (1949) gave an asymptotic expansion for the distribution of a monotone function of the likelihood ratio statistic $\lambda\ [=(U_{p,r,N-s})^{N/2}]$ when H_0 is true. The expansion converges extremely rapidly, and therefore the level of significance derived from it will be quite adequate even for moderate values of N. For large N, the Box result is equivalent to the large sample result of Wilks (1938); that is, under H_0, $-2 \log \lambda$ is distributed as central χ^2_{pr} with pr degrees of freedom as $N \to \infty$. The Box approximation (with $p \leq r$) is, under H_0,

$$P\{-r \log U_{p,r,N-s} \leq z\} = P\{\chi^2_{pr} \leq z\} + (\gamma/r^2)[P\{\chi^2_{pr+4} \leq z\} - P\{\chi^2_{pr} \leq z\}] + O(N^{-4}), \tag{8.121}$$

where $\gamma = pr(p^2 + r^2 - 5)/48$. If just the first term is used, the total error of approximation is $O(N^{-2})$; if both terms are used, the error is $O(N^{-4})$. If $r < p$, we use the result that under H_0, $U_{p,r,N-s}$ is distributed as $U_{r,p,N-s-p+r}$.

For the likelihood ratio criterion, exact tables are available only for $p \leq 4$. The Lawley–Hotelling test criterion cannot be used for small sample sizes and appropriate p, since only a result asymptotic in sample size is available (see Anderson, 1958, p. 224; Pillai, 1954). Morrow (1948) has shown that, under H_0, $N \operatorname{tr}(A'B^{-1})$ has central χ^2_{pr} when $N \to \infty$. The largest and the smallest root criteria of Roy can be used in the general case, although percentage point tables are available only for the restricted values of the parameters. Appropriate tables are given by Foster and Rees (1957), Foster (1957, 1958), Heck (1960), and Pillai (1960). Different criteria for this problem have been compared on the basis of their power functions, in some detail, by Smith *et al.* (1962) and Gabriel (1969).

8.4.4 Optimum Properties of Tests of General Linear Hypotheses

Using the argument that follows Stein's proof of admissibility of Hotelling's T^2-test (a generalization of a result of Birnbaum, 1955) Schwartz (1964a) has shown that for testing $H_0\!:\!\Lambda = 0$ against $H_1\!:\!\Lambda \neq 0$, the test (Pillai, 1955) that rejects H_0 whenever $\operatorname{tr} a(a + b)^{-1} \geq c$, where the constant c depends on the level of significance α of the test, is admissible. He also obtained the following results:

(i) For testing $H_0\!:\!\Lambda = 0$ against the alternatives $\operatorname{tr} \Lambda\Lambda'\Sigma^{-1} = \delta$, Pillai's test is locally minimax in the sense of Giri and Kiefer (1964a) as $\delta \to 0$.

(ii) Among all invariant level α tests of H_0 which depend only on $R_1,\ldots,R_p$ and which therefore have power functions of the form $\alpha + c\ \mathrm{tr}(\Lambda\Lambda'\Sigma^{-1}) + o(\Lambda\Lambda'\Sigma^{-1})$, Pillai's test minimizes the value of c.

Ghosh (1964), using Stein's approach, has shown that the Lawley–Hotelling trace test, which rejects H_0 whenever $\mathrm{tr}(ab^{-1}) \geq c$, and Roy's test based on $\max_i(r_i)$ are admissible for testing $H_0:\Lambda = 0$ against $H_1:\Lambda \neq 0$. Thus as a consequence of the following result of Anderson *et al.* (1964), they are unbiased for this problem.

Anderson *et al.* (1964) gave sufficient conditions on invariant tests (depending only on $R_1, \ldots, R_p$) for the power functions to be monotonically increasing functions of each θ_i, $i = 1, \ldots, p$. Further, they have shown that the likelihood ratio test, the Lawley and Hotelling trace test, and Roy's maximum characteristic root test satisfy these conditions. The monotonicity of the power function of Roy's test has been demonstrated by Roy and Mikhail (1961) using a geometric argument.

Kiefer and Schwartz (1965) have shown, using the Bayes approach, that Pillai's test is admissible Bayes for this problem. The proof proceeds in the same way as that of the admissibility of the T^2- and R^2-tests. The interested reader may consult the original reference for details. This test is fully invariant, similar, and as a consequence of the result given in the preceding paragraph, unbiased. Using the same approach, these authors have also proved the admissibility of the likelihood ratio test under the restriction that $N - s \geq p + r - 1$, although the admissibility of the likelihood ratio test can be proved without this added restriction (see Schwartz, 1964b). Recently Sihna and Giri (1975) proved the Bayes character (and, hence, admissibility) of the likelihood ratio test whenever $N - s > p$. Narain (1950) has shown that the likelihood ratio test is unbiased. We refer to Nandi (1963) for a related admissibility result and to John (1971) for an optimality result.

8.4.5 Multivariate One-Way, Two-Way Classifications

Most of the univariate results in connection with design of experiments can be extended to the multivariate case. We consider here one-way and two-way classifications as examples.

One-Way Classification

Suppose we have r p-variate normal populations with the same positive definite covariance matrix Σ but with different mean vectors $\boldsymbol{\mu}_i$, $i = 1, \ldots, r$. We are interested here in testing the null hypothesis

$$H_0:\boldsymbol{\mu}_1 = \cdots = \boldsymbol{\mu}_r.$$

Let $\mathbf{x}_{ij} = (x_{ij1}, \ldots, x_{ijp})'$, $j = 1, \ldots, N_i$ $(N_i > p)$, $i = 1, \ldots, r$, be a sample of size N_i from the ith p-variate normal population with mean $\boldsymbol{\mu}_i$ and covariance matrix Σ. Define

$$N = \sum_{i=1}^{r} N_i, \qquad N_i\mathbf{x}_{i.} = \sum_{j=1}^{N_i} \mathbf{x}_{ij}, \qquad N\mathbf{x}_{..} = \sum_{i=1}^{r} N_i\mathbf{x}_{i.},$$

$$s_i = \sum_{j=1}^{N_i} (\mathbf{x}_{ij} - \mathbf{x}_{i.})(x_{ij} - \mathbf{x}_{i.})',$$

$$s = \sum_{i=1}^{r} \sum_{j=1}^{N_i} (\mathbf{x}_{ij} - \mathbf{x}_{..})(\mathbf{x}_{ij} - \mathbf{x}_{..})'.$$

A straightforward calculation shows that the likelihood ratio test of H_0 rejects H_0 whenever

$$\lambda = \left[\frac{\det(\sum_{i=1}^{r} s_i)}{\det s}\right]^{N/2} \geq c,$$

where c depends on the level of significance of the test.

Since

$$\sum_{i=1}^{r} \sum_{j=1}^{N_i} (\mathbf{x}_{ij} - \mathbf{x}_{..})(\mathbf{x}_{ij} - \mathbf{x}_{..})'$$
$$= \sum_{i=1}^{r} \sum_{j=1}^{N_i} (\mathbf{x}_{ij} - \mathbf{x}_{i.})(\mathbf{x}_{ij} - \mathbf{x}_{i.})' + \sum_{i=1}^{r} N_i(\mathbf{x}_{i.} - \mathbf{x}_{..})(\mathbf{x}_{i.} - \mathbf{x}_{..})',$$

we obtain

$$\lambda = \left[\frac{\det b}{\det(a + b)}\right]^{N/2}$$

where

$$a = \sum_{i=1}^{r} N_i(\mathbf{x}_{i.} - \mathbf{x}_{..})(\mathbf{x}_{i.} - \mathbf{x}_{..})', \qquad b = \sum_{i=1}^{r} s_i.$$

Under H_0, the corresponding random matrices A, B are independently distributed as $W_p(\Sigma, N - r)$, $W_p(\Sigma, r - 1)$, respectively. Thus under H_0,

$$U = \frac{\det B}{\det(A + B)}$$

is distributed as $U_{p,r-1,N-r}$, and we have discussed its distribution in the context of the general linear hypothesis.

Two-Way Classification

Suppose we have a set of independently normally distributed p-dimensional random vectors $\mathbf{X}_{ij} = (X_{ij1}, \ldots, X_{ijp})'$, $i = 1, \ldots, r$; $j = 1, \ldots, c$, with $E(\mathbf{X}_{ij}) = \boldsymbol{\mu} + \boldsymbol{\alpha}_i + \boldsymbol{\beta}_j$, and the same covariance matrix Σ, where

$$\boldsymbol{\mu} = (\mu_1, \ldots, \mu_p)', \qquad \boldsymbol{\alpha}_i = (\alpha_{i1}, \ldots, \alpha_{ip})', \qquad \boldsymbol{\beta}_j = (\beta_{j1}, \ldots, \beta_{jp})',$$

$$\sum_1^r \boldsymbol{\alpha}_i = \mathbf{0}, \qquad \sum_{j=1}^c \boldsymbol{\beta}_j = \mathbf{0}.$$

We are interested in testing the null hypothesis

$$H_0 : \boldsymbol{\beta}_j = \mathbf{0} \qquad \text{for all} \quad j.$$

In the univariate case, the problem can be treated as a problem of regression by assigning Z suitable values. The same algebra can be used without any difficulty in the multivariate case to reduce the problem to the multiple regression problem. Define

$$\mathbf{X} = \frac{1}{rc} \sum_{i=1}^r \sum_{j=1}^c \mathbf{X}_{ij}, \qquad \mathbf{X}_{i.} = \frac{1}{c} \sum_{j=1}^c \mathbf{X}_{ij}, \qquad \mathbf{X}_{.j} = \frac{1}{r} \sum_{i=1}^r \mathbf{X}_{ij}.$$

The statistic U, analogous to the multiple regression model, is

$$U = \frac{\det B}{\det(A + B)},$$

where

$$B = \sum_{i=1}^r \sum_{j=1}^c (\mathbf{X}_{ij} - \mathbf{X}_{i.} - \mathbf{X}_{.j} + \mathbf{X}_{..})(\mathbf{X}_{ij} - \mathbf{X}_{i.} - \mathbf{X}_{.j} + \mathbf{X}_{..})'$$

$$A = r \sum_{j=1}^c (\mathbf{X}_{.j} - \mathbf{X}_{..})(\mathbf{X}_{.j} - \mathbf{X}_{..})'.$$

Under H_0, U has the distribution $U_{p,r,N-s}$ with $r = c - 1$, $N - s = (r - 1)(c - 1)$. In order for B to be positive definite we need to have $p \leq (r - 1)(c - 1)$.

EXAMPLE 8.4.2 Let us analyze the data in Table 8.1 pertaining to 12 double crosses of barley which were raised during 1971–1972 in Hissar, India. The column indices run over different crosses of barley and the row indices run over four different locations. The observation vector has two components, the first being the height of the barley plant in centimeters and the second the average ear weight in grams. Here

$$b = \begin{pmatrix} 774437.429 & 131452.592 \\ 131452.592 & 22903.067 \end{pmatrix}, \qquad a = \begin{pmatrix} 772958.191 & 131499.077 \\ 131499.077 & 22418.604 \end{pmatrix},$$

TABLE 8.1

Double Crosses of Barley

Location	1	2	3	4	5	6	7	8	9	10	11	12
1	126.60	133.04	113.90	121.52	123.26	133.96	125.42	128.06	137.24	130.50	127.96	129.24
	18.03	23.08	28.56	18.06	20.54	18.78	20.27	27.94	26.74	18.42	20.82	20.75
2	129.26	126.26	115.82	125.10	123.96	127.58	133.74	133.82	140.06	119.36	121.26	130.78
	18.87	22.33	27.70	18.66	19.30	17.42	19.58	26.42	25.85	17.15	20.68	22.46
3	138.76	128.54	107.28	132.56	112.86	118.42	137.08	127.96	129.64	128.04	116.06	137.12
	18.21	24.85	28.16	16.81	20.72	18.40	20.87	25.18	25.90	18.92	22.19	23.46
4	121.40	122.48	118.32	127.64	121.26	133.72	129.56	127.92	134.22	121.98	127.08	132.28
	18.19	24.82	29.32	17.80	19.20	17.09	19.79	25.42	26.35	16.74	23.01	21.92

$\det b/\det(a + b) = 0.4632$. Now

$$\frac{1 - (0.4632)^{1/2}}{(0.4632)^{1/2}}\left(\frac{32}{11}\right) = 1.37$$

is to be compared with $F_{22,64}$ at the 5% level of significance. Thus our data show there is no difference between crosses.

In concluding this section we give some recent developments regarding the complex multivariate general linear hypothesis which is defined for the complex multivariate normal distributions in the same way as that for multivariate normal distributions. The distribution of statistics based on characteristic roots of complex Wishart matrices is also useful in multiple time series analysis (see Hannan, 1970). The joint noncentral distributions of the characteristic roots of complex Wishart matrices associated with the complex multivariate general linear hypothesis model were given explicitly by James (1964) in terms of zonal polynomials, whereas Khatri (1964a) expressed them in the form of integrals. In the case of the central complex Wishart matrices and random matrices connected with the complex multivariate general linear hypothesis, the distributions of extreme characteristic roots were derived by Pillai and Young (1970) and Pillai and Jouris (1972). The noncentral distributions of the individual characteristic roots of the matrices associated with the complex multivariate general linear hypothesis and that of traces, are given by Khatri (1964b, 1970).

8.5 EQUALITY OF SEVERAL COVARIANCE MATRICES

Let $\mathbf{X}_{ij} = (X_{ij1}, \ldots, X_{ijp})'$, $j = 1, \ldots, N_i$, be a random sample of size N_i from a p-variate normal distribution with unknown mean vectors $\boldsymbol{\mu}_i = (\mu_{i1}, \ldots, \mu_{ip})'$ and positive definite covariance matrices Σ_i, $i = 1, \ldots, k$.

We shall consider the problem of testing the null hypothesis

$$H_0:\Sigma_1 = \cdots = \Sigma_k = \Sigma \qquad \text{(say)}$$

when $\boldsymbol{\mu}_i$, $i = 1, \ldots, k$, are unknown. Let $\sum_{i=1}^k N_i = N$ and let

$$S_i = \sum_{j=1}^{N_i} (\mathbf{X}_{ij} - \mathbf{X}_{i.})(\mathbf{X}_{ij} - \mathbf{X}_{i.})',$$

$$S = \sum_{i=1}^{k} S_i, \qquad \mathbf{X}_{i.} = \frac{1}{N_i}\sum_{j=1}^{N_i} \mathbf{X}_{ij}.$$

The parametric space Ω is the space $\{(\boldsymbol{\mu}_1, \ldots, \boldsymbol{\mu}_k, \Sigma_1, \ldots, \Sigma_k)\}$, which reduces to the subspace $\omega = \{(\boldsymbol{\mu}_1, \ldots, \boldsymbol{\mu}_k, \Sigma)\}$ under H_0. The likelihood of the observations $\mathbf{x}_{ij}$ on $\mathbf{X}_{ij}$ is

$$\begin{aligned} L(\Omega) &= (2\pi)^{-Np/2} \prod_{i=1}^{k} (\det \Sigma_i)^{-N_i/2} \\ &\quad \times \exp\left\{-\tfrac{1}{2} \operatorname{tr} \sum_{i=1}^{k} \Sigma_i^{-1}\left(\sum_{j=1}^{N_i} (\mathbf{x}_{ij} - \boldsymbol{\mu}_i)(\mathbf{x}_{ij} - \boldsymbol{\mu}_i)'\right)\right\}. \end{aligned}$$

Using Lemma 5.1.1, a straightforward calculation will yield

$$\max_{\Omega} L(\Omega) = (2\pi)^{-Np/2} \prod_{i=1}^{k} [\det(s_i/N_i)]^{-N_i/2} \exp\{-\tfrac{1}{2}Np\}. \tag{8.122}$$

When H_0 is true the likelihood function reduces to

$$\begin{aligned} L(\omega) &= (2\pi)^{-Np/2}(\det \Sigma)^{-N/2} \\ &\quad \times \exp\left\{-\tfrac{1}{2} \operatorname{tr} \Sigma^{-1}\left(\sum_{i=1}^{k}\sum_{j=1}^{N_i} (\mathbf{x}_{ij} - \boldsymbol{\mu}_i)(\mathbf{x}_{ij} - \boldsymbol{\mu}_i)'\right)\right\}, \end{aligned}$$

and

$$\max_{\omega} L(\omega) = (2\pi)^{-Np/2}[\det(s/N)]^{-N/2} \exp\{-\tfrac{1}{2}Np\}. \tag{8.123}$$

Thus the likelihood ratio test of H_0 rejects H_0 whenever

$$\lambda = \frac{\max_{\omega} L(\omega)}{\max_{\Omega} L(\Omega)} = \frac{N^{pN/2} \prod_{i=1}^{k}(\det s_i)^{N_i/2}}{(\det s)^{N/2} \prod_{i=1}^{k} N_i^{pN_i/2}} \leq c, \tag{8.124}$$

where the constant c is chosen so that the test has the required size α.

From Section 6.3 it follows that the S_i are independently distributed $p \times p$ Wishart random matrices with parameters Σ_i and degrees of freedom $N_i - 1 = n_i$ (say). Bartlett, in the univariate case, suggested modifying λ by replacing N_i by n_i and N by $\sum_{i=1}^k n_i = n$ (say).

In the case of two populations ($k = 2$, $p = 1$) the likelihood ratio test reduces to the F-test, and Bartlett in this case gave an intuitive argument for replacing N_i by n_i. He argued that if N_1 (say) is small, s_1 is given too much weight in λ and other effects may be missed. The modified likelihood ratio test in the general case rejects H_0 whenever

$$\lambda' = \frac{n^{np/2} \prod_{i=1}^{k} (\det s_i)^{n_i/2}}{(\det s)^{n/2} \prod_{i=1}^{k} n_i^{pn_i/2}} \leq c', \tag{8.125}$$

where c' is determined so that the test has the required size α.

For $p = 1$ the modified likelihood ratio test is based on the F-distribution, but for $p > 1$ the distribution is more complicated. The reader is referred to Anderson (1958) for some relevant distribution results based on the moments of λ'.

Define

$$a = 1 - \left(\sum_{i=1}^{k} \frac{1}{n_i} - \frac{1}{n} \right) \frac{2p^2 + 3p - 1}{6(p + 1)(k - 1)}$$

$$b = \frac{p(p + 1)}{48a^3} \left[(p - 1)(p + 2) \left(\sum_{i=1}^{k} \frac{1}{n_i^2} - \frac{1}{n^2} \right) - 6(k - 1)(1 - a)^2 \right] \tag{8.126}$$

$$f = \tfrac{1}{2} p(p + 1)(k - 1).$$

It was shown by Box (1949) that a close approximation to the distribution of $\log \lambda$ under H_0 is given by

$$P\{-2a \log \lambda \leq z\} = P\{\chi_f^2 \leq z\} + b[P\{\chi_{f+4}^2 \leq z\} - P\{\chi_f^2 \leq z\}] + O((N - k)^{-3}). \tag{8.127}$$

From this it follows that in large samples under H_0

$$P\{-2a \log \lambda \leq z\} \simeq P\{\chi_f^2 \leq z\}.$$

Giri (1972) has shown that if $\Sigma_1, \ldots, \Sigma_k$ are such that they can be diagonalized by the same orthogonal matrix [a necessary and sufficient condition for this to be true is that $\Sigma_i \Sigma_j = \Sigma_j \Sigma_i$ for all (i, j)], then the test with rejection region

$$\prod_{i=1}^{k} (\det s_i)^{a_i} / (\det s)^b \leq \text{const}, \tag{8.128}$$

where $b = \sum_{i=1}^{k} a_i = cn$, c being a positive constant, is unbiased for testing H_0 against the alternatives $\det \Sigma_1 \geq \det \Sigma_i$ when $0 < a_i \leq cn_i$ for all i, and against the alternatives $\det \Sigma_1 \leq \det \Sigma_i$ when $a_i > cn_i$ for all i. A special case of this additional restriction, which arises in the analysis of variance components, is the alternatives $H_1':\Sigma_1 = l_2\Sigma_2 = \cdots = l_k\Sigma_k$ where the l_i are unknown scalar constants.

Federer (1951) has pointed out that this type of model is also meaningful in certain genetic problems. From the preceding it follows trivially that for testing H_0 against H_1', the test given in (8.127) is unbiased if $l_i \geq 1$ when $0 < a_i \leq cn_i$ for all i and if $l_i \leq 1$ when $a_i > cn_i$ for all i.

Kiefer and Schwartz (1965) have shown that if $0 < a_i \leq n_i - p$ for all i and b (not necessarily equal to $\sum_{i=1}^k a_i) \leq n - p$, then the test given in (8.127) is admissible Bayes and similar for testing H_0 against the alternatives that not all Σ_i are equal. It is also similar and fully invariant if $\sum_{i=1}^k a_i = b$. Such a test can be obtained from the simplest choice of $a_i = 1$ with $b = k$, provided that $n_i > p$ for all i. The likelihood ratio test (respectively the modified likelihood ratio test) can be obtained in this way by setting $a_i = c_1(n_i + 1)$ (respectively $a_i = c_1 n_i$) and $b = \sum_{i=1}^k a_i$ where $c_1 < 1$.

Some satisfactory solutions to this problem (which cannot be obtained otherwise) can be obtained in the special case $k = 2$. Khatri and Srivastava (1971) have derived the exact nonnull distribution of the modified likelihood ratio test in this case in terms of the H-function. The problem of testing $H_0:\Sigma_1 = \Sigma_2$ against $H_1:\Sigma_1 \neq \Sigma_2$ remains invariant under the group of affine transformations $G = (G_l(p), T)$, where $G_l(p)$ is the full linear group of $p \times p$ real nonsingular matrices and T is the group of translations, transforming

$$\mathbf{X}_{ij} \to g\mathbf{X}_{ij} + \mathbf{b}_i, \qquad j = 1, \ldots, N_i, \quad i = 1, 2,$$

$g \in G_l(p)$, $\mathbf{b}_i = (b_{i1}, \ldots, b_{ip})' \in T$. The induced transformation in the space of the sufficient statistic $(\mathbf{X}_{1.}, S_1; \mathbf{X}_{2.}, S_2)$ is given by

$$(\mathbf{X}_{1.}, S_1; \mathbf{X}_{2.}, S_2) \to (g\mathbf{X}_{1.} + \mathbf{b}_1, gS_1g'; g\mathbf{X}_{2.} + \mathbf{b}_2, gS_2g')$$

and the corresponding induced transformation in the parametric space Ω is given by

$$(\boldsymbol{\mu}_1, \Sigma_1; \boldsymbol{\mu}_2, \Sigma_2) \to (g\boldsymbol{\mu}_1 + \mathbf{b}_1, g\Sigma_1g'; g\boldsymbol{\mu}_2 + \mathbf{b}_2, g\Sigma_2g').$$

A maximal invariant in the space of the sufficient statistic under G is $(R_1, \ldots, R_p)$, the characteristic roots of $S_1S_2^{-1}$. A corresponding set of maximal invariants in the parametric space Ω is $(\theta_1, \ldots, \theta_p)$, the characteristic roots of $\Sigma_1\Sigma_2^{-1}$. In terms of these parameters the null hypothesis can be stated as

$$H_0:\theta_1 = \cdots = \theta_p = 1. \tag{8.129}$$

Several invariant tests have been proposed for this problem:

(1) a test based on $\det(S_1S_2^{-1})$;
(2) a test based on $\mathrm{tr}(S_1S_2^{-1})$;

(3) Roy's test based on the largest and the smallest characteristic roots of $S_1 S_2^{-1}$ (Roy, 1953);

(4) a test based on $\det[(S_1 + S_2)S_2^{-1}]$ (Kiefer and Schwartz, 1965).

We shall now prove some interesting properties of these tests.

Consider two independent random matrices U_1 of dimension $p \times n_1$ and U_2 of dimension $p \times n_2$, such that the column vectors of U_1 are independently and normally distributed with mean $\mathbf{0}$ and covariance matrix Σ_1 and the column vectors of U_2 are independently and normally distributed with mean vector $\mathbf{0}$ and covariance matrix Σ_2. Then

$$S_1 = U_1 U_1', \qquad S_2 = U_2 U_2'.$$

THEOREM 8.5.1 Let ω be a set in the space of $(R_1, \ldots, R_p)$, the characteristic roots of $(U_1 U_1')(U_2 U_2')^{-1}$ such that when a point $(r_1, \ldots, r_p) \in \omega$, so is every point $(\bar{r}_1, \ldots, \bar{r}_p)$ for which $\bar{r}_i \le r_i$, $i = 1, \ldots, p$. Then the probability of the set ω depends on Σ_1 and Σ_2 only through $(\theta_1, \ldots, \theta_p)$ and is a monotonically decreasing function of each θ_i.

Proof Since Σ_1, Σ_2 are positive definite, there exists a $g \in G_l(p)$ such that $\Sigma_1 = g\theta g'$, $\Sigma_2 = gg'$ where θ is a diagonal matrix with diagonal elements $\theta_1, \ldots, \theta_p$. Write $V_1 = g^{-1}U_1$, $V_2 = g^{-1}U_2$. It follows that the column vectors of V_1 are independently normally distributed with mean $\mathbf{0}$ and covariance matrix θ, the column vectors of V_2 are independently normally distributed with mean $\mathbf{0}$ and covariance matrix I, and $(U_1 U_1')(U_2 U_2')^{-1}$ and $(V_1 V_1')(V_2 V_2')^{-1}$ have the same characteristic roots. Let

$$Q(u_1, u_2) = \{(u_1, u_2):(r_1, \ldots, r_p) \in \omega\} \tag{8.130}$$

and let $f_{U_i}(u_i|\Sigma_i)$ be the probability density function of U_i, $i = 1, 2$. Then

$$\int_{Q(u_1,u_2)} f_{U_1}(u_1|\Sigma_1) f_{U_2}(u_2|\Sigma_2)\, du_1\, du_2$$
$$= \int_{Q(v_1,v_2)} f_{V_1}(v_1|\theta) f_{V_2}(v_2|I)\, dv_1\, dv_2 = P\{\omega|\theta\} \qquad \text{(say)}. \tag{8.131}$$

Consider $V_2 = v_2$ fixed and let $(v_2 v_2')^{-1} = TT'$ where T is a $p \times p$ nonsingular matrix. The probability density function of $W = TV_1$ is $f_W(w|T\theta T')$. Obviously $(V_1 V_1')(v_2 v_2')^{-1}$ and WW' have the same characteristic roots. Then for $V_2 = v_2$ we have

$$\int_{R(v_1)} f_{V_1}(v_1|\theta)\, dv_1 = \int_{R(w)} f_W(w|T\theta T')\, dw \tag{8.132}$$

where $R(v_1) = \{v_1 \text{:characteristic roots of } (v_1 v_1')(v_2 v_2')^{-1} \text{ belong to } \omega\}$. Let θ^* be a diagonal matrix such that $\theta^* - \theta$ is positive semidefinite. It now follows from Exercise 8.11 that (denoting Chi as the ith characteristic root)

$$\text{Ch}i(T\theta^* T') = \text{Ch}i(\theta^{*1/2} T' T \theta^{*1/2}) \ge \text{Ch}i(\theta^{1/2} T' T \theta^{1/2}) = \text{Ch}i(T\theta T').$$

From Exercise 8.12 and from (8.132) we get for $V_2 = v_2$ (fixed)

$$\int_{R(v_1)} f_{V_1}(v_1|\theta)\, dv_1 \geq \int_{R(v_1)} f_{V_1}(v_1|\theta^*)\, dv_1. \tag{8.133}$$

Multiplying both sides of (8.133) by $f_{V_2}(v_2|I)$ and integrating with respect to v_2 we obtain $P(\omega|\theta) \geq P(\omega|\theta^*)$ whenever $\theta^* - \theta$ is positive semi-definite. Q.E.D.

From this theorem it now follows that:

COROLLARY 8.5.1 If an invariant test with respect to G has an acceptance region ω' such that if $(r_1, \ldots, r_p) \in \omega'$, so is $(\bar{r}_1, \ldots, \bar{r}_p)$ for $\bar{r}_i \leq r_i$, $i = 1, \ldots, p$, then the power function of the test is a monotonically increasing function of each θ_i.

COROLLARY 8.5.2 The cumulative distribution function of $R_{i_1}, \ldots, R_{i_k}$ where $(i_1, \ldots, i_k)$ is a subset of $(1, \ldots, p)$ is a monotonically decreasing function of each θ_i.

COROLLARY 8.5.3 If $g(r_1, \ldots, r_p)$ is monotonically increasing in each of its arguments, a test with acceptance region $g(r_1, \ldots, r_p) \leq$ const has a monotonically increasing power function in each θ_i.

In particular, Corollary 8.5.3 includes tests with acceptance regions

$$\sum_{i=1}^{k} d_i T_i \leq \text{const}$$

where $d_i \geq 0$ and T_i is the sum of all different products of $r_1, \ldots, r_p$ taken i at a time. Special cases of these regions are

(1) $\prod_{i=1}^{p} r_i = \det(s_1 s_2^{-1}) \leq \text{const}$

(2) $\sum_{i=1}^{p} r_i = \operatorname{tr}(s_1 s_2^{-1}) \leq \text{const}.$

In addition it can be verified that it also includes tests with acceptance region $\sum_{i,j=1}^{p} a_{ij}\omega_{ij} \leq$ const with $a_{ij} \geq 0$ and $\omega_{ij} = T_i/T_j$ $(i > j)$. Roy's tests based on the largest and the smallest characteristic roots with acceptance regions $\max_i r_i \leq$ const and $\min_i r_i \leq$ const, respectively, are also special cases of Corollary 8.5.3.

Sugiura and Nagao (1968) proved the following property of the modified likelihood ratio test.

THEOREM 8.5.2 For testing $H_0:\Sigma_1 = \Sigma_2$ against the alternatives $H_1:\Sigma_1 \neq \Sigma_2$ the modified likelihood ratio test with acceptance region

$$\omega = \left\{(s_1, s_2): \prod_{i=1}^{2} \det(s_i(s_1 + s_2)^{-1})^{n_i/2} \geq c'\right\},$$

where the constant c' is chosen such that the test has size α, is unbiased.

Proof As observed earlier, we can take $\Sigma_2 = I$ and $\Sigma_1 = \theta$, the diagonal matrix with diagonal elements $\theta_1, \ldots, \theta_p$. Now

$$\begin{aligned} P\{\omega|H_1\} &= c_{n_1,p}c_{n_2,p} \int_{(s_1,s_2)\in\omega} (\det s_1)^{(n_1-p-1)/2}(\det s_2)^{(n_2-p-1)/2} \\ &\quad \times (\det \theta)^{-n_1/2} \exp\{-\tfrac{1}{2}\operatorname{tr}(\theta^{-1}s_1 + s_2)\}\, ds_1\, ds_2 \\ &= c_{n_1,p}c_{n_2,p} \int_{(I,u_2)\in\omega} (\det u_1)^{(n-p-1)/2}(\det u_2)^{(n_2-p-1)/2}(\det \theta)^{-n_1/2} \\ &\quad \times \exp\{-\tfrac{1}{2}\operatorname{tr}(\theta^{-1} + u_2)u_1\}\, du_1\, du_2 \\ &= b \int_{(I,u_2)\in\omega} (\det u_2)^{(n_2-p-1)/2}(\det \theta)^{-n_1/2}(\det(\theta^{-1} + u_2))^{-n/2}\, du_2, \end{aligned}$$

where $S_1 = U_1$, $S_2 = U_1^{1/2}U_2U_1^{1/2}$, with $U_1^{1/2}$ a symmetric matrix such that $U_1 = U_1^{1/2}U_1^{1/2}$, and $b = c_{n_1,p}c_{n_2,p}/c_{n,p}$. The Jacobian of the transformation $(s_1, s_2) \to (u_1, u_2)$ is given by

$$\det\left[\frac{\partial(s_1, s_2)}{\partial(u_1, u_2)}\right] = (\det u_1)^{(p+1)/2}.$$

Write $V = \theta^{1/2}U_2\theta^{1/2}$. Let ω^* be the set of all $p \times p$ positive definite matrices v such that $(I, \theta^{-1/2}v\theta^{-1/2}) \in \omega$, and let $\bar{\omega}$ be the set of all $p \times p$ positive definite symmetric matrices v such that $(I, v) \in \omega$. Then

$$\begin{aligned} &P\{\omega|H_0\} - P\{\omega|H_1\} \\ &\quad = b\left\{\int_{\bar{\omega}} - \int_{\omega^*}\right\}(\det v)^{(n_2-p-1)/2}(\det(I + v))^{-n/2}\, dv \\ &\quad = b\left\{\int_{\bar{\omega}-\bar{\omega}\cap\omega^*} - \int_{\omega^*-\bar{\omega}\cap\omega^*}\right\}(\det v)^{n_2/2}(\det(I + v))^{-n/2}(\det v)^{-(p+1)/2}\, dv \\ &\quad \geq bc'\left\{\int_{\bar{\omega}-\bar{\omega}\cap\omega^*} - \int_{\omega^*-\bar{\omega}\cap\omega^*}\right\}(\det v)^{-(p+1)/2}\, dv \\ &\quad = bc'\left\{\int_{\bar{\omega}} - \int_{\omega^*}\right\}(\det v)^{-(p+1)/2}\, dv = 0 \end{aligned}$$

since

$$\int_{\bar{\omega}} (\det v)^{(n_2-p-1)/2}(\det(I+v))^{-n/2}\,dv < \infty,$$

and for any subset ω' of $\bar{\omega}$

$$\int_{\omega'} (\det v)^{(n_2-p-1)/2}(\det(I+v))^{-n/2}\,dv \geq c' \int_{\omega'} (\det v)^{-(p+1)/2}\,dv < \infty.$$

Hence the theorem. Q.E.D.

Subsequently Das Gupta and Giri (1973) considered the following class of rejection regions for testing $H_0:\Sigma_1 = \Sigma_2$:

$$c(a, b) = \left\{(s_1, s_2): \frac{[\det(s_1 s_2^{-1})]^a}{[\det(s_1 s_2^{-1} + I)]^b} \leq k\right\},$$

where k is a constant depending on the size α of the rejection regions. For the likelihood ratio test of this problem $a = N_1$, $b = N_1 + N_2$, and for the modified likelihood ratio test $a = n_1$ $(=N_1 - 1)$ and $b = n_1 + n_2$ $(=N_1 + N_2 - 2)$. Das Gupta (1969) has shown that the likelihood ratio test is unbiased for testing $H_0:\Sigma_1 = \Sigma_2$ against $H_1:\Sigma_1 \neq \Sigma_2$ if and only if $N_1 = N_2$ (it follows trivially from Exercise 8.5b). In what follows we shall assume that $0 < a < b$, in which case the rejection regions $c(a, b)$ are admissible.

THEOREM 8.5.3 (a) The rejection region $c(a, n_1 + n_2)$ is unbiased for testing $\Sigma_1 = \Sigma_2$ against the alternatives $\Sigma_1 \neq \Sigma_2$ for which $(\det \Sigma_1 - \det \Sigma_2)(n_1 - a) \geq 0$.

(b) The rejection region $C(a, b)$ is biased for testing $\Sigma_1 = \Sigma_2$ against the alternatives $\Sigma_1 \neq \Sigma_2$, for which the characteristic roots of $\Sigma_1\Sigma_2^{-1}$ lie in the interval with endpoints d and 1, where $d = a(n_1 + n_2)/bn_1$.

Proof Note that

$$\frac{[\det(s_1 s_2^{-1})]^a}{(\det s_1 s_2^{-1} + 1)^n} = [\det(s_1 s_2^{-1})]^{a-n_1} \frac{\prod_{i=1}^{2}(\det s_i)^{n_i}}{(\det(s_1 + s_2))^n}.$$

Proceeding exactly in the same way as in Theorem 8.5.2 ($\bar{C}$ being the complement of C) we can get

$$\begin{aligned} &P\{\bar{C}(a, n)|\theta = I\} - P\{\bar{C}(a, n)|\theta\} \\ &\quad \geq A(p, n_1, n_2, k)\{1 - (\det \theta)^{(a-n_1)/2}\} \int_{\bar{C}(a,n)} (\det v)^{(a-n_1-p-1)/2}\,dv \\ &\quad \geq 0 \end{aligned}$$

where A is a constant.

To prove part (b), consider a family of regions given by

$$R(a, b) = \{y : y^a(1 + y)^{-b} \geq k, y \geq 0\}.$$

These regions are either intervals or complements of intervals. When $0 < a < b$, $R(a, b)$ is a finite interval not including zero (excluding the trivial extreme case). Consider a random variable Y such that Y/σ $(\sigma > 0)$ is distributed as the ratio of independent $\chi^2_{n_1}$, $\chi^2_{n_2}$ random variables. Let $\beta(\delta) = P\{Y \in R(a, b)\}$. It can be shown by differentiation that $\beta(\delta) > \beta(1)$ if δ lies in the open interval with endpoints d, 1. Define a random variable Z by

$$\frac{(\det S_1)^a(\det S_2)^{b-a}}{[\det(S_1 + S_2)]^b} = \left[\frac{(S_{11}^{(1)})^a(S_{11}^{(2)})^b}{(S_{11}^{(1)} + S_{11}^{(2)})^b}\right] Z \tag{8.134}$$

where $S_k = (S_{ij}^{(k)})$, $k = 1, 2$, and suppose that $\theta_2 = \cdots = \theta_p = 1$. Then the distribution of Z is independent of θ_1 and is independent of the first factor in the right-hand side of (8.134). From Exercise 5b the power of the rejection regions $C(a, b)$ is less than its size if θ_1 lies strictly between d and 1. Q.E.D.

Let R, θ be diagonal matrices with diagonal elements $R_1, \ldots, R_p$ and $\theta_1, \ldots, \theta_p$, respectively, and let $f_R(r|\theta)$ be the probability density function of R. Giri (1968) has shown that

$$\begin{aligned}\frac{f_R(r|\theta)}{f_R(r|I)} &= 1 + \frac{1}{2}\sum_{i=1}^{p} (\theta_i - 1)\left\{N_2 - K\sum_{i=1}^{p} (1 + r_i)^{-1}\right\} + B(\theta, r)\\ &= 1 + \frac{1}{2}\sum_{i=1}^{p} (\theta_i - 1)\{N_2 - K\operatorname{tr}(s_2(s_1 + s_2)^{-1})\} + B(\theta, r),\end{aligned}$$

where $B(\theta, r) = o(\sum_{i=1}^{p}(\theta_i - 1))$ uniformly in r and K is a positive constant. From this he concluded that for testing $H_0 : \theta = I$ against $H_1 : \sum_{i=1}^{p}(\theta_i - 1) = \lambda > 0$, the test which rejects H_0 whenever

$$\operatorname{tr}(s_2(s_1 + s_2)^{-1}) \leq c, \tag{8.135}$$

where the constant c depends on the level of significance α of the test, is locally best invariant when $\lambda \to 0$. It is easy to see that the acceptance region of the test satisfies the condition of Theorem 8.5.1, and hence the power of the test is a monotonically increasing function of each θ_i, i $= 1, \ldots, p$.

For further relevant results in this contest we refer the reader to Brown (1939) and Mikhail (1962).

EXAMPLE 8.5.1 Consider Example 5.3.1. Assume that the data pertaining to 1971, 1972 constitute two independent samples from two six-variate normal populations with mean vectors $\boldsymbol{\mu}_1$, $\boldsymbol{\mu}_2$ and positive definite covariance matrices Σ_1, Σ_2, respectively. We are interested in testing

$H_0:\Sigma_1 = \Sigma_2$ when $\boldsymbol{\mu}_1, \boldsymbol{\mu}_2$ are unknown. Here $N_1 = N_2 = 27$. From (8.126),

$$-2a \log \lambda = 49.7890, \qquad b = 0.0158, \qquad f = 21,$$

since asymptotically

$$P\{-2a \log \lambda \le z\} = P\{\chi_f^2 \le z\} = 1 - \alpha,$$

for

$$\alpha = 0.05, \qquad z = 32.7; \qquad \alpha = 0.01, \quad z = 38.9.$$

Hence we reject the null hypothesis H_0. Since the hypothesis is rejected our method of solution of Example 7.2.1 is not appropriate. It is necessary to test the equality of mean vectors when the covariance matrices are unequal, using the Behrens–Fisher approach.

8.5.1 Test of Equality of Several Multivariate Normal Distributions

Consider the problem as formulated in the beginning of Section 8.5. We are interested in testing the null hypothesis

$$H_0:\Sigma_1 = \cdots = \Sigma_k, \qquad \boldsymbol{\mu}_1 = \cdots = \boldsymbol{\mu}_k.$$

In Section 8.4 we tested the hypothesis $\boldsymbol{\mu}_1 = \cdots = \boldsymbol{\mu}_k$, given that $\Sigma_1 = \cdots = \Sigma_k$, and in this section we tested the hypothesis $\Sigma_1 = \cdots = \Sigma_k$. Let λ_1 be the likelihood ratio test criterion for testing the null hypothesis $\boldsymbol{\mu}_1 = \cdots = \boldsymbol{\mu}_k$ given that $\Sigma_1 = \cdots = \Sigma_k$ and let λ_2 be the likelihood ratio test criterion for testing the null hypothesis $\Sigma_1 = \cdots = \Sigma_k$ when $\boldsymbol{\mu}_1, \ldots, \boldsymbol{\mu}_k$ are unknown. It is easy to conclude that the likelihood ratio test criterion λ for testing H_0 is given by

$$\lambda = \lambda_1 \lambda_2 = \frac{N^{pN/2} \prod_{i=1}^{k} (\det s_i)^{N_i/2}}{(\det b)^{N/2} \prod_{i=1}^{k} N_i^{pN_i/2}}$$

where

$$b = \sum_{i=1}^{k} \sum_{\alpha=1}^{N_i} (\mathbf{x}_{ij} - \mathbf{x}_{..})(\mathbf{x}_{ij} - \mathbf{x}_{..})' = \sum_{i=1}^{k} s_i + \sum_{i=1}^{k} N_i(\mathbf{x}_{i.} - \mathbf{x}_{..})(\mathbf{x}_{i.} - \mathbf{x}_{..})',$$

and the likelihood ratio test rejects H_0 whenever

$$\lambda \le C,$$

where C depends on the level of significance α. The modified likelihood ratio test of H_0 rejects H_0 whenever

$$\omega = \frac{n^{pn/2} \prod_{i=1}^{k} (\det s_i)^{n_i/2}}{(\det b)^{n/2} \prod_{i=1}^{k} n_i^{pn_i/2}} \le C'$$

where C' depends on level α. To determine C' we need to find the probability density function of W under H_0. Using Box (1949) (see Anderson, 1958, p. 256), the distribution of W under H_0 is given by

$$P\{-2\rho \log W \le z\} = P\{\chi_f^2 \le z\} + \omega_2[P\{\chi_{f+4}^2 \le z\} - P\{\chi_f^2 \le z\}] + o(N^{-3}),$$

where

$$f = \tfrac{1}{2}(k-1)p(p+1),$$

$$1 - \rho = \left(\sum_{i=1}^{k} \frac{1}{n_i} - \frac{1}{n}\right)\left(\frac{2p^2 + 3p - 1}{6(k-1)(p+3)}\right) + \frac{p - k + 2}{n(p+3)},$$

$$\begin{aligned}\omega_2 = \frac{p}{288\rho^2}\Bigg[&6\left(\sum_{i=1}^{k} \frac{1}{n_i^2} - \frac{1}{n^2}\right)(p+1)(p+2)(p-1) \\ &- \sum_{i=1}^{k}\left(\frac{1}{n_i} - \frac{1}{n}\right)^2 \frac{(2p^2+3p-1)^2}{(k-1)(p+3)} - 12\left(\sum_{i=1}^{k} \frac{1}{n_i} - \frac{1}{n}\right) \\ &\times \frac{(2p^2+3p-1)(p-k+2)}{n(p+3)} - 36\,\frac{(k-1)(p-k+2)^2}{n^2(p+3)} \\ &- \frac{12(k-1)}{n^2}(-2k^2 + 7k + 3pk - 2p^2 - 6p - 4)\Bigg].\end{aligned}$$

Thus under H_0 in large samples

$$P\{-2\rho \log W \le z\} = P\{\chi_f^2 \le z\}.$$

EXERCISES

1. Prove (8.3).

2. Prove (8.17).

3. Prove (8.18).

4. Show that if η and z are $(p \times m)$ matrices and t is a $p \times p$ positive definite matrix, then (E^{mp}, Euclidean space of dimension mp)

(a) $$\int_{E^{mp}} \exp\{-\tfrac{1}{2}\operatorname{tr}(t\eta\eta' - 2z\eta')\}\, d\eta = C(\det t)^{-m/2} \exp\{-\tfrac{1}{2}\operatorname{tr}(t^{-1}\mathbf{zz}')\}$$

where C is a constant.

(b) Show that

$$\int_{E^{mp}} [\det(I + \eta\eta')]^{-h/2}\, d\eta < \infty$$

if and only if $h > m + p - 1$.

5. (a) Let X be a random variable such that X/θ ($\theta > 0$) has a central chi-square distribution with m degrees of freedom. Then show that for $r > 0$

$$\beta(\theta) = P\{X^r \exp\{-\tfrac{1}{2}X\} \geq C\}$$

satisfies

$$\frac{d\beta(\theta)}{d\theta} \begin{Bmatrix} > \\ = \\ < \end{Bmatrix} 0 \qquad \text{according as} \quad \theta \begin{Bmatrix} < \\ = \\ > \end{Bmatrix} \frac{2r}{m}.$$

(b) Let Y be a random variable such that δY ($\delta > 0$) is distributed as a central F-distribution with $(n_1 - 1, n_2 - 1)$ degrees of freedom, and let

$$\beta(\delta) = P\left\{\frac{Y^{n_1}}{(1 + Y)^{n_1 + n_2}} \geq k \,\middle|\, \delta\right\}.$$

Assuming that $n_1 < n_2$, show that there exists a constant λ (<1) independent of k such that

$$\beta(\delta) > \beta(1)$$

for all δ lying between λ and 1.

6. Prove Theorem 8.3.4.

7. (a) Let A, B be defined as in Section 8.4. Show that the roots of $\det(A - \lambda B) = 0$ comprise a maximal invariant in the space of (A, B) under $G_l(p)$ transforming $(A, B) \to (gAg', gBg')$, $g \in G_l(p)$.

(b) Show that if $r + (N - s) > p$, the $p \times p$ matrix $A + B$ is positive definite with probability 1.

(c) Show that the roots of $\det(A - \lambda(A + B)) = 0$ also comprise a maximal invariant in the space of (A, B) under $G_l(p)$.

8. Show that for the transformation given in (8.93) the Jacobian of the transformation $(A, C) \to (W, V)$ is

$$2^p (\det W)^{p+2} \prod_{i<j} (V_i - V_j).$$

9. (*Two-way classifications with K observations per cell*) Let $\mathbf{X}_{ijk} = (X_{ijk1}, \ldots, X_{ijkp})'$, $i = 1, \ldots, I$; $j = 1, \ldots, J$; $k = 1, \ldots, K$, be independently normally distributed with

$$E(\mathbf{X}_{ijk}) = \boldsymbol{\mu} + \boldsymbol{\alpha}_i + \boldsymbol{\beta}_j + \boldsymbol{\gamma}_{ij}, \qquad \operatorname{cov}(\mathbf{X}_{ijk}) = \Sigma$$

where $\boldsymbol{\mu} = (\mu_1, \ldots, \mu_p)'$, $\boldsymbol{\alpha}_i = (\alpha_{i1}, \ldots, \alpha_{ip})'$, $\boldsymbol{\beta}_j = (\beta_{j1}, \ldots, \beta_{jp})'$, $\boldsymbol{\gamma}_{ij} = (\gamma_{ij1}, \ldots, \gamma_{ijp})'$, Σ is positive definite, and

$$\sum_{i=1}^{I} \boldsymbol{\alpha}_i = \sum_{j=1}^{J} \boldsymbol{\beta}_j = \sum_{i=1}^{I} \boldsymbol{\gamma}_{ij} = \sum_{j=1}^{J} \boldsymbol{\gamma}_{ij} = \mathbf{0}.$$

Assume that $p \leq IJ(K - 1)$.

(a) Show that the likelihood ratio test of $H_0:\boldsymbol{\alpha}_i = 0$ for all i rejects H_0 whenever

$$u = \frac{\det b}{\det(a+b)} \leq C,$$

where C is a constant depending on the level of significance, and

$$b = \sum_{i=1}^{I}\sum_{j=1}^{J}\sum_{k=1}^{K} (\mathbf{x}_{ijk} - \mathbf{x}_{ij.})(\mathbf{x}_{ijk} - \mathbf{x}_{ij.})'$$

$$a = JK \sum_{i=1}^{I} (\mathbf{x}_{i..} - \mathbf{x}_{...})(\mathbf{x}_{i..} - \mathbf{x}_{...})'$$

$$\mathbf{x}_{ij.} = \frac{1}{K}\sum_{k=1}^{K} \mathbf{x}_{ijk}, \qquad \mathbf{x}_{i..} = \frac{1}{JK}\sum_{j,k} \mathbf{x}_{ijk}, \qquad \mathbf{x}_{...} = \frac{1}{IJK}\sum_{i,j,k} \mathbf{x}_{ijk},$$

and so forth.

(b) Find the distribution of the corresponding test statistic U under H_0.

(c) Test the hypothesis $H_0:\boldsymbol{\beta}_1 = \cdots = \boldsymbol{\beta}_J = \mathbf{0}$.

10. Let x_J denote the change in the number of people residing in Montreal, Canada from the year j to the year $j+1$, who would prefer to live in integrated neighborhoods, $j = 1, 2, 3$. Suppose $\mathbf{X} = (X_1, \ldots, X_3)'$ with $E(\mathbf{X}) = z\boldsymbol{\beta}$ where

$$z = \begin{pmatrix} 1 & 0 \\ 0 & 1 \\ 1 & 1 \end{pmatrix}, \qquad \boldsymbol{\beta} = \begin{pmatrix} \beta_1 \\ \beta_2 \end{pmatrix}$$

of unknown quantities β_1, β_2 and

$$\operatorname{cov} \mathbf{X} = \sigma^2 \begin{pmatrix} 1 & \rho & \rho \\ \rho & 1 & \rho \\ \rho & \rho & 1 \end{pmatrix}, \qquad -\tfrac{1}{2} < \rho < 1.$$

Let $\mathbf{x} = (-4, 6, 2)'$.

(a) Estimate $\boldsymbol{\beta}$.

(b) If $\rho = 0$, estimate both $\boldsymbol{\beta}$ and σ^2.

11. Let A be a positive definite matrix of dimension $p \times p$ and D, D^* be two diagonal matrices of dimension $p \times p$ such that $D^* - D$ is positive semidefinite and D is positive definite. Then show that the ith characteristic root satisfies

$$\text{Ch}i(ADA') \leq \text{Ch}i(AD^*A') \qquad \text{for} \quad i = 1, \ldots, p.$$

12. (Anderson and Das Gupta, 1964) Let X be a $p \times n$ $(n \geq p)$ random matrix having probability density function

$$f_X(x|\Sigma) = (2\pi)^{-np/2}(\det \Sigma)^{-n/2} \exp\{-\tfrac{1}{2} \operatorname{tr} \Sigma^{-1}xx'\},$$

where Σ is a symmetric positive definite matrix.

(a) Show that the distribution of the characteristic roots of XX' is the same as the distribution of the characteristic roots of $(\Delta Y)(\Delta Y)'$ where Y is a $p \times n$ random matrix having the probability density function f with $\Sigma = I$, and Δ is a diagonal matrix with diagonal elements $\theta_1, \ldots, \theta_p$, the characteristic roots of Σ.

(b) Let $C_1 \geq C_2 \geq \cdots \geq C_p$ be the characteristic roots of XX' and let ω be a set in the space of $(C_1, \ldots, C_p)$ such that when a point $(C_1, \ldots, C_p)$ is in ω, so is every point $(\bar{C}_1, \ldots, \bar{C}_p)$ for which $\bar{C}_i \leq C_i$ $(i = 1, \ldots, p)$. Then show that the probability of the set ω depends on Σ only through $\theta_1, \ldots, \theta_p$ and is a monotonically decreasing function of each θ_i.

TABLE 8.2

Double Crosses of Barley

Replication	1	2	3	4	5	6	7	8	9	10
1	136.24	121.62	135.52	116.14	115.76	132.58	118.00	124.66	127.82	123.46
	48.2	52.6	64.6	59.8	56.4	49.6	54.4	50.6	48.6	59.8
	72.46	90.26	117.26	123.46	118.94	97.56	109.78	96.22	114.32	84.96
	15.82	18.27	19.94	21.89	22.61	22.79	20.64	20.65	27.09	20.13
2	128.82	138.04	133.62	119.66	119.76	147.56	125.46	121.88	126.72	129.64
	54.2	47.8	56.4	63.6	46.8	52.6	56.8	52.8	47.8	47.4
	82.98	76.76	108.12	135.62	97.06	111.72	107.56	110.38	115.26	91.78
	16.08	17.21	20.10	22.76	21.76	24.18	19.70	22.43	27.84	20.67
3	128.74	125.18	142.00	124.78	132.02	141.76	111.32	115.10	127.76	123.12
	46.6	44.6	56.4	61.6	51.4	48.2	49.8	55.4	48.4	50.2
	69.16	75.12	107.26	135.98	96.44	95.86	92.86	111.62	110.28	89.52
	15.86	17.97	19.78	22.97	20.78	21.70	18.72	21.80	27.57	19.21
4	124.62	134.32	123.06	125.86	121.34	141.26	120.68	126.26	122.22	125.04
	53.2	46.2	54.8	63.2	50.8	40.4	58.2	46.8	39.8	46.2
	80.24	75.02	101.30	140.28	106.48	86.58	112.62	91.78	100.36	85.92
	16.11	19.33	19.04	23.15	21.64	22.43	20.78	20.70	26.69	20.85

13. Analyze the data in Table 8.2 pertaining to 10 double crosses of barley which were raised in Hissar, India during 1972. Column indices run over different crosses of barley; the row indices run over four different locations.

The observation vector has four components $(x_1, \ldots, x_4)$,

x_1 plant height in centimeters,
x_2 average number of grains per ear,
x_3 average yield in grams per plant,
x_4 average ear weight in grams.

14. Let $\mathbf{X}^\alpha = (X_{\alpha 1}, \ldots, X_{\alpha p})'$, $\alpha = 1, \ldots, N$, be independently normally distributed with mean $\boldsymbol{\mu}$ and positive definite covariance matrix Σ. On the basis of observations $\mathbf{x}^\alpha$ on $\mathbf{X}^\alpha$ find the likelihood ratio test of $H_0:\Sigma = \Sigma_0$, $\boldsymbol{\mu} = \boldsymbol{\mu}_0$ where Σ_0 is a fixed positive definite matrix and $\boldsymbol{\mu}_0$ is also fixed. Show that the likelihood ratio test is unbiased for testing H_0 against the alternatives $H_1:\Sigma \neq \Sigma_0, \boldsymbol{\mu} \neq \boldsymbol{\mu}_0$.

15. Let $\mathbf{X}_{ij} = (X_{ij1}, \ldots, X_{ijp})', j = 1, \ldots, N_i$, be a random sample of size N_i from a p-variate normal population with mean $\boldsymbol{\mu}_i$ and positive definite covariance matrix Σ_i, $i = 1, \ldots, k$. On the basis of observations on the $\mathbf{X}_{ij}$, find the likelihood ratio test of $H_0:\Sigma_i = \sigma^2\Sigma_{i0}$, $i = 1, \ldots, k$, when the Σ_{i0} are fixed positive definite matrices and $\boldsymbol{\mu}_1, \ldots, \boldsymbol{\mu}_k$, σ^2 are unknown. Show that both the likelihood ratio test and the modified likelihood ratio test are unbiased for testing H_0 against $H_1:\Sigma_i \neq \sigma^2\Sigma_{i0}$, $i = 1, \ldots, k$.

REFERENCES

Anderson, T. W. (1955). The integral of a symmetric unimodal function over a symmetric convex set and some probability inequalities, *Proc. Am. Math. Soc.* **6**, 170–176.

Anderson, T. W. (1958). "An Introduction to Multivariate Analysis." Wiley, New York.

Anderson, T. W., and Das Gupta, S. (1964a). A monotonicity property of power functions of some tests of the equality of two covariance matrices, *Ann. Math. Statist.* **35**, 1059–1063.

Anderson, T. W., and Das Gupta, S. (1964b). Monotonicity of power functions of some tests of independence between two sets of variates, *Ann. Math. Statist.* **35**, 206–208.

Anderson, T. W., Das Gupta, S., and Mudolkar, G. S. (1964). Monotonicity of power functions of some tests of multivariate general linear hypothesis, *Ann. Math. Statist.* **35**, 200–205.

Bartlett, M. S. (1934). The vector representation of a sample, *Proc. R. Soc. Edinburgh* **53**, 260–283.

Birnbaum, A. (1955). Characterization of complete class of tests of some multiparametric hypotheses with application to likelihood ratio tests, *Ann. Math. Statist.* **26**, 21–36.

Box, G. E. P. (1949). A general distribution theory for a class of likelihood ratio criteria, *Biometrika* **36**, 317–346.

Brown, G. W. (1939). On the power of L_1-test for the equality of variances, *Ann. Math. Statist.* **10**, 119–128.

Constantine, A. G. (1963). Some noncentral distribution problems of multivariate linear hypotheses, *Ann. Math. Statist.* **34**, 1270–1285.

Consul, P. C. (1968). The exact distribution of likelihood ratio criteria for different hypothesis, "Multivariate Analysis" ed., Vol. II. P. R. Krishnaiah, Academic Press, New York.

Das Gupta, S. (1969). Properties of power functions of some tests concerning dispersion matrices, *Ann. Math. Statist.* **40**, 697–702.

Das Gupta, S., and Giri, N. (1973). Properties of tests concerning covariance matrices of normal distributions, *Ann. Statist.* **1**, 1222–1224.

Federer, W. T. (1951). Testing proportionality of covariance matrices, *Ann. Math. Statist.* **22**, 102–106.

Foster, R. G. (1957). Upper percentage point of the generalized beta distribution II, *Biometrika* **44**, 441–453.

Foster, R. G. (1958). Upper percentage point of the generalized beta distribution III, *Biometrika* **45**, 492–503.

Foster, R. G. and Rees, D. D. (1957). Upper percentage point of the generalized beta distribution I, *Biometrika* **44**, 237–247.

Gabriel, K. R. (1969). Comparison of some methods of simultaneous inference in MANOVA, *In* "Multivariate Analysis" (P. R. Krishnaiah, ed.), Vol. II, pp. 67–86. Academic Press, New York.

Ghosh, M. N. (1964). On the admissibility of some tests of Manova, *Ann. Math. Statist.* **35**, 789–794.

Giri, N., Kiefer, J., and Stein, C. (1963). Minimax character of T^2-test in the simplest case. *Ann. Math. Statist.* **35**, 1524–1535.

Giri, N. (1968). On tests of the equality of two covariance matrices, *Ann. Math. Statist.* **39**, 275–277.

Giri, N. (1972). On a class of unbiased tests for the equality of K covariance matrices, "Multivariate Statistical Inference" (D. G. Kabe and H. Gupta, eds.), pp. 57–62. North-Holland Publ., Amsterdam.

Giri, N. (1975). "Introduction to Probability and Statistics," Part 2, Statistics. Dekker, New York.

Giri, N., and Kiefer, J. (1964a). Local and asymptotic minimax property of multivariate tests, *Ann. Math. Statist.* **35**, 21–35.

Giri, N. and Kiefer, J. (1964b). Minimax character of R^2-test in the simplest case, *Ann. Math. Statist.* **35**, 1475–1490.

Glesser, L. J. (1966), A note on sphericity test, *Ann. Math. Statist.* **37**, 464–467.

Hannan, E. J. (1970). "Multiple Time Series." Wiley, New York.

Heck, D. L. (1960). Charts of some upper percentage points of the distribution of the largest characteristic root, *Ann. Math. Statist.* **31**, 625–642.

Hotelling, H. (1951). A generalized T-test and measure of multivariate dispersion, *Proc. Barkeley Symp. Prob. Statist., 2nd* pp. 23–41.

Hsu, P. L. (1940). On generalized analysis of variance (I), *Biometrika* **31**, 221–237.

Hsu, P. L. (1941). Analysis of variance from the power function standpoint. *Biometrika*, **32**, 62–69.

James, A. T. (1964). Distribution of matrix variates and latent roots derived from normal samples, *Ann. Math. Statist.* **35**, 475–501.

John, S. (1971). Some optimal multivariate tests, *Biometrika* **58**, 123–127.

Khatri, C. G. (1964a). Distribution of the generalized multiple correlation matrix in the dual case, *Ann. Math. Statist.* **35**, 1801–1805.

Khatri, C. G. (1964b). Distribution of the largest or the smallest characteristic root under null hypotheses concerning complex multivariate normal populations, *Ann. Math. Statist.* **35**, 1807–1810.

Khatri, C. G. (1970). On the moments of traces of two matrices in three situations for complex multivariate normal populations, Sankhya, **32**, pp. 65–80.

Khatri, C. G., and Srivastava, M. S. (1971). On exact non-null distributions of likelihood ratio criteria for sphericity test and equality of two covariance matrices, *Sankhya* 201–206.

Kiefer, J. and Schwartz, R. (1965). Admissible Bayes character of T^2- and R^2- and other fully invariant tests for classical normal problems, *Ann. Math. Statist.* **36**, 747–760.

Lawley, D. N. (1938). A generalization of Fisher's Z-test, *Biometrika* **30**, 180–187.

Lehmann, E. L. (1959). "Testing Statistical Hypotheses." Wiley, New York.

Lehmann, E. L. and Stein, C. (1948). Most powerful tests of composite hypotheses, I, *Ann. Math. Statist.* **19**, pp. 495–516.

Mauchly, J. W. (1940). Significance test of sphericity of a normal n-variate distribution, *Ann. Math. Statist.* **11**, 204–207.

Mikhail, W. F. (1962). On a property of a test for the equality of two normal dispersion matrices against one sided alternatives, *Ann. Math. Statist.* **33**, 1463–1465.

Morrow, D. J. (1948). On the distribution of the sums of the characteristic roots of a determinantal equation (Abstract), *Bull. Am. Math. Soc.* **54**, 75.

Nandi, H. K. (1963). Admissibility of a class of tests, *Calcutta Statist. Assoc. Bull.* **15**, 13–18.

Narain, R. D. (1950). On the completely unbiased character of tests of independence in multivariate normal system, *Ann. Math. Statist.* **21**, 293–298.

Olkin, I. (1952). Note on the jacobians of certain matrix transformations useful in multivariate analysis, *Biometrika* **40**, 43–46.

Pillai, K. C. S. (1954). On Some Distribution Problems in Multivariate Analysis. Inst. of Statist., Univ. of North Carolina, Chapel Hill, North Carolina.

Pillai, K. C. S. (1955). Some new test criteria in multivariate analysis, *Ann. Math. Statist.* **26**, 117–121.

Pillai, K. C. S. (1960). Statistical Tables for Tests of Multivariate Hypotheses. Manila Statist. Center, Univ. of Phillipines.

Pillai, K. C. S., and Jouris, G. M. (1972). An approximation to the distribution of the largest root of a matrix in the complex Gaussian case, *Ann. Inst. Statist. Math.* **24**, 517–525.

Pillai, K. C. S. and Young, D. L. (1970). An approximation to the distribution of the largest root of a matrix in the complex Gaussian case, *Ann. Inst. Statist. Math.* **22**, 89–96.

Roy, S. N. (1953). On a heuristic method of test construction and its use in multivariate analysis, *Ann. Math. Statist.* **24**, 220–238.

Roy, S. N. (1957). "Some Aspects of Multivariate Analysis." Wiley, New York.

Roy, S. N., and Mikhail, W. F. (1961). On the monotonic character of the power functions of two multivariate tests, *Ann. Math. Statist.* **32**, 1145–1151.

Schwartz, R. (1964a). Properties of tests in Manova (Abstract). *Ann. Math. Statist.* **35**, 939.

Schwartz, R. (1964b). Admissible invariant tests in Manova (Abstract), *Ann. Math. Statist.* **35**, 1398.

Simaika, J. B. (1941). An optimum property of two statistical tests, *Biometrika* **32**, 70–80.

Sinha, B., and Giri, N. (1975). On the optimality and non-optimality of some multivariate normal test procedures (to be published).

Smith, H., Gnanadeshikhan, R. and Huges, J. B. (1962). Multivariate analysis of variance, *Biometrika* **18**, 22–41.

Stein, C. (1956). The admissibility of Hotelling's T^2-test, *Ann. Math. Statist.* **27**, 616–623.

Sugiura, N., and Nagao, H. (1968). Unbiasedness of some test criteria for the equality of one or two covariance matrices, *Ann. Math. Statist.* **39**, 1689–1692.

Wald, A. and Brookner, R. J. (1941). On the distribution of Wilks statistic for testing the independence of several groups of variates, *Ann. Math. Statist.* **12**, 137–152.

Wilks, S. S. (1932). Certain generalizations of analysis of variance, *Biometrika* **24**, 471–494.

Wilks, S. S. (1934). Moment generating operators for determinant product moments in samples from normal system, *Ann. Math. Statist.* **35**, 312–340.

Wilks, S. S. (1938). The large sample distribution of likelihood ratio for testing composite hypotheses, *Ann. Math. Statist.* **9**, 101–112.

CHAPTER IX

Discriminant Analysis

9.0 INTRODUCTION

The basic idea of discriminant analysis consists of assigning an individual or a group of individuals to one of several known or unknown distinct populations, on the basis of observations on several characters of the individual or the group and a sample of observations on these characters from the populations if these are unknown. In scientific literature, discriminant analysis has many synonyms, such as classification, pattern recognition, character recognition, identification, prediction, and selection, depending on the type of scientific area in which it is used. The origin of discriminant analysis is fairly old, and its development reflects the same broad phases as that of general statistical inference, namely, a Pearsonian phase followed by Fisherian, Neyman–Pearsonian, and Waldian phases.

Hodges (1950) prepared an exhaustive list of case studies of discriminant analysis, published in various scientific literatures. In the early work, the problem of discrimination was not precisely formulated and was often viewed as the problem of testing the equality of two or more distributions. Various test statistics which measured in some sense the divergence between two populations were proposed. It was Pearson (see Tildsley, 1921) who first proposed one such statistic and called it the coefficient of racial likeness. Later Pearson (1926) published a considerable amount of theoretical results on this coefficient of racial likeness and proposed the following form for it:

$$\frac{N_1 N_2}{N_1 + N_2}(\bar{\mathbf{X}} - \bar{\mathbf{Y}})' S^{-1} (\bar{\mathbf{X}} - \bar{\mathbf{Y}})$$

on the basis of sample observations $\mathbf{x}^\alpha = (x_{\alpha 1}, \ldots, x_{\alpha p})'$, $\alpha = 1, \ldots, N_1$, from the first distribution, and $\mathbf{y}^\alpha = (y_{\alpha 1}, \ldots, y_{\alpha p})'$, $\alpha = 1, \ldots, N_2$, from the second distribution, where the components characterizing the populations are dependent and

$$\bar{\mathbf{X}} = \frac{1}{N_1} \sum_{\alpha=1}^{N_1} \mathbf{X}^\alpha, \qquad \bar{\mathbf{Y}} = \frac{1}{N_2} \sum_{\alpha=1}^{N_2} \mathbf{Y}^\alpha,$$

$$S = \sum_{\alpha=1}^{N_1} (\mathbf{X}^\alpha - \bar{\mathbf{X}})(\mathbf{X}^\alpha - \bar{\mathbf{X}})' + \sum_{\alpha=1}^{N_2} (\mathbf{Y}^\alpha - \bar{\mathbf{Y}})(\mathbf{Y}^\alpha - \bar{\mathbf{Y}})'.$$

The coefficient of racial likeness for the case of independent components was later modified by Morant (1928) and Mahalanobis (1927, 1930). Mahalanobis called his statistic the D^2-statistic. Subsequently Mahalanobis (1936) also modified his D^2-statistic for the case in which the components are dependent. This form is successfully applied to discrimination problems in anthropological and craniometric studies. For this problem Hotelling (1931) suggested the use of the T^2-statistic which is a constant multiple of Mahalanobis' D^2-statistic in the Studentized form, and obtained its null distribution. For a comprehensive review of this development the reader is referred to Das Gupta (1973).

Fisher (1936) was the first to suggest a linear function of variables representing different characters, hereafter called the linear discriminant function (discriminator) for classifying an individual into one of two populations. Its early applications led to several anthropometric discoveries such as sex differences in mandibles, the extraction from a dated series of the particular compound of cranial measurements showing secular trends and solutions of taxonomic problems in general. The motivation for the use of the linear discriminant function in multivariate populations came from Fisher's own idea in the univariate case.

For the univariate case he suggested a rule which classifies an observation x into the ith univariate population if

$$|x - \bar{x}_i| = \min(|x - \bar{x}_1|, |x - \bar{x}_2|), \qquad i = 1, 2,$$

where $\bar{x}_i$ is the sample mean based on a sample of size N_i from the ith population. For two p-variate populations π_1 and π_2 (with the same covariance matrix) Fisher replaced the vector random variable by an optimum linear combination of its components obtained by maximizing the ratio of the difference of the expected values of a linear combination under π_1 and π_2 to its standard deviation. He then used his univariate discrimination method with this optimum linear combination of components as the random variable.

The next stage of development of discriminant analysis was influenced by Neyman and Pearson's fundamental works (1933, 1936) in the theory of statistical inference. Advancement proceeded with the development of decision theory. Welch (1939) derived the forms of Bayes rules and the minimax Bayes rules for discriminating between two known multivariate populations with the same covariance matrix. This case was also considered by Wald (1944) when the parameters were unknown; he suggested some heuristic rules, replacing the unknown parameters by their corresponding maximum likelihood estimates. Wald also studied the distribution problem of his proposed test statistic. Von Mises (1944) obtained the rule which maximizes the minimum probability of correct classification. The problem of discrimination into two univariate normal populations with different variances was studied by Cavalli (1945) and Penrose (1947). The multivariate analog of this was studied by Smith (1947).

Rao (1946, 1947a,b, 1948, 1949a,b, 1950) studied the problem of discrimination following the approaches of Neyman–Pearson and Wald. He suggested a measure of distance between two populations, and considered the possibility of withholding decision through doubtful regions and preferential decision. Theoretical results on discriminant analysis from the viewpoint of decision theory are given in the book by Wald (1950) and in the paper by Wald and Wolfowitz (1950). Bahadur and Anderson (1962) also considered the problem of discriminating between two unknown multivariate normal populations with different covariance matrices. They derived the minimax rule and characterized the minimal complete class after restricting to the class of discriminant rules based on linear discriminant functions. For a complete bibliography the reader is referred to Das Gupta (1973) and Cacoullos (1973).

9.1 EXAMPLES

The following are some examples in which discriminant analysis can be applied with success.

EXAMPLE 9.1.1 Rao (1948) considered three populations, the Brahmin, Artisan, and Korwa castes of India. He assumed that each of the three populations could be characterized by four characters—stature (x_1), sitting height (x_2), nasal depth (x_3), and nasal height (x_4)—of each member of the population. On the basis of sample observations on these characters from these three populations the problem is to classify an individual with observation $\mathbf{x} = (x_1, \ldots, x_4)'$ into one of the three populations. Rao used a linear discriminator to obtain the solution.

EXAMPLE 9.1.2 On a patient with a diagnosis of myocardial infarction, observations on his systolic blood pressure (x_1), diastolic blood pressure (x_2), heart rate (x_3), stroke index (x_4), and mean arterial pressure (x_5) are taken. On the basis of these observations it is possible to predict whether or not the patient will survive.

EXAMPLE 9.1.3 In developing a certain rural area a question arises regarding the best strategy for this area to follow in its development. This problem can be considered as one of the problems of discriminant analysis. For example, the area can be grouped as catering to recreation users or attractive to industry by means of variables such as distance to the nearest city (x_1), distance to the nearest major airport (x_2), percentage of land under lakes (x_3), and percentage of land under forests (x_4).

EXAMPLE 9.1.4 Admission of students to the state-supported medical program on the basis of examination marks in mathematics (x_1), physics (x_2), chemistry (x_3), English (x_4), and bioscience (x_5) is another example of discriminant analysis.

9.2 FORMULATION OF THE PROBLEM OF DISCRIMINANT ANALYSIS

Suppose we have k distinct populations $\pi_1, \ldots, \pi_k$. We want to classify an individual with observation $\mathbf{x} = (x_1, \ldots, x_p)'$ or a group of N individuals with observations $\mathbf{x}^\alpha = (x_{\alpha 1}, \ldots, x_{\alpha p})'$, $\alpha = 1, \ldots, N$, on p different characters, characterizing the individual or the group, into one of $\pi_1, \ldots, \pi_k$. When considering the group of individuals we make the basic assumption that the group as a whole belongs to only one population among the k given. Furthermore, we shall assume that each of the π_i can be specified by means of the distribution function F_i (or its probability density function f_i with respect to a Lebesgue measure) of a random vector $\mathbf{X} = (X_1, \ldots, X_p)'$, whose components represent random measurements on the p different characters. For convenience we shall treat only the case in which the distribution possesses a density function, although the case of discrete distributions can be treated in almost the same way.

We shall assume that the functional form of F_i, for each i, is known and that the F_i are different for different i. However, the parameters involved in F_i may be known or unknown. If they are unknown, supplementary information about these parameters is obtained through additional samples from these populations. These additional samples are generally called training samples by engineers.

Let us denote by E^p the entire p-dimensional space of values of $\mathbf{X}$. We are interested here in prescribing a rule to divide E^p into k disjoint regions $R_1, \ldots, R_k$ such that if $\mathbf{x}$ (or $\mathbf{x}^\alpha$, $\alpha = 1, \ldots, N$) falls in R_i, we assign the individual (or the group) to π_i. Evidently in using such a classification rule we may make an error by misclassifying an individual to π_i when he really belongs to π_j $(i \neq j)$. As in the case of testing of statistical hypotheses $(k = 2)$, in prescribing a rule we should look for one that controls these errors of misclassification. Let the cost (penalty) of misclassifying an individual to π_j when he actually belongs to π_i be denoted by $C(j|i)$. Generally the $C(j|i)$ are not all equal, and depend on the relative importance of these errors. For example, the error of misclassifying a patient with myocardial infarction to survive is less serious than the error of misclassifying a patient to die. Furthermore, we shall assume throughout that there is no reward (negative penalty) for correct classification. In other words $C(i|i) = 0$ for all i. Let us first consider the case of classifying a single individual with observation $\mathbf{x}$ to one of the π $(=1, \ldots, k)$. Let $\mathbf{R} = (R_1, \ldots, R_k)$. We shall denote a classification rule which divides the space E into disjoint and exhaustive regions $R_1, \ldots, R_k$ by $\mathbf{R}$. The probability of misclassifying an individual with observation $\mathbf{x}$ from π_i as coming from π_j (with the rule $\mathbf{R}$) is

$$P(j|i, \mathbf{R}) = \int_{R_j} f_i(\mathbf{x})\, d\mathbf{x} \tag{9.1}$$

where $d\mathbf{x} = \prod_{i=1}^{p} dx_i$.

The expected cost of misclassifying an observation from π_i (using the rule $\mathbf{R}$) is given by

$$r_i(\mathbf{R}) = \sum_{j=1, j\neq i}^{k} C(j|i)P(j|i, \mathbf{R}), \qquad i = 1, \ldots, k. \tag{9.2}$$

In defining an optimum classification rule we now need to compare the cost vectors $r(\bar{R}) = (r_1(\mathbf{R}), \ldots, r_k(\mathbf{R}))$ for different $\mathbf{R}$.

DEFINITION 9.2.1 Given any two classification rules $\mathbf{R}$, $\mathbf{R}^*$ we say that $\mathbf{R}$ is as good as $\mathbf{R}^*$ if $r_i(\mathbf{R}) \leq r_i(\mathbf{R}^*)$ for all i and $\mathbf{R}$ is better than $\mathbf{R}^*$ if at least one inequality is strict.

DEFINITION 9.2.2 *Admissible rule* A classification rule $\mathbf{R}$ is said to be admissible if there does not exist a classification rule $\mathbf{R}^*$ which is better than $\mathbf{R}$.

DEFINITION 9.2.3 *Complete class* A class of classification rules is said to be complete if for any rule $\mathbf{R}^*$ outside this class, we can find a rule $\mathbf{R}$ inside the class which is better than $\mathbf{R}^*$.

Obviously the criterion of admissibility, in general, does not lead to a unique classification rule. Only in those circumstances in which $r(\mathbf{R})$ for different $\mathbf{R}$ can be ordered can one expect to arrive at a unique classification rule by using this criterion.

DEFINITION 9.2.4 *Minimax rule* A classification rule $\mathbf{R}^*$ is said to be minimax among the class of all rules $\mathbf{R}$ if

$$\max_i r_i(\mathbf{R}^*) = \min_{\mathbf{R}} \max_i r_i(\mathbf{R}). \tag{9.3}$$

This criterion leads to a unique classification rule whenever it exists and it minimizes the maximum expected loss (cost). Thus from a conservative viewpoint this may be considered as an optimum classification rule.

Let p_i denote the proportion of π_i in the population (of which the individual is a member), $i = 1, \ldots, k$. If the p_i are known, we can define the average cost of misclassifying an individual using the classification rule $\mathbf{R}$. Since the probability of drawing an observation from π_i is p_i, the probability of drawing an observation from π_i and correctly classifying it to π_i with the help of the rule $\mathbf{R}$ is given by $p_i P(i|i, \mathbf{R})$, $i = 1, \ldots, k$. Similarly the probability of drawing an observation π_i and misclassifying it to π_j $(i \neq j)$ is $p_i P(j|i, \mathbf{R})$. Thus the quantity

$$\sum_{i=1}^{k} p_i \sum_{j=1, j \neq i}^{k} C(j|i) P(j|i, \mathbf{R}) \tag{9.4}$$

is the average cost of misclassification for the rule $\mathbf{R}$ with respect to the a priori probabilities $\mathbf{p} = (p_1, \ldots, p_k)$.

DEFINITION 9.2.5 *Bayes rule* Given $\mathbf{p}$, a classification rule $\mathbf{R}$ which minimizes the average cost of misclassification is called a Bayes rule with respect to $\mathbf{p}$.

It may be remarked that a Bayes rule may result in a large probability of misclassification, and there have been several attempts to overcome this difficulty (see Anderson, 1969). In cases in which the a priori probabilities p_i are known, the Bayes rule is optimum in the sense that it minimizes the average expected cost. For further results and details about these decision theoretic criteria the reader is referred to Wald (1950), Blackwell and Girshik (1954), and Ferguson (1967).

We shall now evaluate the explicit forms of these rules in cases in which each π_i admits of a probability density function f_i, $i = 1, \ldots, k$. We shall assume that all the classification procedures considered are the same if they differ only on sets of probability measure 0.

THEOREM 9.2.1 *Bayes rule* If the a priori probabilities p_i, $i = 1, \ldots, k$, are known and if π_i admits of a probability density function f_i with respect to a Lebesgue measure, then the Bayes classification rule $\mathbf{R}^* = (R_1^*, \ldots, R_k^*)$ which minimizes the average expected cost is defined by assigning $\mathbf{x}$ to the region R_l^* if

$$\sum_{i=1, i \neq l}^{k} p_i f_i(\mathbf{x}) c(l|i) < \sum_{i=1, i \neq j}^{k} p_i f_i(\mathbf{x}) C(j|i), \qquad j = 1, \ldots, k, \quad j \neq l. \tag{9.5}$$

If the probability of equality between the righthand side and the left-hand side of (9.5) is 0 for each l and j and for each π_i, then the Bayes classification rule is unique except for sets of probability measure 0.

Proof Let

$$h_j(\mathbf{x}) = \sum_{i=1 (i \neq j)}^{k} p_i f_i(\mathbf{x}) c(j|i). \tag{9.6}$$

Then the average expected cost of a classification rule $\mathbf{R} = (R_1, \ldots, R_k)'$ with respect to the a priori probabilities p_i, $i = 1, \ldots, k$, is given by

$$\sum_{j=1}^{k} \int_{R_j} h_j(\mathbf{x})\, d\mathbf{x} = \int h(\mathbf{x})\, d\mathbf{x} \tag{9.7}$$

where

$$h(\mathbf{x}) = h_j(\mathbf{x}) \quad \text{if} \quad \mathbf{x} \in R_j. \tag{9.8}$$

For the Bayes classification rule $\mathbf{R}^*$, $h(x)$ is equal to

$$h^*(\mathbf{x}) = \min_j h_j(\mathbf{x}). \tag{9.9}$$

In other words, $h^*(\mathbf{x}) = h_j(\mathbf{x}) = \min_i h_i(\mathbf{x})$ for $\mathbf{x} \in R_j^*$. The difference between the average expected costs for any classification rules $\mathbf{R}$ and $\mathbf{R}^*$ is

$$\int [h(\mathbf{x}) - h^*(\mathbf{x})]\, d\mathbf{x} = \sum_j \int_{R_j} [h_j(\mathbf{x}) - \min_i h_i(\mathbf{x})]\, d\mathbf{x} \geq 0,$$

and the equality holds if $h_j(\mathbf{x}) = \min_i h_i(\mathbf{x})$ for $\mathbf{x}$ in R_j (for all j). Q.E.D.

Remarks (i) If (9.5) holds for all j $(\neq l)$ except for h indices, for which the inequality is replaced by equality, then $\mathbf{x}$ can be assigned to any one of these $(h + 1)$ π_i terms.

(ii) If $c(i|j) = c$ $(\neq 0)$ for all (i, j), $i \neq j$, then in R_l^* we obtain from (9.5)

$$\sum_{i=1, i \neq l}^{k} p_i f_i(\mathbf{x}) < \sum_{i=1, i \neq j}^{k} p_i f_i(x), \qquad j = 1, \ldots, k, \quad j \neq l,$$

which implies in R_l^*

$$p_j f_j(\mathbf{x}) < p_l f_l(\mathbf{x}), \qquad j = 1, \ldots, k, \quad j \neq l.$$

In other words, the point $\mathbf{x}$ is in R_l^* if l is the index for which $p_i f_i(\mathbf{x})$ is a maximum. If two different indices give the same maximum, it is irrelevant as to which index is selected.

EXAMPLE 9.2.1 Suppose that

$$f_i(x) = \begin{cases} \beta_i^{-1} \exp(-x/\beta_i) & 0 < x < \infty \\ 0 & \text{otherwise,} \end{cases}$$

$i = 1, \ldots, k$, and $\beta_1 < \cdots < \beta_k$ are unknown parameters, and let $p_i = 1/k$, $i = 1, \ldots, k$. If x is observed, the Bayes rule with equal $C(i|j)$ requires us to classify x to π_i if

$$p_i f_i(x) \geq \max_{j(\neq i)} p_j f_j(x),$$

in other words, for $i < j$ if

$$\beta_i^{-1} \exp(-x/\beta_i) \geq \beta_j^{-1} \exp(-x/\beta_j),$$

which holds if and only if

$$x \leq \frac{\beta_i \beta_j}{\beta_j - \beta_i} (\log \beta_j - \log \beta_i).$$

It is easy to show that this is an increasing function of β_j for fixed $\beta_i < \beta_j$ and is an increasing function of β_i for fixed $\beta_j < \beta_i$. Since $f_i(x)$ is decreasing in x for $x > 0$, it implies that we classify x to π_i if

$$x_{i-1} \leq x < x_i$$

where

$$x_0 = 0, \qquad x_k = \infty, \qquad \text{and} \qquad x_i = \frac{\beta_i \beta_{i+1}}{\beta_{i+1} - \beta_i} (\log \beta_{i+1} - \log \beta_i).$$

It is interesting to note that if p_i is proportional to β_i, then the Bayes rule consists of making no observation on the individual and always classifying him to π_k.

EXAMPLE 9.2.2 Let

$$f_i(x) = \begin{cases} \dfrac{1}{(2\pi)^{1/2}} \exp\{-\frac{1}{2}(x - \mu_i)^2\} & -\infty < x < \infty \\ 0 & \text{otherwise,} \end{cases}$$

where the μ_i are unknown parameters, and let $p_i = 1/k$, $i = 1, \ldots, k$. The Bayes rule with equal $C(i|j)$ requires us to classify an observed x to π_j if

$$(x - \mu_j)^2 < \max_{i, i \neq j} \{(x - \mu_i)^2\}. \tag{9.10}$$

For the particular case $k = 2$, the Bayes classification rule against the a priori (p_1, p_2) is given by

$$\text{Assign } x \text{ to} \begin{cases} \pi_1 & \text{if } \dfrac{f_1(x)}{f_2(x)} > \dfrac{C(1|2)p_2}{C(2|1)p_1} \\ \pi_2 & \text{if } \dfrac{f_1(x)}{f_2(x)} < \dfrac{C(1|2)p_2}{C(2|1)p_1} \\ \text{one of } \pi_1 \text{ and } \pi_2 & \text{if } \dfrac{f_1(x)}{f_2(x)} = \dfrac{C(1|2)p_2}{C(2|1)p_1}. \end{cases} \tag{9.11}$$

However, if under π_i, $i = 1, 2$,

$$P\left\{\frac{f_1(x)}{f_2(x)} = \frac{C(1|2)p_2}{C(2|1)p_1} \middle| \pi_i\right\} = 0, \tag{9.12}$$

then the Bayes classification rule is unique except for sets of probability measure 0.

Some Heuristic Classification Rules A likelihood ratio classification rule $\mathbf{R} = (R_1, \ldots, R_k)$ is defined by

$$R_j : C_j f_j(\mathbf{x}) > \max_{i, i \neq j} C_i f_i(\mathbf{x}) \tag{9.13}$$

for positive constants $C_1, \ldots, C_k$. In particular, if the C_i are all equal, the classification rule is called a maximum likelihood rule.

If the distribution F_i is not completely known, supplementary information on it or on the parameters involved in it is obtained through a training sample from the corresponding population. Then assuming complete knowledge of the F_i, a good classification rule $\mathbf{R} = (R_1, \ldots, R_k)$ (i.e., Bayes, minimax, likelihood ratio rule) is chosen. A plug-in classification rule $\mathbf{R}^*$ is obtained from $\mathbf{R}$ by replacing the F_i or the parameters involved in the definition of $\mathbf{R}$ by their corresponding estimates from the training samples.

For other heuristic rules based on the Mahalanobis distance the reader is referred to Das Gupta (1973), who also gives some results in this case and relevant references.

In concluding this section we state without proof some decision theoretic results of the classification rules. For a proof of these results see, for example, Wald (1950, Section 5.1.1), Ferguson (1967), and Anderson (1958).

THEOREM 9.2.2 Every admissible classification rule is a Bayes classification rule with respect to certain a priori probabilities on $\pi_1, \ldots, \pi_k$.

THEOREM 9.2.3 The class of all admissible classification rules is complete.

THEOREM 9.2.4 For every set of a priori probabilities $\mathbf{p} = (p_1, \ldots, p_k)$ on $(\pi_1, \ldots, \pi_k)$, there exists an admissible Bayes classification rule.

THEOREM 9.2.5 For $k = 2$, there exists a unique minimax classification rule $\mathbf{R}$ for which $r_1(\mathbf{R}) = r_2(\mathbf{R})$.

THEOREM 9.2.6 Suppose that $C(j|i) = C > 0$ for all $i \neq j$ and that the distribution functions $F_1, \ldots, F_k$ characterizing the populations $\pi_1, \ldots, \pi_k$ are absolutely continuous. Then there exists a unique minimax classification rule $\mathbf{R}$ for which

$$r_1(\mathbf{R}) = \cdots = r_k(\mathbf{R}). \tag{9.14}$$

It may be cautioned that if either of these two conditions is violated, then (9.14) may not hold.

9.3 CLASSIFICATION INTO ONE OF TWO MULTIVARIATE NORMAL POPULATIONS

Consider the problem of classifying an individual, with observation $\mathbf{x}$ on him, into one of two known p-variate normal populations with means $\boldsymbol{\mu}_1$ and $\boldsymbol{\mu}_2$, respectively, and the same positive definite covariance matrix Σ. Here

$$f_i(\mathbf{x}) = (2\pi)^{-p/2}(\det \Sigma)^{-1/2} \exp\{-\tfrac{1}{2}(\mathbf{x} - \boldsymbol{\mu}_i)'\Sigma^{-1}(\mathbf{x} - \boldsymbol{\mu}_i)\}, \qquad i = 1, 2. \tag{9.15}$$

The ratio of the densities is

$$\begin{aligned} \frac{f_1(\mathbf{x})}{f_2(\mathbf{x})} &= \exp\{-\tfrac{1}{2}(\mathbf{x} - \boldsymbol{\mu}_1)'\Sigma^{-1}(\mathbf{x} - \boldsymbol{\mu}_1) + \tfrac{1}{2}(\mathbf{x} - \boldsymbol{\mu}_2)'\Sigma^{-1}(\mathbf{x} - \boldsymbol{\mu}_2)\} \\ &= \exp\{\mathbf{x}'\Sigma^{-1}(\boldsymbol{\mu}_1 - \boldsymbol{\mu}_2) - \tfrac{1}{2}(\boldsymbol{\mu}_1 + \boldsymbol{\mu}_2)'\Sigma^{-1}(\boldsymbol{\mu}_1 - \boldsymbol{\mu}_2)\}. \end{aligned} \tag{9.16}$$

The Bayes classification rule $\mathbf{R} = (R_1, R_2)$ against the a priori probabilities (p_1, p_2) is given by

$$\begin{aligned} R_1&: (\mathbf{x} - \tfrac{1}{2}(\boldsymbol{\mu}_1 + \boldsymbol{\mu}_2))'\Sigma^{-1}(\boldsymbol{\mu}_1 - \boldsymbol{\mu}_2) \geq k, \\ R_2&: (\mathbf{x} - \tfrac{1}{2}(\boldsymbol{\mu}_1 + \boldsymbol{\mu}_2))'\Sigma^{-1}(\boldsymbol{\mu}_1 - \boldsymbol{\mu}_2) < k, \end{aligned} \tag{9.17}$$

where $k = \log(p_2 C(1|2))/(p_1 C(2|1))$. For simplicity we have assigned the boundary to the region R_1, though we can equally assign it to R_2 also. The linear function $(\mathbf{x} - \frac{1}{2}(\boldsymbol{\mu}_1 + \boldsymbol{\mu}_2))'\Sigma^{-1}(\boldsymbol{\mu}_1 - \boldsymbol{\mu}_2)$ of the components of the observation vector $\mathbf{x}$ is called the discriminant function, and the components of $\Sigma^{-1}(\boldsymbol{\mu}_1 - \boldsymbol{\mu}_2)$ are called discriminant coefficients. It may be noted that if $p_1 = p_2 = \frac{1}{2}$ and $C(1|2) = C(2|1)$, then $k = 0$. Now suppose that we do not have a priori probabilities for the π_i. In this case we cannot use the Bayes technique to obtain the Bayes classification rule given in

(9.17). However, we can find the minimax classification rule by finding k such that the Bayes rule in (9.17) with unknown k satisfies

$$C(2|1)P(2|1, \mathbf{R}) = C(1|2)P(1|2, \mathbf{R}). \tag{9.18}$$

According to Ferguson (1967) such a classification rule is called an equalizer rule.

Let $\mathbf{X}$ be the random vector corresponding to the observed $\mathbf{x}$ and let

$$U = (\mathbf{X} - \tfrac{1}{2}(\boldsymbol{\mu}_1 + \boldsymbol{\mu}_2))'\Sigma^{-1}(\boldsymbol{\mu}_1 - \boldsymbol{\mu}_2). \tag{9.19}$$

On the assumption that $\mathbf{X}$ is distributed according to π_1, U is normally distributed with mean and variance

$$\begin{aligned} E_1(U) &= \tfrac{1}{2}(\boldsymbol{\mu}_1 - \boldsymbol{\mu}_2)'\Sigma^{-1}(\boldsymbol{\mu}_1 - \boldsymbol{\mu}_2) = \tfrac{1}{2}\alpha \\ \operatorname{var}(U) &= E\{(\boldsymbol{\mu}_1 - \boldsymbol{\mu}_2)'\Sigma^{-1}(\mathbf{X} - \boldsymbol{\mu}_1)(\mathbf{X} - \boldsymbol{\mu}_1)'\Sigma^{-1}(\boldsymbol{\mu}_1 - \boldsymbol{\mu}_2)\} \\ &= (\boldsymbol{\mu}_1 - \boldsymbol{\mu}_2)'\Sigma^{-1}(\boldsymbol{\mu}_1 - \boldsymbol{\mu}_2) = \alpha. \end{aligned} \tag{9.20}$$

If $\mathbf{X}$ is distributed according to π_2, then U is normally distributed with mean and variance

$$E_2(U) = -\tfrac{1}{2}\alpha, \qquad \operatorname{var}(U) = \alpha. \tag{9.21}$$

The quantity α is called the Mahalanobis distance between two normal populations with the same covariance matrix. Now the minimax classification rule $\mathbf{R}$ is given by, writing $u = U(\mathbf{x})$,

$$R_1 : u \geq k, \qquad R_2 : u < k, \tag{9.22}$$

where the constant k is given by

$$\begin{aligned} C(2|1) \int_{-\infty}^{k} \frac{1}{(2\pi\alpha)^{1/2}} \exp\left\{-\frac{1}{2\alpha}(u - \alpha/2)^2\right\} du \\ = C(1|2) \int_{k}^{\infty} \frac{1}{(2\pi\alpha)^{1/2}} \exp\left\{-\frac{1}{2\alpha}(u + \alpha/2)^2\right\} du \end{aligned}$$

or, equivalently, by

$$C(2|1)\phi\left(\frac{k - \alpha/2}{\sqrt{\alpha}}\right) = C(1|2)\left(1 - \phi\left(\frac{k + \alpha/2}{\sqrt{\alpha}}\right)\right) \tag{9.23}$$

where $\phi(z) = \int_{-\infty}^{z} (2\pi)^{-1} \exp\{-\frac{1}{2}t^2\}\, dt$.

Suppose we have a group of N individuals, with observations $\mathbf{x}^\alpha$, $\alpha = 1, \ldots, N$, to be classified as a whole to one of the π_i, $i = 1, 2$. Since, writing $\bar{\mathbf{x}} = (1/N) \sum_1^N \mathbf{x}^\alpha$,

$$\prod_{\alpha=1}^{N} \frac{f_1(\mathbf{x}^\alpha)}{f_2(\mathbf{x}^\alpha)} = \exp\{N(\bar{\mathbf{x}} - \tfrac{1}{2}(\boldsymbol{\mu}_1 + \boldsymbol{\mu}_2))'\Sigma^{-1}(\boldsymbol{\mu}_1 - \boldsymbol{\mu}_2)\} \tag{9.24}$$

and $N(\bar{\mathbf{X}} - \frac{1}{2}(\boldsymbol{\mu}_1 + \boldsymbol{\mu}_2))'\Sigma^{-1}(\boldsymbol{\mu}_1 - \boldsymbol{\mu}_2)$ is normally distributed with means $N\alpha/2$, $-N\alpha/2$ and the same variance $N\alpha$ under π_1 and π_2, respectively, the Bayes classification rule $\mathbf{R} = (\mathbf{R}_1, \mathbf{R}_2)$ against the a priori probabilities (p_1, p_2) is given by

$$\begin{aligned} R_1&: N(\bar{\mathbf{x}} - \tfrac{1}{2}(\boldsymbol{\mu}_1 + \boldsymbol{\mu}_2))'\Sigma^{-1}(\boldsymbol{\mu}_1 - \boldsymbol{\mu}_2) \geq k, \\ R_2&: N(\bar{\mathbf{x}} - \tfrac{1}{2}(\boldsymbol{\mu}_1 + \boldsymbol{\mu}_2))'\Sigma^{-1}(\boldsymbol{\mu}_1 - \boldsymbol{\mu}_2) < k. \end{aligned} \tag{9.25}$$

The minimax classification rule $\mathbf{R} = (R_1, R_2)$ is given by (9.25), where k is determined by

$$C(2|1)\phi\left(\frac{k - N\alpha/2}{(N\alpha)^{1/2}}\right) = C(1|2)\left(1 - \phi\left(\frac{k + N\alpha/2}{(N\alpha)^{1/2}}\right)\right). \tag{9.26}$$

If the parameters are unknown, estimates of these parameters are obtained from independent random samples of sizes N_1 and N_2 from π_1 and π_2, respectively. Let $\mathbf{x}_\alpha^{(1)} = (x_{\alpha 1}^1, \ldots, x_{\alpha p}^1)'$, $\alpha = 1, \ldots, N_1$, $\mathbf{x}_\alpha^{(2)} = (x_{\alpha 1}^2, \ldots, x_{\alpha p}^2)'$, $\alpha = 1, \ldots, N_2$, be the sample observations (independent) from π_1, π_2, respectively, and let

$$\begin{aligned} \bar{\mathbf{x}}^{(i)} &= \frac{1}{N_i}\sum_{\alpha=1}^{N_i} \mathbf{x}_\alpha^{(i)}, \qquad i = 1, 2 \\ (N_1 + N_2 - 2)s &= \sum_{i=1}^{2}\sum_{\alpha=1}^{N_i} (\mathbf{x}_\alpha^{(i)} - \bar{\mathbf{x}}^{(i)})(\mathbf{x}_\alpha^{(i)} - \bar{\mathbf{x}}^{(i)})'. \end{aligned} \tag{9.27}$$

We substitute these estimates for the unknown parameters in the expression for U to obtain the sample discriminant function $[v(\mathbf{x})]$

$$v = (\mathbf{x} - \tfrac{1}{2}(\bar{\mathbf{x}}^{(1)} + \bar{\mathbf{x}}^{(2)}))'s^{-1}(\bar{\mathbf{x}}^{(1)} - \bar{\mathbf{x}}^{(2)}), \tag{9.28}$$

which is used in the same way as U in the case of known parameters to define the classification rule $\mathbf{R}$. When classifying a group of N individuals instead of a single one we can further improve the estimate of Σ by taking its estimate as s, defined by

$$\begin{aligned} (N_1 + N_2 + N - 3)s = &\sum_{i=1}^{2}\sum_{\alpha=1}^{N_i} (\mathbf{x}_\alpha^{(i)} - \bar{\mathbf{x}}^{(i)})(\mathbf{x}_\alpha^{(i)} - \bar{\mathbf{x}}^{(i)})' \\ &+ \sum_{\alpha=1}^{N} (\mathbf{x}^\alpha - \bar{\mathbf{x}})(\mathbf{x}^\alpha - \bar{\mathbf{x}})'. \end{aligned}$$

The sample discriminant function in this case is

$$v = N(\bar{\mathbf{x}} - \tfrac{1}{2}(\bar{\mathbf{x}}^{(1)} + \bar{\mathbf{x}}^{(2)}))'s^{-1}(\bar{\mathbf{x}}^{(1)} - \bar{\mathbf{x}}^{(2)}). \tag{9.29}$$

The classification rule based on v is a plug-in rule. To find the cutoff point k it is necessary to find the distribution of V. The distribution of V has been studied by Wald (1944), Anderson (1951), Sitgreaves (1952), Bowker (1960), Kabe (1963), and recently by Sinha and Giri (1975). Okamoto (1963) gave an asymptotic expression for the distribution of V.

Write

$$< \mathbf{Z} = \mathbf{X} - \tfrac{1}{2}(\bar{\mathbf{X}}^{(1)} + \bar{\mathbf{X}}^{(2)}), \qquad \mathbf{Y} = \bar{\mathbf{X}}^{(1)} - \bar{\mathbf{X}}^{(2)}, >$$
$$< (N_1 + N_2 - 2)S = \sum_{i=1}^{2} \sum_{\alpha=1}^{N_i} (\mathbf{X}_\alpha^{(i)} - \bar{\mathbf{X}}^{(i)})(\mathbf{X}_\alpha^{(i)} - \bar{\mathbf{X}}^{(i)})'. > \tag{9.30}$$

Obviously both $\mathbf{Y}$ and $\mathbf{Z}$ are distributed as p-variate normal with

$$E(\mathbf{Y}) = \boldsymbol{\mu}_1 - \boldsymbol{\mu}_2, \qquad \operatorname{cov}(\mathbf{Y}) = \left(\frac{1}{N_1} + \frac{1}{N_2}\right)\Sigma,$$
$$E_1(\mathbf{Z}) = \tfrac{1}{2}(\boldsymbol{\mu}_1 - \boldsymbol{\mu}_2), \qquad E_2(\mathbf{Z}) = \tfrac{1}{2}(\boldsymbol{\mu}_2 - \boldsymbol{\mu}_1), \tag{9.31}$$
$$\operatorname{cov}(\mathbf{Z}) = \left(1 + \frac{1}{4N_1} + \frac{1}{4N_2}\right)\Sigma, \qquad \operatorname{cov}(\mathbf{Y}, \mathbf{Z}) = \left(\frac{1}{2N_2} - \frac{1}{2N_1}\right)\Sigma,$$

and $(N_1 + N_2 - 2)S$ is distributed independently of $\mathbf{Z}, \mathbf{Y}$ as Wishart $W_p(N_1 + N_2 - 2, \Sigma)$ when $N_i > p$, $i = 1, 2$. If $N_1 = N_2$, $\mathbf{Y}$ and $\mathbf{Z}$ are independent. Wald (1944) and Anderson (1951) obtained the distribution of V when $\mathbf{Z}, \mathbf{Y}$ are independent. Sitgreaves (1952) obtained the distribution of $\mathbf{Z}'S^{-1}\mathbf{Y}$ where $\mathbf{Z}, \mathbf{Y}$ are independently distributed normal vectors whose means are proportional and S is distributed as Wishart, independently of $(\mathbf{Z}, \mathbf{Y})$. It may be remarked that the distribution of V is a particular case of this statistic. Giri and Sinha (1975) obtained the distribution of $\mathbf{Z}'S^{-1}\mathbf{Y}$ when the means of $\mathbf{Z}$ and $\mathbf{Y}$ are arbitrary vectors and $\mathbf{Z}, \mathbf{Y}$ are not independent. However, all these distributions are too complicated for practical use.

It is easy to verify that if $N_1 = N_2$, the distribution of V if $\mathbf{X}$ comes from π_1 is the same as that of $-V$ if $\mathbf{X}$ comes from π_2. A similar result holds for V depending on $\bar{\mathbf{X}}$. Thus if $v \geq 0$ is the region R_1 and $v < 0$ is the region R_2 (v is an observed value of V), then the probability of misclassifying $\mathbf{x}$ when it is actually from π_1 is equal to the probability of misclassifying it when it is from π_2. Furthermore, given $\bar{\mathbf{X}}^{(i)} = \bar{\mathbf{x}}^{(i)}$, $i = 1, 2$, $S = s$, the conditional distribution of V is normal with means and variance

$$E_1(V) = (\boldsymbol{\mu}_1 - \tfrac{1}{2}(\bar{\mathbf{x}}^{(1)} + \bar{\mathbf{x}}^{(2)}))'s^{-1}(\bar{\mathbf{x}}^{(1)} - \bar{\mathbf{x}}^{(2)})$$
$$E_2(V) = (\boldsymbol{\mu}_2 - \tfrac{1}{2}(\bar{\mathbf{x}}^{(1)} + \bar{\mathbf{x}}^{(2)}))'s^{-1}(\bar{\mathbf{x}}^{(1)} - \bar{\mathbf{x}}^{(2)}) \tag{9.32}$$
$$\operatorname{var}(V) = (\bar{\mathbf{x}}^{(1)} - \bar{\mathbf{x}}^{(2)})'s^{-1}\Sigma s^{-1}(\bar{\mathbf{x}}^{(1)} - \bar{\mathbf{x}}^{(2)}).$$

However, the unconditional distribution of V is not normal.

9.3.1 Evaluation of the Probability of Misclassification Based on V

As indicated earlier if $N_1 = N_2$, then the classification rule $\mathbf{R} = (R_1, R_2)$ where $v \geq 0$ is the region R_1, $v < 0$ is the region R_2, has equal probabilities of misclassification. Various attempts have been made to evaluate these two probabilities of misclassification for the rule $\mathbf{R}$ in the general case $N_1 \neq N_2$. This classification rule is sometimes referred to as Anderson's rule in literature. As pointed out earlier in this section, the distribution of V, though known, is too complicated to be of any practical help in evaluating these probabilities. Let

$$P_1 = P(2|1, \mathbf{R}), \qquad P_2 = P(1|2, \mathbf{R}). \tag{9.33}$$

We shall now discuss several methods for estimating P_1, P_2. Let us recall that when the parameters are known these probabilities are given by [taking $k = 0$ in (9.23)]

$$P_1 = \phi(-\tfrac{1}{2}\sqrt{\alpha}), \qquad P_2 = 1 - \phi(\tfrac{1}{2}\sqrt{\alpha}). \tag{9.34}$$

Method 1 This method uses the same sample observations $\mathbf{x}_\alpha^{(1)}$, $\alpha = 1, \ldots, N_1$, from π_1, $\mathbf{x}_\alpha^{(2)}$, $\alpha = 1, \ldots, N_2$, from π_2, used to estimate the unknown parameters, to assess the performance of $\mathbf{R}$ based on V. Each of these $N_1 + N_2$ observations $\mathbf{x}_\alpha^{(i)}$, $\alpha = 1, \ldots, N_i$, is substituted in V and the proportions of misclassified observations from among these, using the rule $\mathbf{R}$, are noted. These proportions are taken as the estimates of P_1, P_2. This method, which is sometimes called the resubstitution method, was suggested by Smith (1947). It is obviously very crude and often gives estimates of P_i and P_2 that are too optimistic, as the same observations are used to compute the value of V and also to evaluate its performance.

Method 2 When the population parameters are known, using the analog statistic U, we have observed that the probabilities of misclassification are given by (9.34). Thus one way of estimating P_1 and P_2 is to replace α by its estimates from the samples $\mathbf{x}_\alpha^{(i)}$, $\alpha = 1, \ldots, N_i$, $i = 1, 2$,

$$\hat{\alpha} = (\overline{\mathbf{x}}^{(1)} - \overline{\mathbf{x}}^{(2)})' s^{-1} (\overline{\mathbf{x}}^{(1)} - \overline{\mathbf{x}}^{(2)}), \tag{9.35}$$

as is done to obtain the sample discriminant function V from U. It follows from Theorem 6.8.1 that

$$E(\hat{\alpha}) = \frac{N_1 + N_2 - 2}{N_1 + N_2 - p + 1}\left(\alpha + \frac{pN_1N_2}{N_1 + N_2}\right). \tag{9.36}$$

Thus $\phi(\frac{1}{2}\sqrt{\hat{\alpha}})$ is an underestimate of $\phi(\frac{1}{2}\sqrt{\alpha})$. A modification of this method will be to use an unbiased estimate of α, which is given by

$$\tilde{\alpha} = \frac{N_1 + N_2 - p + 1}{N_1 + N_2 - 2}\hat{\alpha} - \frac{pN_1N_2}{N_1 + N_2}. \tag{9.37}$$

Method 3 This method is similar to the "jackknife technique" used in statistics (see Quenouille, 1956; Tukey, 1958; Schucany *et al.*, 1971). Let $\mathbf{x}_\alpha^{(i)}$, $\alpha = 1, \ldots, N_i$, $i = 1, 2$, be samples of sizes N_1, N_2 from π_1, π_2, respectively. In this method one observation is omitted from either $\mathbf{x}_\alpha^{(1)}$ or $\mathbf{x}_\alpha^{(2)}$, and v is computed by using the omitted observation as $\mathbf{x}$ and estimating the parameters from the remaining $N_1 + N_2 - 1$ observations in the samples. Since the estimates of the parameters are obtained without using the omitted observation, we can now classify the omitted observation which we correctly know to be from π_1 or π_2, using the statistic V and the rule $\mathbf{R}$, and note if it is correctly or incorrectly classified. To estimate P_1 we repeat this procedure, omitting each $\mathbf{x}_\alpha^{(1)}$, $\alpha = 1, \ldots, N_1$. Let m_1 be the number of $\mathbf{x}_\alpha^{(1)}$ that are misclassified. Then m_1/N_1 is an estimate of P_1. To estimate P_2 the same procedure is repeated with respect to $\mathbf{x}_\alpha^{(2)}$, $\alpha = 1, \ldots, N_2$. Intuitively it is felt that this method is not sensitive to the assumption of normality.

Method 4 This method is due to Lachenbruch and Mickey (1968). Let $v(\mathbf{x}_\alpha^{(i)})$ be the value of V obtained from $\mathbf{x}_\alpha^{(i)}$, $\alpha = 1, \ldots, N_i$, $i = 1, 2$, by omitting $\mathbf{x}_\alpha^{(i)}$ as in Method 3, and let

$$u_1 = \frac{1}{N_1} \sum_{\alpha=1}^{N_1} v(\mathbf{x}_\alpha^{(1)}), \qquad u_2 = \frac{1}{N_2} \sum_{\alpha=1}^{N_2} v(\mathbf{x}_\alpha^{(2)}),$$

$$(N_1 - 1){s_1}^2 = \sum_{\alpha=1}^{N_1} (v(\mathbf{x}_\alpha^{(1)}) - u_1)^2,$$

$$(N_2 - 1){s_2}^2 = \sum_{\alpha=1}^{N_2} (v(\mathbf{x}_\alpha^{(2)}) - u_2)^2. \tag{9.38}$$

Lachenbruch and Mickey propose $\phi(-u_1/s_1)$ as the estimate of P_1 and $\phi(u_2/s_2)$ as the estimate of P_2.

When the parameters are known, the probabilities of misclassifications for the classification rule $\mathbf{R} = (R_1, R_2)$ where $R_1 : u \geq 0$, $R_2 : u < 0$ are given by

$$P_i = \phi\left((-1)^i \frac{E_i(U)}{(V(U))^{1/2}}\right), \qquad i = 1, 2. \tag{9.39}$$

In case the parameters $E_i(U)$, $V(U)$ are unknown, for estimating $E_1(U)$ and $V(U)$, we can take $v(\mathbf{x}_\alpha^{(1)})$, $\alpha = 1, \ldots, N_1$, as a sample of N_1 observations on U. So u_1, ${s_1}^2$ are appropriate estimates of $E_1(U)$ and $V(U)$. In other words, an appropriate estimate of P_1 is $\phi(-u_1/s_1)$. Similarly, $\phi(u_2/s_2)$ will be an appropriate estimate of P_2.

It may be added here that since U has the same variance irrespective of whether $\mathbf{X}$ comes from π_1 or π_2, a better estimate of $V(U)$ is

$$\frac{(N_1 - 1){s_1}^2 + (N_2 - 1){s_2}^2}{N_1 + N_2 - 2}.$$

It is worth investigating the effect of replacing $V(U)$ by such an estimate in P_i.

Method 5 Asymptotic case Let

$$\bar{\mathbf{X}}^{(i)} = \frac{1}{N_i} \sum_{\alpha=1}^{N_i} \mathbf{X}_\alpha^{(i)}, \qquad i = 1, 2,$$

$$(N_1 + N_2 - 2)S = \sum_{i=1}^{2} \sum_{\alpha=1}^{N_i} (\mathbf{X}_\alpha^{(i)} - \bar{\mathbf{X}}^{(i)})(\mathbf{X}_\alpha^{(i)} - \mathbf{X}^{(i)})'$$

where $\mathbf{X}_\alpha^{(i)}, \alpha = 1, \ldots, N_1$, and $X_\alpha^{(2)}, \alpha = 1, \ldots, N_2$, are independent random samples from π_1 and π_2, respectively. Since $\bar{\mathbf{X}}^{(i)}$ is the mean of a random sample of size N_1 from a normal distribution with mean $\boldsymbol{\mu}_{(i)}$ and covariance matrix Σ, then as shown in Chapter VI $\bar{\mathbf{X}}^{(i)}$ converges to $\boldsymbol{\mu}_i$ in probability as $N_i \to \infty$, $i = 1, 2$. As also shown S converges to Σ in probability as both N_1 and N_2 tend to ∞. Hence it follows that $S^{-1}(\bar{\mathbf{X}}^{(1)} - \bar{\mathbf{X}}^{(2)})$ converges to $\Sigma^{-1}(\boldsymbol{\mu}_1 - \boldsymbol{\mu}_2)$ and $(\bar{\mathbf{X}}_{(1)} + \bar{\mathbf{X}}_{(2)})' S^{-1}(\bar{\mathbf{X}}^{(1)} - \bar{\mathbf{X}}^{(2)})$ converges to $(\boldsymbol{\mu}_1 + \boldsymbol{\mu}_2)'\Sigma^{-1}(\boldsymbol{\mu}_1 - \boldsymbol{\mu}_2)$ in probability as both N_1, $N_2 \to \infty$. Thus as N_1, $N_2 \to \infty$ the limiting distribution of V is normal with

$$E_1(V) = \tfrac{1}{2}\alpha, \qquad E_2(V) = -\tfrac{1}{2}\alpha, \qquad \text{and} \qquad \operatorname{var}(V) = \alpha.$$

If the dimension p is small, the sample sizes N_1, N_2 occurring in practice will probably be large enough to apply this result. However, if p is not small, we will probably require extremely large sample sizes to make this result relevant for our purpose. In this case one can achieve a better approximation of the probabilities of misclassifications by using the asymptotic results of Okamoto (1963). Okamoto obtained

$$P_1 = \phi(-\tfrac{1}{2}\alpha) + \frac{a_1}{N_1} + \frac{a_2}{N_2} + \frac{a_3}{N_1 + N_2 - 2} + \frac{b_{11}}{N_1{}^2} + \frac{b_{22}}{N_2{}^2} + \frac{b_{12}}{N_1 N_2} + \frac{b_{13}}{N_1(N_1 + N_2 - 2)} + \frac{b_{23}}{N_2(N_1 + N_2 - 2)} + \frac{b_{33}}{(N_1 + N_2 - 2)^2} + O_3 \tag{9.40}$$

where O_3 is $O(1/N_i{}^3)$, and he gave a similar expression for P_2. He gave the values of the a and b in terms of the parameters $\boldsymbol{\mu}_1$, $\boldsymbol{\mu}_2$, and Σ and tabulated the values of the a and b terms for some specific cases. To evaluate P_1 and P_2, α is to be replaced by its unbiased estimate as in (9.37) and the a and b are to be estimated by replacing the parameters by their corresponding estimates.

Lachenbruch and Mickey (1968) made a comparative study of all these methods on the basis of a series of Monte Carlo experiments. They concluded that Methods 1 and 2 give relatively poor results. Methods 3–5 do fairly well overall. If approximate normality can be assumed, Methods 4 and 5 are good. Cochran (1968), while commenting on this study, also reached the conclusion that Method 5 ranks first, with Methods 3 and 4 not far behind. Obviously Method 5 needs sample sizes to be large and cannot be applied for small sample sizes. Methods 3 and 4 can be used for all sample sizes, but perform better for large sample sizes.

For the case of the equal covariance matrix, Kiefer and Schwartz (1965) indicated a method for obtaining a broad class of Bayes classification rules that are admissible. In particular, these authors showed that the likelihood ratio classification rules are admissible Bayes when Σ is unknown.

Rao (1954) derived an optimal classification rule in the class of rules for which P_1, P_2 depend only on α (the Mahalanobis distance) using the following criteria: (i) to minimize a linear combination of derivatives of P_1, P_2 with respect to α at $\alpha = 0$, subject to the condition that P_1, P_2 at $\alpha = 0$ leave a given ratio; (ii) the first criterion with the additional restriction that the derivatives of P_1, P_2 at $\alpha = 0$ bear a given ratio. See Kudo (1959, 1960) also for the minimax and the most stringent properties of the maximum likelihood classification rules.

9.3.2 Penrose's Shape and Size Factors

Let us assume that the common covariance matrix Σ of two p-variate normal populations with mean vectors $\boldsymbol{\mu}_1 = (\mu_{11}, \ldots, \mu_{1p})'$, $\boldsymbol{\mu}_2 = (\mu_{21}, \ldots, \mu_{2p})'$ has the particular form

$$\Sigma = \begin{pmatrix} 1 & \rho & \cdots & \rho \\ \rho & 1 & \cdots & \rho \\ \vdots & \vdots & & \vdots \\ \rho & \rho & \cdots & 1 \end{pmatrix}, \tag{9.41}$$

since

$$\begin{aligned} \mathbf{x}'\Sigma^{-1}(\boldsymbol{\mu}_1 - \boldsymbol{\mu}_2) &= (\boldsymbol{\mu}_1 - \boldsymbol{\mu}_2)'\Sigma^{-1}\mathbf{x} \\ &= \frac{\sum_{i=1}^{p}(\mu_{1i} - \mu_{2i})}{p(1-\rho)} \left\{ \mathbf{b}'\mathbf{x} + \frac{1-\rho}{1+(p-1)\rho} \sum_{i=1}^{b} x_i \right\}, \end{aligned} \tag{9.42}$$

where

$$\mathbf{b}'\mathbf{x} = \frac{p(\boldsymbol{\mu}_1 - \boldsymbol{\mu}_2)'\mathbf{x}}{\sum_{i=1}^{p}(\mu_{1i} - \mu_{2i})} - \sum_{i=1}^{p} x_i. \tag{9.43}$$

Hence the discriminant function depends on two factors, $\mathbf{b}'\mathbf{x}$ and $\sum_{i=1}^{p} x_i$. Penrose (1947) called $\sum_{i=1}^{p} x_i$ the size factor, since it measures the total size, and $\mathbf{b}'\mathbf{x}$ the shape factor. This terminology is more appropriate for biological organs where Σ is of the form just given and $\sum_{i=1}^{p} x_i$, $\mathbf{b}'\mathbf{x}$ measure the size and the shape of an organ. It can be verified that

$$E_i\left(\sum_{j=1}^{p} X_j\right) = \sum_{j=1}^{p} \mu_{ij}, \qquad E_i(\mathbf{b}'\mathbf{X}) = \mathbf{b}'\boldsymbol{\mu}_i, \qquad i = 1, 2,$$

$$\operatorname{cov}\left(\mathbf{b}'\mathbf{X}, \sum_{i=1}^{p} X_i\right) = 0, \qquad \operatorname{var}\left(\sum_{i=1}^{p} X_i\right) = p(1 + p\rho - \rho)$$

$$\operatorname{var}(\mathbf{b}'\mathbf{X}) = p(1-\rho)\left[\frac{p(\boldsymbol{\mu}_1 - \boldsymbol{\mu}_2)'(\boldsymbol{\mu}_1 - \boldsymbol{\mu}_2)}{\sum_{i=1}^{p}(\mu_{1i} - \mu_{2i})^2} - 1\right]. \tag{9.44}$$

Thus the random variables corresponding to the size and the shape factors are independently normally distributed with the means and variances just given. If the covariance matrix has this special form, the discriminant analysis can be performed with the help of two factors only. If Σ does not have this special form, it can sometimes be approximated to this form by first standardizing the variates to have unit variance for each component X_i and then replacing the correlation ρ_{ij} between the components X_i, X_j of $\mathbf{X}$ by ρ, the average correlation among all pairs (i, j). No doubt the discriminant analysis carried out in this fashion is not as efficient as with the true covariance matrix but it is certainly economical. However, if ρ_{ij} for different (i, j) do not differ greatly, such an approximation may be quite adequate.

9.3.3 Unequal Covariance Matrices

The equal covariance assumption is rarely satisfied although in some cases the two covariance matrices are so close that it makes little or no difference in the results to assume equality. When they are quite different we obtain

$$\frac{f_1(\mathbf{x})}{f_2(\mathbf{x})} = \left(\frac{\det(\Sigma_2)}{\det(\Sigma_1)}\right)^{1/2} \exp\{-\tfrac{1}{2}\mathbf{x}'(\Sigma_1^{-1} - \Sigma_2^{-1})\mathbf{x} + \mathbf{x}'(\Sigma_1^{-1}\boldsymbol{\mu}_1 - \Sigma_2^{-1}\boldsymbol{\mu}_2)$$
$$- \tfrac{1}{2}(\boldsymbol{\mu}_1'\Sigma_1^{-1}\boldsymbol{\mu}_1 - \boldsymbol{\mu}_2'\Sigma_2^{-1}\boldsymbol{\mu}_2)\}.$$

The Bayes classification rule $\mathbf{R} = (R_1, R_2)$ against the prior probabilities (p_1, p_2) is given by

$$R_1 : \tfrac{1}{2}\log\left(\frac{\det \Sigma_2}{\det \Sigma_1}\right) - \tfrac{1}{2}\boldsymbol{\mu}_1'\Sigma_1^{-1}\boldsymbol{\mu}_1 + \tfrac{1}{2}\boldsymbol{\mu}_2'\Sigma_2^{-1}\boldsymbol{\mu}_2$$
$$- \tfrac{1}{2}(\mathbf{x}'(\Sigma_1^{-1} - \Sigma_2^{-1})\mathbf{x} - 2\mathbf{x}'(\Sigma_1^{-1}\boldsymbol{\mu}_1 - \Sigma_2^{-1}\boldsymbol{\mu}_2)) \geq k,$$

where $k = \log(p_2 C(1|2)/p_1 C(1|2))$. The quantity

$$\mathbf{x}'(\Sigma_1^{-1} - \Sigma_2^{-1})\mathbf{x} - 2\mathbf{x}'(\Sigma_1^{-1}\boldsymbol{\mu}_1 - \Sigma_2^{-1}\boldsymbol{\mu}_2) \tag{9.45}$$

is called the quadratic discriminant function, and in the case of unequal covariance matrices one has to use a quadratic discriminant function since $\Sigma_1^{-1} - \Sigma_2^{-1}$ does not vanish. For the minimax classification rule **R** one has to find k such that (9.18) is satisfied. Typically this involves the finding of the distribution of the quadratic discriminant function when $\mathbf{x}$ comes from π_i, $i = 1, 2$. It may be remarked that the quadratic discriminant function is also the statistic involved in the likelihood ratio classification rule for this problem. The distribution of this quadratic function is very complicated. It was studied by Cavalli (1945) for the special case $p = 1$; by Smith (1947), Cooper (1963, 1965), and Bunke (1964); by Okamoto (1963) for the special case $\boldsymbol{\mu}_1 = \boldsymbol{\mu}_2$; by Bartlett and Please (1963) for the special case $\boldsymbol{\mu}_1 = \boldsymbol{\mu}_2 = \mathbf{0}$ and

$$\Sigma_i = \begin{pmatrix} 1 & \rho_i & \cdots & \rho_i \\ \rho_i & 1 & \cdots & \rho_i \\ \vdots & \vdots & & \vdots \\ \rho_i & \rho_i & \cdots & 1 \end{pmatrix}; \tag{9.46}$$

and by Han (1968, 1969, 1970) for different special forms of Σ_i. Okamoto (1963) derived the minimax classification rule and the form of a Bayes classification rule when the parameters are known. He also studied some properties of Bayes classification risk function and suggested a method of choosing components. Okamoto also treated the case when the Σ_i are unknown and the common value of $\boldsymbol{\mu}_i$ may be known or unknown. The asymptotic distribution of the sample quadratic discriminant function (plug-in log likelihood statistic) was also obtained by him. Bunke (1964) showed that the plug-in minimax rule is consistent. Following the method of Kiefer and Schwartz (1965), Nishida (1971) obtained a class of admissible Bayes classification rules when the parameters are unknown. Since these results are not very elegant for presentation we shall not discuss them here. The reader is referred to the original references for these results. However, we shall discuss a solution of this problem by Bahadur and Anderson (1962), based on linear discriminant functions only.

Let $\mathbf{b}$ $(\neq \mathbf{0})$ be a p-column vector and c a scalar. An observation $\mathbf{x}$ on an individual is classified as from π_1 if $\mathbf{b}'\mathbf{x} \leq c$ and as from π_2 if $\mathbf{b}'\mathbf{x} > c$. The probabilities of misclassification with this classification rule can be easily evaluated from the fact that $\mathbf{b}'\mathbf{x}$ is normally distributed with mean $\mathbf{b}'\boldsymbol{\mu}_1$ and variance $\mathbf{b}'\Sigma_1\mathbf{b}$ if $\mathbf{X}$ comes from π_1, and with mean $\mathbf{b}'\boldsymbol{\mu}_2$ and variance $\mathbf{b}'\Sigma_2\mathbf{b}$ if $\mathbf{X}$ comes from π_2, and are given by

$$P_1 = P(2|1, \mathbf{R}) = 1 - \phi(z_1), \qquad P_2 = P(1|2, \mathbf{R}) = 1 - \phi(z_2), \tag{9.47}$$

where

$$z_1 = \frac{c - \mathbf{b}'\boldsymbol{\mu}_1}{(\mathbf{b}'\Sigma_1\mathbf{b})^{1/2}}, \qquad z_2 = \frac{\mathbf{b}'\boldsymbol{\mu}_2 - c}{(\mathbf{b}'\Sigma_2\mathbf{b})^{1/2}}. \tag{9.48}$$

We shall assume in this treatment that $C(1|2) = C(2|1)$. Hence each procedure (obtained by varying $\mathbf{b}$) can be evaluated in terms of the two probabilities of misclassification P_1, P_2. Since the transformation by the normal cumulative distribution $\phi(z)$ is strictly monotonic, comparisons of different linear procedures can just as well be made in terms of the arguments z_1, z_2 given in (9.48). For a given z_2, eliminating c, we obtain from (9.48)

$$z_1 = \frac{\mathbf{b}'\boldsymbol{\delta} - z_2(\mathbf{b}'\Sigma_2\mathbf{b})^{1/2}}{(\mathbf{b}'\Sigma_1\mathbf{b})^{1/2}},$$

where $\boldsymbol{\delta} = (\boldsymbol{\mu}_2 - \boldsymbol{\mu}_1)$. Since z_1 is homogeneous in $\mathbf{b}$ of degree 0, we can restrict $\mathbf{b}$ to lie on an ellipse, say $\mathbf{b}'\Sigma_1\mathbf{b} = \text{const}$, and on this bounded closed domain z_1 is continuous and hence has a maximum. Thus among the linear procedures with a specified z_2 coordinate (equivalently, with a specified P_2) there is at least one procedure which maximizes the z_1 coordinate (equivalently, minimizes P_1).

LEMMA 9.3.1 The maximum z_1 coordinate is a decreasing function of z_2.

Proof Let $z_2^* > z_2$ and let $\mathbf{b}^*$ be a vector maximizing z_1^* for given z_2^*. Then

$$\max z_1 = \max_{\mathbf{b}} \frac{\mathbf{b}'\boldsymbol{\delta} - z_2(\mathbf{b}'\Sigma_2\mathbf{b})^{1/2}}{(\mathbf{b}'\Sigma_1\mathbf{b})^{1/2}}$$

$$\geq \frac{\mathbf{b}^{*\prime}\boldsymbol{\delta} - z_2(\mathbf{b}^{*\prime}\Sigma_2\mathbf{b}^*)^{1/2}}{(\mathbf{b}^{*\prime}\Sigma_1\mathbf{b}^*)^{1/2}} > \frac{\mathbf{b}^*\boldsymbol{\delta} - z_2^*(\mathbf{b}^{*\prime}\Sigma_2\mathbf{b}^*)^{1/2}}{(\mathbf{b}^{*\prime}\Sigma_1\mathbf{b}^*)^{1/2}} = \max z_1^*. \quad (9.49)$$

The set of z_2 with corresponding maximum z_1 is thus a curve in the (z_1, z_2) plane running downward and to the right. Since $\boldsymbol{\delta} \neq \mathbf{0}$, the curve lies above and to the right of the origin. Q.E.D.

THEOREM 9.3.1 A linear classification rule $\mathbf{R}$ with $P_1 = 1 - \phi(z_1)$, $P_2 = 1 - \phi(z_2)$, where z_1 is maximized with respect to $\mathbf{b}$ for a given z_2, is admissible.

Proof Suppose $\mathbf{R}$ is not admissible. Then there is a linear classification rule $\mathbf{R}^* = (R_1^*, R_2^*)$ with arguments (z_1^*, z_2^*) such that $z_1^* \geq z_1$, $z_2^* \geq z_2$ with at least one inequality being strict. If $z_2^* = z_2$, then $z_1^* > z_1$, which contradicts the fact that z_1 is a maximum. If $z_2^* > z_2$, the maximum coordinate corresponding to z_2^* must be less than z_1, which contradicts $z_1^* \geq z_1$. Q.E.D.

Furthermore, it can be verified that the set of admissible linear classification rules is complete in the sense that for any linear classification rule outside this set there is a better one in the set.

We now want to characterize analytically the admissible linear classification rules. To achieve this the following lemma will be quite helpful.

LEMMA 9.3.2 If a point (α_1, α_2) with $\alpha_i > 0$, $i = 1, 2$, is admissible, then there exists $t_i > 0$, $i = 1, 2$, such that the corresponding linear classification rule is defined by

$$\mathbf{b} = (t_1\Sigma_1 + t_2\Sigma_2)^{-1}\boldsymbol{\delta} \tag{9.50}$$

$$c = \mathbf{b}'\boldsymbol{\mu}_1 + t_1\mathbf{b}'\Sigma_1\mathbf{b} = \mathbf{b}'\boldsymbol{\mu}_2 - t_2\mathbf{b}'\Sigma_2\mathbf{b}. \tag{9.51}$$

Proof Let the admissible linear classification rule be defined by the vector $\boldsymbol{\beta}$ and the scalar γ. The line

$$z_1 = \frac{s - \boldsymbol{\beta}'\boldsymbol{\mu}_1}{(\boldsymbol{\beta}'\Sigma_1\boldsymbol{\beta})^{1/2}}, \qquad z_2 = \frac{\boldsymbol{\beta}'\boldsymbol{\mu}_2 - s}{(\boldsymbol{\beta}'\Sigma_2\boldsymbol{\beta})^{1/2}}, \tag{9.52}$$

with s as parameter, has negative slope with the point (α_1, α_2) on it. Hence there exist positive numbers t_1, t_2 such that the line (9.52) is tangent to the ellipse

$$\frac{z_1^2}{t_1} + \frac{z_2^2}{t_2} = k \tag{9.53}$$

at the point (α_1, α_2). Consider the line defined by an arbitrary vector $\mathbf{b}$ and all scalars c. This line is tangent to an ellipse similar or concentric to (9.53) at the point (z_1, z_2) if c in (9.48) is chosen so that $-z_1t_2/z_2t_1$ is equal to the slope of this line. For a given $\mathbf{b}$, the values of c and the resulting z_1, z_2 are

$$\begin{gathered} c = \frac{t_1\mathbf{b}'\Sigma_1\mathbf{b}\mathbf{b}'\boldsymbol{\mu}_2 + t_2\mathbf{b}'\Sigma_2\mathbf{b}\mathbf{b}'\boldsymbol{\mu}_1}{t_1\mathbf{b}'\Sigma_1\mathbf{b} + t_2\mathbf{b}'\Sigma_2\mathbf{b}}, \\ z_1 = \frac{t_2(\mathbf{b}'\Sigma_1\mathbf{b})^{1/2}\mathbf{b}'\boldsymbol{\delta}}{t_1\mathbf{b}'\Sigma_1\mathbf{b} + t_2\mathbf{b}'\Sigma_2\mathbf{b}}, \qquad z_2 = \frac{t_2(\mathbf{b}'\Sigma_2\mathbf{b})^{1/2}\mathbf{b}'\boldsymbol{\delta}}{t_1\mathbf{b}'\Sigma_1\mathbf{b} + t_2\mathbf{b}'\Sigma_2\mathbf{b}}. \end{gathered} \tag{9.54}$$

This point (z_1, z_2) is on the ellipse

$$\frac{z_1^2}{t_1} + \frac{z_2^2}{t_2} = \frac{(\mathbf{b}'\boldsymbol{\delta})^2}{\mathbf{b}'(t_1\Sigma_1 + t_2\Sigma_2)\mathbf{b}}. \tag{9.55}$$

The maximum of the right side of (9.55) with respect to $\mathbf{b}$ occurs when $\mathbf{b}$ is given by (9.50). However, the maximum must correspond to the admissible procedure, for if there were a $\mathbf{b}$ such that the constant in (9.55) were larger than k, the point (α_1, α_2) would be within the ellipse with the constant in (9.55) and would be nearer the origin than the line tangent at (z_1, z_2). Then some points on this line (corresponding to procedures with $\mathbf{b}$ and scalar c) would be better. The expressions for the value of c in (9.54) and (9.51) are the same if we use the value of $\mathbf{b}$ as given in (9.50). Q.E.D.

Remark Since Σ_1, Σ_2 are positive definite and $t_i > 0, i = 1, 2, t_1\Sigma_1 + t_2\Sigma_2$ is positive definite, any multiples of (9.50) and (9.51) are equivalent solutions. When $\mathbf{b}$ in (9.50) is normalized so that

$$\mathbf{b}'\boldsymbol{\delta} = \mathbf{b}'(t_1\Sigma_1 + t_2\Sigma_2)\mathbf{b} = \boldsymbol{\delta}'(t_1\Sigma_1 + t_2\Sigma_2)^{-1}\boldsymbol{\delta}, \tag{9.56}$$

then from (9.54) we get

$$z_1 = t_1(\mathbf{b}'\Sigma_1\mathbf{b})^{1/2}, \qquad z_2 = t_2(\mathbf{b}'\Sigma_2\mathbf{b})^{1/2}. \tag{9.57}$$

Since these are homogeneous of degree 0 in t_1 and t_2 for $\mathbf{b}$ given by (9.50) we shall find it convenient to take $t_1 + t_2 = 1$ when $t_i > 0, i = 1, 2, t_1 - t_2 = 1$ when $t_1 > 0, t_2 < 0$, and $t_2 - t_1 = 1$ when $t_2 > 0, t_1 < 0$.

THEOREM 9.3.2 A linear classification rule with

$$\mathbf{b} = (t_1\Sigma_1 + t_2\Sigma_2)^{-1}\boldsymbol{\delta}, \tag{9.58}$$

$$c = \mathbf{b}'\boldsymbol{\mu}_1 + t_1\mathbf{b}'\Sigma_1\mathbf{b} = \mathbf{b}'\boldsymbol{\mu}_2 - t_2\mathbf{b}'\Sigma_2\mathbf{b} \tag{9.59}$$

for any t_1, t_2 such that $t_1\Sigma_1 + t_2\Sigma_2$ is positive definite is admissible.

Proof If $t_i > 0$, $i = 1, 2$, the corresponding z_1, z_2 are also positive. If this linear classification rule is not admissible, there would be a linear admissible classification rule that would be better (as the set of all linear admissible classification rules is complete) and both arguments for this rule would also be positive. By Lemma 9.3.2 the rule would be defined by

$$\boldsymbol{\beta} = (\tau_1\Sigma_1 + \tau_2\Sigma_2)^{-1}\boldsymbol{\delta}$$

for $\tau_i > 0$, $i = 1, 2$, such that $\tau_1 + \tau_2 = 1$. However, by the monotonicity properties of z_1, z_2 as functions of t_1, one of the coordinates corresponding to τ_1 would have to be less than one of the coordinates corresponding to t_1. This shows that the linear classification rule corresponding to $\boldsymbol{\beta}$ is not better than the rule defined by $\mathbf{b}$. Hence the theorem is proved for $t_i > 0$, $i = 1, 2$.

If $t_1 = 0$, then $z_1 = 0$, $\mathbf{b} = \Sigma_2^{-1}\boldsymbol{\delta}$, $z_2 = (\boldsymbol{\delta}'\Sigma_2^{-1}\boldsymbol{\delta})^{1/2}$. However, for any $\mathbf{b}$ if $z_1 = 0$, then $z_2 = \mathbf{b}'\boldsymbol{\delta}(\mathbf{b}'\Sigma_2\mathbf{b})^{-1/2}$, and z_2 is maximized if $\mathbf{b} = \Sigma_2^{-1}\boldsymbol{\delta}$. Similarly if $t_2 = 0$, the solution assumed in the theorem is optimum.

Now consider $t_1 > 0$, $t_2 < 0$, and $t_1 - t_2 = 1$. Any hyperbola

$$\frac{{z_1}^2}{t_1} + \frac{z_2}{t_2} = k \tag{9.60}$$

for $k > 0$ cuts the z_1 axis at $\pm(t_1k)^{1/2}$. The rule assumed in the theorem has $z_1 > 0$ and $z_2 < 0$. From (9.48) we get

$$\frac{(c - \mathbf{b}'\boldsymbol{\mu}_1)^2}{t_1\mathbf{b}'\Sigma_1\mathbf{b}} + \frac{(\mathbf{b}'\boldsymbol{\mu}_2 - c)^2}{t_2\mathbf{b}'\Sigma_2\mathbf{b}} = k. \tag{9.61}$$

The maximum of this expression with respect to c for given $\mathbf{b}$ is attained for c as given in (9.54). Then z_1, z_2 are of the form (9.54), and (9.61) reduces to (9.55). The maximum of (9.61) is then given by $\mathbf{b} = (t_1\Sigma_1 + t_2\Sigma_2)^{-1}\boldsymbol{\delta}$. It is easy to argue that this point is admissible because otherwise there would be a better point which would lie on a hyperbola with greater k.

The case $t_1 < 0$, $t_2 > 0$ can be similarly treated. Q.E.D.

Given t_1, t_2 so that $t_1\Sigma_1 + t_2\Sigma_2$ is positive definite, one would compute the optimum $\mathbf{b}$ such that

$$(t_1\Sigma_1 + t_2\Sigma_2)\mathbf{b} = \boldsymbol{\delta} \tag{9.62}$$

and then compute c as given in (9.51). Usually t_1, t_2 are not given. A desired solution can be obtained as follows. For another solution the reader is referred to Bahadur and Anderson (1962).

Minimization of One Probability of Misclassification Given the Other

Suppose z_2 is given and let $z_2 > 0$. Then if the maximum $z_1 > 0$, we want to find $t_2 = 1 - t_1$ such that $z_2 = t_2(\mathbf{b}'\Sigma_2\mathbf{b})^{1/2}$ with $\mathbf{b}$ given by (9.62). The solution can be approximated by trial and error. For $t_2 = 0$, $z_2 = 0$ and for $t_2 = 1$, $z_2 = (\mathbf{b}'\Sigma_2\mathbf{b})^{1/2} = (\mathbf{b}'\boldsymbol{\delta})^{1/2} = \boldsymbol{\delta}'\Sigma_2^{-1}\boldsymbol{\delta}$, where $\Sigma_2\mathbf{b} = \boldsymbol{\delta}$. One could try other values of t_2 successively by solving (9.62) and inserting the solution in $\mathbf{b}'\Sigma_2\mathbf{b}$ until $t_2(\mathbf{b}'\Sigma_2\mathbf{b})^{1/2}$ agrees closely enough with the desired z_2.

For $t_2 > 0$, $t_1 < 0$, and $t_2 - t_1 = 1$, z_2 is a decreasing function of t_2 ($t_2 \leq 1$) and at $t_2 = 1$, $z_2 = (\boldsymbol{\delta}'\Sigma_2^{-1}\boldsymbol{\delta})^{1/2}$. If the given z_2 is greater than $(\boldsymbol{\delta}'\Sigma_2\boldsymbol{\delta})^{1/2}$, then $z_1 < 0$ and we look for a value of t_2 such that $z_2 = t_2(\mathbf{b}'\Sigma_2\mathbf{b})^{1/2}$. We require that t_2 be large enough so that $t_1\Sigma_1 + t_2\Sigma_2 = (t_2 - 1)\Sigma_1 + t_2\Sigma_2$ is positive definite.

The Minimax Classification

The minimax linear classification rule is the admissible rule with $z_1 = z_2$. Obviously in this case $z_1 = z_2 > 0$ and $t_i > 0$, $i = 1, 2$. Hence we want to find $t_1 = 1 - t_2$ such that

$$0 = z_1{}^2 - z_2{}^2 = \mathbf{b}'(t_1{}^2\Sigma_1 - (1 - t_1)^2\Sigma_2)\mathbf{b}. \tag{9.63}$$

The values of $\mathbf{b}$ and t_1 satisfying (9.63) and (9.62) are obtained by the trial and error method.

Since Σ_1, Σ_2 are positive definite there exists a nonsingular matrix C such that $\Sigma_1 = C'\Delta C$, $\Sigma_2 = C'C$ where Δ is a diagonal matrix with diagonal elements $\lambda_1, \ldots, \lambda_p$, the roots of $\det(\Sigma_1 - \lambda\Sigma_2) = 0$.

Let $\mathbf{b}^* = (b_1{}^*, \ldots, \mathbf{b}_p{}^*)' = C\mathbf{b}$. Then (9.63) can be written as

$$\sum_{i=1}^{p} (\lambda_i - \theta)b_i^{*2} = 0 \tag{9.64}$$

where $\theta = (1 - t_1^2)/t_1^2$. If $\lambda_i - \theta$ are all positive or all negative, (9.64) will not have a solution for $\mathbf{b}^*$. To obtain a solution θ must lie between the minimum and the maximum of $\lambda_1, \ldots, \lambda_p$. This treatment is due to Banerjee and Marcus (1965), and it provides a valuable tool for obtaining $\mathbf{b}$ and t_1 for the minimax solution.

9.3.4 Test Concerning Discriminant Coefficients

As we have observed earlier, for discriminating between two multivariate normal populations with means $\boldsymbol{\mu}_1$, $\boldsymbol{\mu}_2$ and the same positive definite covariance matrix Σ, the optimum classification rule depends on the linear discriminant function $\mathbf{x}'\Sigma^{-1}(\boldsymbol{\mu}_1 - \boldsymbol{\mu}_2) - \frac{1}{2}(\boldsymbol{\mu}_1 + \boldsymbol{\mu}_2)\Sigma^{-1}(\boldsymbol{\mu}_1 - \boldsymbol{\mu}_2)$. The elements of $\Sigma^{-1}(\boldsymbol{\mu}_1 - \boldsymbol{\mu}_2)$ are called discriminant coefficients. In the case in which $\Sigma, \boldsymbol{\mu}_1, \boldsymbol{\mu}_2$ are unknown we can consider estimation and testing problems concerning these coefficients on the basis of sample observations $\mathbf{x}_\alpha^{(1)}$, $\alpha = 1, \ldots, N_1$, from π_1, and $\mathbf{x}_\alpha^{(2)}$, $\alpha = 1, \ldots, N_2$, from π_2. We have already tackled the problem of estimating these coefficients, here we will consider testing problems concerning them.

For testing hypotheses about these coefficients, the sufficiency consideration leads us to restrict our attention to the set of sufficient statistics $(\bar{\mathbf{X}}^{(1)}, \bar{\mathbf{X}}^{(2)}, S)$ as given in (9.27), where $\bar{\mathbf{X}}^{(1)}, \bar{\mathbf{X}}^{(2)}$ are independently distributed p-dimensional normal random vectors and $(N_1 + N_2 - 2)S$ is distributed independently of $(\bar{\mathbf{X}}^{(1)}, \bar{\mathbf{X}}^{(2)})$ as a Wishart random matrix with parameter Σ and $N_1 + N_2 - 2$ degrees of freedom. Further, invariance and sufficiency considerations permit us to consider the statistics $(\bar{\mathbf{X}}^{(1)} - \bar{\mathbf{X}}^{(2)}, S)$ instead of the random samples (independent) $\mathbf{X}_\alpha^{(1)}$, $\alpha = 1, \ldots, N_1$, from π_1, and $\mathbf{X}_\alpha^{(2)}$, $\alpha = 1, \ldots, N_2$, from π_2. Since $(1/N_1 + 1/N_2)^{-1/2}(\bar{\mathbf{X}}^{(1)} - \bar{\mathbf{X}}^{(2)})$ is distributed as a p-dimensional normal random vector with mean $(1/N_1 + 1/N_2)^{-1/2}(\boldsymbol{\mu}_1 - \boldsymbol{\mu}_2)$ and positive definite covariance matrix Σ, by relabeling variables we can consider the following canonical form where $\mathbf{X}$ is distributed as a p-dimensional normal random vector with mean $\boldsymbol{\mu} = (\mu_1, \ldots, \mu_p)'$ and positive definite covariance matrix Σ, and S is distributed (independent of $\mathbf{X}$) as Wishart with parameter Σ, and consider testing problems concerning the components of $\boldsymbol{\Gamma} = \Sigma^{-1}\boldsymbol{\mu}$. Equivalently this problem can be stated as follows: Let $\mathbf{X}^\alpha = (X_{\alpha 1}, \ldots, X_{\alpha p})'$, $\alpha = 1, \ldots, N$, be a random sample of size N $(>p)$ from a p-dimensional normal population with mean $\boldsymbol{\mu}$ and covariance matrix Σ. Write

$$\bar{\mathbf{X}} = \frac{1}{N}\sum_{\alpha=1}^{N} \mathbf{X}^\alpha, \qquad S = \sum_{\alpha=1}^{N} (\mathbf{X}^\alpha - \bar{\mathbf{X}})(\mathbf{X}^\alpha - \bar{\mathbf{X}})'.$$

(Note that we have changed the definition of S to be consistent with the notation of Chapter VII.) Let $\boldsymbol{\Gamma} = (\Gamma_1, \ldots, \Gamma_p)' = \Sigma^{-1}\boldsymbol{\mu}$. We shall now consider the following testing problems concerning $\boldsymbol{\Gamma}$, using the notation of Section 7.2.2. We refer to Giri (1964, 1965) for further details.

A. To test the null hypothesis $H_0:\boldsymbol{\Gamma} = \mathbf{0}$ against the alternatives $H_1:\boldsymbol{\Gamma} \neq \mathbf{0}$ when $\boldsymbol{\mu}$, Σ are unknown. Since Σ is nonsingular this problem is equivalent to testing $H_0:\boldsymbol{\mu} = \mathbf{0}$ against the alternatives $H_1:\boldsymbol{\mu} \neq \mathbf{0}$, which we have discussed in Chapter VII. This case does not seem to be of much interest in the context of linear discriminant functions but is included for completeness.

B. Let $\boldsymbol{\Gamma} = (\boldsymbol{\Gamma}_{(1)}, \boldsymbol{\Gamma}_{(2)})'$, where the $\Gamma_{(i)}$ are subvectors of dimension $p_i \times 1$, $i = 1, 2$, with $p_1 + p_2 = p$. We are interested in testing the null hypothesis $H_0:\boldsymbol{\Gamma}_{(1)} = \mathbf{0}$ against the alternatives $H_1:\boldsymbol{\Gamma}_{(1)} \neq \mathbf{0}$ when it is given that $\boldsymbol{\Gamma}_{(2)} = \mathbf{0}$ and $\boldsymbol{\mu}$, Σ are unknown. Let $S^* = S + N\bar{\mathbf{X}}\bar{\mathbf{X}}'$, and let S^*, S, $\bar{\mathbf{X}}$, $\boldsymbol{\mu}$, and Σ be partitioned as in (7.21) and (7.22) with $k = 2$. Let Ω be the parametric space of $((\boldsymbol{\Gamma}_{(1)}, \mathbf{0}), \Sigma)$ and $\omega = (\mathbf{0}, \Sigma)$ be the subspace of Ω when H_0 is true. The likelihood of the observations $\mathbf{x}^\alpha$ on $\mathbf{X}^\alpha$, $\alpha = 1, \ldots, N$, is

$$L(\boldsymbol{\Gamma}_{(1)}, \Sigma) = (2\pi)^{-Np/2}(\det \Sigma)^{-N/2} \times \exp\{-\tfrac{1}{2}\operatorname{tr}(\Sigma^{-1}s^* - 2N\boldsymbol{\Gamma}_{(1)}\bar{\mathbf{x}}'_{(1)} + N\Sigma_{(11)}\boldsymbol{\Gamma}_{(1)}\boldsymbol{\Gamma}'_{(1)})\}.$$

LEMMA 9.3.3

$$\max_{\Omega} L(\boldsymbol{\Gamma}_{(1)}, \Sigma) = (2N\pi)^{-Np/2}(\det \Sigma)^{-N/2} \times (1 - N\bar{\mathbf{x}}'_{(1)}(s_{(11)} + N\bar{\mathbf{x}}_{(1)}\bar{\mathbf{x}}'_{(1)})^{-1}\bar{\mathbf{x}}_{(1)})^{-N/2} \exp\{-\tfrac{1}{2}Np\}.$$

Proof

$$\begin{aligned}
\max_{\Omega} L(\boldsymbol{\Gamma}_{(1)}, \Sigma) &= \max_{\Sigma,\boldsymbol{\Gamma}_{(1)}} (2\pi)^{-Np/2}(\det \Sigma)^{-N/2} \exp\{-\tfrac{1}{2}\operatorname{tr}(\Sigma^{-1}s^* + N\Sigma_{(11)}^{-1}(\bar{\mathbf{x}}_{(1)} - \Sigma_{(11)}\boldsymbol{\Gamma}_{(1)}) \\
&\quad \times (\bar{\mathbf{x}}_{(1)} - \Sigma_{(11)}\boldsymbol{\Gamma}_{(1)})' - N\Sigma_{(11)}^{-1}\bar{\mathbf{x}}_{(1)}\bar{\mathbf{x}}'_{(1)})\} \\
&= \max_{\Sigma} (2\pi)^{-Np/2}(\det s^*)^{-N/2} \exp\{-\tfrac{1}{2}\operatorname{tr}(\Sigma^{-1}s^* - N\Sigma_{(11)}^{-1}\bar{\mathbf{x}}_{(1)}\bar{\mathbf{x}}'_{(1)})\}. \qquad (9.65)
\end{aligned}$$

Since Σ and s^* are positive definite there exist nonsingular upper triangular matrices K and T such that

$$\Sigma = KK', \qquad s^* = TT'.$$

Partition K and T as

$$K = \begin{pmatrix} K_{(11)} & K_{(12)} \\ 0 & K_{(22)} \end{pmatrix}, \qquad T = \begin{pmatrix} T_{(11)} & T_{(12)} \\ 0 & T_{(22)} \end{pmatrix}$$

where $K_{(11)}$, $T_{(11)}$ are (upper triangular) submatrices of K, T, respectively, of dimension $p_1 \times p_1$. Now

$$K^{-1} = \begin{pmatrix} K_{(11)}^{-1} & -(K_{(11)}^{-1}K_{(12)}K_{(22)}^{-1}) \\ 0 & K_{(22)}^{-1} \end{pmatrix}, \qquad T^{-1} = \begin{pmatrix} T_{(11)}^{-1} & -(T_{(11)}^{-1}T_{(12)}T_{(22)}^{-1}) \\ 0 & T_{(22)}^{-1} \end{pmatrix}$$

and $\Sigma_{(11)} = K_{(11)}K'_{(11)}$, $s^*_{(11)} = T_{(11)}T'_{(11)}$. Let $L = T^{-1}K$ and $\Sigma^* = LL'$. Let L, Σ^* be partitioned in the same way as K into submatrices $L_{(ij)}$, $\Sigma^*_{(ij)}$, respectively. Obviously $K_{(11)} = T_{(11)}L_{(11)}$. Writing $\mathbf{z}'_{(1)} = \mathbf{x}'_{(1)}T^{-1}_{(11)}$, from (9.65) we obtain

$$
\begin{aligned}
\max_{\Omega} L(\mathbf{\Gamma}_{(1)}, \Sigma)
&= \max_{K} (2\pi)^{-Np/2}(\det K)^{-N} \\
&\quad \times \exp\{-\tfrac{1}{2}\operatorname{tr}(K^{-1}(K')^{-1}T'T - NK^{-1}_{(11)}(K'_{(11)})^{-1}\bar{\mathbf{x}}_{(1)}\bar{\mathbf{x}}'_{(1)})\} \\
&= \max_{K} (2\pi)^{-Np/2}(\det s^*)^{-N/2}(\det \Sigma^*)^{-N/2} \\
&\quad \times \exp\{-\tfrac{1}{2}\operatorname{tr}(\Sigma^{*-1} - N\Sigma^{*-1}_{(11)}\mathbf{z}_{(1)}\mathbf{z}'_{(1)})\} \\
&= \max_{\Lambda} (2\pi)^{-Np/2}(\det s^*)^{-N/2}(\det \Lambda_{(22)})^{N/2}(\det(\Lambda_{(11)} - \Lambda_{(12)}\Lambda^{-1}_{(22)}\Lambda_{(21)}))^{N/2} \\
&\quad \times \exp\{-\tfrac{1}{2}\operatorname{tr}(\Lambda_{(11)} + \Lambda_{(22)} - (\Lambda_{(11)} - \Lambda_{(12)}\Lambda^{-1}_{(22)}\Lambda_{(21)})(N\mathbf{z}_{(1)}\mathbf{z}'_{(1)}))\} \\
&= (2\pi/N)^{-Np/2}(\det s^*)^{-N/2}(\det(I - N\mathbf{z}_{(1)}\mathbf{z}'_{(1)}))^{-N/2}\exp\{-\tfrac{1}{2}Np\} \\
&= (2\pi/N)^{-Np/2}(\det s^*)^{-N/2}(1 - N\bar{\mathbf{x}}'_{(1)}(s_{(11)} + N\bar{\mathbf{x}}_{(1)}\bar{\mathbf{x}}_{(1)})^{-1}\bar{\mathbf{x}}_{(1)})^{-N/2} \\
&\quad \times \exp\{-\tfrac{1}{2}Np\}, \qquad (9.66)
\end{aligned}
$$

where $(\Sigma^*)^{-1} = \Lambda$ and Λ is partitioned into submatrices $\Lambda_{(ij)}$ similar to those of Σ^*. The next to last step in (9.66) follows from the fact that the maximum likelihood estimates of $\Lambda_{(22)}$, $\Lambda_{(11)}$ are I/N, $(I - N\mathbf{z}_{(1)}\mathbf{z}'_{(1)})/N$ (see Lemma 5.1.1) and that of $\Lambda_{(12)}$ is $\mathbf{0}$. Q.E.D.

Since

$$\max_{\omega} L(\mathbf{\Gamma}_{(1)}, \Sigma) = (2\pi N)^{-Np/2}(\det s^*)^{-N/2}\exp\{-\tfrac{1}{2}Np\},$$

the likelihood ratio criterion for testing H_0 is given by

$$
\begin{aligned}
\lambda = \frac{\max_{\omega} L(\mathbf{\Gamma}_{(1)}, \Sigma)}{\max_{\Omega} L(\mathbf{\Gamma}_{(1)}, \Sigma)} &= (1 - N\bar{\mathbf{x}}'_{(1)}(s_{(11)} + N\bar{\mathbf{x}}_{(1)}\bar{\mathbf{x}}'_{(1)})^{-1}\bar{\mathbf{x}}_{(1)})^{N/2} \\
&= (1 - r_1)^{N/2}, \qquad (9.67)
\end{aligned}
$$

where r_1 is given in Section 7.2.2. (We have used the same notation for the classification regions $\mathbf{R}$ and the statistic $\mathbf{R}$.) Thus the likelihood ratio test of H_0 rejects H_0 whenever

$$r_1 \geq C, \qquad (9.68)$$

where the constant C depends on the level of significance α of the test. From Chapter VI the probability density function of R_1 under H_1 is given by

$$
\begin{aligned}
f_{R_1}(r_1|\delta_1^2) &= \frac{\Gamma(\frac{1}{2}N)}{\Gamma(\frac{1}{2}p_1)\Gamma(\frac{1}{2}(N - p_1))} r_1^{p_1/2-1}(1 - r_1)^{(N-p_1)/2-1} \\
&\quad \times \exp\{-\tfrac{1}{2}\delta_1^2\}\phi(\tfrac{1}{2}N, \tfrac{1}{2}p_1; \tfrac{1}{2}(r_1\delta_1^2)) \qquad (9.69)
\end{aligned}
$$

provided $r_1 \geq 0$ and is zero elsewhere, where $\delta_1^2 = N\mathbf{\Gamma}'_{(1)}\Sigma_{(11)}\mathbf{\Gamma}_{(1)}$. Obviously under H_0, $\delta_1^2 = 0$ and R_1 is distributed as central beta with parameter $(\frac{1}{2}p_1, \frac{1}{2}(N - p_1))$.

Let G_{BT} (as defined in Section 7.2.2 with $k = 2$) be the multiplicative group of lower triangular matrices

$$g = \begin{pmatrix} g_{(11)} & 0 \\ g_{(21)} & g_{(22)} \end{pmatrix}$$

of dimension $p \times p$. The problem of testing H_0 against H_1 remains invariant under G_{BT} with $k = 2$ operating as $\mathbf{X}_\alpha \to g\mathbf{X}_\alpha$, $\alpha = 1, \ldots, N$, $g \in G_{BT}$. The induced transformation in the space of $(\bar{\mathbf{X}}, S)$ is given by $(\bar{\mathbf{X}}, S) \to (g\bar{\mathbf{X}}, gSg')$ and in the space of $(\boldsymbol{\mu}, \Sigma)$ is given by $(\boldsymbol{\mu}, \Sigma) \to (g\boldsymbol{\mu}, g\Sigma g')$. A set of maximal invariants in the space of $(\bar{\mathbf{X}}, S)$ under G_{BT} is (R_1, R_2) as defined in (6.63) with $k = 2$. A corresponding maximal invariant in the parametric space of $(\boldsymbol{\mu}, \Sigma)$ is given by (δ_1^2, δ_2^2), where

$$\begin{aligned} \delta_1^2 &= N(\Sigma_{(11)}\mathbf{\Gamma}_{(1)} + \Sigma_{(12)}\mathbf{\Gamma}_{(2)})'\Sigma_{(11)}^{-1}(\Sigma_{(11)}\mathbf{\Gamma}_{(1)} + \Sigma_{(12)}\mathbf{\Gamma}_{(2)}) \\ \delta_1^2 + \delta_2^2 &= N\mathbf{\Gamma}'\Sigma\mathbf{\Gamma}. \end{aligned} \tag{9.70}$$

Since $\mathbf{\Gamma}_{(2)} = 0$ in this case, we get $\delta_2^2 = 0$ and $\delta_1^2 = N\mathbf{\Gamma}'_{(1)}\Sigma_{(11)}\mathbf{\Gamma}_{(1)}$. Hence under $H_0{:}\delta_1^2 = 0$ and under $H_1{:}\delta_1^2 > 0$, the joint probability density function of (R_1, R_2) under H_1 is given by (6.73). The ratio of the density of (R_1, R_2) under H_1 to its density under H_0 is given by

$$\exp\{-\tfrac{1}{2}\delta_1^2\} \sum_{j=0}^{\infty} \left(\frac{r_1\delta_1^2}{2}\right)^j \frac{\Gamma(\frac{1}{2}N + j)\Gamma(\frac{1}{2}p_1)}{j!\Gamma(\frac{1}{2}p_1 + j)\Gamma(\frac{1}{2}N)}. \tag{9.71}$$

Hence we have the following theorem.

THEOREM 9.3.3 For testing H_0 against H_1, the likelihood ratio test which rejects H_0 for large values of R_1 is uniformly most powerful invariant.

C. To test the null hypothesis $H_0{:}\mathbf{\Gamma}_{(2)} = \mathbf{0}$ against the alternatives $H_1{:}\mathbf{\Gamma}_{(2)} \neq \mathbf{0}$ when $\boldsymbol{\mu}$ and Σ are unknown and $\mathbf{\Gamma}_{(1)}$, $\mathbf{\Gamma}_{(2)}$ are defined as in case B. The likelihood of the observations $\mathbf{x}_\alpha$ on $\mathbf{X}_\alpha$, $\alpha = 1, \ldots, N$, is

$$\begin{aligned} L(\mathbf{\Gamma}, \Sigma) &= (2\pi)^{-Np/2}(\det \Sigma)^{-N/2} \\ &\quad \times \exp\{-\tfrac{1}{2}\operatorname{tr}(\Sigma^{-1} s^* - 2N\mathbf{\Gamma}'\bar{\mathbf{x}} + N\Sigma\mathbf{\Gamma}\mathbf{\Gamma}')\}. \end{aligned} \tag{9.72}$$

Proceeding exactly in the same way as in Lemma 9.3.3, we obtain

$$\begin{aligned} \max_{\Omega} (\mathbf{\Gamma}, \Sigma) &= (2\pi N)^{-Np/2}(\det s^*)^{-N/2}(1 - N\bar{\mathbf{x}}'(s + N\bar{\mathbf{x}}\bar{\mathbf{x}}')^{-1}\bar{\mathbf{x}})^{-N/2} \\ &\quad \times \exp\{-\tfrac{1}{2}Np\}, \end{aligned} \tag{9.73}$$

where $\Omega = \{(\boldsymbol{\Gamma}, \Sigma)\}$. From Lemma 9.3.3 and (9.73) the likelihood ratio criterion for testing H_0 is given by

$$\lambda = \frac{\max_\omega L(\boldsymbol{\Gamma}, \Sigma)}{\max_\Omega L(\boldsymbol{\Gamma}, \Sigma)} = \left(\frac{1 - r_1 - r_2}{1 - r_1}\right)^{N/2}, \tag{9.74}$$

where $\omega = ((\boldsymbol{\Gamma}_{(1)}, \mathbf{0}), \Sigma)$ and r_1, r_2 are as defined in case B. Thus the likelihood ratio test for testing H_0 rejects H_0 whenever

$$z = \frac{1 - r_1 - r_2}{1 - r_1} \leq C, \tag{9.75}$$

where the constant C depends on the level of significance α of the test. From (6.73) the joint probability density function of Z and R_1 under H_0 is given by

$$\exp\{-\tfrac{1}{2}\delta_1^2\} \sum_{j=0}^{\infty} \Gamma(\tfrac{1}{2}N + j)(\tfrac{1}{2}r_1\delta_1^2)^j (r_1)^{p_1/2-1}$$
$$\times \frac{(1 - r_1)^{(N-p_1)/2-1} z^{(N-p_1)/2-1}(1 - z)^{(p-p_1)/2-1}}{j!\Gamma(\tfrac{1}{2}p_1 + j)\Gamma(\tfrac{1}{2}(N - p)\Gamma(\tfrac{1}{2}(p - p_1))}. \tag{9.76}$$

From this it follows that under H_0, Z is distributed as a central beta random variable with parameter $(\tfrac{1}{2}(N - p_1), \tfrac{1}{2}p_2)$ and is independent of R_1.

The problem of testing H_0 against H_1 remains invariant under the group of transformations G_{BT} with $k = 2$, operating as $\mathbf{X}^\alpha \to g\mathbf{X}^\alpha$, $g \in G_{BT}$, $\alpha = 1, \ldots, N$. A set of maximal invariants in the space of $(\bar{\mathbf{X}}, S)$ under G_{BT} is (R_1, R_2) of case B and the corresponding maximal invariant in the parametric space of $(\boldsymbol{\mu}, \Sigma)$ is (δ_1^2, δ_2^2) of (9.70). Under H_0, $\delta_2^2 = 0$ and under H_1, $\delta_2^2 > 0$ (δ_1^2 is unknown). The joint probability density function of (R_1, R_2) is given by (6.73). From this we conclude that R_1 is sufficient for δ_1^2 when H_0 is true, and the marginal probability density function of R_1 when H_0 is true is given by (9.69). This is also the probability density function of R_1 when H_1 is true.

LEMMA 9.3.4 The family of probability density functions $\{f_{R_1}(r_1|\delta_1^2), \delta_1^2 \geq 0\}$ is boundedly complete.

Proof Let $\Psi(r_1)$ be any real valued function of r_1. Then

$$E_{\delta_1^2}(\Psi(R_1)) = \exp\{-\tfrac{1}{2}\delta_1^2\} \sum_{j=0}^{\infty} (\tfrac{1}{2}\delta_1^2)^j a_j \int_0^1 \Psi(r_1) r_1^{p_1/2+j-1}(1 - r_1)^{(N-p_1)/2-1}\, dr_1$$
$$= \exp\{-\tfrac{1}{2}\delta_1^2\} \sum_{j=0}^{\infty} (\tfrac{1}{2}\delta_1^2)^j a_j \int_0^1 \Psi^*(r_1) r_1^{\,j}\, dr_1,$$

where

$$a_j = \frac{\Gamma(\frac{1}{2}N + j)}{j!\Gamma(\frac{1}{2}(N - p_1))\Gamma(\frac{1}{2}p_1 + j)}, \qquad \Psi^*(r_1) = r_1^{p_1/2-1}(1 - r_1)^{(N-p_1)/2-1}\psi(r_1).$$

Hence $E_{\delta_1^2}(\Psi(R_1)) = 0$ identically in δ_1^2 implies that

$$\sum_{j=0}^{\infty} (\tfrac{1}{2}\delta_1^2)^j a_j \int_0^1 \Psi^*(r_1) r_1^j \, dr_1 = 0 \tag{9.77}$$

identically in δ_1^2. Since the left-hand side of (9.77) is a polynomial in δ_1^2, all its coefficients must be zero. In other words,

$$\int_0^1 \Psi^*(r_1) r_1^j \, dr_1 = 0, \qquad j = 1, 2, \ldots, \tag{9.78}$$

which implies that $\Psi^{*+}(r_1) = \Psi^{*-}(r_1)$ for all r_1, except possibly for a set of values of r_1 of probability measure 0. Hence $\Psi^*(r_1) = 0$ almost everywhere, which implies that $\Psi(r_1) = 0$ almost everywhere. Q.E.D.

THEOREM 9.3.4 The likelihood ratio test of $H_0 : \boldsymbol{\Gamma}_{(2)} = \mathbf{0}$ when $\boldsymbol{\mu}$, Σ are unknown is uniformly most powerful invariant similar against the alternatives $H_1 : \boldsymbol{\Gamma}_{(2)} \neq \mathbf{0}$.

Proof Since R_1 is sufficient for δ_1^2 when H_0 is true and the distribution of R_1 is boundedly complete, it is well known that (see, e.g., Lehmann, 1959, p. 134) any level α invariant test $\phi(r_1, r_2)$ has Neyman structure with respect to R_1, i.e.,

$$E_{\delta_1^2}(\phi(R_1, R_2) | R_1 = r_1) = \alpha. \tag{9.79}$$

Now to find the uniformly most powerful test among all similar invariant tests we need the ratio of the conditional probability density function of R_2 given $R_1 = r_1$ under H_1 to that under H_0, and this ratio is given by

$$\exp\{-\tfrac{1}{2}\delta_1^2(1 - r_1)\} \sum_{j=0}^{\infty} \frac{(\frac{1}{2}r_2\delta_2^2)^j \Gamma(\frac{1}{2}(N - p_1) + j)\Gamma(\frac{1}{2}p_2)}{j!\Gamma(\frac{1}{2}p_2 + j)\Gamma(\frac{1}{2}(N - p_1))}. \tag{9.80}$$

Since the distribution of R_2 on each surface $R_1 = r_1$ is independent of δ_1^2, condition (9.79) reduces the problem to that of testing a simple hypothesis $\delta_2^2 = 0$ against the alternatives $\delta_2^2 > 0$ on each surface $R_1 = r_1$. In this conditional situation, by Neyman and Pearson's fundamental lemma, the uniformly most powerful level α invariant test of $\delta_2^2 = 0$ against the alternatives $\delta_2^2 > 0$ [from (9.80)] rejects H_0 whenever

$$\sum_{j=0}^{\infty} \frac{(\frac{1}{2}r_2\delta_2^2)^j \Gamma(\frac{1}{2}(N - p_1) + j)\Gamma(\frac{1}{2}p_2)}{j!\Gamma(\frac{1}{2}p_2 + j)\Gamma(\frac{1}{2}(N - p_1))} \geq C(r_1), \tag{9.81}$$

where $C(r_1)$ is a constant such that the test has level α on each surface $R_1 = r_1$.

Since the left-hand side of (9.81) is an increasing function of r_2 and $r_2 = (1 - r_1)(1 - z)$, this reduces to rejecting H_0 on each surface $R_1 = r_1$ whenever $z \leq C$, where the constant C is chosen such that the test has level α. Since, under H_0, z is independent of R_1, the constant C does not depend on r_1. Hence the theorem. Q.E.D.

D. Let $\boldsymbol{\Gamma} = (\boldsymbol{\Gamma}_{(1)}, \boldsymbol{\Gamma}_{(2)}, \boldsymbol{\Gamma}_{(3)})'$, where $\boldsymbol{\Gamma}_{(i)}$ is $p_i \times 1, i = 1, 2, 3$, and $\sum_1^3 p_i = p$. We are interested in testing the null hypothesis $H_0 : \boldsymbol{\Gamma}_{(2)} = \mathbf{0}$ against the alternatives $H_1 : \boldsymbol{\Gamma}_{(2)} \neq \mathbf{0}$ when it is given that $\boldsymbol{\Gamma}_{(3)} = \mathbf{0}$ and $\boldsymbol{\Gamma}_{(1)}$ is unknown. Here

$$\Omega = \{(\boldsymbol{\Gamma}_{(1)}, \boldsymbol{\Gamma}_{(2)}, \mathbf{0}), \Sigma), \omega = \{(\boldsymbol{\Gamma}_{(1)}, \mathbf{0}, \mathbf{0}), \Sigma\}.$$

Let S^*, S, $\bar{\mathbf{X}}$, $\boldsymbol{\mu}$, and Σ be partitioned as in (7.21) and (7.22) with $k = 3$. Using Lemma 9.3.3 we get from (9.72)

$$\frac{\max_\omega L(\boldsymbol{\Gamma}, \Sigma)}{\max_\Omega L(\boldsymbol{\Gamma}, \Sigma)} = \left(\frac{1 - r_1 - r_2}{1 - r_1}\right)^{N/2}, \tag{9.82}$$

where r_1, r_2, r_3 are given in Section 7.2.2 with $k = 3$. The likelihood ratio test of H_0 rejects H_0 whenever

$$z = \frac{1 - r_1 - r_2}{1 - r_1} \leq C, \tag{9.83}$$

where C is a constant such that the test has size α. The joint probability density function of R_1, R_2, R_3 (under H_1) is given in (6.73) with $k = 3$, where

$$\begin{aligned}
\delta_1^2 &= N(\Sigma_{(11)}\boldsymbol{\Gamma}_{(1)} + \Sigma_{(12)}\boldsymbol{\Gamma}_{(2)})'\Sigma_{(11)}^{-1}(\Sigma_{(11)}\boldsymbol{\Gamma}_{(1)} + \Sigma_{(12)}\boldsymbol{\Gamma}_{(2)}) \\
\delta_1^2 + \delta_2^2 &= N\begin{pmatrix} \Sigma_{(11)}\boldsymbol{\Gamma}_{(1)} + \Sigma_{(12)}\boldsymbol{\Gamma}_{(2)} \\ \Sigma_{(21)}\boldsymbol{\Gamma}_{(1)} + \Sigma_{(22)}\boldsymbol{\Gamma}_{(2)} \end{pmatrix}' \begin{pmatrix} \Sigma_{(11)} & \Sigma_{(12)} \\ \Sigma_{(21)} & \Sigma_{(11)} \end{pmatrix}^{-1} \\
&\quad \times \begin{pmatrix} \Sigma_{(11)}\boldsymbol{\Gamma}_{(1)} + \Sigma_{(12)}\boldsymbol{\Gamma}_{(2)} \\ \Sigma_{(21)}\boldsymbol{\Gamma}_{(1)} + \Sigma_{(22)}\boldsymbol{\Gamma}_{(2)} \end{pmatrix} \\
\delta_3^2 &= N\boldsymbol{\Gamma}_{(3)}'\left(\Sigma_{(33)} - \begin{pmatrix} \Sigma_{(13)} \\ \Sigma_{(23)} \end{pmatrix}' \begin{pmatrix} \Sigma_{(11)} & \Sigma_{(12)} \\ \Sigma_{(21)} & \Sigma_{(22)} \end{pmatrix}^{-1} \begin{pmatrix} \Sigma_{(13)} \\ \Sigma_{(23)} \end{pmatrix}\right)\boldsymbol{\Gamma}_{(3)} = 0,
\end{aligned} \tag{9.84}$$

and under H_0, $\delta_2^2 = 0$. From this it follows that the joint probability density function of Z and R_1 under H_0 is given by (9.75) with p replaced by $p_1 + p_2$. Hence under H_0, Z is distributed as central beta with parameters $(\frac{1}{2}(N - p_1), \frac{1}{2}p_2)$ and is independent of R_1.

The problem of testing H_0 against H_1 remains invariant under G_{BT} with $k = 3$ operating as $\mathbf{X}_\alpha \to g\mathbf{X}_\alpha$, $g \in G_{\mathrm{BT}}$, $\alpha = 1, \ldots, N$. A set of maximal invariants in the space of $(\bar{\mathbf{X}}, S)$ under G_{BT} with $k = 3$ is (R_1, R_2, R_3), and the corresponding maximal invariants in the parametric space is $(\delta_1^2, \delta_2^2, \delta_3^2)$

as given in (9.83). Under H_0, $\delta_2^2 = 0$ and under H_1, $\delta_1^2 > 0$, and it is given that $\delta_3^2 = 0$. As we have proved in case C, R_1 is sufficient for δ_1^2 under H_0 and the distribution of R_1 is boundedly complete. Now arguing in the same way as in case C we prove the following theorem.

THEOREM 9.3.5 For testing $H_0:\boldsymbol{\Gamma}_{(2)} = \mathbf{0}$ the likelihood ratio test which rejects H_0 whenever $z \leq C$, C depending on the level α of the test, is uniformly most powerful invariant similar against $H_1:\boldsymbol{\Gamma}_{(2)} \neq \mathbf{0}$ when it is given that $\boldsymbol{\Gamma}_{(3)} = \mathbf{0}$.

Tests depending on the Mahalanobis distance statistic are also used for testing hypotheses concerning discriminant coefficients. The reader is referred to Rao (1965) or Kshirsagar (1972) for an account of this. Recently Sinha and Giri (1976) have studied the optimum properties of the likelihood ratio tests of these problems from the point of view of Isaacson's type D and type E property (see Isaacson, 1951).

9.4 CLASSIFICATION INTO MORE THAN TWO MULTIVARIATE NORMAL POPULATIONS

As pointed out in connection with Theorem 9.2.1 if $C(i|j) = C$ for all $i \neq j$, then the Bayes classification rule $\mathbf{R}^* = (R_1^*, \ldots, R_k^*)$ against the a priori probabilities $(p_1, \ldots, p_k)$ classifies an observation $\mathbf{x}$ to R_l^* if

$$\frac{f_l(\mathbf{x})}{f_j(\mathbf{x})} \geq \frac{p_j}{p_l} \qquad \text{for} \quad j = 1, \ldots, k, \quad j \neq l. \tag{9.85}$$

In this section we shall assume that $f_i(\mathbf{x})$ is the probability density function of a p-variate normal random vector with mean $\boldsymbol{\mu}_i$ and the same positive definite covariance matrix Σ. Most known results in this area are straightforward extensions of the results for the case $k = 2$. In this case the Bayes classification rule $\mathbf{R}^* = (R_1^*, \ldots, R_k^*)$ classifies $\mathbf{x}$ to R_l^* whenever

$$u_{lj} = \log \frac{f_l(\mathbf{x})}{f_j(\mathbf{x})} = (\mathbf{x} - \tfrac{1}{2}(\boldsymbol{\mu}_l + \boldsymbol{\mu}_j))'\Sigma^{-1}(\boldsymbol{\mu}_l - \boldsymbol{\mu}_j) \geq \log \frac{p_j}{p_l}. \tag{9.86}$$

Each u_{lj} is the linear discriminant function related to the jth and the lth populations and obviously $u_{lj} = -u_{jl}$.

In the case in which the a priori probabilities are unknown the minimax classification rule $\mathbf{R} = (R_1, \ldots, R_k)$ classifies $\mathbf{x}$ to R_l if

$$u_{lj} \geq C_l - C_j, \qquad j = 1, \ldots, k, \quad j \neq l, \tag{9.87}$$

where the C_j are nonnegative constants and are determined in such a way that all $P(i|i, \mathbf{R})$ are equal. Let us now evaluate $P(i|i, \mathbf{R})$. First observe that the random variable

$$U_{ij} = (\mathbf{X} - \tfrac{1}{2}(\boldsymbol{\mu}_i + \boldsymbol{\mu}_j))'\Sigma^{-1}(\boldsymbol{\mu}_i - \boldsymbol{\mu}_j) \tag{9.88}$$

satisfies $U_{ij} = -U_{ji}$. Thus we use $k(k-1)/2$ linear discriminant functions U_{ij} if the mean vectors $\boldsymbol{\mu}_i$ span a $(k-1)$-dimensional hyperplane. Now the U_{ij} are normally distributed with

$$\begin{aligned} E_i(U_{ij}) &= \tfrac{1}{2}(\boldsymbol{\mu}_i - \boldsymbol{\mu}_j)'\Sigma^{-1}(\boldsymbol{\mu}_i - \boldsymbol{\mu}_j), \\ E_j(U_{ij}) &= -\tfrac{1}{2}(\boldsymbol{\mu}_i - \boldsymbol{\mu}_j)'\Sigma^{-1}(\boldsymbol{\mu}_i - \boldsymbol{\mu}_j) \\ \operatorname{var}(U_{ij}) &= (\boldsymbol{\mu}_i - \boldsymbol{\mu}_j)'\Sigma^{-1}(\boldsymbol{\mu}_i - \boldsymbol{\mu}_j) \\ \operatorname{cov}(U_{ij}, U_{ij'}) &= (\boldsymbol{\mu}_i - \boldsymbol{\mu}_j)'\Sigma^{-1}(\boldsymbol{\mu}_i - \boldsymbol{\mu}_{j'}), \qquad j \neq j', \end{aligned} \tag{9.89}$$

where $E_i(U_{ij})$ denotes the expectation of U_{ij} when $\mathbf{X}$ comes from π_i. For a given j let us denote the joint probability density function of U_{ji}, $i = 1, \ldots, k$; $i \neq j$, by p_j. Then

$$P(j|j, \mathbf{R}) = \int_{C_j - C_k}^{\infty} \cdots \int_{C_j - C_1}^{\infty} p_j \prod_{i \neq j} du_{ji}.$$

Note that the sets of regions given by (9.87) form an admissible class.

If the parameters are unknown, they are replaced by their appropriate estimates from training samples from these populations to obtain sample discriminant functions as discussed in the case of two populations. We discussed earlier the problems associated with the distribution of sample discriminant functions and different methods of evaluating the probabilities of misclassification. For some relevant results the reader is referred to Das Gupta (1973) and the references therein.

The problem of unequal covariance matrices can be similarly resolved by using the results presented earlier for the case of two multivariate normal populations with unequal covariance matrices. For further discussions in this case the reader is referred to Fisher (1938), Brown (1947), Rao (1952, 1963), and Cacoullos (1965). Das Gupta (1962) considered the problems where $\boldsymbol{\mu}_1, \ldots, \boldsymbol{\mu}_k$ are linearly restricted and showed that the maximum likelihood classification rule is admissible Bayes when the common covariance matrix Σ is known. Following Kiefer and Schwartz (1965), Srivastava (1964) obtained similar results when Σ is unknown.

EXAMPLE 9.4.1 Consider two populations π_1 and π_2 of plants of two distinct varieties of wheat. The measurements for each member of these two populations are

x_1 plant height (cm),

x_2 number of effective tillers,

x_3 length of ear (cm),

x_4 number of fertile spikelets per 10 ears,

x_5 numbers of grains per 10 ears,

x_6 weight of grains per 10 ears (gm).

Assuming that these are six-dimensional normal populations with different unknown mean vectors $\boldsymbol{\mu}_1, \boldsymbol{\mu}_2$ and with the same unknown covariance matrix Σ we shall consider here the problem of classifying an individual with observation $\mathbf{x} = (x_1, \ldots, x_6)'$ on him to one of these populations. Since the parameters are unknown we obtained two training samples (Table 9.1) (of size 27 each) from them (these data were collected from the Indian Agricultural Research Institute, New Delhi, India). The sample mean vectors

TABLE 9.1

Samples from Populations

	π_1						π_2					
Observation	x_1	x_2	x_3	x_4	x_5	x_6	x_1	x_2	x_3	x_4	x_5	x_6
1	77.60	136	9.65	12.6	322	14.7	65.55	166	9.29	11.3	323	13.1
2	83.45	177	9.76	13.1	321	14.5	67.10	132	9.52	11.7	319	13.6
3	76.20	164	10.52	13.9	384	17.1	66.25	173	9.88	12.1	319	13.6
4	80.30	185	9.76	12.5	259	15.4	80.45	155	11.19	13.8	394	17.6
5	82.30	187	9.77	13.4	314	14.4	78.30	202	10.78	13.3	376	16.7
6	86.00	171	9.25	13.0	278	13.0	77.80	155	10.86	14.0	401	18.2
7	90.50	211	9.75	12.9	308	13.6	79.20	161	10.68	14.3	417	17.8
8	81.50	158	10.38	13.6	258	14.8	82.65	158	10.64	12.2	382	17.4
9	79.75	176	9.31	12.0	307	13.2	79.85	156	10.83	13.7	366	16.1
10	86.85	175	10.23	14.2	330	14.6	67.30	157	9.78	11.8	354	14.0
11	72.90	139	10.29	12.9	346	15.5	70.65	173	9.97	12.2	310	12.5
12	73.50	124	9.68	12.0	308	14.1	67.15	159	9.99	12.3	325	11.9
13	86.85	149	10.33	13.5	337	15.1	80.85	160	10.47	12.7	358	15.5
14	89.15	224	9.70	13.0	317	14.7	81.80	162	10.87	13.9	403	18.3
15	78.05	149	9.63	12.6	285	12.4	81.15	178	11.07	13.8	401	16.2
16	81.95	200	9.28	12.8	272	12.5	82.95	177	11.04	13.5	366	16.6
17	81.70	187	9.46	12.6	276	12.3	81.20	172	11.14	14.1	412	19.3
18	89.65	200	9.58	11.1	285	12.5	83.85	192	11.24	14.1	372	17.2
19	79.90	152	9.49	13.2	275	11.7	67.60	164	10.07	11.9	305	11.8
20	71.15	144	9.55	12.0	292	11.9	64.35	170	9.34	11.0	303	11.6
21	83.05	147	10.30	13.3	326	14.2	66.40	158	9.71	11.9	326	12.9
22	87.25	231	10.32	13.1	332	14.7	79.10	162	10.49	12.9	395	17.0
23	78.65	183	9.90	14.1	324	14.6	81.65	171	11.31	14.1	403	17.2
24	79.95	165	9.34	12.5	290	12.1	79.35	162	10.43	12.6	390	15.9
25	86.65	198	10.07	12.7	293	12.3	78.90	166	11.14	14.0	432	18.4
26	92.05	212	9.81	13.1	304	13.9	80.45	172	11.32	14.3	306	18.7
27	76.80	193	9.80	13.1	288	13.4	83.75	202	10.38	13.4	343	13.8

TABLE 9.2

Sample Means

	π_1	π_2
x_1	81.98704	76.13333
x_2	175.44444	167.22222
x_3	9.81148	10.49741
x_4	12.91852	12.99630
x_5	305.22222	363.00000
x_6	13.82222	15.66296

Sample Covariance Matrix s

3.13548	90.44476	6.54646	38.81429	71.91165	1.97064
2.61154	−41.76262	0.37415	71.91165	498.06410	0.98669
0.37533	11.89829	0.82986	1.97064	0.98669	0.26782
0.75635	18.28440	1.27375	3.13548	2.61158	0.37533
18.28440	1214.74359	51.04744	90.44476	−41.76282	11.89829
1.27375	51.04744	3.73134	6.54646	0.37415	0.82986

and the sample covariance matrix are given in Table 9.2 [see Eq. (9.27) for the notation]. Using the sample discriminant function

$$v = (\mathbf{x} - \tfrac{1}{2}(\overline{\mathbf{x}}^{(1)} + \overline{\mathbf{x}}^{(2)}))' s^{-1}(\overline{\mathbf{x}}^{(1)} - \overline{\mathbf{x}}^{(2)}),$$

we classify x to π_1 if $v \geq 0$ and to π_2 if $v < 0$. Since $v \geq 0$ implies that

$$\mathbf{x}' s^{-1}\overline{\mathbf{x}}^{(1)} - \tfrac{1}{2}\overline{\mathbf{x}}^{(1)\prime} s^{-1}\overline{\mathbf{x}}^{(1)} \geq \mathbf{x}' s^{-1}\overline{\mathbf{x}}^{(2)} - \tfrac{1}{2}\overline{\mathbf{x}}^{(2)\prime} s^{-1}\overline{\mathbf{x}}^{(2)},$$

writing

$$d_1(\mathbf{x}) = \mathbf{x}' s^{-1}\overline{\mathbf{x}}^{(1)} - \tfrac{1}{2}\overline{\mathbf{x}}^{(1)\prime} s^{-1}\overline{\mathbf{x}}^{(1)}, \qquad d_2(\mathbf{x}) = \mathbf{x}' s^{-1}\overline{\mathbf{x}}^{(2)} - \tfrac{1}{2}\overline{\mathbf{x}}^{(2)\prime} s^{-1}\overline{\mathbf{x}}^{(2)},$$

we classify $\mathbf{x}$

$$\begin{array}{lll} \text{to } \pi_1 & \text{if } & d_1(\mathbf{x}) \geq d_2(\mathbf{x}) \\ \text{to } \pi_2 & \text{if } & d_1(\mathbf{x}) < d_2(\mathbf{x}). \end{array}$$

Now

$$\begin{aligned} d_1(\mathbf{x}) &= 0.10070x_1 + 0.20551x_2 + 75.13581x_3 + 1.69460x_4 + 0.16121x_5 \\ &\quad - 15.98724x_6 - 315.81156 \\ d_2(\mathbf{x}) &= -0.49307x_1 + 0.28011x_2 + 84.84069x_3 - 1.88664x_4 \\ &\quad + 0.22783x_5 - 16.30691x_6 - 351.33860. \end{aligned}$$

To verify the efficacy of this plug-in classification rule we now classify the observed sample observations using the proposed criterion. The results are given in Table 9.3.

TABLE 9.3

Evaluation of the Classification Rule for Sample Observation s

Observation	Population π_1 Classified to:	Population π_2 Classified to:
1	π_1	π_2
2	π_1	π_2
3	π_2	π_2
4	π_1	π_2
5	π_1	π_2
6	π_1	π_2
7	π_1	π_2
8	π_1	π_2
9	π_1	π_2
10	π_1	π_2
11	π_2	π_2
12	π_1	π_2
13	π_1	π_2
14	π_1	π_2
15	π_1	π_2
16	π_1	π_2
17	π_1	π_2
18	π_1	π_2
19	π_1	π_2
20	π_1	π_2
21	π_1	π_2
22	π_2	π_2
23	π_1	π_2
24	π_1	π_2
25	π_1	π_2
26	π_1	π_2
27	π_1	π_2

9.5 CONCLUDING REMARKS

We have limited our discussions mainly to the case of multivariate normal distributions. The cases of nonnormal and discrete distributions are equally important in practice and have been studied by various workers. For multinomial distributions the works of Matusita (1956), Chernoff (1956), Wesler (1956, 1959), Cochran and Hopkins (1961), Bunke (1966), and Glick (1969) are worth mentioning. For multivariate Bernouilli distributions we refer to Bahadur (1961), Solomon (1960, 1961), Hills (1966), Martin and Bradly (1972), Cooper (1963, 1965), Bhattacharya and Das Gupta (1964), and Anderson (1972). The works of Kendall (1966) and Marshall and Olkin (1968) are equally important for related results in connection with discrete distributions. The reader is also referred to the book edited by Cacoullos (1973) for an up-to-date account of research work in the area of discriminant analysis.

EXERCISES

1. Let π_1, π_2 be two p-variate normal populations with means $\boldsymbol{\mu}_1, \boldsymbol{\mu}_2$ and the same covariance matrix Σ. Let $\mathbf{X} = (X_1, \ldots, X_p)'$ be a random vector distributed according to π_1 or π_2 and let $\mathbf{b} = (b_1, \ldots, b_p)'$ be a real vector. Show that

$$\frac{[E_1(\mathbf{b}'\mathbf{X}) - E_2(\mathbf{b}'\mathbf{X})]^2}{\operatorname{var}(\mathbf{b}'\mathbf{X})}$$

is maximum for all choices of $\mathbf{b}$ whenever $\mathbf{b} = \Sigma^{-1}(\boldsymbol{\mu}_1 - \boldsymbol{\mu}_2)$. [$E_i(\mathbf{b}'\mathbf{X})$ is the expected value of $\mathbf{b}'\mathbf{X}$ under π_i.)

2. Let $\mathbf{x}_\alpha^{(i)}$, $\alpha = 1, \ldots, N_i$, $i = 1, 2$. Define dummy variables $y_\alpha^{(i)}$

$$y_\alpha^{(i)} = \frac{N_i}{N_1 + N_2}, \qquad \alpha = 1, \ldots, N_i, \quad i = 1, 2.$$

Find the regression on the variables $\mathbf{x}_\alpha^{(i)}$ by choosing $\mathbf{b} = (b_1, \ldots, b_p)'$ to minimize

$$\sum_{i=1}^{2} \sum_{\alpha=1}^{N_i} (y_\alpha^{(i)} - \mathbf{b}'(\mathbf{x}_\alpha^{(i)} - \bar{\mathbf{x}}))^2,$$

where

$$\bar{\mathbf{x}} = \frac{N_1\bar{\mathbf{x}}^{(1)} + N_2\bar{\mathbf{x}}^{(2)}}{N_1 + N_2}, \qquad N_i\bar{\mathbf{x}}^{(i)} = \sum_{\alpha=1}^{N_i} \mathbf{x}_\alpha^{(i)}.$$

Show that the minimizing $\mathbf{b}$ is proportional to $s^{-1}(\bar{\mathbf{x}}^{(1)} - \bar{\mathbf{x}}^{(2)})$, where

$$(N_1 + N_2 - 2)s = \sum_{i=1}^{2} \sum_{\alpha=1}^{N_i} (\mathbf{x}_\alpha^{(i)} - \bar{\mathbf{x}}^{(i)})(\mathbf{x}_\alpha^{(i)} - \bar{\mathbf{x}}^{(i)})'.$$

3. (a) For discriminating between two p-dimensional normal distributions with unknown means $\boldsymbol{\mu}_1, \boldsymbol{\mu}_2$ and the same unknown covariance matrix Σ, show that the sample discriminant function v can be obtained from

$$\mathbf{b}'(\mathbf{x} - \tfrac{1}{2}(\bar{\mathbf{x}}^{(1)} + \bar{\mathbf{x}}^{(2)}))$$

by finding $\mathbf{b}$ to maximize the ratio

$$\frac{[\mathbf{b}'(\bar{\mathbf{x}}^{(1)} - \bar{\mathbf{x}}^{(2)}]^2}{(\mathbf{b}'s\mathbf{b})}$$

where $\bar{\mathbf{x}}^{(i)}$, s are given in (9.27).

(b) In the analysis of variance terminology (a) amounts to finding **b** to maximize the ratio of the between-population sum of squares to the within-population sum of squares. With this terminology show that the sample discriminant function obtained by finding **b** to maximize the ratio of the between-population sum of squares to the total sum of squares is proportional to v.

4. For discriminating between two-p-variate normal populations with known mean vectors $\boldsymbol{\mu}_1, \boldsymbol{\mu}_2$ and the same known positive definite covariance matrix Σ show that the linear discriminant function u is also good for any p-variate normal population with mean $a_1\boldsymbol{\mu}_1 + a_2\boldsymbol{\mu}_2$, where $a_1 + a_2 = 1$, and the same covariance matrix Σ.

5. Prove Theorems 9.2.2 and 9.2.3.

6. Consider the problem of classifying an individual into one of two populations π_1, π_2 with probability density functions f_1, f_2, respectively.

(a) Show that if $P(f_2(\mathbf{x}) = 0|\pi_1) = 0$, $P(f_1(\mathbf{x}) = 0|\pi_2) = 0$, then every Bayes classification rule is admissible.

(b) Show that if $P(f_1(\mathbf{x})/f_2(\mathbf{x}) = k|\pi_i) = 0$, $i = 1, 2$, $0 \le k \le \infty$, then every admissible classification rule is a Bayes classification rule.

7. Let $v = v(\mathbf{x})$ be defined as in (9.28). Show that for testing the equality of mean vectors of two p-variate normal populations with the same positive definite covariance matrix Σ, Hotelling's T^2-test on the basis of sample observations $\mathbf{x}_\alpha^{(1)}$, $\alpha = 1, \ldots, N_1$, from the first population and $\mathbf{x}_\alpha^{(2)}$, $\alpha = 1, \ldots, N_2$, from the second population, is proportional to $v(\overline{\mathbf{x}}^{(1)})$ and $v(\overline{\mathbf{x}}^{(2)})$.

8. Consider the problem of classifying an individual with observation $\mathbf{x} = (x_1, \ldots, x_p)'$ between two p-dimensional normal populations with the same mean vector $\mathbf{0}$ and positive definite covariance matrices Σ_1, Σ_2.

(a) Given $\Sigma_1 = \sigma_1{}^2 I$, $\Sigma_2 = \sigma_2{}^2 I$, where $\sigma_1{}^2, \sigma_2{}^2$ are known positive constants and $C(2|1) = C(1|2)$, find the minimax classification rule.

(b) (i) Let

$$\Sigma_1 = \begin{pmatrix} 1 & \rho_1 & \cdots & \rho_1 \\ \rho_1 & 1 & \cdots & \rho_1 \\ \vdots & \vdots & & \vdots \\ \rho_1 & \rho_1 & \cdots & 1 \end{pmatrix}, \qquad \Sigma_2 = \sigma_2 \begin{pmatrix} 1 & \rho_2 & \cdots & \rho_2 \\ \rho_2 & 1 & \cdots & \rho_2 \\ \vdots & \vdots & & \vdots \\ \rho_2 & \rho_2 & \cdots & 1 \end{pmatrix}.$$

Show that the likelihood ratio classification rule leads to $aZ_1 - bZ_2 = C$ as the boundary separating the regions R_1, R_2 where

$$Z_1 = \mathbf{x}'\mathbf{x}, \qquad Z_2 = \left(\sum_1^p x_i\right)^2$$

$$a = (1 - \rho_1)^{-1} - (\sigma^2(1 - \rho_2))^{-1},$$

$$b = \frac{\rho_1}{(1 - \rho_1)(1 + (p - 1)\rho_1)} - \frac{\rho_2}{(1 - \rho_2)\sigma^2(1 + (p - 1)\rho_2)}.$$

(ii) (Bartlett and Please, 1963) Suppose that $\rho_1 = \rho_2$ in (a). Then the classification rule reduces to: Classify $\mathbf{x}$ to π_1 if $u \geq c'$ and to π_2 if $u < c'$ where c' is a constant and

$$U = Z_1 - \frac{\rho}{1 + (p - 1)\rho} Z_2.$$

Show that the corresponding random variable U has a $((1 - \rho)\sigma_i^2)\chi^2$ distribution with p degrees of freedom where $\sigma_i^2 = 1$ if $\mathbf{X}$ comes from π_1 and $\sigma_i^2 = \sigma^2$ if $\mathbf{X}$ comes from π_2.

9. Show that the likelihood ratio tests for cases C and D in Section 9.3 are uniformly most powerful similar among all tests whose power depends only on δ_1^2 and δ_2^2.

10. Giri (1973) Let $\boldsymbol{\xi} = (\xi_1, \ldots, \xi_p)', \boldsymbol{\eta} = (\eta_1, \ldots, \eta_p)'$ be two p-dimensional independent complex Gaussian random vectors with complex means $E(\boldsymbol{\xi}) = \boldsymbol{\alpha}$, $E(\boldsymbol{\eta}) = \boldsymbol{\beta}$ and with the same Hermitian positive definite covariance matrix Σ.

(a) Find the likelihood ratio rule for classifying an observation into one of these two populations.

(b) Let $\boldsymbol{\xi}$ be distributed as a p-dimensional complex Gaussian random vector with mean $E(\xi) = \boldsymbol{\alpha}$ and Hermitian positive definite covariance matrix Σ. Let $\Gamma = \Sigma^{-1}\boldsymbol{\alpha}$. Find the likelihood ratio tests for problems analogous to B, C, and D in Section 9.3.

REFERENCES

Anderson, J. A. (1969). Discrimination between k populations with constraints on the probabilities of misclassification, *J. R. Statist. Soc. B* **31**, 123–139.

Anderson, J. A. (1972). Separate sample logistic discrimination, *Biometrika* **59**, 19–36.

Anderson, T. W. (1951). Classification by multivariate analysis, *Psychometrika* **16**, 631–650.

Anderson, T. W. (1958). "An Introduction to Multivariate Statistical Analysis." Wiley, New York.

Bahadur, R. R. (1961). On classification based on response to N dichotomus items, *In* "Studies in item analysis and prediction" (H. Solomon, ed.), pp. 177–186. Stanford Univ. Press, Stanford, California.

Bahadur, R. R., and Anderson, T. W. (1962). Classification into two multivariate normal distributions with different covariance matrices, *Ann. Math. Statist.* **33**, 420–431.

Banerjee, K. S., and Marcus, L. F. (1965). Bounds in minimax classification procedures, *Biometrika* **52**, 153–654.

Bartlett, M. S., and Please, N. W. (1963). Discrimination in the case of zero mean differences, *Biometrika* **50**, 17–21.

Bhattacharya, P. K., and Das Gupta, S. (1964). Classification into exponential populations, *Sankhya* **A26**, 17–24.

Blackwell, D., and Girshik, M. A. (1954). "Theory of Games and Statistical Decisions." Wiley, New York.

Bowker, A. H. (1960). A representation of Hotelling's T^2 and Anderson's classification statistic, "Contribution to Probability and Statistics" (Hotelling's vol.). Stanford Univ. Press, Stanford, California.

Brown, G. R. (1947). Discriminant functions, *Ann. Math. Statist.* **18**, 514–528.
Bunke, O. (1964). Uber optimale verfahren der discriminazanalyse, *Abl. Deutsch. Akad. Wiss. Klasse. Math. Phys. Tech.* **4**, 35–41.
Bunke, O. (1966). Nichparametrische klassifikations verfahren für qualitative und quantitative Beobachtunger, *Berlin Math. Naturwissensch. Reihe* **15**, 15–18.
Cacoullos, T. (1965). Comparing Mahalanobis distances, I and II, *Sankhya* **A27**, 1–22, 27–32.
Cacoullos, T. (1973). "Discriminant Analysis and Applications." Academic Press, New York.
Cavalli, L. L. (1945). Alumi problemi dela analyse biometrica di popolazioni naturali, *Mem. Inst. Indrobiol.* **2**, 301–323.
Chernoff, H. (1956). A Classification Problem. Tech. rep. no 33. Stanford Univ., Stanford, California.
Cochran, W. G. (1968). Commentary of estimation of error rates in discriminant analysis, *Technometrics* **10**, 204–210.
Cochran, W. G., and Hopkins, C. E. (1961). Some classification problems with multivariate quantitative data, *Biometrics* **17**, 10–32.
Cooper, D. W. (1963). Statistical classifications with quadratic forms, *Biometrika* **50**, 439–448.
Cooper, D. W. (1965). Quadratic discriminant function in pattern recognition, *IEEE Trans. Informat. II* **11**, 313–315.
Das Gupta, S. (1962). On the optimum properties of some classification rules, *Ann. Math. Statist.* **33**, 1504.
Das Gupta, S. (1973), Classification procedures, a review, "Discriminant Analysis and Applications" T. Cacoullos, ed.). Academic Press, New York.
Ferguson, T. S. (1967). "Mathematical Statistics." Academic Press, New York.
Fisher, R. A. (1936). Use of multiple measurements in Taxonomic problems, *Ann. Eug.* **7**, 179–184.
Fisher, R. A. (1938). The statistical utilization of multiple measurements, *Ann. Eug.* **8**, 376–386.
Giri, N. (1964). On the likelihood ratio test of a normal multivariate testing problem, *Ann. Math. Statist.* **35**, 181–189.
Giri, N. (1965). On the likelihood ratio test of a normal multivariate testing problem, II, *Ann. Math. Statist.* **36**, 1061–1065.
Giri, N. (1973). On discriminant decision functions in complex Gaussian distributions, "Probability and Information Theory" (M. Behara, K. Krickeberg, and J. Wolfowitz, eds.), pp. 139–148. Springer Verlag No. 296, Berlin and New York.
Glick, N. (1969). Estimating Unconditional Probabilities of Correct Classification. Stanford Univ., Dept. Statist. Tech. Rep. No. 3.
Han, Chien Pai (1968). A note on discrimination in the case of unequal covariance matrices, *Biometrika* **55**, 586–587.
Han, Chien Pai (1969). Distribution of discriminant function when covariance matrices are proportional, *Ann. Math. Statist.* **40**, 979–985.
Han, Chien Pai (1970). Distribution of discriminant function in Circular models, *Ann. Inst. Statist. Math.* **22**, 117–125.
Hills, M. (1966). Allocation rules and their error rates, *J. R. Statist. Soc. Ser. B* **28**, 1–31.
Hodges, J. L. (1950). Survey of discriminant analysis, USAF School of Aviation Medicine, rep. no. 1, Randolph Field, Texas.
Hotelling, H. (1931). The generalization of Student's ratio, *Ann. Math. Statist.* **2**, 360–378.
Isaacson, S. L. (1951). On the theory of unbiased tests of simple statistical hypotheses specifying the values of two or more parameters, *Ann. Math. Statist.* **22**, 217–234.
Kabe, D. G. (1963). Some results on the distribution of two random matrices used in classification procedures, *Ann. Math. Statist.* **34**, 181–185.
Kendall, M. G. (1966). Discrimination and classification, *Proc. Int. Symp. Multv. Anal.* (P. R. Krishnaiah, ed.), pp. 165–185. Academic Press. New York.

Kiefer, J., and Schwartz, R. (1965). Admissible bayes character of T^2-, R^2-, and fully invariant tests for classical multivariate normal problem, *Ann. Math. Statist.* **36**, 747–770.

Kshirsagar, A. M. (1972). "Multivariate Analysis." Dekker, New York.

Kudo, A. (1959). The classification problem viewed as a two decision problem, I, *Mem. Fac. Sci. Kyushu Univ.* **A13**, 96–125.

Kudo, A. (1960). The classification problem viewed as a two decision problem, II, *Mem. Fac. Sci. Kyushu Univ.* **A14**, 63–83.

Lachenbruch, P. A., and Mickey, M. R. (1968). Estimation of error rates in discriminant analysis, *Technometries* **10**, 1–11.

Lehmann, E. (1959). "Testing Statistical Hypotheses." Wiley, New York.

Mahalanobis, P. C. (1927). Analysis of race mixture in Bengal, *J. Proc. Asiatic Soc. Bengal* **23**, No 3.

Mahalanobis, P. C. (1930). On tests and measurements of group divergence, *Proc. Asiatic Soc. Bengal* **26**, 541–589.

Mahalanobis, P. C. (1936). On the generalized distance in statistics, *Proc. Nat. Inst. Sci. India* **2**, 49–55.

Marshall, A. W., and Olkin, I. (1968). A general approach to some screening and classification problems, *J. R. Statist. Soc.* **B30**, 407–435.

Martin, D. C., and Bradly, R. A. (1972). Probability models, estimation and classification for multivariate dichotomous populations, *Biometrika* **28**, 203–222.

Morant, G. M. (1928). A preliminary classification of European races based on cranial measurements, *Biometrika*, **20**, 301–375.

Neyman, J., and Pearson, E. S. (1933). The problem of most efficient tests of statistical hypotheses. *Phil. Trans. R. Soc.* **231**.

Neyman, J., and Pearson, E. S. (1936). Contribution to the theory of statistical hypotheses, I. *Statist. Res. Memo. I* 1–37.

Matusita, K. (1956). Decision rules based on the distance for the classification problem, *Ann. Inst. Statist. Math.* **8**, 67–77.

Nishida, N. (1971). A note on the admissible tests and classification in multivariate analysis, *Hiroshima Math. J.* **1**, 427–434.

Okamoto, M. (1963). An asymptotic expansion for the distribution of linear discriminant function, *Ann. Math. Statist.* **34**, 1286–1301, correction vol. **39**, 1358–1359.

Pearson, K. (1926). On the coefficient of racial likeness, *Biometrika* **18**, 105–117.

Penrose, L. S. (1947). Some notes on discrimination, *Ann. Eug.* **13**, 228–237.

Quenouille, M. (1956). Notes on bias in estimation, *Biometrika* **43**, 353–360.

Rao, C. R. (1946). Tests with discriminant functions in multivariate analysis, *Sankhya* **7**, 407–413.

Rao, C. R. (1947a). The problem of classification and distance between two populations, *Nature (London)* **159**, 30–31.

Rao, C. R. (1947b). Statistical criterion to determine the group to which an individual belongs, *Nature (London)* **160**, 835–836.

Rao, C. R. (1948). The utilization of multiple measurements in problem of biological classification, *J. R. Statist. Soc. B* **10**, 159–203.

Rao, C. R. (1949a). On the distance between two populations, *Sankhya* **9**, 246–248.

Rao, C. R. (1949b). On some problems arising out of discrimination with multiple characters, *Sankhya* **9**, 343–366.

Rao, C. R. (1950). Statistical inference applied to classification problems, *Sankhya* **10**, 229–256.

Rao, C. R. (1952). "Advanced Statistical Methods in Biometric Research." Wiley, New York.

Rao, C. R. (1954). A general theory of discrimination when the information about alternative population is based on samples. *Ann. Math. Statist.* **25**, 651–670.

Rao, M. M. (1963). Discriminant analysis, *Ann. Inst. Statist. Math.* **15**, 15–24.

Rao, C. R. (1965). "Linear Statistical Inference and its Applications." Wiley, New York.

Schucany, W. R., Gray, H. L., and Owen, D. B. (1971). On bias reduction in estimation, *Biometrika* **43**, 353.

Sinha, B. K., and Giri, N. (1975). On the distribution of a random matrix, *Commun. Statist.* **4**, 1057–1063.

Sinha, B. K., and Giri, N. (1976). On the optimality and non-optimality of some multivariate normal test procedures, *Sankhya* (to appear).

Sitgreaves, R. (1952). On the distribution of two random matrices used in classification procedures, *Ann. Math. Statist.* **23**, 263–270.

Smith, C. A. B. (1947). Some examples of discrimination *Ann. Eug.* **13**, 272–282.

Solomon, H. (1960). Classification procedures based on dichotomous response vectors, "Contributions to Probability and Statistics" (Hotelling's volume), pp. 414–423. Stanford Univ. Press, Stanford, California.

Solomon, H. (1961). Classification procedures based on dichotomous response vectors, *in* "Studies in Item Analysis and Predictions, (H. Solomon, ed.), pp. 177–186. Stanford Univ. Press, Stanford, California.

Srivastava, M. S. (1964). Optimum procedures for Classification and Related Topics. Tech. rep. no. 11, Dept. Statist., Stanford Univ.

Tildesley, M. L. (1921). A first study of the Burmese skull, *Biometrika* **13**, 247–251.

Tukey, J. W. (1958). Bias and confidence in not quite large samples, *Ann. Math. Statist.* **20**, 618.

von Mises, R. (1945). On the classification of observation data into distinct groups, *Ann. Math. Statist.* **16**, 68–73.

Wald, A. (1944). On a statistical problem arising in the classification of an individual into one of two groups, *Ann. Math. Statist.* **15**, 145–162.

Wald, A. (1950). "Statistical Decision Function." Wiley, New York.

Wald, A., and Wolfowitz, J. (1950). Characterization of minimum complete class of decision function when the number of decisions is finite, *Proc. Berkeley Symp. Prob. Statist., 2nd, California.*

Welch, B. L. (1939). Note on discriminant functions, *Biometrika* **31**, 218–220.

CHAPTER X

Multivariate Covariance Models

10.0 INTRODUCTION

In this chapter we deal mainly with three interrelated concepts concerning multivariate covariance models: principal components, factor models, and canonical correlations. All these concepts deal with the covariance structure of the multivariate normal distribution and aim at reducing the dimension of the observable vector variables. We shall also include a brief treatment of time series analysis.

10.1 PRINCIPAL COMPONENTS

Let $\mathbf{X} = (X_1, \ldots, X_p)'$ be a random vector with

$$E(\mathbf{X}) = \boldsymbol{\mu}, \qquad \operatorname{cov}(\mathbf{X}) = \Sigma = (\sigma_{ij}),$$

where $\boldsymbol{\mu}$ is a real p-vector and Σ is a real positive semidefinite matrix. In multivariate analysis the dimension of $\mathbf{X}$ often causes problems in obtaining suitable statistical techniques to analyze a set of repeated observations (data) on $\mathbf{X}$. For this reason it is natural to look for methods for rearranging the data so that with as little loss of information as possible, the dimension of the problem is considerably reduced. We have seen one such attempt in connection with discriminant analysis in Chapter IX.

This notion is motivated by the fact that in early stages of research interest was usually focused on those variables that tend to exhibit greatest variation from observation to observation. Since variables which do not change much from observation to observation can be treated as constants, by discarding low variance variables and centering attention on high variance variables, one can more conveniently study the problem of interest in a subspace of lower dimension. No doubt some information on the relationship among variables is lost by such a method; nevertheless, in many practical situations there is much more to gain than to lose by this approach.

The principal component approach was first introduced by Karl Pearson (1901) for nonstochastic variables. Hotelling (1933) generalized this concept to random vectors. Principal components of $\mathbf{X}$ are normalized linear combinations of the components of $\mathbf{X}$ which have special properties in terms of variances. For example, the first principal component of $\mathbf{X}$ is the normalized linear combination

$$Z_1 = \mathbf{L}'\mathbf{X}, \qquad \mathbf{L} = (l_1, \ldots, l_p)' \in E^p,$$

where $\mathbf{L}$ is chosen so that $\operatorname{var}(\mathbf{L}'\mathbf{X})$ is maximum with respect to $\mathbf{L}$. Obviously each weight l_i is a measure of the importance to be placed on the component X_i. We require the condition $\mathbf{L}'\mathbf{L} = 1$ in order to obtain a unique solution for the principal components. We shall assume that components of $\mathbf{X}$ are measured in the same units; otherwise the requirement $\mathbf{L}'\mathbf{L} = 1$ is not a sensible one. It will be seen that estimates of principal components are sensitive to units used in the analysis so that different sets of weights are obtained for different sets of units. Sometimes the sample correlation matrix is used instead of the sample covariance matrix to estimate these weights, thereby avoiding the problem of units, since the principal components are then invariant to changes in units of measurement. The use of the correlation matrix amounts to standardizing the variables to unit sample variance. However, since the new variables are not really standardized relative to the population, there is then introduced the problem of interpreting what has actually been computed. In practice such a technique is not recommended unless the sample size is large.

The second principal component is a linear combination that has maximum variance among all normalized linear combinations uncorrelated with Z_1 and so on up to the pth principal component of $\mathbf{X}$. The original vector $\mathbf{X}$ can thus be transformed to the vector of its principal components by means of a rotation of the coordinate axes that has inherent statistical properties. The choosing of such a type of coordinate system is to be contrasted with previously treated problems where the coordinate system in which the original data are expressed is irrelevant. The weights in the principal components associated with the random vector **X** are exactly the normalized characteristic vectors of the covariance matrix Σ of **X**, whereas the characteristic roots of Σ are the variances of the principal components, the largest root being the variance of the first principal component.

It may be cautioned that sample observations should not be indiscriminately subjected to principal component analysis merely to obtain fewer variables with which to work. Rather, principal component analysis should be used only if it complements the overall objective. For example, in problems in which correlation rather than variance is of primary interest or in which there are likely to be important nonlinear functions of observations that are of interest, most of the information about such relationships may be lost if all but the first few principal components are dropped.

10.1.1 Population Principal Components

Let $\mathbf{X} = (X_1, \ldots, X_p)'$ be a p-variate random vector with $E(\mathbf{X}) = \boldsymbol{\mu}$ and known covariance matrix Σ. We shall consider cases in which Σ is a positive semidefinite matrix or cases in which Σ has multiple roots. Since we shall only be concerned with variances and covariances of **X** we shall assume that $\boldsymbol{\mu} = \mathbf{0}$. The first principal component of **X** is the normalized linear combination (say) $Z_1 = \boldsymbol{\alpha}'\mathbf{X}$, $\boldsymbol{\alpha} = (\alpha_1, \ldots, \alpha_p)' \in E^p$ with $\alpha'\alpha = 1$ such that

$$\operatorname{var}(\boldsymbol{\alpha}'\mathbf{X}) = \max_{\mathbf{L}} \operatorname{var}(\mathbf{L}'\mathbf{X}) \tag{10.1}$$

for all $\mathbf{L} \in E^p$ satisfying $\mathbf{L}'\mathbf{L} = 1$. Now

$$\operatorname{var}(\mathbf{L}'\mathbf{X}) = \mathbf{L}'\Sigma\mathbf{L}.$$

Thus to find the first principal component $\boldsymbol{\alpha}'\mathbf{X}$ we need to find the $\boldsymbol{\alpha}$ that maximizes $\mathbf{L}'\Sigma\mathbf{L}$ for all choices of $\mathbf{L} \in E^p$, subject to the restriction that $\mathbf{L}'\mathbf{L} = 1$. Using the Lagrange multiplier λ, we need to find the $\boldsymbol{\alpha}$ that maximizes

$$\phi_1(\mathbf{L}) = \mathbf{L}'\Sigma\mathbf{L} - \lambda(\mathbf{L}'\mathbf{L} - 1) \tag{10.2}$$

for all choices of **L** satisfying $\mathbf{L}'\mathbf{L} = 1$. Since $\mathbf{L}'\Sigma\mathbf{L}$ and $\mathbf{L}'\mathbf{L}$ have derivatives everywhere in a region containing $\mathbf{L}'\mathbf{L} = 1$, we conclude that the vector $\boldsymbol{\alpha}$ which maximizes ϕ_1 must satisfy

$$2\Sigma\boldsymbol{\alpha} - 2\lambda\boldsymbol{\alpha} = \mathbf{0}, \tag{10.3}$$

or

$$(\Sigma - \lambda I)\boldsymbol{\alpha} = \mathbf{0}. \tag{10.4}$$

Since $\boldsymbol{\alpha} \neq \mathbf{0}$ (as a consequence of $\boldsymbol{\alpha}'\boldsymbol{\alpha} = 1$), Eq. (10.4) has a solution if

$$\det(\Sigma - \lambda I) = 0. \tag{10.5}$$

That is, λ is a characteristic root of Σ and $\boldsymbol{\alpha}$ is the corresponding characteristic vector. Since Σ is of dimension $p \times p$, there are p values of λ which satisfy (10.5). Let

$$\lambda_1 \geq \lambda_2 \geq \cdots \geq \lambda_p \tag{10.6}$$

denote the ordered characteristic roots of Σ and let

$$\boldsymbol{\alpha}_1 = (\alpha_{11}, \ldots, \alpha_{1p})', \ldots, \boldsymbol{\alpha}_p = (\alpha_{p1}, \ldots, \alpha_{pp})' \tag{10.7}$$

denote the corresponding characteristic vectors of Σ. Note that since Σ is positive semidefinite some of the characteristic roots may be zeros; in addition, some of the roots may have multiplicities greater than unity. From (10.4)

$$\boldsymbol{\alpha}'\Sigma\boldsymbol{\alpha} = \lambda\boldsymbol{\alpha}'\boldsymbol{\alpha} = \lambda. \tag{10.8}$$

Thus we conclude that if $\boldsymbol{\alpha}$ with $\boldsymbol{\alpha}'\boldsymbol{\alpha} = 1$ satisfies (10.4), then

$$\operatorname{var}(\boldsymbol{\alpha}'\mathbf{X}) = \boldsymbol{\alpha}'\Sigma\boldsymbol{\alpha} = \lambda, \tag{10.9}$$

where λ is the characteristic root of Σ corresponding to $\boldsymbol{\alpha}$. Thus to maximize $\operatorname{var}(\boldsymbol{\alpha}'\mathbf{X})$ we need to choose $\lambda = \lambda_1$, the largest characteristic root of Σ, and $\boldsymbol{\alpha} = \boldsymbol{\alpha}_1$, the characteristic vector of Σ corresponding to λ_1. If the rank of $\Sigma - \lambda_1 I$ is $p - 1$, then there is only one solution to

$$(\Sigma - \lambda_1 I)\boldsymbol{\alpha}_1 = \mathbf{0} \qquad \text{with} \quad \boldsymbol{\alpha}_1'\boldsymbol{\alpha}_1 = 1.$$

DEFINITION 10.1.1 *First principal component* The normalized linear function $\boldsymbol{\alpha}_1'\mathbf{X} = \sum_{i=1}^{p} \alpha_{1i}X_i$, where $\boldsymbol{\alpha}_1$ is the normalized characteristic vector of Σ corresponding to its largest characteristic root λ_1, is called the first principal component of $\mathbf{X}$.

We have assumed no distributional form for $\mathbf{X}$. If $\mathbf{X}$ has a p-variate normal distribution with positive definite covariance matrix Σ, then the surfaces of constant probability density are concentric ellipsoids and $Z_1 = \boldsymbol{\alpha}_1'\mathbf{X}$ represents the major principal axis of these ellipsoids. In general under the assumption of normality of $\mathbf{X}$, the principal components will represent a rotation of coordinate axes of its components to the principal axes of these ellipsoids. If there are multiple roots, the axes are not uniquely defined.

The second principal component is the normalized linear function $\boldsymbol{\alpha}'\mathbf{X}$ having maximum variance among all normalized linear functions $\mathbf{L}'\mathbf{X}$ that are uncorrelated with Z_1. If any normalized linear function $\mathbf{L}'\mathbf{X}$ is uncorrelated with Z_1, then

$$\begin{aligned} E(\mathbf{L}'\mathbf{X}Z_1) &= E(\mathbf{L}'\mathbf{X}Z_1') = E(\mathbf{L}'\mathbf{X}\mathbf{X}'\boldsymbol{\alpha}_1) \\ &= \mathbf{L}'\Sigma\boldsymbol{\alpha}_1 = \mathbf{L}'\lambda_1\boldsymbol{\alpha}_1 = \lambda_1\mathbf{L}'\boldsymbol{\alpha}_1 = 0. \end{aligned} \tag{10.10}$$

This implies that the vectors $\mathbf{L}$ and $\boldsymbol{\alpha}_1$ are orthogonal. We now want to find a linear combination $\boldsymbol{\alpha}'\mathbf{X}$ that has maximum variance among all normalized linear combinations $\mathbf{L}'\mathbf{X}$, $\mathbf{L} \in E^p$, which are uncorrelated with Z_1. Using Lagrange multipliers λ, v, we want to find the $\boldsymbol{\alpha}$ that maximizes

$$\phi_2(\mathbf{L}) = \mathbf{L}'\Sigma\mathbf{L} - \lambda(\mathbf{L}'\mathbf{L} - 1) + 2v(\mathbf{L}'\Sigma\boldsymbol{\alpha}_1). \tag{10.11}$$

Since

$$\frac{\partial\phi_2}{\partial\mathbf{L}} = 2\Sigma\mathbf{L} - 2\lambda\mathbf{L} - 2v\Sigma\boldsymbol{\alpha}_1, \tag{10.12}$$

the maximizing $\boldsymbol{\alpha}$ must satisfy

$$\boldsymbol{\alpha}_1'\Sigma\boldsymbol{\alpha} - \lambda\boldsymbol{\alpha}_1'\boldsymbol{\alpha} - v\boldsymbol{\alpha}_1'\Sigma\boldsymbol{\alpha}_1 = 0. \tag{10.13}$$

Since from (10.10) $\boldsymbol{\alpha}_1'\Sigma\boldsymbol{\alpha} = 0$ and $\boldsymbol{\alpha}_1'\Sigma\boldsymbol{\alpha}_1 = \lambda_1$, we get from (10.13),

$$v\lambda_1 = 0. \tag{10.14}$$

Since $\lambda_1 \neq 0$, we conclude that $v = 0$, and therefore from (10.12) we conclude that λ and $\boldsymbol{\alpha}$ must satisfy (10.3) and (10.4). Thus it follows that the coefficients of the second principal component of $\mathbf{X}$ are the elements of the normalized characteristic vector $\boldsymbol{\alpha}_2$ of Σ, corresponding to its second largest characteristic root λ_2. The second principal component of Σ is

$$Z_2 = \boldsymbol{\alpha}_2'\mathbf{X}.$$

This is continued to the rth ($r < p$) principal component Z_r. For the $(r + 1)$th principal component we want to find a linear combination $\boldsymbol{\alpha}'\mathbf{X}$ that has maximum variance among all normalized linear combinations $\mathbf{L}'\mathbf{X}$, $\mathbf{L} \in E^p$, which are uncorrelated with $Z_1, \ldots, Z_r$. So, with $Z_i = \boldsymbol{\alpha}_i'\mathbf{X}$,

$$\operatorname{cov}(\mathbf{L}'\mathbf{X}, Z_i) = \mathbf{L}'\Sigma\boldsymbol{\alpha}_i = \mathbf{L}'\lambda_i\boldsymbol{\alpha}_i = \lambda_i\mathbf{L}'\boldsymbol{\alpha}_i = 0, \qquad i = 1, \ldots, r. \tag{10.15}$$

To find $\boldsymbol{\alpha}$ we need to maximize

$$\phi_{r+1}(\mathbf{L}) = \mathbf{L}'\Sigma\mathbf{L} - \lambda(\mathbf{L}'\mathbf{L} - 1) - 2\sum_{i=1}^{r} v_i\mathbf{L}'\Sigma\boldsymbol{\alpha}_i, \tag{10.16}$$

where $\lambda, v_1, \ldots, v_r$ are Lagrange multipliers. Setting the vector of partial derivatives

$$\frac{\partial\phi_{r+1}}{\partial\mathbf{L}} = 0,$$

the vector $\boldsymbol{\alpha}$ that maximizes $\phi_{r+1}(\mathbf{L})$ is given by

$$2\Sigma\boldsymbol{\alpha} - 2\lambda\boldsymbol{\alpha} - 2\sum_{i=1}^{r} v_i\Sigma\boldsymbol{\alpha}_i = 0. \tag{10.17}$$

Since from this

$$\boldsymbol{\alpha}_i'\Sigma\boldsymbol{\alpha} - \lambda\boldsymbol{\alpha}_i'\boldsymbol{\alpha} - 2\sum_{i=1}^{r} v_i\lambda_i = 0 \tag{10.18}$$

and $\boldsymbol{\alpha}_i'\Sigma\boldsymbol{\alpha}_i = \lambda_i$, we conclude from (10.17) and (10.18) that if $\lambda_i \neq 0$,

$$v_i\lambda_i = 0, \tag{10.19}$$

that is, $v_i = 0$. If $\lambda_i = 0$, $\Sigma\boldsymbol{\alpha}_i = \lambda_i\boldsymbol{\alpha}_i = \mathbf{0}$, so that the factor $\mathbf{L}'\Sigma\boldsymbol{\alpha}_i$ in (10.16) vanishes. This argument holds for $i = 1, \ldots, r$, so we conclude from (10.17) that the maximizing $\boldsymbol{\alpha}$ [satisfying (10.4)] is the characteristic vector of Σ, orthogonal to $\boldsymbol{\alpha}_i$, $i = 1, \ldots, r$, corresponding to its characteristic root λ. If $\lambda_{r+1} \neq 0$, taking $\lambda = \lambda_{r+1}$ and $\boldsymbol{\alpha}$ for the normalized characteristic vector $\boldsymbol{\alpha}_{r+1}$, corresponding to the $(r+1)$th largest characteristic root λ_{r+1}, the $(r+1)$th principal component is given by

$$Z_{r+1} = \boldsymbol{\alpha}_{r+1}'\mathbf{X}.$$

However, if $\lambda_{r+1} = 0$ and $\lambda_i = 0$ for $1 \leq i \leq r$, then

$$\boldsymbol{\alpha}_i'\Sigma\boldsymbol{\alpha}_{r+1} = 0$$

does not imply that $\boldsymbol{\alpha}_i'\boldsymbol{\alpha}_{r+1} = 0$. In such cases replacing $\boldsymbol{\alpha}_{r+1}$ by a linear combination of $\boldsymbol{\alpha}_{r+1}$ and the $\boldsymbol{\alpha}_i$ for which $\boldsymbol{\alpha}_i = \mathbf{0}$, we can make the new $\boldsymbol{\alpha}_{r+1}$ orthogonal to all $\boldsymbol{\alpha}_i$, $i = 1, \ldots, r$. We continue in this way to the mth step such that at the $(m+1)$th step we cannot find a normalized vector $\boldsymbol{\alpha}$ such that $\boldsymbol{\alpha}'\mathbf{X}$ is uncorrelated with all $Z_1, \ldots, Z_m$. Since Σ is of dimension $p \times p$, obviously $m = p$ or $m < p$. We now show that $m = p$ is the only solution. Assume $m < p$. There exist $p - m$ normalized orthogonal vectors $\boldsymbol{\beta}_{m+1}, \ldots, \boldsymbol{\beta}_p$ such that

$$\boldsymbol{\alpha}_i'\boldsymbol{\beta}_j = 0, \qquad i = 1, \ldots, m, \quad j = m+1, \ldots, p. \tag{10.20}$$

Write $B = (\boldsymbol{\beta}_{m+1}, \ldots, \boldsymbol{\beta}_p)$. Consider a root of $\det(B'\Sigma B - \lambda I) = 0$ and the corresponding $\boldsymbol{\beta} = (\beta_{m+1}, \ldots, \beta_p)'$ satisfying

$$(B'\Sigma B - \lambda I)\boldsymbol{\beta} = \mathbf{0}. \tag{10.21}$$

Since

$$\boldsymbol{\alpha}_i'\Sigma B\boldsymbol{\beta} = \lambda_i\boldsymbol{\alpha}_i'\sum_{j=m+1}^{p}\boldsymbol{\beta}_j\beta_j = \lambda_i\sum_{j=m+1}^{p}\beta_j\boldsymbol{\alpha}_i'\boldsymbol{\beta}_j = 0,$$

the vector $\Sigma B\boldsymbol{\beta}$ is orthogonal to $\boldsymbol{\alpha}_i$, $i = 1, \ldots, r$. It is therefore a vector in the space spanned by $\boldsymbol{\beta}_{m+1}, \ldots, \boldsymbol{\beta}_p$, and can be written as

$$\Sigma B\boldsymbol{\beta} = B\mathbf{C},$$

where $\mathbf{C}$ is a $(p - m)$-component vector. Now

$$B'\Sigma B\boldsymbol{\beta} = B'B\mathbf{C} = C.$$

Thus from (10.21)

$$\lambda\boldsymbol{\beta} = \mathbf{C}, \qquad \Sigma(B\boldsymbol{\beta}) = \lambda B\boldsymbol{\beta}.$$

Then $(B\boldsymbol{\beta})'\mathbf{X}$ is uncorrelated with $\boldsymbol{\alpha}_j'\mathbf{X}$, $j = 1, \ldots, m$, and it leads to a new $\boldsymbol{\alpha}_{m+1}$. This contradicts the assumption that $m < p$, and we must have $m = p$. Let

$$A = (\boldsymbol{\alpha}_1, \ldots, \boldsymbol{\alpha}_p), \qquad \Lambda = \begin{pmatrix} \lambda_1 & 0 & \cdots & 0 \\ 0 & \lambda_2 & \cdots & 0 \\ \vdots & \vdots & & \vdots \\ 0 & 0 & \cdots & \lambda_p \end{pmatrix}, \tag{10.22}$$

where $\lambda_1 \geq \lambda_2 \geq \cdots \geq \lambda_p$ are the ordered characteristic roots of Σ and $\boldsymbol{\alpha}_1, \ldots, \boldsymbol{\alpha}_p$ are the corresponding normalized characteristic vectors. Since $AA' = I$ and $\Sigma A = A\Lambda$ we conclude that $A'\Sigma A = \Lambda$. Thus with $\mathbf{Z} = (Z_1, \ldots, Z_p)'$ we have the following theorem.

THEOREM 10.1.1 There exists an orthogonal transformation

$$\mathbf{Z} = A'\mathbf{X}$$

such that $\operatorname{cov}(\mathbf{Z}) = \Lambda$, a diagonal matrix with diagonal elements $\lambda_1 \geq \cdots \geq \lambda_p \geq 0$, the ordered roots of $\det(\Sigma - \lambda I) = 0$. The ith column $\boldsymbol{\alpha}_i$ of A satisfies $(\Sigma - \lambda_i I)\boldsymbol{\alpha}_i = \mathbf{0}$. The components of $\mathbf{Z}$ are uncorrelated and Z_i has maximum variance among all normalized linear combinations uncorrelated with $Z_1, \ldots, Z_{i-1}$.

The vector $\mathbf{Z}$ is called the vector of principal components of $\mathbf{X}$.

In the case of multiple roots suppose that

$$\lambda_{r+1} = \cdots = \lambda_{r+m} = \lambda \qquad \text{(say)}.$$

Then $(\Sigma - \lambda I)\boldsymbol{\alpha}_i = \mathbf{0}$, $i = r + 1, \ldots, r + m$. That is, $\boldsymbol{\alpha}_i$, $i = r + 1, \ldots, r + m$, are m linearly independent solutions of $(\Sigma - \lambda I)\boldsymbol{\alpha} = \mathbf{0}$. They are the only linearly independent solutions. To show that there cannot be another linearly independent solution of

$$(\Sigma - \lambda I)\boldsymbol{\alpha} = 0, \tag{10.23}$$

take $\sum_{i=1}^{p} a_i\boldsymbol{\alpha}_i$, where the a_i are scalars. If it is a solution of (10.23), we must have

$$\lambda \sum_{i=1}^{p} a_i\boldsymbol{\alpha}_i = \Sigma\left(\sum_{i=1}^{p} a_i\boldsymbol{\alpha}_i\right) - \sum_{i=1}^{p} a_i\Sigma\boldsymbol{\alpha}_i = \sum_{i=1}^{p} a_i\lambda_i\boldsymbol{\alpha}_i.$$

Since $\lambda a_i = \lambda_i a_i$, we must have $a_i = 0$ unless $i = r+1, \ldots, r+m$. Thus the rank of $(\Sigma - \lambda I)\boldsymbol{\alpha}$ is $p - m$.

Obviously if $(\boldsymbol{\alpha}_{r+1}, \ldots, \boldsymbol{\alpha}_{r+m})$ is a solution of (10.23), then for any nonsingular matrix C,

$$(\boldsymbol{\alpha}_{r+1}, \ldots, \boldsymbol{\alpha}_{r+m})C$$

is also a solution of (10.23). But from the condition of orthonormality of $\boldsymbol{\alpha}_{r+1}, \ldots, \boldsymbol{\alpha}_{r+m}$, we easily conclude that C is an orthogonal matrix. Hence we have the following theorem.

THEOREM 10.1.2 If $\lambda_{r+1} = \cdots = \lambda_{r+m} = \lambda$, then $(\Sigma - \lambda I)$ is a matrix of rank $p - m$. Furthermore, the corresponding characteristic vector $(\boldsymbol{\alpha}_{r+1}, \ldots, \boldsymbol{\alpha}_{r+m})$ is uniquely determined except for multiplication from the right by an orthogonal matrix.

From Theorem 10.1.1 it follows trivially that

$$\det \Sigma = \det \Lambda, \qquad \operatorname{tr} \Sigma = \operatorname{tr} \Lambda, \tag{10.24}$$

and we conclude that the generalized variance of the vector $\mathbf{X}$ and its principal component vector $\mathbf{Z}$ are equal, and the same is true for the sum of variances of components of $\mathbf{X}$ and $\mathbf{Z}$. Sometimes $\operatorname{tr} \Sigma$ is called the total system variance.

10.1.2 Sample Principal Components

In practice the covariance matrix Σ is usually unknown. So the population principal components will be of no use and the decision as to which principal components have sufficiently small variances to be ignored must be made from sample observations on $\mathbf{X}$. In the preceding discussion on population principal components we do not need the specific form of the distribution of $\mathbf{X}$. To deal with the problem of an unknown covariance matrix we shall assume that $\mathbf{X}$ has a p-variate normal distribution with mean $\boldsymbol{\mu}$ and unknown positive definite covariance matrix Σ. In most applications of principal components all the characteristic roots of Σ are different, although the possibility of multiple roots cannot be entirely ruled out. For an interesting case in which Σ has only one root of multiplicity p see Exercise 10.1.

Let $\mathbf{x}^\alpha = (x_{\alpha 1}, \ldots, x_{\alpha p})'$, $\alpha = 1, \ldots, N$ $(N > p)$, be a sample of size N from the distribution of the random vector $\mathbf{X}$ which is assumed to be normal with unknown mean $\boldsymbol{\mu}$ and unknown covariance matrix Σ. Let

$$\bar{\mathbf{x}} = \frac{1}{N} \sum_{\alpha=1}^{N} \mathbf{x}^\alpha, \qquad s = \sum_{\alpha=1}^{N} (\mathbf{x}^\alpha - \bar{\mathbf{x}})(\mathbf{x}^\alpha - \bar{\mathbf{x}})'.$$

The maximum likelihood estimate of Σ is s/N and that of $\boldsymbol{\mu}$ is $\bar{\mathbf{x}}$.

THEOREM 10.1.3 The maximum likelihood estimates of the ordered characteristic roots $\lambda_1, \ldots, \lambda_p$ of Σ and the corresponding normalized characteristic vectors $\boldsymbol{\alpha}_1, \ldots, \boldsymbol{\alpha}_p$ of Σ are, respectively, the ordered characteristic roots $r_1 > \cdots > r_p$ of s/N and the corresponding normalized characteristic vectors $\mathbf{a}_1, \ldots, \mathbf{a}_p$ of s/N.

Proof Since the characteristic roots of Σ are all different, the normalized characteristic vectors $\boldsymbol{\alpha}_1, \ldots, \boldsymbol{\alpha}_p$ are uniquely determined except for multiplication by ± 1. To remove this arbitrariness we impose the condition that the first nonzero component of each $\boldsymbol{\alpha}_i$ is positive. Now since $(\boldsymbol{\mu}, \Lambda, A)$ is a single-valued function of $\boldsymbol{\mu}, \Sigma$, by Lemma 5.1.3, the maximum likelihood estimates of $\lambda_1, \ldots, \lambda_p$ are given by the ordered characteristic roots $r_1 > r_2 > \cdots > r_p$ of s/N, and that of $\boldsymbol{\alpha}_i$, is given by $\mathbf{a}_i$ satisfying

$$(s/N - r_i I)\mathbf{a}_i = \mathbf{0}, \qquad \mathbf{a}_i'\mathbf{a}_i = 1, \tag{10.25}$$

with the added restriction that the first nonzero element of $\mathbf{a}_i$ is positive. Note that since $\det(\Sigma) \neq 0$ and $N > p$, the characteristic roots of S/N are all different with probability 1. Since $\Sigma = A\Lambda A'$, that is,

$$\Sigma = \sum_{i=1}^{p} \lambda_i \boldsymbol{\alpha}_i \boldsymbol{\alpha}_i', \tag{10.26}$$

we obtain

$$\frac{s}{N} = \sum_{i=1}^{p} r_i \mathbf{a}_i \mathbf{a}_i'. \tag{10.27}$$

Obviously replacing $\mathbf{a}_i$ by $-\mathbf{a}_i$ does not change this expression for s/N. Hence the maximum likelihood estimate of $\boldsymbol{\alpha}_i$ is given by any solution of $(s/N - r_i I)\mathbf{a}_i = \mathbf{0}$ with $\mathbf{a}_i'\mathbf{a}_i = 1$. Q.E.D.

The estimate of the total system variance is given by

$$\operatorname{tr}\left(\frac{s}{N}\right) = \sum_{i=1}^{p} r_i, \tag{10.28}$$

and is called the total sample variance. The importance of the ith principal component is measured by

$$\frac{r_i}{\sum_{i=1}^{p} r_i}, \tag{10.29}$$

which, when expressed in percentage, will be called the percentage of contribution of the ith principal component to the total sample variance.

If the estimates of the principal components are obtained from the sample correlation matrix

$$r = (r_{ij}), \qquad r_{ij} = \frac{s_{ij}}{(s_{ii}s_{jj})^{1/2}}, \tag{10.30}$$

with $s = (s_{ij})$, then the estimate of the total sample variance will be $p = \operatorname{tr}(r)$.

If the first k principal components explain a large amount of total sample variance, they may be used in future investigations in place of the original vector $\mathbf{X}$. For the computation of characteristic roots and vectors standard programs are now available.

EXAMPLE 10.1.1 Consider once again Example 9.1.1. We have two groups with 27 observations in each group. For *group 1* the sample covariance matrix s/N and the sample correlation matrix r are given by

$$\frac{s}{27} = \begin{pmatrix} 30.58 & & & & & \\ 108.70 & 781.8 & & & & \\ 0.1107 & -0.7453 & 0.1381 & & & \\ 0.4329 & 0.8684 & 0.1465 & 0.4600 & & \\ -10.72 & -98.22 & 6.302 & 7.992 & 840.2 & \\ -0.2647 & -1.910 & 0.3519 & 0.4715 & 25.76 & 1.761 \end{pmatrix},$$

$$r = \begin{pmatrix} 1.0000 & & & & & \\ 0.7025 & 1.0000 & & & & \\ 0.0539 & -0.0717 & 1.0000 & & & \\ 0.1154 & 0.0458 & 0.5812 & 1.0000 & & \\ -0.0669 & -0.1212 & 0.5851 & 0.4065 & 1.0000 & \\ -0.0361 & -0.0515 & 0.7137 & 0.5238 & 0.6698 & 1.0000 \end{pmatrix}.$$

(i) The ordered characteristic roots of $s/27$ along with the corresponding percentages of contribution to the total sample variance (given within parentheses) are

920.312	717.984	15.1837	1.0756	0.3016	0.0533
(55.61%)	(43.39%)	(0.92%)	(0.06%)	(0.02%)	(0%)

(ii) The characteristic vectors $\mathbf{a}_i$ (column vectors) of $s/27$ are

1	2	3	4	5	6
0.0851	−0.1122	0.9898	−0.0029	0.0208	0.0100
0.6199	−0.7720	−0.1408	−0.0017	−0.0012	−0.0020
−0.0058	−0.0047	0.0126	0.1772	−0.1074	−0.9782
−0.0062	−0.0080	0.0186	0.3196	−0.9336	0.1607
−0.7797	−0.6253	−0.0040	−0.0332	−0.0007	0.0017
−0.0232	−0.0204	−0.0061	0.9302	0.3413	0.1312

(iii) The ordered characteristic roots of r along with the corresponding percentages of contribution to the total sample variance (given within parentheses) are

2.7578	1.7284	0.5892	0.3700	0.3277	0.2270
(45.96%)	(28.81%)	(9.82%)	(6.16%)	(5.46%)	(3.79%)

(iv) The characteristic vectors $\mathbf{a}_i$ (column vectors) of r are

1	2	3	4	5	6
0.0093	−0.7012	0.0619	−0.1850	0.5296	0.4356
0.0628	−0.6933	0.1615	0.1894	−0.5158	−0.4329
−0.5274	−0.0403	−0.0713	−0.6615	0.1488	−0.5454
−0.4436	−0.1455	−0.7751	0.4236	0.0237	0.0364
−0.4861	0.0695	0.5602	0.5342	0.3616	−0.1700
−0.5336	0.0035	0.2245	−0.1660	−0.5478	0.5807

For *group 2* the sample covariance matrix and the sample correlation matrix are given by

$$\frac{s}{27} = \begin{pmatrix} 47.05 & & & & & \\ 35.21 & 214.3 & & & & \\ 3.831 & 2.719 & 0.3976 & & & \\ 5.838 & 4.355 & 0.6042 & 1.053 & & \\ 191.6 & 14.69 & 17.49 & 28.58 & 1598 & \\ 13.76 & 2.656 & 1.308 & 2.076 & 76.33 & 5.702 \end{pmatrix},$$

$$r = \begin{pmatrix} 1.0000 & & & & & \\ 0.3506 & 1.0000 & & & & \\ 0.8857 & 0.2945 & 1.0000 & & & \\ 0.8295 & 0.2899 & 0.9339 & 1.0000 & & \\ 0.7007 & 0.0252 & 0.6960 & 0.6987 & 1.0000 & \\ 0.8155 & 0.0761 & 0.8686 & 0.8474 & 0.8019 & 1.0000 \end{pmatrix}.$$

(i) The ordered characteristic roots of $s/27$ along with the corresponding percentages of contribution to the total sample variance (given in parentheses) are

1617.46	219.829	19.0293	1.2743	0.2262	0.0283
(87.06%)	(11.83%)	(1.02%)	(0.07%)	(0.02%)	(0%)

(ii) The characteristic vectors $\mathbf{a}_i$ (column vectors) of $s/27$ are

1	2	3	4	5	6
0.1217	0.1641	0.9476	0.2412	0.0396	−0.0236
0.0136	0.9855	−0.1672	−0.0225	0.0117	−0.0002
0.0111	0.0124	0.0692	−0.1416	−0.3243	0.9326
0.0181	0.0196	0.0924	−0.2749	−0.8873	−0.3576
0.9911	−0.0347	−0.1268	0.0218	−0.0010	0.0011
0.0480	0.0104	0.2114	−0.9194	0.3253	−0.0429

(iii) The ordered characteristic roots of r along with the corresponding percentages of contribution to the total sample variance are

4.3060	1.0439	0.3147	0.1685	0.1129	0.0529
(71.77%)	(17.40%)	(5.24%)	(2.81%)	(1.88%)	(0.90%)

(iv) The characteristic vectors $\mathbf{a}_i$ (column vectors) of r are

1	2	3	4	5	6
0.4477	0.1098	0.0619	−0.8278	−0.1882	0.2508
0.1418	0.9174	−0.3001	0.1309	0.1752	−0.0174
0.4624	0.0481	0.3534	0.0791	−0.1589	−0.7922
0.4543	0.0414	0.3311	0.5189	−0.3519	0.5378
0.3980	−0.3067	−0.8174	0.1376	−0.2282	−0.0914
0.4481	−0.2195	0.0588	0.0558	0.8560	0.1083

We shall now investigate the distribution of the ordered characteristic roots $R_1, \ldots, R_p$ of the random (sample) covariance matrix S/N and the corresponding normalized characteristic vector $\mathbf{A}_i$ given by

$$(S/N - R_i I)\mathbf{A}_i = \mathbf{0}, \qquad i = 1, \ldots, p, \tag{10.31}$$

with $\mathbf{A}_i'\mathbf{A}_i = 1$. In Chapter VIII we derived the joint distribution of $R_1, \ldots, R_p$ when $\Sigma = I$ (identity matrix). We now give the large sample distribution of these statistics, the initial derivation of which was performed by Girshik (1936, 1939). Subsequently this was extended by Anderson (1951, 1963), Bartlett (1954), and Lawley (1956, 1963). In what follows we shall assume that the characteristic roots of Σ are different and N is large. These distribution results are summarized in the following theorems which we state without proof.

THEOREM 10.1.4 (Girshik, 1939) If Σ is positive definite and all its characteristic roots are distinct so that $\lambda_1 > \lambda_2 > \cdots > \lambda_p > 0$, then

(a) as $N \to \infty$, the ordered characteristic roots $R_1, \ldots, R_p$ are independent, unbiased, and approximately normally distributed with

$$E(R_i) = \lambda_i, \qquad \operatorname{var}(R_i) = 2\lambda_i^2/(N-1); \tag{10.32}$$

(b) as $N \to \infty$, $(N-1)^{1/2}(\mathbf{A}_i - \boldsymbol{\alpha}_i)$ has a p-variate normal distribution with mean $\mathbf{0}$ and covariance matrix

$$\lambda_i \sum_{j=1, j \neq i}^{p} \frac{\lambda_j}{(\lambda_j - \lambda_i)^2} \boldsymbol{\alpha}_i \boldsymbol{\alpha}_i'. \tag{10.33}$$

THEOREM 10.1.5 (Anderson, 1963) Under the assumption of Theorem 10.1.4, the likelihood ratio test of

$$H_0 : \lambda_{i+1} = \cdots = \lambda_{i+k}, \qquad i + k \leq p,$$

rejects H_0 whenever

$$q = k(N-1) \log\left\{\frac{1}{k} \sum_{j=1}^{k} r_{i+j}\right\} - (N-1) \sum_{j=1}^{k} (r_{i+j}) \geq C, \tag{10.34}$$

where the constant C is determined so that the test has level of significance α. Under H_0, using Box (1949), the statistic Q (with values q) has approximately the chi-square distribution with $k(k+1)/2 - 1$ degrees of freedom as $N \to \infty$.

10.2 FACTOR ANALYSIS

Factor analysis is a multivariate technique which attempts to account for the correlation pattern present in the distribution of an observable random vector $\mathbf{X} = (X_1, \ldots, X_p)'$ in terms of a minimal number of unobservable random variables, called factors. In this approach each component X_i is examined to see if it could be generated by a linear function involving a minimum number of unobservable random variables, called common factor variates, and a single variable, called the specific factor variate.

The common factors will generate the covariance structure of $\mathbf{X}$ where the specific factor will account for the variance of the component X_i.

Though, in principle, the concept of latent factors seems to have been suggested by Galton (1888), the formulation and early development of factor analysis have their genesis in psychology and are generally attributed to Spearman (1904). He first hypothesized that the correlations among a set of intelligence test scores could be generated by linear functions of a single latent factor of general intellective ability and a second set of specific factors representing the unique characteristics of individual tests. Thurston (1945) extended Spearman's model to include many latent factors and proposed a method, known as the centroid method, for estimating the coefficients of different factors (usually called factor loadings) in the linear model from a given correlation matrix. Lawley (1940), assuming normal distribution for the random vector **X**, estimated these factor loadings by using the method of maximum likelihood.

Factor analysis models are widely used in behavioral and social sciences. We refer to Armstrong (1967) for a complete exposition of factor analysis for an applied viewpoint, to Anderson and Rubin (1956) for a theoretical exposition, and to Thurston (1945) for a general treatment. We refer to Lawley (1949, 50, 53), Morrison (1967), Rao (1955) and Solomon (1960) for further relevent results in factor analysis.

10.2.1 Orthogonal Factor Model

Let $\mathbf{X} = (X_1, \ldots, X_p)'$ be an observable random vector with $E(\mathbf{X}) = \boldsymbol{\mu}$ and $\operatorname{cov}(\mathbf{X}) = \Sigma = (\sigma_{ij})$, a positive definite matrix. Assuming that each component X_i can be generated by a linear combination of m ($m < p$) mutually uncorrelated (orthogonal) unobservable variables $Y_1, \ldots, Y_m$ upon which a set of errors may be superimposed, we write

$$\mathbf{X} = \Lambda \mathbf{Y} + \boldsymbol{\mu} + \mathbf{U}, \tag{10.35}$$

where $\mathbf{Y} = (Y_1, \ldots, Y_m)'$, $\mathbf{U} = (U_1, \ldots, U_p)'$ denotes the error vector and $\Lambda = (\lambda_{ij})$ is a $p \times m$ matrix of unknown coefficients λ_{ij} which is usually called a factor loading matrix. The elements of **Y** are called common factors. We shall assume that **U** is distributed independently of **Y** with $E(\mathbf{U}) = \mathbf{0}$ and $\operatorname{cov}(\mathbf{U}) = D$, a diagonal matrix with diagonal elements $\sigma_1^2, \ldots, \sigma_p^2$; $\operatorname{var}(U_i) = \sigma_i^2$ is called the specific factor variance of X_i. The vector **Y** in some cases will be a random vector and in other cases will be an unknown parameter which varies from observation to observation. A component of **U** is made up of the error of measurement in the test plus specific factors representing the unique character of the individual test. The model (10.35) is similar to the multivariate regression model except that the independent variables **Y** in this case are not observable.

When **Y** is a random vector we shall assume that $E(\mathbf{Y}) = \mathbf{0}$ and $\operatorname{cov}(\mathbf{Y}) = I$, the identity matrix. Since

$$E(\mathbf{X} - \boldsymbol{\mu})(\mathbf{X} - \boldsymbol{\mu})' = E(\Lambda \mathbf{Y} + \mathbf{U})(\Lambda \mathbf{Y} + \mathbf{U})' = \Lambda\Lambda' + D, \tag{10.36}$$

we see that **X** has a p-variate normal distribution with mean $\boldsymbol{\mu}$ and covariance matrix $\Sigma = \Lambda\Lambda' + D$, so that Σ is positive definite. Furthermore, since

$$E(\mathbf{X}\mathbf{Y}') = E(\Lambda \mathbf{Y} + \mathbf{U})\mathbf{Y}') = \Lambda, \tag{10.37}$$

the elements λ_{ij} of Λ are correlations of X_i, $\mathbf{Y}_j$. In behavioral science the term loading is used for correlation. The diagonal elements of $\Lambda\Lambda'$ are called communalities of the components. The purpose of factor analysis is the determination of Λ with elements of D such that

$$\Sigma - D = \Lambda\Lambda'. \tag{10.38}$$

If the errors are small enough to be ignored, we can take $\Sigma = \Lambda\Lambda'$. From this point of view factor analysis is outwardly similar to finding the principal components of Σ since both procedures start with a linear model and end up with matrix factorization. However, the model for principal component analysis must be linear by the very fact that it refers to a rigid rotation of the original coordinate axes, whereas in the factor analysis model the linearity is as much a part of our hypothesis about the dependence structure as the choice of exactly m common factors. The linear model in factor analysis allows us to interpret λ_{ij} as correlation coefficients but if the covariances reproduced by the m-factor linear model fail to fit the linear model adequately, it is as proper to reject linearity as to advance the more usual finding that m common factors are inadequate to explain the correlation structure.

Existence Since a necessary and sufficient condition that a $p \times p$ matrix A be expressed as BB', with B a $p \times m$ matrix, is that A is a positive semidefinite matrix of rank m, we see that the question of existence of a factor analysis model can be resolved if there exists a diagonal matrix D with nonnegative diagonal elements such that $\Sigma - D$ is a positive semidefinite matrix of rank m. So the question is how to tell if there exists such a diagonal matrix D, and we refer to Anderson and Rubin (1956) for an answer to this question.

10.2.2 Oblique Factor Model

This is obtained from the orthogonal factor model by replacing $\mathrm{cov}(\mathbf{Y}) = I$ by $\mathrm{cov}(\mathbf{Y}) = R$, where R is a positive definite correlation matrix; that is, all its diagonal elements are equal to unity. In other words, all factors in the oblique factor model are assumed to have mean 0 and variance 1 but are correlated. In this case $\Sigma = \Lambda R \Lambda' + D$.

10.2.3 Estimation of Factor Loadings

We shall assume that m is fixed beforehand and that $\mathbf{X}$ has the p-variate normal distribution with mean $\boldsymbol{\mu}$ and covariance matrix Σ (positive definite). We are interested in the maximum likelihood estimates of these parameters. Let $\mathbf{x}^\alpha = (x_{\alpha 1}, \ldots, x_{\alpha p})'$, $\alpha = 1, \ldots, N$, be a sample of size N on $\mathbf{X}$. The maximum likelihood estimates of $\boldsymbol{\mu}$ and Σ are given by

$$\hat{\boldsymbol{\mu}} = \bar{\mathbf{x}} = \frac{1}{N}\sum_{\alpha=1}^{N} \mathbf{x}^{\alpha}, \qquad \hat{\Sigma} = \frac{s}{N} = \sum_{\alpha=1}^{N} (\mathbf{x}^{\alpha} - \bar{\mathbf{x}})(\mathbf{x}^{\alpha} - \bar{\mathbf{x}})'.$$

Orthogonal Factor Model

Here $\Sigma = \Lambda\Lambda' + D$. The likelihood of $\mathbf{x}^{\alpha}$, $\alpha = 1, \ldots, N$, is given by

$$L(\Lambda, D, \boldsymbol{\mu}) = (2\pi)^{-Np/2}[\det(\Lambda\Lambda' + D)]^{-N/2} \times \exp\{-\tfrac{1}{2}\operatorname{tr}[(\Lambda\Lambda' + D)^{-1}(s + N(\bar{\mathbf{x}} - \boldsymbol{\mu})(\bar{\mathbf{x}} - \boldsymbol{\mu})')]\}. \tag{10.39}$$

Observe that changing Λ to ΛO, where O is an $m \times m$ orthogonal matrix, does not change $L(\Lambda, D, \boldsymbol{\mu})$. Thus if $\hat{\Lambda}$ is a maximum likelihood estimate of Λ, then $\hat{\Lambda}O$ is also a maximum likelihood estimate of Λ. To obtain uniqueness we impose the restriction that

$$\Lambda' D^{-1}\Lambda = \Gamma \tag{10.40}$$

is a diagonal matrix with distinct diagonal elements $\gamma_1, \ldots, \gamma_p$. We are now interested in obtaining the maximum likelihood estimates $\hat{\boldsymbol{\mu}}$, $\hat{\Lambda}$, $\hat{D}$ of $\boldsymbol{\mu}$, Λ, D, respectively, subject to (10.40).

To maximize the likelihood function the term

$$\operatorname{tr}\{\Lambda\Lambda' + D)^{-1}N(\bar{\mathbf{x}} - \boldsymbol{\mu})(\bar{\mathbf{x}} - \boldsymbol{\mu})'\}$$

may be put equal to zero in (10.39) since it vanishes when $\hat{\boldsymbol{\mu}} = \bar{\mathbf{x}}$. With this in mind let us find $\hat{\Lambda}$, $\hat{D}$.

Note Λ will not depend on the units in which $Y_1, \ldots, Y_m$ are expressed. Suppose that $\mathbf{Y}$ has an m-dimensional normal distribution with mean $\mathbf{0}$ and covariance matrix $\theta\theta'$, where θ is a diagonal matrix with diagonal elements $\theta_1, \ldots, \theta_m$ such that $\theta_i^2 = \operatorname{var}(Y_i)$. Hence

$$\operatorname{cov}(\mathbf{X}) = (\Lambda\theta)(\Lambda\theta)' + D = \Lambda^*\Lambda^{*\prime} + D,$$

where $\Lambda^* = \Lambda\theta$. Thus for the estimation of factor loadings, without any loss of generality we can assume that the Y_i have unit variance and $\operatorname{cov}(\mathbf{Y}) = R$, a correlation matrix. For the orthogonal factor model $R = I$ and for the oblique factor model $R = R$.

THEOREM 10.2.1 The maximum likelihood estimates $\hat{\Lambda}$, $\hat{D}$ of Λ, D, respectively, in the orthogonal factor model are given by

$$\operatorname{diag}(\hat{D} + \hat{\Lambda}\hat{\Lambda}') = \operatorname{diag}((1/N)s), \tag{10.41}$$

$$(s/N)\hat{D}^{-1}\hat{\Lambda} = \hat{\Lambda}(I + \hat{\Lambda}' D^{-1}\hat{\Lambda}). \tag{10.42a}$$

Proof Let

$$L(\Lambda, D) = (2\pi)^{-Np/2}[\det(\Lambda\Lambda' + D)]^{-N/2} \exp\{-\tfrac{1}{2} \operatorname{tr}(\Lambda\Lambda' + D)^{-1}s\}. \quad (10.42b)$$

Then for $i = 1, \ldots, p$,

$$\frac{\partial \log L(\Lambda, D)}{\partial \sigma_i^2} = -\tfrac{1}{2}N \frac{(\Lambda\Lambda' + D)_{ii}}{\det(\Lambda\Lambda' + D)} + \tfrac{1}{2} tr(\Lambda\Lambda' + D)^{-1}s(\Lambda\Lambda' + D)^{-1} \frac{\partial D}{\partial \sigma_i^2}, \quad (10.43)$$

where $\partial D/\partial \sigma_i^2$ is the $p \times p$ matrix with unity in the ith diagonal position and zero elsewhere, and $(\Lambda\Lambda' + D)_{ii}$ is the cofactor of the ith diagonal element of $\Lambda\Lambda' + D$. Note that for any symmetric matrix $A = (a_{ij})$,

$$\frac{\partial \det A}{\partial a_{ii}} = A_{ii}, \qquad \frac{\partial \det A}{\partial a_{ij}} = 2A_{ij}$$

where the A_{ij} are the cofactors of the elements a_{ij}. For $L(\Lambda, D)$ to be maximum it is necessary that each of the p derivatives in (10.43) equal zero at $\Lambda = \hat{\Lambda}$, $D = \hat{D}$. This reduces to the condition that the diagonal elements of $(\hat{D} + \hat{\Lambda}\hat{\Lambda}')^{-1}[I - (s/N)(\hat{D} + \hat{\Lambda}\hat{\Lambda}')^{-1}]$ are zeros, that is,

$$\operatorname{diag}\{[\hat{D} + \hat{\Lambda}\hat{\Lambda}')^{-1}[I - (s/N)(\hat{D} + \hat{\Lambda}\hat{\Lambda}')^{-1}]\} = 0. \quad (10.44)$$

Now differentiating $L(\Lambda, D)$ with respect to λ_{ij}, we get, with $\Lambda\Lambda' + D = \Sigma = (\sigma_{ij})$,

$$\begin{aligned}\frac{\partial \log L(\Lambda, D)}{\partial \lambda_{ij}} &= -\frac{N}{2} \det(\Lambda\Lambda' + D)^{-1} \sum_{g,h=1}^{p} (\Lambda\Lambda' + D)_{gh} \frac{\partial \sigma_{gh}}{\partial \lambda_{ij}} \\ &\quad + \tfrac{1}{2} \operatorname{tr}(\Lambda\Lambda' + D)^{-1}\left(\frac{\partial \Sigma}{\partial \lambda_{ij}}\right)(\Lambda\Lambda' + D)^{-1}s \\ &= -\tfrac{1}{2}N \operatorname{tr} \Sigma^{-1}\left(\frac{\partial \Sigma}{\partial \lambda_{ij}}\right) \\ &\quad + \tfrac{1}{2} \operatorname{tr}(\Lambda\Lambda' + D)^{-1}\left(\frac{\partial \Sigma}{\partial \lambda_{ij}}\right)(\Lambda\Lambda' + D)^{-1}s. \quad (10.45)\end{aligned}$$

Denoting $\Sigma^{-1} = (\sigma^{ij})$, from Exercise 4, we obtain

$$\operatorname{tr} \Sigma^{-1}\left(\frac{\partial \Sigma}{\partial \lambda_{ij}}\right) = 2(\boldsymbol{\sigma}^i)'\boldsymbol{\lambda}_j, \quad (10.46)$$

where $(\boldsymbol{\sigma}^i)'$ is the ith row of Σ^{-1} and $\boldsymbol{\lambda}_j$ is the jth column of Λ. Thus the first term in $\partial \log L(\Lambda, D)/\partial\Lambda$ is $-\frac{1}{2}N(\Lambda\Lambda' + D)^{-1}\Lambda$. Making two cyclic permutations of matrices within the trace symbol, we get

$$\operatorname{tr}(\Lambda\Lambda' + D)^{-1}\left(\frac{\partial\Sigma}{\partial\lambda_{ij}}\right)(\Lambda\Lambda' + D)^{-1}s$$
$$= \operatorname{tr}(\Lambda\Lambda' + D)^{-1}s(\Lambda\Lambda' + D)^{-1}\left(\frac{\partial\Sigma}{\partial\lambda_{ij}}\right). \tag{10.47}$$

Write

$$Z = (\Lambda\Lambda' + D)^{-1}s(\Lambda\Lambda' + D)^{-1} = (Z_{ij})$$

and let the ith row of Z be $\mathbf{Z}_i'$. From Exercise 5,

$$\operatorname{tr} Z\left(\frac{\partial\Sigma}{\partial\lambda_{ij}}\right) = 2\mathbf{Z}_i'\boldsymbol{\lambda}_j. \tag{10.48}$$

Thus the second term in $\partial \log L(\Lambda, D)/\partial\Lambda$ is $Z\Lambda$. From (10.45)–(10.48) we get

$$[N(\hat\Lambda\hat\Lambda' + \hat D)^{-1} - (\hat\Lambda\hat\Lambda' + \hat D)^{-1}s(\hat\Lambda\hat\Lambda' + \hat D)^{-1}]\hat\Lambda = 0$$

or, equivalently,

$$N\hat\Lambda = s(\hat\Lambda\hat\Lambda' + \hat D)^{-1}\hat\Lambda. \tag{10.49}$$

Since

$$(\hat D + \hat\Lambda\hat\Lambda')^{-1}\hat\Lambda = D^{-1}\hat\Lambda(I + \hat\Lambda'\hat D^{-1}\hat\Lambda)^{-1},$$

from (10.49) we get

$$N\hat\Lambda = s\hat D^{-1}\hat\Lambda(I + \hat\Lambda'\hat D^{-1}\hat\Lambda)^{-1} \tag{10.50}$$

or

$$(s/N)\hat D^{-1}\hat\Lambda = \hat\Lambda(I + \hat\Lambda'\hat D^{-1}\hat\Lambda),$$

which yields (10.42). From

$$(\hat D + \hat\Lambda\hat\Lambda')^{-1} = \hat D^{-1} - \hat D^{-1}\hat\Lambda\hat\Lambda'(\hat D + \hat\Lambda\hat\Lambda')^{-1},$$
$$(\hat D + \hat\Lambda\hat\Lambda')^{-1}\hat D = I - (\hat D + \hat\Lambda\hat\Lambda')^{-1}\hat\Lambda\hat\Lambda',$$

we get

$$\hat D(\hat D + \hat\Lambda\hat\Lambda')^{-1}\hat D = \hat D - \hat\Lambda\hat\Lambda'(\hat D + \hat\Lambda\hat\Lambda')^{-1}\hat D$$
$$= \hat D - \hat\Lambda\hat\Lambda' + \hat\Lambda\hat\Lambda'(\hat D + \hat\Lambda\hat\Lambda')^{-1}\hat\Lambda\hat\Lambda'. \tag{10.51}$$

Similarly,

$$\hat D(\hat D + \hat\Lambda\hat\Lambda')^{-1}(s/N)(\hat D + \hat\Lambda\hat\Lambda')^{-1}\hat D$$
$$= (s/N) - (s/N)(\hat D + \hat\Lambda\hat\Lambda')^{-1}\hat\Lambda\hat\Lambda'$$
$$- \hat\Lambda\hat\Lambda'(\hat D + \hat\Lambda\hat\Lambda')^{-1}(s/N)$$
$$+ \hat\Lambda\hat\Lambda'(\hat D + \hat\Lambda\hat\Lambda')^{-1}(s/N)(\hat D + \hat\Lambda\hat\Lambda')^{-1}\hat\Lambda\hat\Lambda'. \tag{10.52}$$

Using (10.49) and (10.51)–(10.52), we get from (10.44),

$$\operatorname{diag}(\hat{D} + \hat{\Lambda}\hat{\Lambda}') = \operatorname{diag}(s/N),$$

which yields (10.41). It can be verified that these estimates yield a maximum for $L(\Lambda, D)$. Q.E.D.

Oblique Factor Model

Similarly for the oblique factor model with $\operatorname{cov}(\mathbf{Y}) = R$ (correlation matrix) we obtain the following theorem.

THEOREM 10.2.2 The maximum likelihood estimates $\hat{\Lambda}, \hat{R}, \hat{D}$ of Λ, R, D, respectively, for the oblique factor model are given by

(1) $\hat{D} = \operatorname{diag}(s/N - \hat{\Lambda}\hat{R}\hat{\Lambda}')$,

(2) $\hat{R}\hat{\Lambda}\hat{D}^{-1}\hat{\Lambda} + I = (\hat{\Lambda}\hat{D}^{-1}\hat{\Lambda}')^{-1}(\hat{\Lambda}'\hat{D}^{-1}(s/N)\hat{D}^{-1}\hat{\Lambda})$,

(3) $\hat{R}\hat{\Lambda}(\hat{\Lambda}\hat{\Lambda}' + \hat{D}^{-1}(I - (s/N)(\hat{\Lambda}\hat{\Lambda}' + \hat{D})^{-1})$
$\qquad = \hat{R}\hat{\Lambda}'[I - (\hat{\Lambda}\hat{\Lambda}' + \hat{D})^{-1}(s/N)]\hat{D}^{-1}$.

For numerical evaluation of these estimates, standard computer programs are now available (see Press, 1971). Anderson and Rubin (1956) have shown that as $N \to \infty$, $\sqrt{N}(\hat{\Lambda} - \Lambda)$ has mean 0 but the covariance matrix is extremely complicated.

Identification For the orthogonal factor analysis model we want to represent the population covariance matrix as

$$\Sigma = \Lambda\Lambda' + D.$$

For any orthogonal matrix O of dimension $p \times p$

$$\Sigma = \Lambda\Lambda' + D = \Lambda OO'\Lambda' + D = (\Lambda O)(\Lambda O)' + D = \Lambda^*\Lambda^{*\prime} + D.$$

Thus, regardless of the value of Λ used, it is always possible to transform Λ by an orthogonal matrix O to get a new Λ^* which gives the same representation for Λ. Furthermore, since Σ is symmetric, there are $p(p+1)/2$ distinct elements in Σ, and in the factor representation model there is generally a greater number, $p(m+1)$, of distinct parameters. So in general a unique estimate of Λ is not possible and there remains the problem of identification in the factor analysis model. We refer to Anderson and Rubin (1956) for a detailed treatment of this topic.

10.2.4 Tests of Hypothesis in Factor Models

Let $\mathbf{x}^\alpha = (x_{\alpha 1}, \ldots, x_{\alpha p})'$, $\alpha = 1, \ldots, N$, be a sample of size N from a p-variate normal population with positive definite covariance matrix Σ. On the basis of these observations we are interested in testing, with the orthogonal factor model, the null hypothesis $H_0{:}\Sigma = \Lambda\Lambda' + D$ against the alternatives H_1 that Σ is a symmetric positive definite matrix. (The corresponding hypothesis in the oblique factor model is $H_0{:}\Sigma = \Lambda R \Lambda' + D$.) The likelihood of the observations $\mathbf{x}^\alpha$, $\alpha = 1, \ldots, N$, is

$$L(\Sigma, \boldsymbol{\mu}) = (2\pi)^{-Np/2}(\det \Sigma)^{-N/2} \exp\left\{-\tfrac{1}{2} \operatorname{tr} \Sigma^{-1}\left[\sum_{\alpha=1}^{N} (\mathbf{x}^\alpha - \boldsymbol{\mu})(\mathbf{x}^\alpha - \boldsymbol{\mu})'\right]\right\}$$

and hence

$$\max_{H_1} L(\Sigma, \boldsymbol{\mu}) = (2\pi)^{-Np/2}(\det(s/N))^{-N/2} \exp\{-\tfrac{1}{2}Np\}.$$

Under H_0, $L(\Sigma, \boldsymbol{\mu})$ reduces to (for the orthogonal factor model)

$$L(\Lambda, D, \boldsymbol{\mu}) = (2\pi)^{-N/2}(\det(\Lambda\Lambda' + D))^{-N/2} \times \exp\left\{-\tfrac{1}{2} \operatorname{tr}(\Lambda\Lambda' + D)^{-1}\left[\sum_{\alpha=1}^{N} (\mathbf{x}^\alpha - \boldsymbol{\mu})(\mathbf{x}^\alpha - \boldsymbol{\mu})'\right]\right\}$$

and

$$\max_{H_0} L(\Lambda, D, \boldsymbol{\mu}) = (2\pi)^{-Np/2}(\det(\hat{\Lambda}\hat{\Lambda}' + \hat{D}))^{-N/2} \exp\{-\tfrac{1}{2} \operatorname{tr}(\hat{\Lambda}\hat{\Lambda} + \hat{D})^{-1}s\},$$

where $\hat{\Lambda}$, $\hat{D}$ are as given in Theorem 10.2.1. Hence the modified likelihood ratio test of H_0 rejects H_0 whenever, with $N - 1 = n$ (say),

$$\lambda = \left[\frac{\det(s/N)}{\det(\hat{\Lambda}\hat{\Lambda}' + \hat{D})}\right]^{-n/2} \exp\{\tfrac{1}{2} \operatorname{tr}(\hat{\Lambda}\hat{\Lambda}' + \hat{D})^{-1}s - \tfrac{1}{2}np\} \geq C, \quad (10.53)$$

where C depends on the level of significance α of the test. In large samples under H_0, using Box (1949),

$$P\{-2 \log \lambda \leq z\} = P\{\chi_f^{\,2} \leq z\},$$

where

$$f = \tfrac{1}{2}p(p + 1) - [mp + p - \tfrac{1}{2}m(m + 1) + m]. \quad (10.54)$$

The modification needed for the oblique factor model is obvious and the value of degrees of freedom f for the chi-square approximation in this case is

$$f = \tfrac{1}{2}p(p - 2m + 1). \quad (10.55)$$

Bartlett (1954) has pointed out that if $N - 1 = n$ is replaced by n_0, where

$$n_0 = n - \tfrac{1}{6}(2p + 5) - \tfrac{2}{3}m, \quad (10.56)$$

then under H_0, the convergence of $-2 \log \lambda$ to chi-square distribution is more rapid.

10.3 CANONICAL CORRELATIONS

Suppose we have two sets of variates and we wish to study their interrelations. If the dimensions of both sets are large, one may wish to consider only a few linear combinations of each set and study those linear combinations which are highly correlated. The admission of students into a medical program is highly competitive. For an efficient selection one may wish to predict a linear combination of scores in the medical program for each candidate from certain linear combinations of scores obtained by the candidate in high school. Economists may find it useful to use a linear combination of easily available economic quantities to study the behavior of the prices of a group of stocks.

The canonical model was first developed by Hotelling (1936b). It selects linear combinations of variables from each of the two sets, so that the correlations between the new variables in different sets are maximized subject to the restriction that the new variables in each set are uncorrelated with mean 0 and variance 1. In developing the concepts and the algebra we do not need a specific assumption of normality, though these will be necessary in making statistical inference.

10.3.1 Population Canonical Correlations

Consider a random vector $\mathbf{X} = (X_1, \ldots, X_p)'$ with mean $\boldsymbol{\mu}$ and positive definite covariance matrix Σ. Since we shall be interested only in the covariances of the components of $\mathbf{X}$, we shall take $\boldsymbol{\mu} = \mathbf{0}$. Let

$$\mathbf{X} = \begin{pmatrix} \mathbf{X}_{(1)} \\ \mathbf{X}_{(2)} \end{pmatrix},$$

where $\mathbf{X}_{(1)}$, $\mathbf{X}_{(2)}$ are subvectors of $\mathbf{X}$ of p_1, p_2 components, respectively. Assume that $p_1 < p_2$. Let Σ be similarly partitioned as

$$\Sigma = \begin{pmatrix} \Sigma_{(11)} & \Sigma_{(12)} \\ \Sigma_{(21)} & \Sigma_{(22)} \end{pmatrix},$$

where $\Sigma_{(ij)}$ is $p_i \times p_j$, $i, j = 1, 2$. Recall that if $p_1 = 1$, then the multiple correlation coefficient is the largest correlation attainable between $\mathbf{X}_{(1)}$ and a linear combination of the components of $\mathbf{X}_{(2)}$. For $p_1 > 1$, a natural generalization of the multiple correlation coefficient is the largest correlation coefficient ρ_1 (say), attainable between linear combinations of $\mathbf{X}_{(1)}$ and linear combinations of $\mathbf{X}_{(2)}$.

Consider arbitrary linear combinations

$$U_1 = \boldsymbol{\alpha}'\mathbf{X}_{(1)}, \qquad V_1 = \boldsymbol{\beta}'\mathbf{X}_{(2)},$$

where $\boldsymbol{\alpha} = (\alpha_1, \ldots, \alpha_{p_1})' \in E^{p_1}$, $\boldsymbol{\beta} = (\beta_1, \ldots, \beta_{p_2})' \in E^{p_2}$. Since the coefficient of correlation between U_1 and V_1 remains invariant under affine transformations

$$U_1 \to aU_1 + b, \qquad V_1 \to cV_1 + d,$$

where a, b, c, d are real constants and $a \neq 0$, $c \neq 0$, we can make an arbitrary normalization of $\boldsymbol{\alpha}, \boldsymbol{\beta}$ to study the correlation. We shall therefore require that

$$\operatorname{var}(U_1) = \boldsymbol{\alpha}'\Sigma_{(11)}\boldsymbol{\alpha} = 1, \qquad \operatorname{var}(V_1) = \boldsymbol{\beta}'\Sigma_{(22)}\boldsymbol{\beta} = 1, \tag{10.57}$$

and maximize the coefficient of correlation between U_1 and V_1. Since $E(\mathbf{X}) = \mathbf{0}$, using (10.57),

$$\begin{aligned}\rho(U_1, V_1) &= \frac{E(U_1 V_1)}{(\operatorname{var}(U_1)\operatorname{var}(V_1))^{1/2}} = \frac{\boldsymbol{\alpha}'\Sigma_{(12)}\boldsymbol{\beta}}{((\boldsymbol{\alpha}'\Sigma_{(11)}\boldsymbol{\alpha})(\boldsymbol{\beta}'\Sigma_{(22)}\boldsymbol{\beta}))^{1/2}} \\ &= \boldsymbol{\alpha}'\Sigma_{(12)}\boldsymbol{\beta} = \operatorname{cov}(U_1, V_1).\end{aligned} \tag{10.58}$$

Thus we want to find $\boldsymbol{\alpha}, \boldsymbol{\beta}$ to maximize $\operatorname{cov}(U_1, V_1)$ subject to (10.57). Let

$$\phi_1(\boldsymbol{\alpha}, \boldsymbol{\beta}) = \boldsymbol{\alpha}'\Sigma_{(12)}\boldsymbol{\beta} - \tfrac{1}{2}\rho(\boldsymbol{\alpha}'\Sigma_{(11)}\boldsymbol{\alpha} - 1) + \tfrac{1}{2}v(\boldsymbol{\beta}'\Sigma_{(22)}\boldsymbol{\beta} - 1)$$

where ρ, v are Lagrange multipliers. Differentiating ϕ_1 with respect to the elements of $\boldsymbol{\alpha}, \boldsymbol{\beta}$ separately and setting the results equal to zero, we get

$$\frac{\partial \phi_1}{\partial \boldsymbol{\alpha}} = \Sigma_{(12)}\boldsymbol{\beta} - \rho\Sigma_{(11)}\boldsymbol{\alpha} = \mathbf{0}, \qquad \frac{\partial \phi_1}{\partial \boldsymbol{\beta}} = \Sigma_{(21)}\boldsymbol{\alpha} - v\Sigma_{(22)}\boldsymbol{\beta} = \mathbf{0}. \tag{10.59}$$

From (10.57) and (10.59) we obtain

$$\rho = v = \boldsymbol{\alpha}\Sigma_{(12)}\boldsymbol{\beta}, \qquad \begin{pmatrix} -\rho\Sigma_{(11)} & \Sigma_{(12)} \\ \Sigma_{(21)} & -\rho\Sigma_{(22)} \end{pmatrix}\begin{pmatrix} \boldsymbol{\alpha} \\ \boldsymbol{\beta} \end{pmatrix} = \mathbf{0}. \tag{10.60}$$

In order that there be a nontrivial solution of (10.60) it is necessary that

$$\det\begin{pmatrix} -\rho\Sigma_{(11)} & \Sigma_{(12)} \\ \Sigma_{(21)} & -\rho\Sigma_{(22)} \end{pmatrix} = 0. \tag{10.61}$$

The left-hand side of (10.61) is a polynomial of degree p in ρ and hence has p roots (say) $\rho_1 \geq \cdots \geq \rho_p$ and $\rho = \boldsymbol{\alpha}'\Sigma_{(12)}\boldsymbol{\beta}$ is the correlation between U_1 and V_1 subject to the restriction (10.57). From (10.60)–(10.61) we get

$$\det(\Sigma_{(12)}\Sigma_{(22)}^{-1}\Sigma_{(21)} - \rho^2\Sigma_{(11)}) = 0, \tag{10.62}$$

$$(\Sigma_{(12)}\Sigma_{(22)}^{-1}\Sigma_{(21)} - \rho^2\Sigma_{(11)})\boldsymbol{\alpha} = \mathbf{0}, \tag{10.63}$$

which has p_1 solutions for ρ^2, $\rho_1{}^2 \geq \cdots \geq \rho_{p_1}^2$ (say), and p_1 solutions for

$\boldsymbol{\alpha}$, and

$$\det(\Sigma_{(21)}\Sigma_{(11)}^{-1}\Sigma_{(12)} - \rho^2\Sigma_{(22)}) = 0, \tag{10.64}$$

$$(\Sigma_{(21)}\Sigma_{(11)}^{-1}\Sigma_{(12)} - \rho^2\Sigma_{(22)})\boldsymbol{\beta} = \mathbf{0}, \tag{10.65}$$

which has p_2 solutions for ρ^2 and p_2 solutions for $\boldsymbol{\beta}$. Now (10.62) implies that

$$\det(\Lambda\Lambda' - \rho^2 I) = 0, \qquad \text{where} \quad \Lambda = \Sigma_{(11)}^{-1/2}\Sigma_{(12)}\Sigma_{(22)}^{-1/2}. \tag{10.66}$$

Since

$$\det(\Lambda\Lambda' - \rho^2 I) = \det(\Lambda'\Lambda - \rho^2 I) = \det(\Sigma_{(22)}^{-1/2}\Sigma_{(21)}\Sigma_{(11)}^{-1}\Sigma_{(12)}\Sigma_{(22)}^{-1/2} - \rho^2 I),$$

we conclude that (10.62) and (10.64) have the same solutions. Thus (10.61) has p roots of which $p_2 - p_1$ are zeros, and the remaining $2p_1$ nonzero roots are of the form $\rho = \pm\rho_i$, $i = 1, \ldots, p_1$. The ordered p roots of (10.61) are thus $(\rho_1, \ldots, \rho_{p_1}, 0, \ldots, 0, -\rho_{p_1}, \ldots, -\rho_1)$. We shall show later that $\rho_i \geq 0$, $i = 1, \ldots, p_1$.

To get the maximum correlation of U_1, V_1 we take $\rho = \rho_1$. Let $\boldsymbol{\alpha}^{(1)}$, $\boldsymbol{\beta}^{(1)}$ be the solution (10.60) when $\rho = \rho_1$. Thus $U_1 = \boldsymbol{\alpha}^{(1)'}\mathbf{X}_{(1)}$, $V_1 = \boldsymbol{\beta}^{(1)'}\mathbf{X}_{(2)}$ are normalized (with respect to variance) linear combinations of $\mathbf{X}_{(1)}$, $\mathbf{X}_{(2)}$, respectively, with maximum correlation ρ_1.

DEFINITION 10.3.1 $U_1 = \boldsymbol{\alpha}^{(1)'}\mathbf{X}_{(1)}$, $V_1 = \boldsymbol{\beta}^{(1)'}\mathbf{X}_{(2)}$ are called the first canonical variates and ρ_1 is called the first canonical correlation between $\mathbf{X}_{(1)}$ and $\mathbf{X}_{(2)}$.

Next we define

$$U_2 = \boldsymbol{\alpha}'\mathbf{X}_{(1)}, \qquad V_2 = \boldsymbol{\beta}'\mathbf{X}_{(2)},$$

$\boldsymbol{\alpha} \in E^{p_1}$, $\boldsymbol{\beta} \in E^{p_2}$, so that $\operatorname{var}(U_2) = \operatorname{var}(V_2) = 1$, U_2, V_2 are uncorrelated with U_1, V_1, respectively, and the coefficient of correlation $\rho(U_2, V_2)$ is as large as possible. It is now left as an exercise to establish that $\rho(U_2, V_2) = \rho_2$, the second largest root of (10.61). Let $\boldsymbol{\alpha}^{(2)}$, $\boldsymbol{\beta}^{(2)}$ be the solution of (10.60) when $\rho = \rho_2$.

DEFINITION 10.3.2 $U_2 = \boldsymbol{\alpha}^{(2)'}\mathbf{X}_{(1)}$, $V_2 = \boldsymbol{\beta}^{(2)'}\mathbf{X}_{(2)}$ are called the second canonical variates and ρ_2 is called the second canonical correlation.

This procedure is continued and at each step we define canonical variates as normalized variates, which are uncorrelated with all previous canonical variates, having maximum correlation. Because of (10.62) and (10.64) the maximum number of pairs (U_i, V_i) of positively correlated canonical variates is p_1.

Let

$$\begin{aligned} \mathbf{U} &= (U_1, \ldots, U_{p_1})' = A'\mathbf{X}_{(1)}, \qquad & A &= (\boldsymbol{\alpha}^{(1)}, \ldots, \boldsymbol{\alpha}^{(p_1)}), \\ \mathbf{V}_{(1)} &= (V_1, \ldots, V_{p_1})' = B_1'\mathbf{X}_{(2)}, \qquad & B_1 &= (\boldsymbol{\beta}^{(1)}, \ldots, \boldsymbol{\beta}^{(p_1)}), \end{aligned} \tag{10.67}$$

and let D be a diagonal matrix with diagonal elements $\rho_1, \ldots, \rho_{p_1}$. Since (U_i, V_i), $i = 1, \ldots, p_1$, are canonical variates,

$$\begin{gathered} \operatorname{cov}(\mathbf{U}) = A'\Sigma_{(11)}A = I, \qquad \operatorname{cov}(\mathbf{V}_{(1)}) = B_1'\Sigma_{(22)}B_1 = I, \\ \operatorname{cov}(\mathbf{U}, \mathbf{V}_{(1)}) = A'\Sigma_{(12)}B_1 = \Lambda. \end{gathered} \tag{10.68}$$

Let $B_2 = (\boldsymbol{\beta}^{(p_1+1)}, \ldots, \boldsymbol{\beta}^{(p_2)})$ be a $p_2 \times (p_2 - p_1)$ matrix satisfying

$$B_2'\Sigma_{(22)}B_1 = 0, \qquad B_2'\Sigma_{(22)}B_2 = I,$$

and formed one column at a time in the following way: $\boldsymbol{\beta}^{(p_1+1)}$ is a vector orthogonal to $\Sigma_{(22)}B_1$ and $\boldsymbol{\beta}^{(p_1+1)'}\Sigma_{(22)}\boldsymbol{\beta}^{(p_1+1)} = 1$; $\boldsymbol{\beta}^{(p_1+2)}$ is a vector orthogonal to $\Sigma_{(22)}(B_1, \boldsymbol{\beta}^{(p_1+1)})$ and $\boldsymbol{\beta}^{(p_1+2)'}\Sigma_{(22)}\boldsymbol{\beta}^{(p_1+2)} = 1$; and so on. Let $B = (B_1, B_2)$. Since $B'\Sigma_{(22)}B = I$, we conclude that B is nonsingular. Now

$$\begin{aligned} &\det\left[\begin{pmatrix} A' & 0 \\ 0 & B_1' \\ 0 & B_2' \end{pmatrix}\begin{pmatrix} -\rho\Sigma_{(11)} & \Sigma_{(12)} \\ \Sigma_{(21)} & -\rho\Sigma_{(22)} \end{pmatrix}\begin{pmatrix} A & 0 & 0 \\ 0 & B_1 & B_2 \end{pmatrix}\right] \\ &= \det\begin{pmatrix} -\rho I & D & 0 \\ D & -\rho I & 0 \\ 0 & 0 & \rho I \end{pmatrix} = (-\rho)^{p_2-p_1}\det\begin{pmatrix} -\rho I & D \\ D & -\rho I \end{pmatrix} \\ &= (-\rho)^{p_2-p_1}\det(\rho^2 I - DD) \\ &= (-\rho)^{p_2-p_1}\prod_{i=1}^{p_1}(\rho^2 - \rho_i^2). \end{aligned} \tag{10.69}$$

Hence the roots of the equation obtained by setting (10.69) equal to zero are the roots of (10.61).

Observe that for $i = 1, \ldots, p_1$ [from (10.60)]

$$\Sigma_{(12)}\boldsymbol{\beta}^{(i)} = -\rho_i\Sigma_{(11)}(-\boldsymbol{\alpha}^{(i)}), \tag{10.70}$$

$$\Sigma_{(21)}(-\boldsymbol{\alpha}^{(i)}) = -\rho_i\Sigma_{(22)}(\boldsymbol{\beta}^{(i)}). \tag{10.71}$$

Thus, if ρ_i, $\boldsymbol{\alpha}^{(i)}$, $\boldsymbol{\beta}^{(i)}$ is a solution so is $-\rho_i$, $-\boldsymbol{\alpha}^{(i)}$, $\boldsymbol{\beta}^{(i)}$. Hence if the ρ_i were negative, then $-\rho_i$ would be nonnegative and $-\rho_i \geq \rho_i$. But since ρ_i was to be a maximum, we must have $\rho_i \geq -\rho_i$ and therefore $\rho_i \geq 0$.

The components of $\mathbf{U}$ are one set of canonical variates, the components of $(\mathbf{V}_{(1)}, \mathbf{V}_{(2)}) = B_2\mathbf{X}_{(2)}$ are other sets of canonical variates, and

$$\operatorname{cov}\begin{pmatrix}\mathbf{U}\\ \mathbf{V}_{(1)}\\ \mathbf{V}_{(2)}\end{pmatrix} = \begin{pmatrix} I & \Lambda & 0\\ \Lambda & I & 0\\ 0 & 0 & I\end{pmatrix}.$$

DEFINITION 10.3.3 The ith pair of canonical variates, $i = 1, \ldots, p_1$, is the pair of linear combinations $U_i = \boldsymbol{\alpha}^{(i)\prime}\mathbf{X}_{(1)}$, $\mathbf{V}_i = \boldsymbol{\beta}^{(i)\prime}\mathbf{X}_{(2)}$, each of unit variance and uncorrelated with the first $(i-1)$ pairs of canonical variates (U_j, V_j), $j = 1, \ldots, i-1$, and having maximum correlation. The coefficient of correlation between U_i and V_i is called the ith canonical correlation.

Hence we have the following theorem.

THEOREM 10.3.1 The ith canonical correlation between $\mathbf{X}_{(1)}$ and $\mathbf{X}_{(2)}$ is the ith largest root ρ_i of (10.61) and is positive. The coefficients $\boldsymbol{\alpha}^{(i)}$, $\boldsymbol{\beta}^{(i)}$ of the normalized ith canonical variates $U_i = \boldsymbol{\alpha}^{(i)\prime}\mathbf{X}_{(1)}$, $V_i = \boldsymbol{\beta}^{(i)\prime}\mathbf{X}_{(2)}$ satisfy (10.60) for $\rho = \rho_i$.

In applications the first few pairs of canonical variates usually have appreciably large correlations, so that a large reduction in the dimension of two sets can be achieved by retaining these variates only.

10.3.2 Sample Canonical Correlations

In practice $\boldsymbol{\mu}$, Σ are unknown. We need to estimate them on the basis of sample observations from the distribution of $\mathbf{X}$. In what follows we shall assume that $\mathbf{X}$ has a p-variate normal distribution with mean $\boldsymbol{\mu}$ and positive definite covariance matrix Σ (in the case of nonnormality see Rao, 1965). Let $\mathbf{x}^\alpha = (x_{\alpha 1}, \ldots, x_{\alpha p})$, $\alpha = 1, \ldots, N$, be a sample of N observations on $\mathbf{X}$ and let

$$\bar{\mathbf{x}} = \frac{1}{N}\sum_{\alpha=1}^{N} \mathbf{x}^\alpha, \qquad s = \sum_{\alpha=1}^{N} (\mathbf{x}^\alpha - \bar{\mathbf{x}})(\mathbf{x}^\alpha - \bar{\mathbf{x}})'.$$

Partition s, similarly to Σ, as

$$s = \begin{pmatrix} s_{(11)} & s_{(12)}\\ s_{(21)} & s_{(22)}\end{pmatrix},$$

where $s_{(ij)}$ is $p_i \times p_j$, $i, j = 1, 2$. The maximum likelihood estimates of the $\Sigma_{(ij)}$ are $s_{(ij)}/N$. The maximum likelihood estimates $\hat{\boldsymbol{\alpha}}^{(i)}$, $i = 1, \ldots, p_1$, $\hat{\boldsymbol{\beta}}^{(j)}$,

$j = 1, \ldots, p_2$ and $\hat{\rho}_i$, $i = 1, \ldots, p_1$, of $\boldsymbol{\alpha}^{(i)}$, $\boldsymbol{\beta}^{(j)}$, and ρ_i, respectively, are obtained from (10.60) and (10.61) by replacing $\Sigma_{(ij)}$ by $s_{(ij)}/N$. Standard programs are available for the computation of $\hat{\boldsymbol{\alpha}}^{(i)}$, $\hat{\boldsymbol{\beta}}^{(j)}$, $\hat{\rho}_i$, and we refer to Press (1971) for details. We define the squared sample canonical correlation R_i^2 (with values $r_i^2 = \hat{\rho}_i^2$) by the roots of

$$\det(S_{(12)}S_{(22)}^{-1}S_{(21)} - r^2 S_{(11)}) = 0, \tag{10.72}$$

which can be written as

$$\det(B - r^2(A + B)) = 0, \tag{10.73}$$

where $B = S_{(12)}S_{(22)}^{-1}S_{(21)}$, $A = S_{(11)} - S_{(12)}S_{(22)}^{-1}S_{(21)}$. From Theorem 6.4.1, A, B are independently distributed, A is distributed as Wishart

$$W_{p_1}(\Sigma_{(11)} - \Sigma_{(12)}\Sigma_{(22)}^{-1}\Sigma_{(21)}, N - 1 - p_2),$$

and the conditional distribution of $S_{(12)}S_{(22)}^{-1/2}$, given that $S_{(22)} = s_{(22)}$, is normal in the sense of Example 6.4.1 with mean $\Sigma_{(12)}\Sigma_{(22)}^{-1/2}$ and covariance matrix

$$(\Sigma_{(22)} - \Sigma_{(21)}\Sigma_{(11)}^{-1}\Sigma_{(12)}) \otimes s_{(22)}^{-1}.$$

Hence if $\Sigma_{(12)} = 0$, then A, B are independently distributed as $W_{p_1}(\Sigma_{(11)}, N - 1 - p_2)$, $W_{p_2}(\Sigma_{(11)}, p_2)$, respectively. Thus, in the case $\Sigma_{(12)} = 0$, the squared sample canonical correlation coefficients are the roots of the equation

$$\det(B - r^2(A + B)) = 0, \tag{10.74}$$

where A, B are independent Wishart matrices with the same parameter $\Sigma_{(11)}$. The distribution of these ordered roots $R_1^2 > R_2^2 > \cdots > R_p^2$ (say) was derived in Chapter VIII and is given by

$$K \prod_{i=1}^{p} (r_i^2)^{(p_2 - p_1 - 1)/2}(1 - r_i^2)^{(N-1-p_1-1)/2} \prod_{i<j} (r_i^2 - r_j^2), \tag{10.75}$$

where

$$K = \Pi^{p/2}\left[\prod_{i=1}^{p} \Gamma(\tfrac{1}{2}(p_2 - i + 1))\Gamma\left(\frac{i}{2}\right)\right]^{-1} \prod_{i=1}^{p} \frac{\Gamma(\frac{1}{2}(N - p_1 + p_2 - i))}{\Gamma(\frac{1}{2}(N - p_1 - i))}. \tag{10.76}$$

These roots are maximal invariants in the space of the random Wishart matrix S under the transformations $S \to ASA'$ where

$$A = \begin{pmatrix} A_1 & 0 \\ 0 & A_2 \end{pmatrix} \qquad \text{with} \quad A_i = p_i \times p_i, \qquad i = 1, 2.$$

10.3.3 Tests of Hypotheses

Let us now consider the problem of testing the null hypothesis $H_{10}:\Sigma_{(12)} = 0$ against the alternatives $H_1:\Sigma_{(12)} \neq 0$ on the basis of sample observations $\mathbf{x}^\alpha$, $\alpha = 1, \ldots, N$ $(N \geq p)$. In other words, H_{10} is the hypothesis of joint nonsignificance of the first p_1 canonical correlations as a set. It can be easily calculated that the likelihood ratio test of H_{10} rejects H_{10} whenever

$$\lambda_1 = \frac{\det s}{\det(s_{(11)})\det(s_{(22)})} \leq c, \tag{10.77}$$

where the constant c is chosen so that the test has level of significance α. Narain (1950) showed that the likelihood ratio test for testing H_{10} is unbiased against H_1 (see Section 8.3). The exact distribution of

$$\lambda_1 = \frac{\det S}{\det(S_{(11)})\det(S_{(22)})}$$

was studied by Hotelling (1936b), Girshik (1939), and Anderson (1958, p. 237). These forms are quite complicated. Bartlett (1938, 1939, 1941) gave an approximate large sample distribution of λ_1. Since

$$\begin{aligned}\det(S) &= \det(S_{(22)})\det(S_{(11)} - S_{(12)}S_{(22)}^{-1}S_{(21)}) \\ &= \det S_{(22)} \det(S_{(11)})\det(I - S_{(11)}^{-1}S_{(12)}S_{(22)}^{-1}S_{(21)}),\end{aligned}$$

we can write λ_1 as

$$\lambda_1 = \det(I - S_{(11)}^{-1}S_{(12)}S_{(22)}^{-1}S_{(21)}) = \prod_{i=1}^{p_1}(1 - R_i^2).$$

Using Box (1949), as $N \to \infty$ and under H_{10}

$$P\{-v \log \lambda_1 \leq z\} = P\{\chi_f^2 \leq z\},$$

where $v = N - \frac{1}{2}(p_1 + p_2 + 1)$, $f = p_1 p_2$.

Now suppose that H_{10} is rejected; that is, the likelihood ratio test accepts $H_1:\Sigma_{(12)} \neq 0$. Bartlett (1941) suggested testing the hypothesis H_{20}: (the joint nonsignificance of $\rho_2, \ldots, \rho_{p_1}$ as a set), and proposed the test of rejecting H_{20} whenever

$$\lambda_2 = \prod_{i=2}^{p_1}(1 - r_i^2) \leq c,$$

where c depends on the level of significance α of the test, and under H_{20} for large N

$$P\{-v \log \lambda_2 \leq z\} = P\{\chi_{f_1}^2 \leq z\},$$

where $f_1 = (p_1 - 1)(p_2 - 1)$. That is, for large N, Bartlett suggested the possibility of testing the joint nonsignificance of $\rho_2, \ldots, \rho_{p_1}$. If H_{10} is rejected and H_{20} is accepted, then ρ_1 is the only significant canonical correlation. If H_{20} is also rejected, the procedure should be continued to test H_{30}: (the joint nonsignificance of $\rho_3, \ldots, \rho_{p_1}$ as a set), and then if necessary to test H_{40}, and so on. For H_{r0}: (the joint nonsignificance of $\rho_r, \ldots, \rho_{p_1}$ as a set), the test rejects H_{r0} whenever

$$\lambda_r = \prod_{i=r}^{p_1} (1 - r_i^2) \leq c,$$

where the constant c depends on the level of significance α of the test, and for large N under H_{r0}

$$P\{-v \log \lambda_r \leq z\} = P\{\chi^2_{f_r} \leq z\},$$

where $f_r = (p_1 - r)(p_2 - r)$.

EXAMPLE 10.3.1 Measurements on 12 different characters $\mathbf{x}' = (x_1, \ldots, x_{12})$ for each of 27 randomly selected wheat plants of a particular variety grown at the Indian Agricultural Research Institute, New Delhi, are taken. The sample correlation matrix is given by

$$\begin{pmatrix}
1.0000 \\
0.7025 & 1.0000 \\
0.0539 & -0.0717 & 1.0000 \\
0.1154 & 0.0458 & 0.5811 & 1.0000 \\
-0.0669 & -0.1212 & 0.5851 & 0.4065 & 1.0000 \\
-0.0361 & -0.0515 & 0.7137 & 0.5238 & 0.6698 & 1.0000 \\
0.4381 & 0.6109 & -0.2064 & -0.1113 & -0.4702 & -0.2029 & 1.0000 \\
-0.1332 & 0.1667 & -0.0708 & -0.1186 & -0.0686 & -0.1693 & 0.3503 & 1.0000 \\
0.4611 & 0.5927 & -0.2545 & -0.1213 & -0.4649 & -0.2284 & 0.8857 & 0.2945 & 1.0000 \\
0.5139 & 0.6633 & -0.3099 & -0.1602 & -0.3441 & -0.2141 & 0.8295 & 0.2899 & 0.9339 & 1.0000 \\
0.4197 & 0.5148 & -0.1491 & -0.0216 & -0.3475 & -0.1929 & 0.7007 & 0.0252 & 0.6960 & 0.6987 & 1.0000 \\
0.6601 & 0.7129 & -0.1652 & -0.0121 & -0.3632 & -0.1119 & 0.8155 & 0.0761 & 0.8686 & 0.8474 & 0.8015 & 1.0000
\end{pmatrix}$$

We are interested in finding the canonical correlations between the set of the first six characters and the set of the remaining six characters. Ordered sample canonical correlations r_i^2 and the corresponding normalized coefficients $\boldsymbol{\alpha}^{(i)}, \boldsymbol{\beta}^{(i)}$ of the canonical variates are given by:

$$\begin{aligned}
r_1^2 &= 0.86018 \\
\boldsymbol{\alpha}^{(1)} &= (0.32592, 0.24328, -0.20063, -0.02249, -0.18167, 0.27907)' \\
\boldsymbol{\beta}^{(1)} &= (0.20072, -0.10910, -0.49652, 0.39825, -0.26955, 0.68558)' \\
r_2^2 &= 0.64546 \\
\boldsymbol{\alpha}^{(2)} &= (-0.09184, 0.02063, 0.22661, 0.05155, -0.36115, -0.01726)'
\end{aligned}$$

$$\boldsymbol{\beta}^{(2)} = (0.17221, -0.01490, 0.66931, -0.69215, 0.12967, -0.16206)'$$
$$r_3^2 = 0.51725$$
$$\boldsymbol{\alpha}^{(3)} = (-0.63826, 0.71640, 0.14862, -0.10546, 0.21983, -0.48038)'$$
$$\boldsymbol{\beta}^{(3)} = (0.06472, 0.46436, -0.30851, 0.55171, 0.36140, -0.50000)'$$
$$r_4^2 = 0.30779$$
$$\boldsymbol{\alpha}^{(4)} = (-0.04954, 0.17225, 0.35083, 0.13124, 0.12220, -0.22384)'$$
$$\boldsymbol{\beta}^{(4)} = (-0.13684, 0.30706, -0.39213, -0.42461, 0.11164, 0.73523)'$$
$$r_5^2 = 0.17273$$
$$\boldsymbol{\alpha}^{(5)} = (0.19760, -0.18106, -0.09297, 0.28382, 0.12898, -0.43494)'$$
$$\boldsymbol{\beta}^{(5)} = (-0.68243, -0.01088, 0.51799, -0.03772, 0.48737, -0.16404)'$$
$$r_6^2 = 0.00413$$
$$\boldsymbol{\alpha}^{(6)} = (0.19967, -0.11996, 0.32842, -0.33865, 0.04872, -0.19203)'$$
$$\boldsymbol{\beta}^{(6)} = (0.51365, -0.17532, -0.67766, 0.27563, 0.28516, -0.29818)'.$$

Bartlett's Test of Significance

i	Sample Canonical Correlations r_i	Likelihood Ratio λ_i	Chi-Square $-v \log \lambda_i$	Degrees of Freedom f_i
1	0.86018	0.09764	47.70060	36
2	0.64327	0.37527	20.09220	25
3	0.51725	0.64327	9.04426	16
4	0.30779	0.87824	2.66151	9
5	0.17273	0.97015	0.62126	4
6	0.00413	0.99998	0.00035	1

10.4 TIME SERIES ANALYSIS

A time series is a sequence of observations usually ordered in time. The main distinguishing feature of time series analysis is its explicit recognition of the importance of the order in which the observations are made. Although in general statistical investigation the observations are independent, in a time series successive observations are dependent. Consider a stochastic process $X(t)$ as a random variable indexed by the continuous parameter t. Let t be a time scale and the process be observed at the particular p points $t_1 < t_2 < \cdots < t_p$. The random vector

$$\mathbf{X}(t) = (X(t_1), \ldots, X(t_p))'$$

is called a time series. It has a multidimensional distribution characterizing the process. In most situations we assume that $\mathbf{X}(t)$ has a p-variate normal distribution specified by the mean $E(\mathbf{X}(t))$ and covariance matrix with general elements

$$\operatorname{cov}(X(t_i), X(t_j)) = \sigma_i \sigma_j \rho(t_i, t_j),$$

where $\operatorname{var}(X(t_i)) = \sigma_i^2$. The term $\rho(t_i, t_j)$ is called the correlation function of the time series. The analysis of the time series data depends on the specific form of $\rho(t_i, t_j)$. A general model of time series can be written as

$$X(t) = f(t) + U(t),$$

where $\{f(t)\}$ is a completely determined sequence, often called the systematic part, and $\{U(t)\}$ is the random sequence having different probability laws. They are sometimes called signal and noise sequences, respectively. The sequence $\{f(t)\}$ may depend on unknown coefficients and known quantities depending on time. This model is analogous to the regression model discussed in Chapter VIII. If $f(t)$ is a slowly moving function of t, for example a polynomial of lower degree, it is called a trend; if it is exemplified by a finite Fourier series, it is called cyclical. The effect of time may be present both in $f(t)$ (i.e., trend in time or cyclical) and in $U(t)$ as a stochastic process. When $f(t)$ has a given structure involving a finite number of parameters, we consider the problem of inference about these parameters. When the stochastic process is specified in terms of a finite number of parameters we want to estimate and test hypotheses about these parameters. We refer to Anderson (1971) for an explicit treatment of this topic.

EXERCISES

1. Let $\mathbf{X} = (X_1, \ldots, X_p)'$ be normally distributed with mean $\boldsymbol{\mu}$ and covariance matrix Σ, and let Σ have one characteristic root λ_1 of multiplicity p.

(a) On the basis of observations $\mathbf{x}^\alpha = (x_{\alpha 1}, \ldots, x_{\alpha p})'$, $\alpha = 1, \ldots, N$, show that the maximum likelihood estimate $\hat{\lambda}_1$ of λ_1 is given by

$$\hat{\lambda}_1 = \frac{1}{pN} \sum_{i=1}^{p} \sum_{\alpha=1}^{N} (x_{\alpha i} - \bar{x}_i)^2 \qquad \text{where} \quad \bar{x}_i = \frac{1}{N} \sum_{\alpha=1}^{N} x_{\alpha i}.$$

(b) Show that the principal component of $\mathbf{X}$ is given by $O\mathbf{X}$ where O is any $p \times p$ orthogonal matrix.

2. Let $\mathbf{X} = (X_1, \ldots, X_p)'$ be a random p-vector with covariance matrix

$$\Sigma = \sigma^2 \begin{pmatrix} 1 & \rho & \cdots & \rho \\ \rho & 1 & \cdots & \rho \\ \vdots & \vdots & & \vdots \\ \rho & \rho & \cdots & 1 \end{pmatrix}, \qquad 0 < \rho \leq 1.$$

(a) Show that the largest characteristic root of Σ is

$$\lambda_1 = \sigma^2(1 + (p - 1)\rho).$$

(b) Show that the first principal component of $\mathbf{X}$ is

$$Z_1 = \frac{1}{\sqrt{p}} \sum_{i=1}^{p} X_i.$$

3. Let $\mathbf{X} = (X_1, \ldots, X_4)'$ be a random vector with covariance matrix

$$\Sigma = \begin{pmatrix} \sigma^2 & \sigma_{12} & \sigma_{13} & \sigma_{14} \\ & \sigma^2 & \sigma_{14} & \sigma_{13} \\ & & \sigma^2 & \sigma_{12} \\ & & & \sigma^2 \end{pmatrix}.$$

Show that the principal components of $\mathbf{X}$ are

$$Z_1 = \tfrac{1}{2}(X_1 + X_2 + X_3 + X_4), \qquad Z_2 = \tfrac{1}{2}(X_1 + X_2 - X_3 - X_4),$$
$$Z_3 = \tfrac{1}{2}(X_1 - X_2 + X_3 - X_4), \qquad Z_4 = \tfrac{1}{2}(X_1 - X_2 - X_3 + X_4).$$

4. Consider the orthogonal factor analysis model of Section 10.3. Let $\Sigma = \Lambda\Lambda' + D$. Show that

$$\frac{\partial \Sigma}{\partial \lambda_{ij}} = \begin{pmatrix} 0 & \cdots & 0 & \lambda_{1,j} & 0 & \cdots & 0 \\ \vdots & & \vdots & \vdots & \vdots & & \vdots \\ 0 & \cdots & 0 & \lambda_{i-1,j} & 0 & \cdots & 0 \\ \lambda_{1j} & \cdots & \lambda_{i-1,j} & 2\lambda_{ij} & \lambda_{i+1,j} & \cdots & \lambda_{pj} \\ \vdots & & \vdots & \vdots & \vdots & & \vdots \\ 0 & \cdots & 0 & \lambda_{pj} & 0 & \cdots & 0 \end{pmatrix}.$$

Hence show that

$$\operatorname{tr} \Sigma^{-1} \frac{\partial \Sigma}{\partial \lambda_{ij}} = 2(\boldsymbol{\sigma}^i)'\boldsymbol{\lambda}_j.$$

5. For the data given in Example 5.3.1, find the ordered characteristic roots and the corresponding normalized characteristic vectors.

REFERENCES

Anderson, T. W. (1951). Classification by multivariate analysis, *Psychometrika* **16**, 31–50.

Anderson, T. W. (1958). "An Introduction to Multivatiate Statistical Analysis." Wiley, New York.

Anderson, T. W. (1963). Asymptotic theory for principal components analysis, *Ann. Math. Statist.* **3**, 122–148.

Anderson, T. W. (1971). "The Statistical Analysis of Time Series." Wiley, New York.

Anderson, T. W., and Rubin, H. (1956). Statistical inference in factor analysis, *Proc. Berkeley Symp. Math. Statist. Prob., 3rd* **5**, 111–150. Univ. of California, Berkeley, California.

Armstrong, J. S. (1967). Derivation of theory by means of factor analysis or Tom Swift and his electric factor analysis machine, The American Statistician, December, pp. 17–21.

Bartlett, M. S. (1938). Further aspects of the theory of multiple regression, *Proc. Cambridge Phil. Soc.* **34**, 33–40.

Bartlett, M. S. (1939). A note on test of significance in multivariate analysis, *Proc. Cambridge Phil. Soc.* **35**, 180–185.

Bartlett, M. S. (1941). The statistical significance of canonical correlation, *Biometrika* **32**, 29–38.

Bartlett, M. S. (1954). A note on the multiplying factors for various chi-square approximation, *J. R. Statist. Soc. B* **16**, 296–298.

Box, A. E. P. (1949). A general distribution theory for a class of likelihood ratio criteria, *Biometrika* **36**, 317–346.

Galton, F. (1888). Co-relation and their measurements, chiefly from anthropometric data, *Proc. R. Soc.* **45**, 135–140.

Girshik, M. A. (1936). Principal components, *J. Am. Statist. Assoc.* **31**, 519–528.

Girshik, M. A. (1939). On the sampling theory of roots of determinantal equation, *Ann. Math. Statist.* **10**, 203–224.

Hotelling, H. (1933). Analysis of a complex of statistical variables into principal components, *J. Educ. Psychol.* **24**, 417–441.

Hotelling, H. (1936a). Simplified calculation of principal components, *Psychometrika* **1**, 27–35.

Hotelling, H. (1936b). Relation between two sets of variates, *Biometrika* **28**, 321–377.

Lawley, D. N. (1940). The estimation of factor loadings by the method of maximum likelihood, *Proc. R. Soc. Edinburgh A* **60**, 64–82.

Lawley, D. N. (1949). Problems in factor analysis, *Proc. R. Soc. Edinburgh* **62**, 394–399.

Lawley, D. N. (1950). A further note on a problem in factor analysis. *Proc. R. Soc. Edinburgh* **63**, 93–94.

Lawley, D. N. (1953). A modified method of estimation in factor analysis and some large sample results, *Uppsala Symp. Psycho Factor Anal.* 35–42. Almqvist and Wiksell, Sweden.

Lawley, D. N. (1956). Tests of significance for the latent roots of covariance and correlation matrices, *Biometrika* **43**, 128–136.

Lawley, D. N. (1963). On testing a set of correlation coefficients for equality, *Ann. Math. Statist.* **34**, 149–151.

Morrison, D. F. (1967). "Multivariate Statistical Method." McGraw-Hill, New York.

Narain, R. D. (1950). On the completely unbiased character of tests of independence in multivariate normal system, *Ann. Math. Statist.* **21**, 293–298.

Pearson, K. (1901). On lines and planes of closest fit to system of points in space, *Phil. Mag.* **2**, 559–572.

Press, J. (1971). "Applied Multivariate Analysis." Holt, New York.

Rao, C. R. (1955). Estimation and tests of significance in factor analysis, *Psychometrika* **20**, 93–111.

Rao, C. R. (1965). "Linear Statistical Inference and its Applications." Wiley, New York.

Solomon, H. (1960). A survey of mathematical models in factor analysis, "Mathematical thinking in the Measurement of Behaviour." Free Press, Glenco, New York.

Spearman, C. (1904). General intelligence objectively determined and measured, *Am. J. Psychol.* **15**, 201–293.

Thurston, L. (1945). "Multiple Factor Analysis." Univ. of Chicago Press, Chicago, Illinois.

Author Index

Subject Index

A
B 7
C 8
D 9
E 0
F 1
G 2
H 3
I 4
J 5